PLEASE STAMP DATE DUE, BOTH BEL

Springer Series in
CHEMICAL PHYSICS

96

Springer Series in
CHEMICAL PHYSICS

Series Editors: A. W. Castleman, Jr.   J. P. Toennies   K. Yamanouchi   W. Zinth

The purpose of this series is to provide comprehensive up-to-date monographs in both well established disciplines and emerging research areas within the broad fields of chemical physics and physical chemistry. The books deal with both fundamental science and applications, and may have either a theoretical or an experimental emphasis. They are aimed primarily at researchers and graduate students in chemical physics and related fields.

Please view available titles in *Springer Series in Chemical Physics* on series homepage http://www.springer.com/series/676

Astrid Gräslund
Rudolf Rigler
Jerker Widengren

Editors

Single Molecule Spectroscopy in Chemistry, Physics and Biology

Nobel Symposium

With 223 Figures

Springer

Editors

Professor Astrid Gräslund

Stockholm University
Department of Biophysics
10691 Stockholm, Sweden
E-Mail: astrid@dbb-su.se

Professor Jerker Widengren

Royal Institute or Technology (KTH)
Department of Biomolecular Physics
10691 Stockholm, Sweden
E-Mail: jerker@biomolphysics.kth.se

Professor Rudolf Rigler

Swiss Federal Institute of Technology Lausanne (EPFL)
1015 Lausanne, Switzerland
E-Mail: rudolf.rigler@epfl.ch

Series Editors:

Professor A.W. Castleman, Jr.

Department of Chemistry, The Pennsylvania State University
152 Davey Laboratory, University Park, PA 16802, USA

Professor J.P. Toennies

Max-Planck-Institut für Strömungsforschung
Bunsenstrasse 10, 37073 Göttingen, Germany

Professor K. Yamanouchi

University of Tokyo, Department of Chemistry
Hongo 7-3-1, 113-0033 Tokyo, Japan

Professor W. Zinth

Universität München, Institut für Medizinische Optik
Öttingerstr. 67, 80538 München, Germany

Springer Series in Chemical Physics   ISSN 0172-6218
ISBN 978-3-642-02596-9   e-ISBN 978-3-642-02597-6
DOI 10.1007/978-3-642-02597-6
Springer Heidelberg Dordrecht London New York

Library of Congress Control Number: 2009934497

Cover design: SPi Publisher Services

Printed on acid-free paper

Springer is a part of Springer Science+Business Media (www.springer.com)

Nobel Symposium, June 2008, at the Sånga Säby Conference

Foreword

By selecting the first week of June 2008 for the Nobel Symposium "Single Molecular Spectroscopy in Chemistry, Physics and Biology", Rudolf Rigler, Jerker Widengren and Astrid Gräslund have once again won the top prize for Meeting Organizers, providing us with a Mediterranean climate on top of the warm hospitality that is unique to Sweden. The Sånga Säby Conference Center was an ideal place to spend this wonderful week, and the comfort of this beautiful place blended perfectly with the high calibre of the scientific programme. It was a special privilege for me to be able to actively participate in this meeting on a field that is in many important ways complementary to my own research. I was impressed by the interdisciplinary ways in which single molecule spectroscopy has evolved and is currently pursued, with ingredients originating from physics, all branches of chemistry and a wide range of biological and biomedical research. A beautiful concert by Semmy Stahlhammer and Johan Ullén further extended the interdisciplinary character of the symposium. I would like to combine thanks to Rudolf, Jerker and Astrid with a glance into a future of other opportunities to enjoy top-level science combined with warm hospitality in the Swedish tradition.

Zürich,   *Kurt Wüthrich*
April 2009

Participants of the Nobel-Symposium 138: First row: Sarah Unterkofler, Anders Liljas, Xiao-Dong Su, Birgitta Rigler, Carlos Busta-mante, Toshio Yanagida, Steven Block, Xiaowei Zhuang, Sunney Xie.
Second row: Ivan Scheblykin, Lars Thelander, Petra Schwille, Watt W. Webb, Rudolf Rigler, Jerker Widengren, Peter Lu, Shimon Weiss, William E Moerner, David Bensimon.
Third row: Anders Ehrenberg, Yu Ming, Fredrik Elinder, Kazuhiko Kinosita, Vladana Vukojevic, Masataka Kinjo, May D Wang, Yu Ohsugi, Shuming Nie, Andreas Engel, Peter G Wolynes, Michel Orrit, Hans Blom, Johan Hofkens.
Fourth row: Claus Seidel, Heike Hevekerl, Taekjip Ha, Evangelos Sisamakis, Per Ahlberg, Joseph Nordgren, Kurt Wthrich, Sune Svanberg, Bengt Nordn, Paul Alivisatos, Per Thyberg, Richard Keller, Andriy Chmyrov, Johan Elf, Per Rigler, Kai Hassler, Gustav Persson, Jürgen Köhler, Eric Betzig, Thomas Schmidt, Christoph Bräuchle, Elliot Elson, Måns Ehrenberg, Dimitrios K Papadopoulos, Ingemar Lundström, Horst Vogel, Stefan Wennmalm, Hermann Gaub, Håkan Wennerström, Yosif Klafter, Julio Fernandez.

Preface

The development of Single Molecule Detection and Spectroscopy started in the late eighties. The developments came from several areas. Fluorescence-based single molecule spectroscopy developed in particular from (i) holeburning and zero phonon spectroscopy of organic molecules at cryo temperatures and (ii) confocal fluctuation spectroscopy of emitting molecules at elevated temperatures. Of crucial importance for these approaches was the ability to suppress background radiation to the point where signals of single molecules could be detected. Today, confocal single molecule analysis is the dominating approach, particularly in chemistry and in biosciences, but attempts to combine analysis at low and high temperatures are being pursued.

In parallel with this development, significant progress has been made in the field of single molecule force spectroscopy. Approaches based on atomic force microscopy, optical trapping, microneedles or magnetic beads have made it possible to investigate mechanical properties, and not least, the interplay between mechanics and chemistry on a single molecule level.

In June 1999 the first Nobel Conference on Single Molecule Spectroscopy was organized in Södergarn Mansion, Lidingö (Sweden) and a comprehensive presentation of the results obtained in the first decade of single molecule analysis was given (Orrit, Rigler, Basche (eds.) 1999)

Now after almost another decade, it was of interest to find out whether the developments and promises presented at the Södergarn Conference were still valid or had even exceeded our expectations.

The contributions to this volume come from the pioneers of the early period of single molecule spectroscopy as well as from other laboratories which have made important contributions to demonstrate the importance of SM analysis in various applications in Chemistry, Physics and BioSciences.

The Nobel Symposium No. 138 dedicated to Single Molecule Spectroscopy in Chemistry, Physics and Biosciences was held at the Mansion Sånga-Säby situated at the island of Ekerö in Lake Mälaren outside Stockholm, from June 1–6, 2008. The Conference was blessed with pleasant weather and sunshine all the days. Together with the wonderful surroundings this contributed to many stimulating opportunities for individual discussions, in parallel with outdoor excursions including swimming in the lake, jogging tours, walks in the forests and sauna.

The Symposium started with an evening session on molecules and dynamic processes by Kurt Wüthrich and Martin Karplus. The program of the next days included the presentation of the fields which initiated single molecule analysis in cryo temperatures (Moerner,Orrit) followed by confocal analysis of molecular fluctuations at room temperature (Keller, Rigler, Elson, Webb, Widengren, Schwille). Major topics in the following sessions included quantum dots (Alivasatos, Nie), the analysis of conformational dynamics (Weiss, Ha, Seidel), the motion of molecular motors (Yanagida, Kinosita) and replicating assemblies (Bustamante, Block). A special session was devoted to the analysis of forces operating on single molecules (Gaub, Fernandez) as well as to high resolution imaging of single molecules (Hell, Betzig, Zhuang, Engel). Stochastic single molecule events at the cellular level were another important topic (Xie, Schmidt, Vogel, Wolynes) as well as single molecule enzymology (Lu, Xie, Rigler, Hofkens, Klafter, Köhler), which together with atomic force microscopy formed the basis for intense discussions. Several presentations brought the single molecule methodologies and perspectives to a sub-cellular and cellular context (Rigler, Schwille, Weiss, Bensimon, Axner, Hell, Betzig, Zhuang, Schmidt, Xie, Orwar, Bräuchle), which seems to form one of several exciting future directions of this field.

A special event was the evening concert with Semmy Stalhammer on the violin and Johan Ullén on the piano. The violin sonata of Cesar Franck and its masterly performance matched perfectly the level and tension of the scientific sessions.

As organizers we would like to thank all the invited speakers for their excellent contributions to this symposium, as well as all those who contributed with a chapter to this book. We would also like to thank Margareta Klingberg and colleagues at the conference site of Sånga-Säby for the prerequisites and support of an excellent venue, and not the least the Nobel Foundation for supporting this Symposium.

Stockholm,   *Astrid Gräslund*
July 2009   *Rudolf Rigler*
                                                                   Jerker Widengren

Contents

Part X Fields and Outlook

Contributors

Kengo Adachi
Department of Physics
Faculty of Science and Engineering
Waseda University, Okubo
Shinjuku-ku, Tokyo 169-8555
Japan

A. P. Alivisatos
Department of Chemistry
University of California
Berkeley, USA

and

Materials Science Division
Lawrence Berkeley National Lab
Berkeley, USA
alivis@berkeley.edu

Magnus Andersson
Department of Physics
Umeå University
901 87 Umeå, Sweden

Ove Axner
Department of Physics
Umeå University
901 87 Umeå, Sweden
ove.axner@physics.umu.se

Jürgen Baier
Experimental Physics IV
and Bayreuth Institute for
Macromolecular Research

Universität Bayreuth
Universtitätsstrasse 30
95440 Bayreuth, Germany

Mark Bates
School of Engineering and Applied
Sciences
29 Oxford Street, Cambridge
MA 02138, USA

David Bensimon
Laboratoire de Physique Statistique
UMR 8550
Ecole Normale Supérieure
Paris, France

and

Department of Chemistry and
Biochemistry
University of California at Los
Angeles
Los Angeles, CA, USA
david.bensimon@ens.fr,
david@lps.ens.fr

Oscar Björnham
Department of Applied Physics
and Electronics
Umeå University
901 87 Umeå, Sweden

Kerstin Blank
Department of Chemistry
Katholieke Universiteit Leuven
Leuven, Belgium

Christoph Bräuchle
Department of Chemistry und
Biochemistry and Center for
Nanoscience (CeNS)
Ludwig-Maximilians-Universität
München
Butenandtstrasse 11
81377 München, Germany
Christoph.Braeuchle@
cup.uni-muenchen.de

Carlos Bustamante
Jason L. Choy Laboratory of Single
Molecule Biophysics and Department
of Physics
University of California, Berkeley
CA 94720, USA

and

Departments of Chemistry and
Molecular and Cell Biology
Howard Hughes Medical Institute
University of California, Berkeley
CA 94720, USA
carlos@alice.berkeley.edu

Mickaël Castelain
Department of Physics
Umeå University
901 87 Umeå, Sweden

Richard J. Cogdell
Division of Biochemistry and
Molecular Biology
Institute of Biomedical and Life
Sciences
Biomedical Research Building
University of Glasgow
120 University Place
Glasgow G12 8TA, UK

Gert De Cremer
Department of Microbial and
Molecular Systems
Katholieke Universiteit Leuven
Leuven, Belgium

Ilja Czolkos
Department of Physical Chemistry
Chalmers University of Technology
412 96 Gothenburg, Sweden

Graham T. Dempsey
Program in Biophysics
Harvard University, Cambridge
MA 02138, USA

Lorna Dougan
Department of Biological Sciences
Columbia University
New York, NY 10027, USA

Yuval Ebenstein
Department of Chemistry and
Biochemistry
and DOE Institute for Genomics and
Proteomics
UCLA, Germany

Elliot L. Elson
Department of Biochemistry and
Molecular Biophysics
Washington University
St. Louis, MO 63110, USA
elson@wustl.edu

A. Engel
Maurice E. Müller Institute for
Structural Biology
Biozentrum, University of Basel
Klingelbergstrasse 70, 4056 Basel
Switzerland

and

Department of Pharmacology
Case Western Reserve University
10900 Euclid Avenue
Wood Bldg 321D, Cleveland
OH 44106, USA
andreas.engel@unibas.ch

Erik Fällman
Department of Physics
Umeå University
901 87 Umeå, Sweden

Julio M Fernandez
Department of Biological Sciences
Columbia University, New York
NY 10027, USA
jf2120@columbia.edu

Shou Furuike
Department of Physics
Faculty of Science and Engineering
Waseda University, Okubo
Shinjuku-ku, Tokyo 169-8555
Japan

Sergi Garcia-Manyes
Department of Biological Sciences
Columbia University
New York, NY 10027, USA

Natalie Gassman
Department of Chemistry and
Biochemistry
and DOE Institute for Genomics and
Proteomics
UCLA, Germany

Hermann E. Gaub
Lehrstuhl für Angewandte Physik
LMU Munich, Amalienstr. 54
80799 Munich, Germany

and

Center for Nanoscience (CENS)
Nanosystems Initiative Munich
(NIM) and Center for Integrated
Protein Science Munich (CIPSM)
Germany
gaub@physik.uni-muenchen.de

Stefan W. Hell
Department of NanoBiophotonics
Max Planck Institute for Biophysical
Chemistry
37070 Göttingen, Germany
hell@nanoscopy.de,shell@gwdg.de

Johan Hofkens
Department of Chemistry
Katholieke Universiteit Leuven
Leuven, Belgium
johan.hofkens@chem.kuleuven.
ac.be

Mohammad Delawar Hossain
Department of Physics
Faculty of Science and Engineering
Waseda University, Okubo
Shinjuku-ku, Tokyo 169-8555
Japan

and

Department of Physics
School of Physical Sciences
Shahjalal University of Science and
Technology
Sylhet-3114, Bangladesh

Bo Huang
Department of Chemistry and
Chemical Biology
Cornell University, Ithaca
NY, USA

and

Howard Hughes Medical Institute
Harvard University, Cambridge
MA 02138, USA
zhuang@chemistry.harvard.edu

Hiroyasu Itoh
Tsukuba Research Laboratory
Hamamatsu Photonics KK
Tokodai, Tsukuba 300-2635
Japan

Aldo Jesorka
Department of Physical Chemistry
Chalmers University of Technology
412 96 Gothenburg, Sweden

Ludovic Jullien
Département de Chimie UMR 8640
Ecole Normale Supérieure
Paris, France
ludovic.jullien@ens.fr

Martin Karplus
Department of Chemistry and
Chemical Biology
Harvard University, Cambridge
MA 02138, USA
marci@tammy.harvard.edu

Petronella Kettunen
Department of Physiological Science
University of California at Los
Angeles
Los Angeles, CA, USA
kettunen@ucla.edu

Kazuhiko Kinosita, Jr
Department of Physics
Faculty of Science and Engineering
Waseda University, Okubo
Shinjuku-ku, Tokyo 169-8555
Japan
kazuhiko@waseda.jp
http://www.k2.phys.waseda.ac.jp

Jürgen Köhler
Experimental Physics IV
and Bayreuth Institute for
Macromolecular Research
Universität Bayreuth
Universtitätsstrasse 30
95440 Bayreuth, Germany
juergen.koehler@uni-bayreuth.de

Zoran Konkoli
Microtechnology and Nanoscience
Center
Chalmers University of Technology
412 96 Gothenburg
Sweden

Efstratios Koutris
Department of Physics
Umeå University
901 87 Umeå, Sweden

Ludvig Lizana
Department of Physical Chemistry
Chalmers University of Technology
412 96 Gothenburg, Sweden

H. Peter Lu
Department of Chemistry
Center for Photochemical Sciences
Bowling Green State University
Bowling Green
OH 43403, USA
hplu@bgsu.edu

William E. Moerner
Departments of Chemistry and
(by Courtesy) of Applied Physics
Stanford University, Stanford
CA 94305, USA
wmoerner@stanford.edu

Jeffrey R. Moffitt
Jason L. Choy Laboratory of Single
Molecule Biophysics and Department
of Physics
University of California
Berkeley
CA 94720, USA

Jörg Mütze
Biophysics group
Biotechnologisches Zentrum
Technische Universität Dresden
Tatzberg 47-51
01307 Dresden
Germany,
petra.schwille@
biotec.tu-dresden.de

Pierre Neveu
Kavli Institute for Theoretical
Physics
University of California at Santa
Barbara
Santa Barbara
CA, USA
neveu@kitp.ucsb.edu

Shuming Nie
Departments of Biomedical
Engineering and Chemistry
Emory University and Georgia
Institute of Technology
101 Woodruff Circle
Suite 2001, Atlanta
GA 30322, USA
snie@emory.edu

Silke Oellerich
Experimental Physics IV
and Bayreuth Institute for
Macromolecular Research
Universität Bayreuth
Universtitätsstrasse 30
95440 Bayreuth
Germany

Thomas Ohrt
Biophysics group
Biotechnologisches Zentrum
Technische Universität Dresden
Tatzberg 47-51, 01307 Dresden
Germany,
petra.schwille@
biotec.tu-dresden.de

Yasuhiro Onoue
Department of Physics
Faculty of Science and Engineering
Waseda University, Okubo
Shinjuku-ku, Tokyo 169-8555
Japan
and

Department of Functional Molecular
Science
The Graduate University for
Advanced Studies (Sokendai)
Okazaki, Aichi 444-8585
Japan

Michel Orrit
MoNOS, LION
Postbox 9504, Leiden University
2300 RA Leiden
The Netherlands
orrit@molphys.leidenuniv.nl

Owe Orwar
Department of Physical Chemistry
Chalmers University of Technology
412 96 Gothenburg, Sweden
orwar@chalmers.se

Zdeněk Petrášek
Biophysics group
Biotechnologisches Zentrum
Technische Universität Dresden
Tatzberg 47-51
01307 Dresden, Germany
petra.schwille@
biotec.tu-dresden.de

Jingzhi Pu
Laboratoire de Chimie Biophysique
ISIS, Université Louis Pasteur
67000 Strasbourg, France

Elias M. Puchner
Lehrstuhl für Angewandte Physik
LMU Munich, Amalienstr. 54
80799 Munich, Germany
and

Center for Nanoscience (CENS)
Nanosystems Initiative Munich
(NIM) and Center for Integrated
Protein Science Munich (CIPSM)
Germany

Hong Qian
Department of Applied Mathematics
University of Washington
Seattle, WA 98195, USA

Martin F. Richter
Experimental Physics IV
and Bayreuth Institute for
Macromolecular Research
Universität Bayreuth
Universtitätsstrasse 30
95440 Bayreuth, Germany

Susana Rocha
Department of Chemistry
Katholieke Universiteit Leuven
Leuven, Belgium

Maarten B. J. Roeffaers
Department of Chemistry
Katholieke Universiteit Leuven
Leuven, Belgium

Michael J. Rust
Department of Physics Harvard
University
Cambridge, MA 02138
USA

Staffan Schedin
Department of Applied Physics and
Electronics
Umeå University
901 87 Umeå, Sweden

Petra Schwille
Biophysics group
Biotechnologisches Zentrum
Technische Universität Dresden
Tatzberg 47-51, 01307 Dresden
Germany
petra.schwille@
biotec.tu-dresden.de

Rieko Shimo-Kon
Department of Physics
Faculty of Science and Engineering
Waseda University
Okubo, Shinjuku-ku
Tokyo 169-8555, Japan

Deepak Kumar Sinha
Laboratoire de Physique Statistique
UMR 8550
Ecole Normale Supérieure
Paris, France
deepak@lps.ens.fr

Andrew M. Smith
Departments of Biomedical
Engineering and Chemistry
Emory University and Georgia
Institute of Technology
101 Woodruff Circle
Suite 2001, Atlanta
GA 30322, USA

Hiroshi Uji-i
Department of Chemistry
Katholieke Universiteit Leuven
Leuven, Belgium

Sophie Vriz
Inserm U770
Hémostase et Dynamique Cellulaire
Vasculaire
Le Kremlin-Bicêtre
France
vriz@univ-paris-diderot.fr

May D. Wang
Departments of Biomedical
Engineering
Georgia Institute of Technology
313 Ferst Drive
UA Whitaker Building 4106
Atlanta, GA 30332, USA
and

Department of Electrical and
Computer Engineering
Georgia Institute of Technology
313 Ferst Drive
UA Whitaker Building 4106
Atlanta, GA 30332, USA

Wenqin Wang
Department of Physics
Harvard University
Cambridge
MA 02138, USA

Watt W. Webb
Cornell University
School of Applied and Engineering
Physics
212 Clark Hall
Ithaca, NY 14853-2501, USA
www2@cornell.edu

Shimon Weiss
Department of Chemistry and
Biochemistry
and DOE Institute for Genomics and
Proteomics
UCLA, University of California
Los Angeles, CA, USA
sweiss@chem.ucla.edu

Mary M. Wen
Departments of Biomedical
Engineering and Chemistry
Emory University and Georgia
Institute of Technology
101 Woodruff Circle
Suite 2001, Atlanta
GA 30322, USA

Jerker Widengren
Exp.Biomol.Physics Dept. Appl.
Physics
Royal Institute of Technology (KTH)
Albanova University Center
106 91 Stockholm, Sweden
jwideng@kth.se

Peter G. Wolynes
Department of Chemistry and
Biochemistry
University of California at San Diego
9500 Gilman Drive
La Jolla, CA 92093
USA
pwolynes@ucsd.edu

X. Sunney Xie
Department of Chemistry and
Chemical Biology
Harvard University
Cambridge
MA 02138 USA
xie@chemistry.harvard.edu

Toshio Yanagida
Graduate School of Frontier
Biosciences
Osaka University, 1-3 Yamadaoka,
Suita, Osaka
565-0871 Japan

and

Formation of soft nano-machines
CREST 1-3 Yamadaoka
Suita, Osaka
565-0871 Japan
yanagida@phys1.med.
osaka-u.ac.jp
http://www.phys1.med.
osaka-u.ac.jp/

Xiaowei Zhuang
Department of Chemistry and
Chemical Biology
Howard Hughes Medical Institute
Harvard University
Cambridge, MA 02138
USA

and

Department of Physics
Program in Biophysics
Harvard University, Cambridge,
MA 02138, USA
zhuang@chemistry.harvard.edu

Part I

Introductory Lecture: Molecular Dynamics of
Single Molecules

How Biomolecular Motors Work: Synergy Between Single Molecule Experiments and Single Molecule Simulations

Martin Karplus and Jingzhi Pu

Summary. Cells are a collection of machines with a wide range of functions. Most of these machines are proteins. To understand their mechanisms, a synergistic combination of experiments and computer simulations is required. Some underlying concepts concerning proteins involved in such machines and their motions are presented. An essential element is that the conformational changes required for machine function are built into the structure by evolution. Specific biomolecular motors (kinesin and F_1-ATPase) are considered and how they work is described.

On the basis of my lecture at Nobel Symposium 138 on Single Molecule Spectroscopy, I shall present studies of proteins that illustrate how single molecule experiments and single molecule simulations complement each other to provide insights not available from either one by itself. I will focus particularly on molecular motors and how they work. Before considering specific examples, I shall describe some general properties of the protein free energy surface and how evolution encodes the required information in protein structures so that they can perform their motor functions. Figure 1.1a shows a schematic picture of the free energy of a polypeptide chain under native conditions of temperature and solvent environment, as a function of an order parameter, such as the radius of gyration, R_g. We see that at large values of R_g, the chain has a high free energy and forms what is often referred to as a random coil, though it is now known that, even in a denaturing environment, there is considerable residual structure. As solution conditions are changed to stabilize the native state, the coil state condenses to a compact globule. This can still be disorganized (i.e., no more native structural features than in the "random" coil) or it can be organized in what is called a molten globule, which has much of the secondary structural elements (α-helices and β-strands) of the native protein, but the tertiary structure has not yet formed and the sidechains are disordered. As R_g continues to decrease, there is usually a free energy barrier before the collapse to the native state, which is a deep minimum (on the order of $10\,\text{kcal}\,\text{mol}^{-1}$) and narrow on the length scale of Fig. 1.1a. As the native state is the one in which most, but not all proteins,

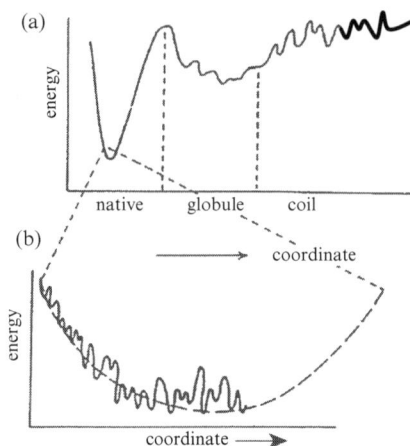

Fig. 1.1. (**a**) Schematic free energy surface for a polypeptide that folds to form a stable protein. The energy is shown as a function of a size coordinate, such as the radius of gyration R_g. (**b**) Details of native state energy surface at approximately constant R_g (see text)

are active, it is useful to examine it at higher resolution. To do so, we choose a coordinate "perpendicular" to R_g; by perpendicular we mean that the size is essentially constant on the scale of Fig. 1.1a. The contributing structures have very similar values of R_g, but differ in the detailed arrangement of the atoms, in accord with the fluctuations demonstrated by native state molecular dynamics simulations [1] or measured by X-ray thermal parameters [2]. Figure 1.1b shows that the surface along the perpendicular direction corresponds overall to a "broad" minimum with a complex multiminimum character. This multiminimum character was demonstrated by quenched molecular dynamics simulations of myoglobin [3]. They showed that the smallest barriers separating two minima are such that they are crossed in 0.1 ps and that there is a whole hierarchy of barriers of increasing height that may require nanosecond, microsecond, or even longer to cross. The quenching simulations were stimulated by the experiments of Frauenfelder and coworkers [2], who studied the rebinding of CO to myoglobin after photodissociation over a temperature range of 40–300 K and times ranging from 10^{-7} to 10^3 s. What made such studies possible in an ensemble system is that the photodissociation reaction provides a "trigger", which synchronizes the initial state of the molecules. The rebinding reaction was shown to be "complex" [4]; that is, the rebinding reaction is stretched exponential or power law, rather than exponential in time, and the rate of the reaction decreases faster than expected from the Arrhenius equation as the temperature is lowered. To interpret both of these observations Frauenfelder et al. postulated a surface such as that shown in Fig. 1.1b. The nonexponential time dependence was explained by the ensemble average over myoglobin molecules trapped in different minima, each of which has a different activation energy for rebinding. The non-Arrhenius temperature dependence was rationalized by the "glassy" nature of the protein at low temperatures. It will be interesting to have single molecule experiments for myoglobin to confirm the Frauenfelder model.

Recent advances in room-temperature fluorescence spectroscopy have made possible the real-time observation of single biomolecules, thus circumventing the problem of synchronization. Of particular interest are distance-sensitive probes based on fluorescence resonance energy transfer (FRET) [5] or electron transfer (ET) [6], which provide information on conformational fluctuations. In the electron transfer experiments I consider here, the Fre/FAD protein complex was used and the quenching of the fluorescent chromophore FAD by electron transfer from an excited Tyr was studied. The observed variation in the quenching rate of a single molecule was interpreted in terms of distance fluctuations between the FAD and a nearby Tyr, on the basis of the exponential distance dependence of the ET rate [7,8]. A stretched exponential decay of the distance autocorrelation function was observed and shown to be consistent with an anomalous diffusion-based model [9,10]; also, a one-dimensional generalized Langevin equation (GLE) model with a power-law memory kernel was found to provide an interpretation of the results [11]. Such formulations provided compact descriptions of the experiments, but they do not determine the underlying molecular mechanism that results in the wide distribution of relaxation times.

There are three tyrosine residues in Fre, Tyr 35, Tyr 72, and Tyr 116, close to the flavin-binding pocket (Fig. 1.2). Fluorescence lifetime measurements of the wild-type and mutant Fre/flavin complexes showed that electron transfer from Tyr 35 to the excited FAD isoalloxazine is responsible for the fluorescence quenching [6]. The average positions of the bound FAD and the three tyrosine residues of the protein are shown in the figure. To study the fluctuations in the

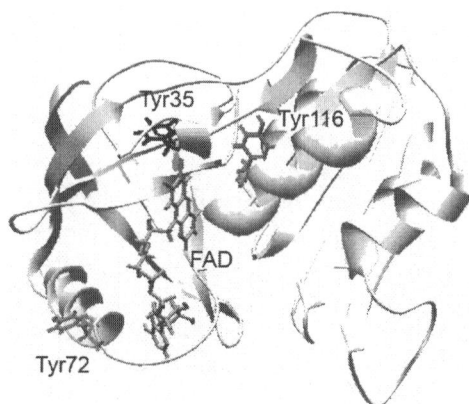

Fig. 1.2. Positions of the Tyr residues and FAD in Fre: The average positions of the three nearby Tyr[35], Tyr[72], and Tyr[116] plus FAD in a 5 ns simulation are shown

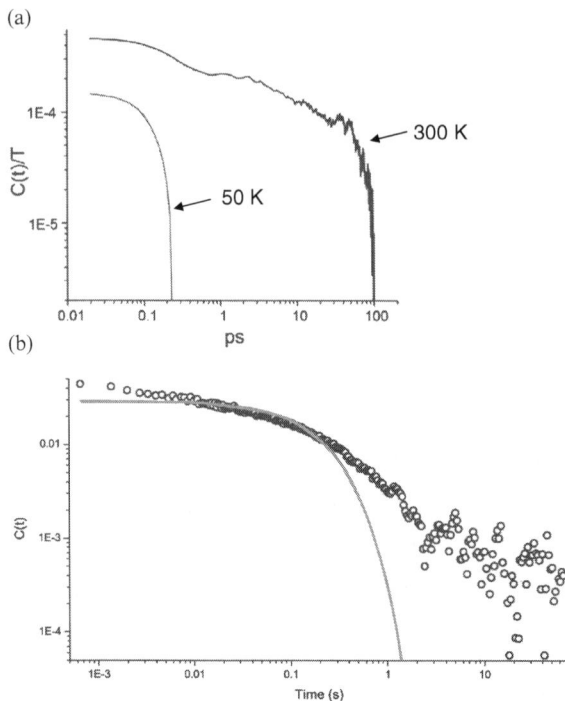

Fig. 1.3. Distance autocorrelation function, C(t); see text. (**a**) Calculated values at 300 K and 50 K; (**b**) Experimental values from [6]

distance, all-atom stimulations were performed [11]. In a 5 ns simulation, Tyr 35 (with an average distance of 7.8 Å) is always nearest to the isoalloxazine; Tyr 116 is slightly further away (9.4 Å), and Tyr 72 is the furthest (15.9 Å). Thus, the calculated relative distances (Tyr72 > Tyr116 > Tyr35) are in accord with the lifetime measurement results and the static X-ray structure. Figure 1.3 shows log–log plots of Tyr 35–isoalloxazine distance autocorrelation functions, $C(t)$, defined by $C(t) =< \delta d(\tau)\delta d(\tau + t) >$ [12], where δd is the deviation of the distance from its average value and $\langle \ldots \rangle$ denotes the time average. The results from a 5 ns trajectory at 300 K are shown in Fig. 1.3a; for times greater than 100 ps, the statistics are such that the results are not meaningful. The calculated decay corresponds to a stretched exponential $C(t) = C(0) \exp(-t/\tau)^{\beta}$, with $\beta = 0.33$ and $\tau = 306$ ps; the value of β is very close to that found experimentally ($\beta = 0.30$), in spite of the very different time scales of the experiments (the experimental τ is 54 ms); see Fig. 1.3b. The importance of the simulations is that they make possible a determination of the origin of the stretched exponential behavior, which is not available from experiment. From earlier work [13], it is known that at 50 K, the system is trapped in a single well over the simulation time scale. Figure 1.3a shows that, as expected, the autocorrelation function from a 50 K trajectory decays exponentially; that is, in this well, the dynamics is simple and can be modeled approximately as the Brownian motion of a harmonic oscillator. This indicates that a trapping mechanism with "jumps" involving a range of barriers is responsible for the stretched exponential behavior observed in the 300 K molecular dynamics simulation. That a corresponding trapping mechanism applies to the actual electron transfer experiment has not been demonstrated. However, one suggestive result is the agreement between the potential of mean force (PMF) for the Tyr–isoalloxazine distances estimated from the single molecule experiments and that calculated from umbrella sampling simulations used to extend the range of distances visited in the dynamics (Fig. 1.4). Figure 1.4a shows the statistics of the Tyr 35–isoalloxazine distances sampled with different umbrella potentials. The potential of mean force as a function of the distance obtained by combining the simulation results, using the weighted histogram analysis method (WHAM) [14–16], is shown in Fig. 1.4b, which also shows the experimental values. As can be seen from the figure, the calculated PMF is in good agreement with the experimental estimate [6]. This agreement provides support for the use of the simulations to examine the nonexponential relaxation, even though the time scales are not commensurate.

Motor Proteins. The above examples provide information about certain aspects of the free energy surface of native proteins, but are not concerned with their function per se. It has been proposed that living cells can be regarded as a collection of machines, which carry out many of the functions essential for their existence, differentiation, and reproduction [17]. The terms "motors" or "machines" are used to describe these molecules because they transduce one form of energy (say, chemical binding) to another (say, mechanical). Each protein machine possesses its specific function and it often forms an element of the chemical network of which the cell is composed [18]. The biomolecular motors range from single subunit proteins (e.g., some DNA polymerases [19]) through the smallest rotary motor F_1-ATPase, composed of nine subunits in mitochondria [20], to the flagella motors of bacteria, which can be composed of several hundred molecules of a number of different proteins [21]. Molecular motors make use of chemical energy from a variety of sources, of which the most common is the differential binding energy of ATP, H_2O, and its hydrolysis products ADP, $H_2PO_4^-$. Proton and ion gradients, as well as redox potential differences, also serve as the energy source in certain cases. The motors have a wide range of functions, including chemical (e.g., ATP) synthesis, organelle transport, muscle contraction, protein folding, and translocation along DNA/RNA, as well as their role in cellular signaling, cell division and cellular motion.

In an insightful chapter in his textbook *"The Feynman Lectures in Physics"*, which includes a description of the relation of physics to other sciences, Feynman pointed out the importance of motion in the function of proteins (Fig. 1.5a). Figure 1.5b is a more poetic description of the atomic fluctuations by my friend, the late Claude Poyart. However, the existence of

(a)

(b)

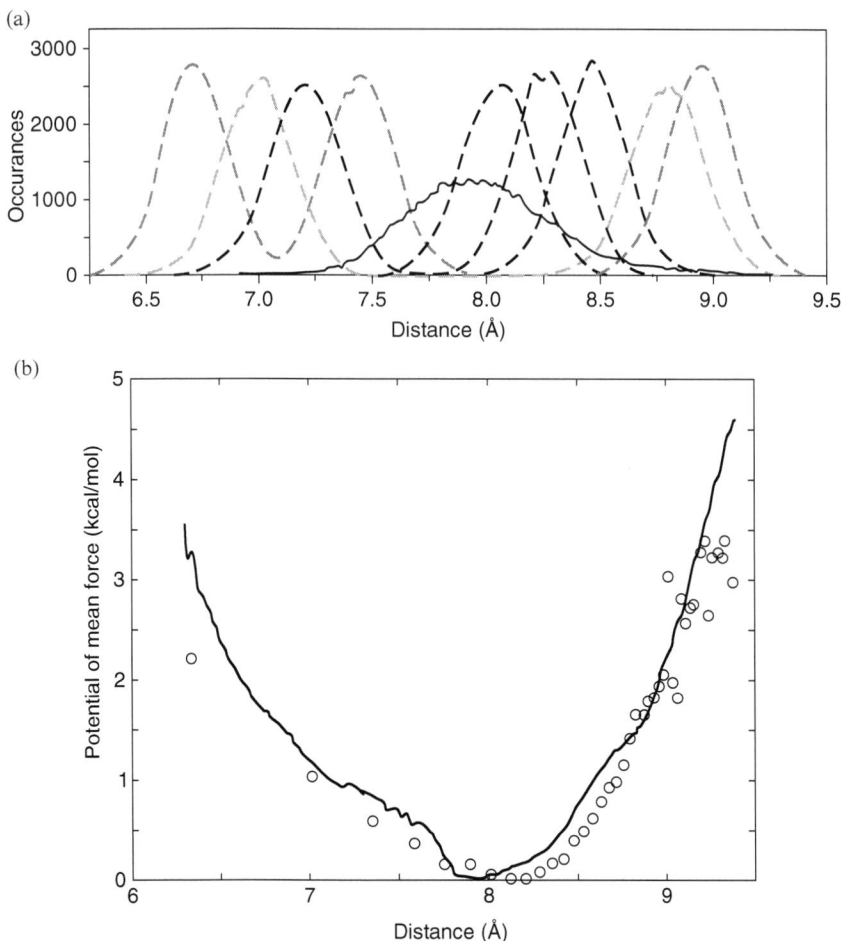

Fig. 1.4. The potential of mean force (PMF) for the Tyr[35]-isoalloxazine center-to-center distance. (**a**) Sampling histograms of the Tyr[35]-isoalloxazine distance; the solid line is the histogram with a restraint-free simulation. The dashed lines are histograms obtained with umbrella potentials. (**b**) The PMF generated with the data from (**a**) is shown as a solid line. Also shown is the experimental PMF (*open circles*) from reference [6]

"jigglings and wigglings of atoms" leaves unanswered the question of how such motion leads to function. One essential point is that the native free energy surface of motor proteins deviates from that shown in Fig. 1.1b. Instead, it has the form shown schematically in Fig. 1.6a; that is, there are at least two major minima on the surface and their relative free energies can be varied by the binding of different ligands. The introduction of such multiple conformations is the key evolutionary development that serves to put the jigglings and

(a)   "...everything that living things do can be
understood in terms of the jigglings and
wigglings of atoms."

Richard Feynman

(b)   "The X-ray structures of proteins are like
trees in winter, beautiful in their stark
outline but lifeless in appearance. Molecular
dynamics gives life to these structures by
clothing the branches with leaves that flutter
in the thermal winds."

Claude Poyart

Fig. 1.5. (a) Exerpt from Feynman RP, Leighton RB, Sands M (1963) *The Feynman Lectures in Physics* (Addison-Wesley, Reading), Vol. I, Chap. 3. (b) Quote from Claude Poyart (private communication)

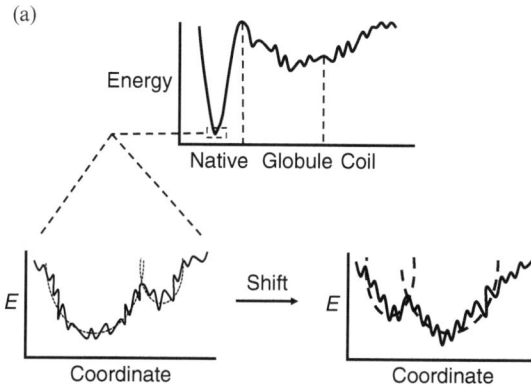

(b)
Putting the "Jigglings and Wigglings" to work

(i)   Semirigid domains with hinges (GroEL, F_1-ATPase)
Disorder to order transition (Kinesin)
Semirigid subunits (Hemoglobin)

(ii) Binding of ligand to change equilibria amongst
conformations (ATP/ADP, P_i in GroEL, F_1-ATPase,
and Kinesin) (O_2 or NO in Hemoglobin)

Fig. 1.6. Putting the "jigglings" and "wigglings" to work. (a) Schematic double minimum potential for a molecular motor with two important conformations. The change in relative stabilities of the two minima is induced by differential ligand binding; (b) Some examples of types of protein functional conformational changes and the ligands involved

wigglings to work. Figure 1.6b indicates some possible structural mechanisms and the ligands involved; they range from two or more different conformations of semi-rigid domains connected by flexible hinges, to disorder-to-order transitions and different quaternary conformations of semi-rigid subunits of oligomers. In what follows, I shall describe two motor systems (kinesin and ATP synthase), in whose understanding single molecule experiments and simulations have played an important role. They represent two different classes in terms of Fig. 1.6b: kinesin is a linear motor that uses ATP to walk on microtubules, while ATP synthase is a rotary motor that synthesizes ATP.

Kinesins. Kinesins, which are the smallest processive motors, generally function as dimers. Each monomer consists of a motor domain and an α-helical stalk; the latter forms a coiled-coil in the dimer, at the end of which is the globular element that transport cargo. From single molecule experiments [22], kinesins seem to walk on microtubules in an asymmetric hand-over-hand manner. There is a 12 residue "neck linker" (NL), which connects the motor domain and the α-helical stalk. Although the neck linker has been shown to be important for walking (i.e., mutations in the neck linker impair motility), it appears to be too flexible to provide the measured directional force [23]. In looking at a series of structures of kinesins with ADP or ATP bound, it became evident that, in fact, these structures did not have a free neck linker but that it forms a two-stranded sheet with the N-terminal β strand (see Fig. 1.7) [24]. Clearly, such a two-stranded β sheet would be considerably stiffer than a single strand, which interacted very weakly, if at all, with the rest of the motor domain. The name "cover strand" (CS) was introduced for the N-terminal β-strand and "cover neck bundle" (CNB) for the two-stranded β sheet [25]. In nucleotide-free motor domains, the CS is not interacting with the undocked neck-linker and appears to be disordered (Fig. 1.8a). In this state, the α4 helix (corresponding to myosin's relay helix) prevents the α6 helix from forming an extra helical turn at the N-terminal end of the NL, which renders the NL out-of-register with the CS (Fig. 1.8a). When ATP binds to the motor head, conformational changes in the switch II cluster lead to retraction of α4 [26, 27], the subsequent formation of an extra helical turn in α6, followed by the shortening of the NL. This places the CS and NL in-register to form the CNB (Fig. 1.8b). The CNB was shown by simulations [25] to possess a forward conformational bias and generate forces consistent with 2D force clamp motility measurements [23]. In contrast, the NL alone exhibited little forward bias and generated much smaller forces, which explains why its role as a force-generating element has been under debate. The simulations thus suggest a force generation mechanism triggered by this dynamic disorder-to-order transition (i.e., formation of the CNB from the NL and CS). Simulations are now being done for a dimer interacting with a microtubule to elaborate this proposal.

To test the model based on the simulations, optical trapping experiments in the presence of an external force were performed for a wild-type kinesin and for two mutants [27]. One mutant introduces two Gly (G2), which are

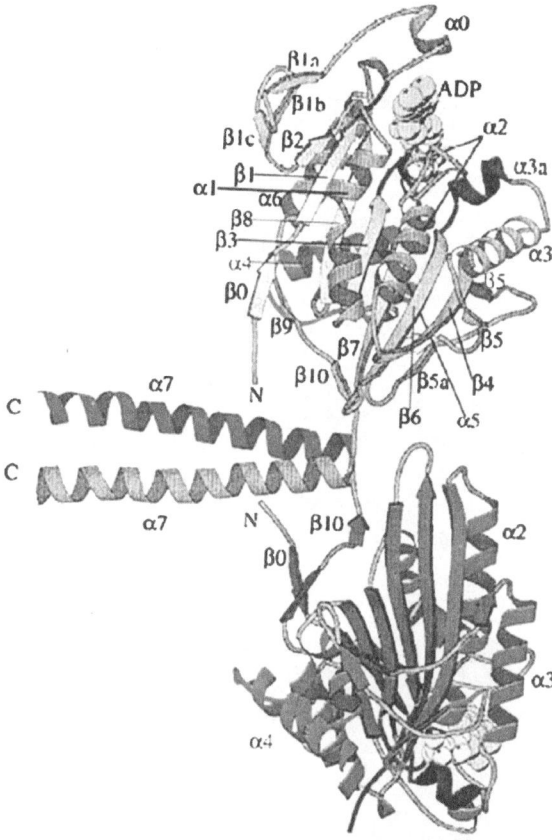

Fig. 1.7. Dimeric kinesin structure with ADP bound showing the β strands involved in the CNB: β_{10} is part of NL and β_0 is the CS. From [24]

expected to make the CNB more flexible and the other completely deletes the CS (DEL) (Fig. 1.9a). Figure 1.9b presents one set of results, namely the decrease of the stall force required for the G2 mutant and the almost zero stall force required for DEL, which appears at best to "limp" along the microtubule [28]; more details of the experimental studies that support the CNB model are described separately [27].

ATP Synthase. The motor enzyme, F_oF_1-ATP synthase, of mitochondria uses the proton-motive force across the mitochondrial membranes to make ATP from ADP and Pi $(H_2PO_4^-)$ [29–32]. It does so under cellular conditions that favor the hydrolysis reaction by a factor of 2×10^5. As a result of the activity of the F_oF_1-ATP synthase, the concentration ratio (ATP:ADP/Pi) is close to unity in mitochondria [33]. This remarkable property is based on the essential difference between an ordinary enzyme, which increases the rate of reaction without shifting the equilibrium, and a catalytic motor like F_oF_1-ATP

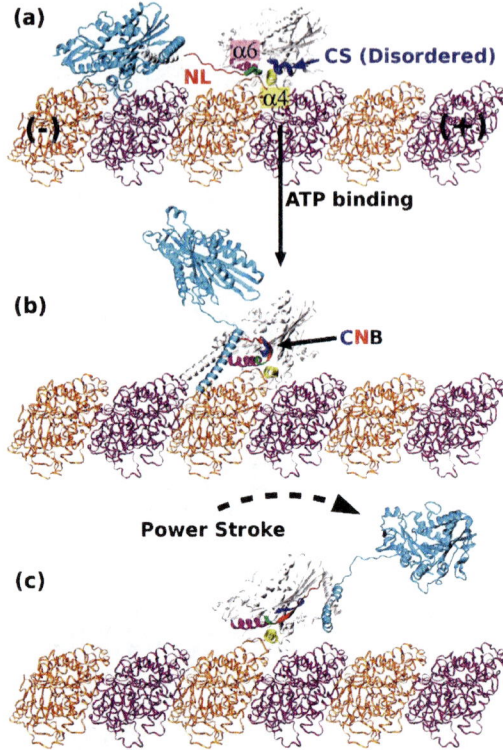

Fig. 1.8. Model for the power stroke based on the simulations (see text). (**a**) Prior to ATP binding, the NL (*red*) and N-terminal CS (*blue*, thick S-shaped tube) of the leading motor head are out-of-register due to the unwound portion (*green*, thick tube) of the α6 helix (*magenta*). (**b**) ATP binding results in retraction of the α4 helix (*yellow*), allowing the extra helical turn of α6 to form and bringing the NL and CS into a favorable position to form a β-sheet, known as the CNB. (**c**) The "kinked" CNB possesses the forward bias to deliver a power stroke and propel the trailing head forward. Kinesin dimers were constructed using PDB 1MKJ (with CNB) and PDB 1BG2 (without CNB). The neck coiled-coil stalk was extended based on PDB 3KIN

synthase, which can drive a reaction away from equilibrium by harnessing an external energy source. Given that a sedentary adult uses (and, therefore, synthesizes) about 40 kg of ATP per day [34], an understanding of the detailed mechanism of F_oF_1-ATP synthase is essential for a molecular explanation of the biology of living cells.

F_oF_1-ATP synthase is composed of two domains (Fig. 1.10a): a transmembrane portion (F_o), the rotation of which is induced by a proton gradient, and a globular catalytic moiety (F_1) that synthesizes and hydrolyzes ATP. The F_1−ATPase moiety, for which several high-resolution structures with the different ligands are available (e.g., [30, 35, 36]), can synthesize, as well as hydrolyze, ATP. Synthesis has been demonstrated by applying an external

(a)   (b)   (c)

WT
MSAEREIPAEDSI

2G
MSAEREIPGEDGI

DEL

(d)

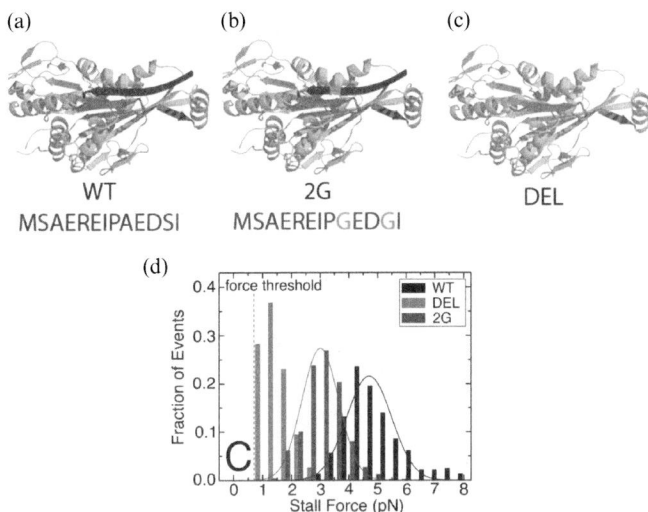

Fig. 1.9. Kinesin mutant design and single-molecule motility results based on an optical trapping assay. (a) WT: full CS. (b) 2G: CS with mutated residues (light area in CS). (c) DEL: CS is absent. The structure is based on PDB 2KIN, modified to incorporate the *Drosophila* CS (SwissProt ID P17210). (d) Stall force histogram. Solid lines: Gaussian fits for WT and 2G; a DEL histogram was not fitted because of the unknown number of stalls below the minimum detection force threshold. See [27] for details

torque to the γ subunit, causing it to rotate in the reverse direction from that observed during ATP hydrolysis [37]. Thus, the primary function of the proton-motive force acting on F_oF_1-ATP synthase is to provide the torque required to rotate the γ subunit in the direction of ATP synthesis. In what follows, my focus is on ATP synthesis, as Kinosita is describing his experiments on ATP hydrolysis at this meeting. However, some brief comments on our simulations of hydrolysis are given at the end, emphasizing their relation to his recent experiments.

F_1-ATPase has three α and three β subunits arranged in alternation around the γ subunit, which has a globular base and an extended coiled-coil domain (Fig. 1.10). All of the α and β subunits bind nucleotides, but only the three β subunits are catalytically active. The crystal structures of F_1-ATPase provide views of distinct conformational states of the catalytic β subunits [30, 35, 36]. The centrally located and asymmetric γ subunit forms a shaft, and it has been proposed that its orientation determines the conformations of the β subunits. The original crystal structure [30] of F_1-ATPase from bovine heart mitochondria led to the identification of three conformations of the β subunits: β_E (empty), β_{TP} (ATP analog bound), and β_{DP} (ADP bound). In a more recent high-resolution crystal structure [35], the β_{TP} and β_{DP} subunits contain an ATP analog (ADP plus AlF_4^-) and the third catalytic

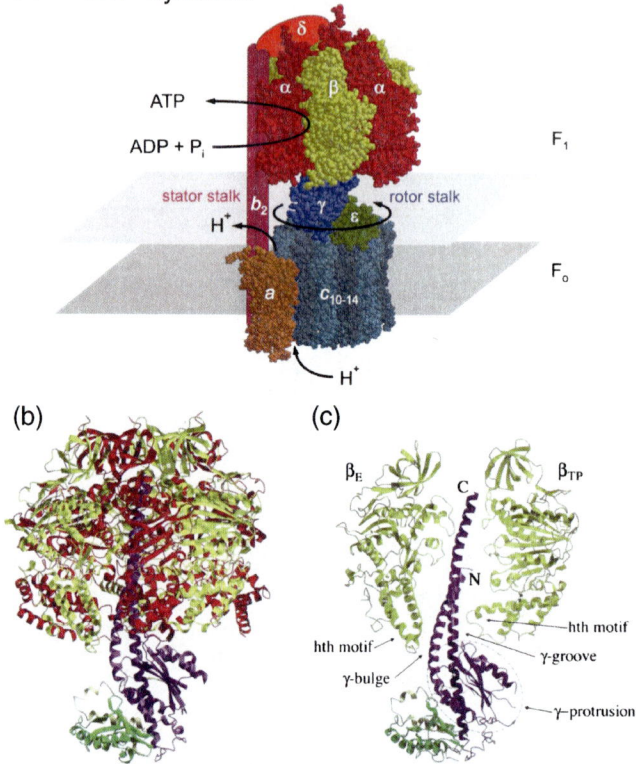

Fig. 1.10. (a) Structural model of F_oF_1-ATP synthase; the figure is from [31]. The three conformations of the β subunits in the 1BMF crystal structure [30] are called β_E (empty) β_{TP} (bound with AMP-PNP) and β_{DP} (bound with ADP), and the three conformations of the β subunits in the 1H8E crystal structure [35] are called β_{HC} (bound with ADP/SO$_4^{2-}$), β_{TP} (bound with ADP/AlF$_4^-$), and β_{DF} (bound with ADP/AlF$_4^-$). (b) Ribbon structure of F_1-ATPase synthase showing $\alpha_3\beta_3\gamma\delta\varepsilon$ based on [30]. α subunits, red; β subunits, yellow; γ subunit, purple; δ subunit, green; ε subunit, light yellow. (c) Cut away diagram showing β_E and β_{TP}, plus γ, ε, and δ, from the same structure as in (b) with labels, including γ bulge, γ groove, and γ protrusion

subunit has a half-closed conformation, called β_{HC}, containing ADP plus SO$_4^-$ (an analog of Pi). The open β_E and half-closed β_{HC} conformations of the third β subunit are both very different from those of the β_{TP} and β_{DP} subunits, which are both closed and very similar to each other in all structures. From his insightful analysis of kinetic data, in advance of detailed structural information, Paul Boyer proposed a "binding change mechanism" by which rotary catalysis could operate [29]. In a modern interpretation of the mechanism, which differs in some details from Boyer's original proposal, ATP synthesis proceeds by the cyclical conversion of each of the β subunits into different conformational states, mainly related to those observed in the crystal structures (see below). The first crystal structure [30] clearly supported the general features of the binding-change mechanism, as did the single-molecule experiments that visualized the rotation of the γ-subunit, which occurs when ATP was provided to F_1-ATPase [38–40]. Researchers were reluctant to believe in the rotary mechanism until its explicit demonstration, but it is now generally accepted and forms the conceptual basis of the quantitative model I describe here [41]. Given the remarkable properties of F_1-ATPase, this rotary-motor enzyme has been the subject of many experimental studies. Nevertheless, a detailed understanding of the mechanism by which it carries out its functions is not available. This has been due primarily to the lack of a quantitative description of how the thermodynamics and kinetics of the enzyme are related to the known crystal structures. Recent molecular dynamics simulations have supplied the "missing link" between the high-resolution crystal structures of F_1-ATPase and measurements of ATP affinities in solution. On the basis of these results, a consistent structure-based model for ATP synthesis can be formulated. Details are given in [41]. Here I present mainly the contributions made to the formulation of the model by molecular dynamics simulations.

Molecular dynamics simulations [42, 43] have shown how rotation of the γ shaft induced either by the proton-motive force [29] or an external force [37] alter the conformations of the subunits. Both calculations [42, 43] were performed with rotation of the γ-subunit enforced in the direction predicted for synthesis. The timescale for one 360° rotation of the γ shaft is in the microsecond-to-millisecond range [44, 45] and is therefore not directly accessible to the nanosecond timescales probed by standard molecular dynamics simulations. To overcome this problem, the conformational transitions in F_1-ATPase were obtained by simulations in the presence of biasing forces, which were applied to either the γ subunit alone or to the entire structure in a procedure that drives the system from one state to the other on the nanosecond timescale without explicitly constraining the nature of the transition path. The implicit assumption in such studies is that meaningful information concerning the mechanism can be obtained even though the time scale of the forced rotational transition is several orders of magnitude faster than the actual rotation rate. This is tested, in part, by doing simulations over a range of times, say 500 ps–10 ns, and determining what aspects of the transition are "robust." The simulation results demonstrated how the rotation of the γ subunit can induce the observed structural changes in the catalytic β subunits and explained why there is much less movement in the catalytically inactive α subunits that are bound to ligand. Both van der Waals (steric) and electrostatic interactions contribute to the coupling between the β and the γ subunits. The dominant electrostatic interactions occur between positive residues of both the coiled-coil portion and the globular region of the γ subunit (the "ionic track") and the negatively charged residues of the β subunits (see Figs. 1.4 and 1.5 of [42]). This ionic track leads to a smooth rotation pathway without

large jumps in the coupling energy and is likely to contribute to the high efficiency of the chemo-mechanical energy transduction. An experimental paper
subsequently confirmed that certain of the ionic track residues play a role
[46]. The simulations show how the rotation of the γ subunit, based primarily on its asymmetric coiled-coil shaft (see Fig. 1.10c), induces the opening
motion of the β subunits. In contrast, the closing motion of the β subunits
appears to be spontaneous once ligand is bound to the active site and there
are no steric restrictions caused by the γ subunit. This has been confirmed for
an isolated β subunit in solution by nuclear magnetic resonance (NMR) and
by thermodynamic measurements [47, 48]. Interestingly, the conformational
changes observed in the β subunits have been shown to correspond to their
lowest frequency normal modes [49]. This is in accord with the concept that
the structure of the protein, as designed by evolution, is such that the motions
required for its function involve relatively small energies.

As different β subunit conformations are involved in ADP/Pi binding,
ATP synthesis, and ATP release, an essential element of the structure-based
mechanism of F_1-ATPase is the standard free-energy difference between ATP,
H_2O, and ADP/Pi at the various catalytic sites. Four different binding constants for ATP have been measured for F_1-ATPase in solution. The values
for the *E. coli* enzyme are 0.2nM, 2μM, 25μM, and 5m [50]. It is generally
agreed that the open (β_E) and half-closed (β_{HC}) subunits have the two weak-
est binding affinities and that the two other subunits (β_{TP} and β_{DP}) contain
the tightest site and the second highest affinity site. To resolve the uncertainty concerning the β_{TP} and β_{DP} binding affinities, free-energy difference
simulations were performed [51]. The standard free-energy change (ΔG^o) of
the hydrolysis reaction $\left(\text{ATP} + H_2O \rightarrow \text{ADP} + H_2PO_4^-\right)$ in the various β subunit conformations was calculated [52]. In the β_{TP} site, bound ATP/H_2O was
found to have a free energy similar to that of ADP/Pi (1.4–1.5 kcal mol^{-1}
difference), whereas the free energy in the β_{DP} site favors ADP/Pi relative
to ATP/H_2O by 9 kcal mol^{-1}. Experiments have shown that under unisite
hydrolysis conditions (i.e., at ATP concentrations so low that only one ATP
is bound to the enzyme), the free energies of ATP/H_2O and ADP/Pi in the
occupied site are nearly the same; the measured ΔG^o value is 0.4 kcal mol^{-1}
in the mitochondrial enzyme and -0.6 in the *E. coli* enzyme [52]. Because
unisite hydrolysis takes place in the site that has the highest affinity for ATP,
the simulation results can be used to identify the β_{TP} and β_{DP} subunits as
the ones with the highest and second highest affinity for ATP, respectively.
Recently [53], an ensemble FRET study confirmed the identification of the
β_{TP} site as the strong binding site for ATP. Further, free energy simulations
showed that the β_{HC} subunit binding affinity of ATP agreed with the values
measured when the proton-motive force is present [33]. Having matched the
ATP binding constants with specific β subunit conformations, it was possible
to determine the binding constants for ADP and Pi for each of the β sites [50].

Given these results, a mechanism was proposed that shows how F_1-ATPase
can synthesize ATP from ADP/Pi against a strong thermodynamic driving

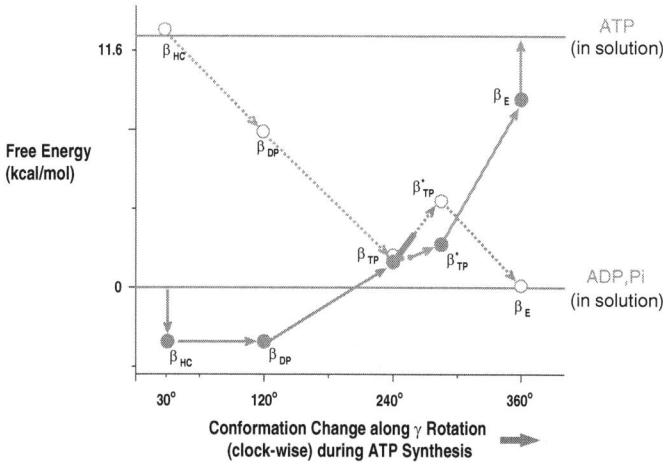

Fig. 1.11. The changing chemical potentials of ATP and ADP/Pi in a β subunit as a function of γ subunit rotation; a full 360° cycle for a single catalytic subunit is shown. The rotation angle of the γ subunit is shown on the abscissa, and the corresponding β subunit conformations are indicated. In addition to the known structures, an intermediate state, β^*_{TP}, favoring ATP, is shown between β_{TP} and β_E sites (see text). Dark gray is used for ATP and light gray for ADP/Pi. The thermodynamically favored states are represented by filled circles and the unfavored states by open circles. The dominant transitions are represented by solid arrows and the unimportant ones by dotted arrows. The chemical reaction is indicated by arrows that are half solid light (ADP/Pi) and half solid dark (ATP). The solution state of ATP is represented by an dark line at $11.3\,\mathrm{kcal\,mol}^{-1}$ (the potential of ATP relative to ADP/Pi at cellular concentrations), and the solution state of ADP/Pi is represented by a light line, which is set to zero

force biased toward hydrolysis [41]. Figure 1.11 shows the thermodynamic properties of the different conformations of the catalytic β subunits involved in the ATP synthesis reaction, which is driven by a clockwise rotation of the γ subunit. The β_{HC} subunit binds ADP/Pi because it has a lower free energy in the β_{HC} subunit than in the solution. Because the potential of ATP in the β_{HC} site is higher than that in the solution (see Fig. 1.11), there is little interference (inhibition) from ATP binding, even when it is present at a concentration similar to that of ADP/Pi. Rotation of the γ subunit by 90°, after ADP and Pi are bound, transforms β_{HC} into β_{DP}. This conformational change does not induce synthesis because the reaction free energy still strongly favors ADP/Pi. A further rotation of 120° changes β_{DP} to β_{TP}, the high-affinity site for ATP. The free energies of bound ATP and ADP/Pi are approximately equal in β_{TP}, and synthesis can begin, but the rate is much slower ($0.04\,\mathrm{s}^{-1}$ in *E. coli*) than the observed maximal rate (10–$100\,\mathrm{s}^{-1}$ in *E. coli*) [31]. A conformational change is required to shift the equilibrium toward ATP and increase the rate of ATP synthesis. Free-energy simulations (unpublished data) suggest that this occurs in a conformation similar to β_{TP} but with local structural changes induced by rotation of the γ subunit to about 300°; we refer to this structure, which has not been observed experimentally, as β_{TP}^* (Fig. 1.11). Rotation of the γ subunit to complete the 360° cycle creates the β_E site, which binds to ATP less strongly, as required for product release. As the β_E site is approached, the free energy of ATP becomes higher than that of ADP/Pi. A lower value of the free energy of ATP in the β_E site, relative to that in solution, is required for optimization of hydrolysis, as well as synthesis, because β_E is the binding site for the ATP substrate in hydrolysis, as well as the release site for ATP after synthesis. The openness of the β_E site makes the release and binding rate of ATP rapid enough such that it is not rate limiting under normal conditions. Hydrolysis of ATP is prevented as β_E is approached, because the catalytic residues, particularly Arg α373 and Arg β189, are no longer in a position to accelerate the reaction [41].

What I have described so far is concerned with the synthesis of ATP by F_1−ATPase, with the γ subunit rotating in the synthesis direction. Because ATP hydrolysis has been discussed by Kinosita at this meeting (see Lecture 17) based on a series of single molecule experiments, I thought it would be useful to briefly describe our studies of hydrolysis. We have developed a coarse-grained structural model [54], which made possible the simulation of the full rotation cycle involved in hydrolysis. The $\alpha_3\beta_3$-crown and the γ-stalk are represented by separate plastic network models (PNMs) [55], and they interact by a repulsive van der Waals-type interaction. The PNM represents each entity ($\alpha_3\beta_3$-crown in a given conformation, γ-stalk) by an elastic network (EN), whose energy minimum corresponds to a known crystal structure. A modified targeted molecular dynamics method [56] was applied to the $\alpha_3\beta_3$-crown to gradually transform the conformation of the catalytic β-subunits (and their neighboring α-subunits) from the EN representing one structure to the other, and the response of the γ-stalk was monitored. The model shows how the conformational changes of the catalytic β-subunits, particularly the in/out motions of the helix-turn-helix motifs, induced by binding of ATP and product release, produce a torque that leads to the rotation of the γ-subunit. The simulations reproduce the 85°/35° rotational substeps observed in single molecule experiments (80°/40° to 90°/30°) [40]; see Fig. 1.12a for a typical set of simulation results. Details of this work are described in [54]. Analysis of the simulation shows what residues of the γ-subunit play an important role in the coupling between the in/out motion of the β subunits and the γ-rotation. Particularly for the 85° portion of the rotation, the inward motion of the β_E subunit on ATP binding has the dominant effect, although there are contributions from other subunits. Figure 1.12b shows the residues involved in the torque generation, while Fig. 1.12c shows the distribution of the torque over the γ-stalk residues as a function of time during the conformational transitions that produce the 85° and 35° substeps, respectively. Four clusters of γ-residues are identified in Fig. 1.12c; they are the ones primarily responsible for the two stages of the 85°/35° rotation of the γ-stalk. The parts of the

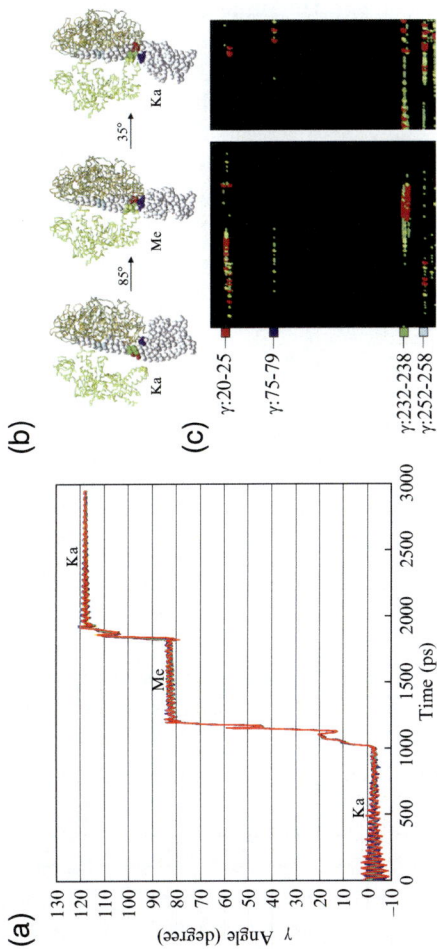

Fig. 1.12. Simulation of γ-rotation due to β subunit motion. (**a**) Rotational angles of the γ-stalk during the molecular dynamics simulations. The system is first equilibrated for 1 ns with an elastic network representation of the "Ka" structure (PDB 2HLD), then the crown conformation is gradually transformed to that of the "Me" structure (PDB 1H8E) by a TMD simulation (220 ps). The first transition yields an 85° γ-rotation. After a 600-ps further equilibration with the Me crown, the α₃β₃ subcomplex is gradually transformed back to the Ka structure over a period of 130 ps by a TMD simulation. The second transition introduces a 35° γ-rotation. The system was then equilibrated for another 1 ns until a plateau of the γ-rotational angle was obtained. Results for 10 independent simulations are plotted; they superpose so well that they cannot be distinguished. (**b**) Residues involved in torque generation in the γ-subunit, shown as colored spheres. (**c**) Torque distribution over the residues (residue number on the y axis) of the γ-subunit. Torque profiles during the 85° (*Left*) and 35° (*Right*) substep rotations, respectively; the torque is scaled to lie between 0 and 1 on each map. The torque (τ) is then colored based on a relative scale, $\tau < 0.1$ (*black*), $0.1 \leq \tau < 0.2$ (*green*), $0.2 \leq \tau < 0.3$ (*yellow*), and $\tau \geq 0.3$ (*red*). Mitochondrial residue numbering is used; the corresponding yeast number is given in parenthesis. Four torque generation "hot spot" clusters on the γ-stalk are labeled by using the same color scheme for identifying residues as in (**b**): γ:20–25 (20–25) (*red*), γ:75–79 (81–85) (*dark blue*), γ:232–238 (237–243) (*green*), and γ:252–258 (257–263) (*cyan*). The torque generation for the 85° substep rotation displays a relay pattern, where the first set of torques is generated primarily on the N-terminal helix of the coiled-coil (γ:20–25) and the second set of torques is generated on the C-terminal helix (γ:232–238). The program VMD was used for making this illustration

β_E-subunit that generate the torque are indicated in Fig. 1.12b; the latter goes from β_E to β_{TP} during the 85° rotation and remains in the β_{TP} conformation during the remaining 35° rotation, which completes the 120° rotation cycle. Structurally, the portion of the γ-subunit that inserts into the $\alpha_3\beta_3$-subunits consists of a left-handed coiled-coil, formed by its N-terminal helix (short) and C-terminal helix (long) with the N and C helices antiparallel. Two of the torque-generating clusters are located in the "neck" region; that is, the most convex curved part of the coiled-coil just above the globular base of the γ-subunit where close contacts with the surrounding γ-subunits occur. They are γ:20–25 (red) on the N-terminal helix and γ:232–238 (green) on the C-terminal helix. The third cluster, γ:252–258 (cyan), is located on the upper part of the C-terminal α helix, and the last torque-generation cluster (dark blue) is located at γ:75–79. Several of these residues have been shown to be important for torque generation by mutation experiments; for a discussion, see [54].

In his lecture, Kinosita emphasized studies in which he deleted part of the γ shaft of the F_1-ATPase from a thermophilic bacterium (TF_1) and demonstrated that rotation still occurs [57], albeit in what might be called a "limping" mode, in analogy to the terminology used for certain kinesin mutants. Specifically Kinosita et al. created F_1-ATPase constructs in which portions of the N and/or C helices of the γ-stalk coiled-coil were deleted. They found that rotation still took place, although there was a general tendency to slow the rotation rate; the γ-rotation also became more erratic as the extent of the deletion increased. In these constructs, some residues of the γ-subunit found to be important for the γ-rotation in the simulations [54] are still present. In unpublished experiments reported at the meeting, the Kinosita group showed that even more drastic reductions of the γ-subunit can lead to limited rotation; an example cited at the meeting was a construct with the entire N helix deleted, which showed some rotation. In Fig. 1.12c, it is evident that there are torque contributions involving the C-helix that could explain the observation. To obtain a more precise model, structural data showing the changes in the modified γ-stalk orientation are needed; that is, different stable positions of the γ-stalk may be involved, so that an analysis based on the wild type structure may not be valid.

Because of the importance of ATP synthesis for life, it is likely that the F_1-ATPase has evolved into a very robust, highly efficient machine. The simulation results I have reported show how F_oF_1-ATP synthase and F_1-ATPase function in present-day living systems. The Kinosita laboratory results on various reduced γ-constructs make clear that the rotational motion induced by ATP hydrolysis in F_1-ATPase is "over designed," in agreement with this concept; that is, the entire γ-stalk is not essential for rotation, although the γ-stalks in all known F_oF_1-ATP synthases from bacteria to humans are structurally very similar. It is possible that some of the reduced γ-subunits are similar to a hypothetical precursor, which is only marginally effective (Kinosita (private communication) agrees with this viewpoint). One proposal is that ATP synthase has evolved from helicases, which resemble the $\alpha_3\beta_3$

crown but do not have a rotary motor function. In fact, a DNA transporter, TrwB, active in bacterial conjugation, has a structure suggestive of the $\alpha_3\beta_3$ crown and uses ATP as an energy source to pump DNA through its central channel [58]. An analogue of such a helicase could have recruited a reduced γ-subunit, like the one of Kinosita's constructs, to form the "Ur-ATP synthase," for which there is, as yet, no evolutionary evidence.

The above overview has tried to show how single molecule experiments and single molecule simulations act in a complementary fashion. For F_oF_1-ATP synthase and many molecular motors, such a synergy can be expected to continue to aid in our understanding of these wonderful machines.

Acknowledgments

Past members of my research group who participated in the work described here are Qiang Cui, Ron Elber, Yi Qin Gao, Ron Levy, Jianpeng Ma, Paul Maragakis, Arjan van der Vaart, and Wei Yang. The study of kinesin is a collaboration with Wonmuk Hwang and the group of Matthew Lang; see [25] and [27]. The research described here was supported in part by the National Institutes of Health (USA) and the Department of Energy (USA).

References

1. J.A. McCammon, M. Karplus, Nature **268**, 765–766 (1977)
2. H. Frauenfelder, G.A. Petsko, D. Tsernoglou, Nature (London) **280**, 558–563 (1979)
3. R. Elber, M. Karplus, Science **235**, 318–321 (1987)
4. M. Karplus, J. Phys. Chem. B **104**, 11–27 (2000)
5. X. Zhuang, H. Kim, M.J.B. Pereira et al, Science **296**, 1473–1476 (2002)
6. H. Yang, G. Luo, P. Karnchanaphanurach et al, Science **302**, 262–266 (2003)
7. H.B. Gray, J.R. Winkler, Annu. Rev. Biochem. **65**, 537–561 (1996)
8. C.C. Moser, J.M. Keske, K. Warncke et al, Nature (London) **355**, 796–802 (1992)
9. R. Metzler, E. Barkai, J. Klafter, Phys. Rev. Lett. **82**, 3563–3567 (1999)
10. S.C. Kou, X.S. Xie, Phys. Rev. Lett. **93**, 180603/1–180603/4 (2004)
11. G. Luo, I. Andricioaei, X.S. Xie, M. Karplus, J Phys Chem B **110**, 9363–9367 (2006)
12. H. Yang, X.S. Xie, J Chem Phys **117**, 10965–10979 (2002)
13. K. Kuczera, J. Kuriyan, M. Karplus, J Mol Biol **213**, 351–373 (1990)
14. S. Kumar, D. Bouzida, R.H. Swendsen et al, J Comput Chem **13**, 1011–1021 (1992)
15. E.M. Boczko, C.L. Brooks III, J. Phys. Chem. **97**, 4509–4513 (1993)
16. M. Souaille, B. Roux, Comput Phys Comm **135**, 40–57 (2001)
17. B. Alberts, Cell **92**, 291–294 (1998)
18. M. Schliwa, *Molecular Motors*. (Wiley, Weinheim, 2003)
19. P.H. Patel, L.A. Loeb, Nat Struct Biol **8**, 656–659 (2001)
20. J.P. Abrahams, A.G.W. Leslie, R. Lutter, J.E. Walker, Nature **370**, 621–628 (1994)

21. H.C. Berg, Ann Rev of Biochem **72**, 19–54 (2003)
22. C.L. Asbury, Curr Opinion in Cell Biology **17**, 89–97 (2005)
23. S.M. Block, C.L. Asbury, J.W. Shaevitz, M.J. Lang, Proc Natl Acad Sci USA **100**, 2351–2356 (2003)
24. F. Kozielski, S. Sack, A. Marx et al, Cell **91**, 985–994 (1997)
25. W. Hwang, M.J. Lange, M. Karplus, Structure **16**, 62–71 (2008)
26. C.V. Sindelar, M.J. Budny, S. Rice et al, Nat Struct Biol **9**, 844–848 (2002)
27. A.S. Khalil, D.C. Appleyard, A.K. Labno et al., Proc Natl Acad Sci USA **105**, 19246–19251 (2008)
28. S.M. Block, Biophysical J **92**, 2986–2995 (2007)
29. P.D. Boyer, Annu Rev Biochem **66**, 717–749 (1997)
30. J.P. Abrahams, A.G.W. Leslie, R. Lutter, J.E. Walker, Nature **370**, 621–628 (1994)
31. A.E. Senior, S. Nadanaciva, J. Weber, Biochim Biophys Acta **1553**, 188–211 (2002)
32. M. Karplus, Y.Q. Gao, Current Opinion in Structural Biology **14**, 250–259 (2004)
33. R.K. Nakamoto, C.J. Ketchum, P. Kuo et al, Biochim Biophys Acta **1458**, 289–299 (2000)
34. D. Voet, J.G. Voet, *Biochemistry*, 2nd edn. (Wiley, New York, 1995)
35. R.I. Menz, J.E. Walker, A.G.W. Leslie, Cell **106**, 331–341 (2001)
36. V. Kabaleeswaran, N. Puri, J.E. Walker et al, EMBO J **25**, 5433–5442 (2006)
37. H. Itoh, A. Takahashi, K. Adachi et al, Nature **427**, 465–468 (2004)
38. H. Noji, R. Yasuda, M. Roshida et al, Nature **386**, 299–302 (1997)
39. R. Yasuda, H. Noji, K. Kinosita, M. Yoshida, Cell **93**, 1117–1124 (1998)
40. R. Yasuda, H. Noji, M. Yoshida, K. Kinosita, Nature **410**, 898–904 (2001)
41. Y.Q. Gao, W. Yang, M. Karplus, Cell **123**, 195–205 (2005)
42. J. Ma, T.C. Flynn, Q. Cui et al, Structure **10**, 921–931 (2002)
43. R.A. Böckmann, H. Grubmüller, Nat Struct Biol. **9**, 198–202 (2002)
44. K. Kinosita Jr, K. Adachi, H. Itoh, Annu Rev Biophys Biomol Struct **33**, 245–268 (2004)
45. D. Spetzler, J. York, D. Daniel et al, Biochemistry **45**, 3117–3124 (2006)
46. S. Bandyopadhyay, W.S. Allison, Biochemistry **43**, 2533–2540 (2004)
47. H. Yagi, T. Tsujimoto, T. Tamazaki et al, J Am Chem Soc **126**, 16632–16638 (2004)
48. G. Perez-Hernandez, E. Garcia-Hernandez, R.A. Zubillaga, M. Tuena de Gomez-Puyou, Arch Biochem and Biophysics **408**, 177–183 (2002)
49. Q. Cui, G. Li, J. Ma, M. Karplus, J Mol Biol **340**, 345–372 (2004)
50. Y.Q. Gao, W. Yang, R.A. Marcus, M. Karplus, Proc Natl Acad Sci USA **100**, 11339–11344 (2003)
51. T. Simonson, G. Archontis, M. Karplus, Accts Chem Res **35**, 430–437 (2002)
52. W. Yang, Y.Q. Gao, Q. Cui et al, Proc Natl Acad Sci USA **100**, 874–879 (2003)
53. H.Z. Mao, J. Weber, Proc Natl Acad Sci USA **104**, 18478–18483 (2007)
54. J. Pu, M. Karplus, Proc Natl Acad Sci USA **105**, 1192–1197 (2008)
55. P. Maragakis, M. Karplus, J Mol Biol **352**, 807–822 (2005)
56. A. van der Vaart, M. Karplus, J Chem Phys **122**, 114903.1–114903.6 (2005)
57. S. Furuike, M.D. Hossain, Y. Maki et al, Science **319**, 955–958 (2008)
58. E. Cabezon, F. de la Cruz, Research in Microbiology **157**, 299–305 (2005)

Part II

Detection of Single Molecules and Single
Molecule Processes

2

Single-Molecule Optical Spectroscopy and Imaging: From Early Steps to Recent Advances

William E. Moerner

Summary. The initial steps toward optical detection and spectroscopy of single molecules arose out of the study of spectral hole-burning in inhomogeneously broadened optical absorption profiles of molecular impurities in solids at low temperatures. Spectral signatures relating to the fluctuations of the number of molecules in resonance led to the attainment of the single-molecule limit in 1989. In the early 1990s, many fascinating physical effects were observed for individual molecules such as spectral diffusion, optical switching, vibrational spectra, and magnetic resonance of a single molecular spin. Since the mid-1990s when experiments moved to room temperature, a wide variety of biophysical effects have been explored, and a number of physical phenomena from the low temperature studies have analogs at high temperature. Recent advances worldwide cover a huge range, from in vitro studies of enzymes, proteins, and oligonucleotides, to observations in real time of a single protein performing a specific function inside a living cell. Because each single fluorophore acts a light source roughly 1 nm in size, microscopic observation of individual fluorophores leads naturally to localization beyond the optical diffraction limit. Combining this with active optical control of the number of emitting molecules leads to superresolution imaging, a new frontier for optical microscopy beyond the optical diffraction limit and for chemical design of photoswitchable fluorescent labels. Finally, to study one molecule in aqueous solution without surface perturbations, a new electrokinetic trap is described (the ABEL trap) which can trap single small biomolecules without the need for large dielectric beads.

2.1 Introduction

Nobel Symposium Number 138 on Single-Molecule Spectroscopy (SMS) in Chemistry, Physics and Biology was held at the Sånga-Säby Conference Center, Stockholm, Sweden on June 1–6, 2008. This meeting gathered researchers from all over the globe utilizing studies of individual molecules to explore a wide range of problems spanning numerous fields of natural science. In the 9 years since the earlier Nobel Conference on the same subject in June 1999 [1], the continuing growth of interest, increase in the number of researchers, and wide range of new ideas based on the study of single molecules have been spectacular. At present, the impact of SMS and imaging spans several fields, from chemistry to physics and to biology (Fig. 2.1A). Thanks are due to the Nobel Foundation for making this Symposium possible, and in particular to the organizers, Professors Rudolf Rigler, Jerker Widengren, and Astrid Gräslund for preparing a most stimulating event. Single-molecule forces and positions can be precisely measured in great detail in vitro using attachment to a large dielectric bead and optical tweezers [2]; however, this paper concentrates on direct optical methods which probe individual molecules, one at a time.

It has now been roughly two decades since the first experiment demonstrating optical detection and spectroscopy of single molecules in a condensed phase [3]. SMS allows *exactly one* molecule hidden deep within a crystal, polymer, or cell to be observed via optical excitation of the molecule of interest (Fig. 2.1B). This represents detection and spectroscopy at the ultimate

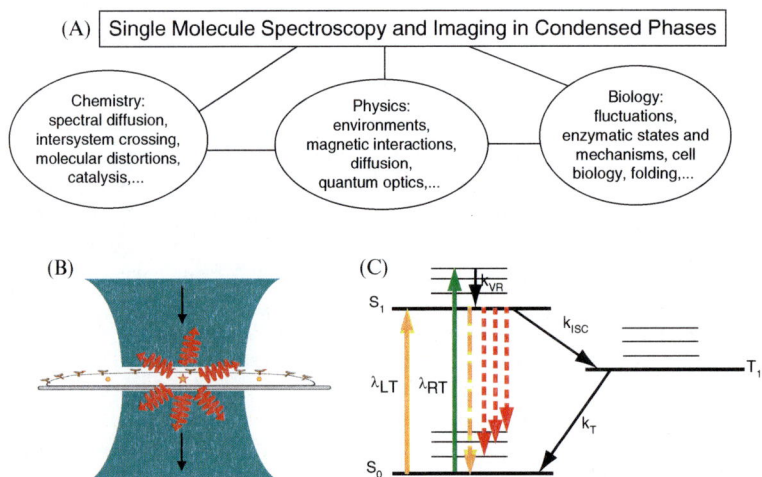

Fig. 2.1. (**A**) The impact of single-molecule spectroscopy and imaging spans areas of chemistry, physics, and biology. (**B**) Schematic of a focused optical beam pumping a single resonant molecule in a cell or other condensed phase sample. The molecule may emit fluorescence or its presence may be detected by carefully measuring the transmitted beam. (**C**) Typical energy level scheme for single-molecule spectroscopy showing the interaction with the pumping light. S_0, ground singlet state; S_1, first excited singlet; T_1, lowest triplet state or other intermediate state. For each electronic state, several levels in the vibrational progression are shown. Typical low-temperature studies use wavelength λ_{LT} to pump the dipole-allowed (0–0) transition, while at room temperature shorter wavelengths λ_{RT} which pump vibronic sidebands are more common. Fluorescence emission shown as dotted lines originates from S_1 and terminates on various vibrationally excited levels of S_0 or S_0 itself. Molecules are typically chosen to minimize entry into dark states such as the triplet state (illustrated). The intersystem crossing or intermediate production rate is k_{ISC}, and the triplet decay rate is k_T

sensitivity level of $\sim 1.66 \times 10^{-24}$ moles of the molecule of interest (1.66 yoc-tomole), or a quantity of moles equal to the inverse of Avogadro's number. Detection of the single molecule of interest must be done in the presence of billions to trillions of solvent or host molecules. To achieve this, a light beam (typically a laser) is used to pump an electronic transition of the one molecule resonant with the optical wavelength (Fig. 2.1C), and it is the interaction of this optical radiation with the molecule that allows the single molecule to be detected. Successful experiments must meet the requirements of (a) guaranteeing that only one molecule is in resonance in the volume probed by the laser and (b) providing a signal-to-noise ratio (SNR) for the single-molecule signal that is greater than unity for a reasonable averaging time.

    Why are single-molecule studies now regarded as a critical part of modern physical chemistry, chemical physics, and biophysics? By removing ensemble averaging, it is now possible to directly measure distributions of behavior to explore hidden heterogeneity. In the time domain, the ability to optically sense internal states of one molecule and the transitions among them allows measurement of hidden kinetic pathways and the detection of rare intermediates. Because typical single-molecule labels behave like tiny light sources roughly 1–2 nm in size and can report on their immediate local environment, single-molecule studies provide a new window into the nanoscale with intrinsic access to time-dependent changes.

    The basic principles of single-molecule optical spectroscopy and imaging have been the subject of many reviews [4–15] and books [16–19]. In this paper, selected milestones achieved in the Moerner laboratory will be summarized, starting from the early steps at low temperatures arising from explorations of spectral hole-burning as a method to achieve frequency domain optical storage (Sects. 2.2 and 2.3). Some of the intriguing physical effects first observed at low temperatures will then be described (Sect. 2.4), as many of these have counterparts in the modern studies of single molecules at room temperature. Finally, recent milestones in the study of single biomolecules in and out of living cells as well as new methods for trapping single molecules at room temperature will be described (Sect. 2.5).

2.2 Early Steps: Statistical Fine Structure in Inhomogeneous Lines

Our first steps toward the single-molecule regime arose from work at IBM Research in the early 1980s on persistent spectral hole-burning effects in the optical transitions of impurities in solids (for a review, see [20]). Briefly, if a molecule with a strong zero-phonon transition and minimal Franck–Condon distortion is doped into a solid and cooled to liquid helium temperatures, the optical absorption becomes inhomogeneously broadened (Fig. 2.2A). The width of the lowest electronic transition for any one molecule (homogeneous width, γ_H) becomes very small because few phonons are present, while at the

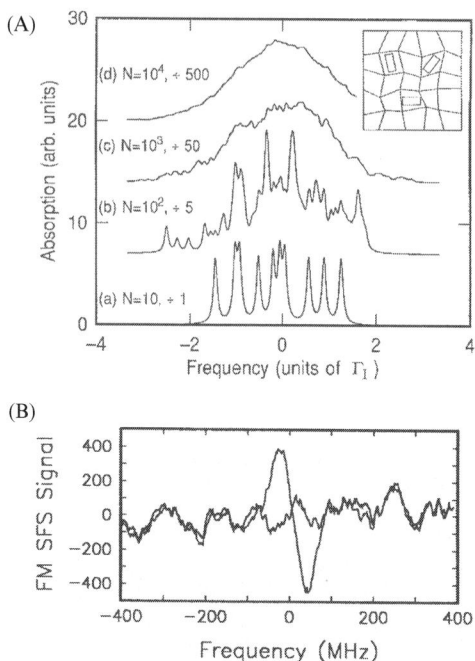

Fig. 2.2. (**A**) Illustration of the source of statistical fine structure (SFS) using simulated absorption spectra with different total numbers of absorbers N, where a Gaussian random variable provides center frequencies for the inhomogeneous distribution. Traces (a) through (d) correspond to N values of 10, 100, 1,000, and 10,000, respectively, and the traces have been divided by the factors shown. For clarity, $\gamma_H = \Gamma_I/10$. Inset: several guest impurity molecules are sketched as rectangles with different local environments produced by strains, local electric fields, and other imperfections in the host matrix. (**B**) SFS detected by FM spectroscopy for pentacene in p-terphenyl at 1.4 K, with a spectral hole at zero relative frequency for one of the two scans. Note the repeatable fine structure

same time different copies of the impurity molecule acquire slightly different absorption energies (resonant optical frequencies) due to local strains and other defects in the solid. Spectral hole-burning occurs when light-driven physical or chemical changes are produced only in those molecules resonant with the light, which yields a dip or "spectral hole" in the overall absorption profile that may be used for optical recording of information in the optical frequency domain.

One goal of the research in the Moerner group at IBM was the exploration of ultimate limits to the spectral hole-burning optical storage process. A particularly interesting limit on the SNR of a spectral hole results from the finite number of molecules that contribute to the absorption profile near the hole. Due to unavoidable number of fluctuations in the density of molecules in any spectral interval, there should exist a "spectral noise" on an inhomogeneous absorption profile scaling as the square root of the number of molecules in resonance (i.e., the number of molecules per homogeneous width, N_H). We named this effect "statistical fine structure" (SFS), and Fig. 2.2A shows how the *relative* size of SFS scales as $1/\sqrt{N_H}$, while the absolute root-mean-square (rms) size of the fine structure scales as $\sqrt{N_H}$. Surprisingly, prior to the late 1980s, SFS had not been detected.

In 1987, SFS was observed for the first time [21, 22], using a powerful zero-background optical absorption technique, laser frequency-modulation (FM) spectroscopy [23, 24]. FM spectroscopy probes the sample with a phase-modulated laser beam; when a narrow spectral feature is present, the imbalance in the laser sidebands leads to amplitude modulation in the detected photocurrent at the modulation frequency. A key feature of the method is that it senses only the deviations of the absorption from the average value, so that detection of SFS could be easily accomplished, but only if a test sample was chosen with minimal spectral hole-burning so that the shape of the inhomogeneous profile could be measured without disturbance from the scanning. The choice of sample was critical: pentacene dopant molecules in a p-terphenyl crystal (Fig. 2.3A), in which spectral hole-burning was weak. In Fig. 2.2B, SFS is the repeatable spectral structure (which looks like noise, but is not time-dependent) over the entire range of the two scans, one of which includes a spectral hole burned at the center. This shows directly that the size of a spectral hole must be larger than the SFS in order to be detected [25]. SFS is clearly an unusual spectral feature, in that its size depends not upon the

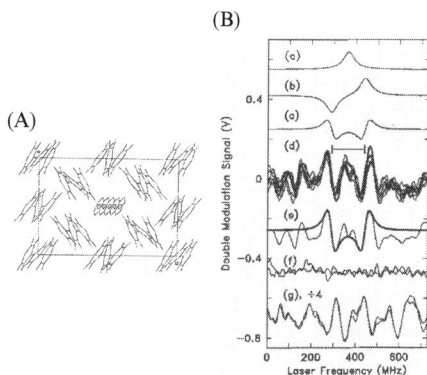

Fig. 2.3. (**A**) Illustration of the crystal structure of p-terphenyl, with a single substitutional impurity of pentacene. (**B**) Single molecules of pentacene in p-terphenyl detected by FM–Stark optical absorption spectroscopy. (a–c) Simulated traces for absorption, FM and FM-Stark, respectively. The "W"-shaped structure in the center of trace (d) is the absorption from a single pentacene molecule, acquired multiple times to show repeatability. (e) Average of traces in (d) with expected lineshape. Trace (f) is the signal from a region of the spectrum with no molecules, while trace (g) is a region with many molecules showing SFS. For details, see [3, 26]

total number of resonant molecules, but rather upon the square root of the number.

2.3 A Scaling Argument Led the way to the First Single-Molecule Detection and Spectroscopy

A simple scaling argument provoked by the observation of SFS suggested that single-molecule detection would be possible. When SFS due to $\sim 1,000$ molecules is detectable (roughly the case in Fig. 2.2B), that means that the measured signal (rms amplitude) is due to ~ 32 molecules in resonance (i.e., $1,000^{1/2}$). This means that in terms of improving the SNR of FM spectroscopy, it was only necessary to work 32 times harder to observe a single molecule, not 1,000 times harder! This realization, combined with two additional facts: (1) FM absorption spectroscopy was insensitive to any scattering background from imperfect samples and (2) FM allows quantum-limited detection sensitivity, led the author to push FM spectroscopy to the single-molecule limit. It is also true that the particularly low quantum efficiency for spectral hole-burning made pentacene in p-terphenyl (Fig. 2.3A) an obvious first choice for single-molecule detection.

The first SMS experiments in 1989 utilized either of two powerful double-modulation FM absorption techniques, laser frequency-modulation with Stark secondary modulation (FM-Stark) or frequency-modulation with ultrasonic strain secondary modulation (FM-US) [3,26]. The secondary modulation was required in order to remove the effects of residual amplitude modulation produced by the imperfect phase modulator. In contrast to fluorescence methods, Rayleigh and Raman scattering were unimportant. Figure 2.3B (specifically trace d) shows examples of the optical absorption spectrum from a single molecule of pentacene in p-terphenyl using the FM-Stark method.

Although this early observation and similar data from the FM-US method served to stimulate much further work, there was one important limitation to the general use of FM methods for SMS. As was shown in the early papers on FM spectroscopy ([23,27], extremely low fractional absorption changes as small as 10^{-7} can be detected in a 1-s averaging time, but only if large laser powers on the order of several milliWatts can be delivered to the detector to drive down the relative size of the shot noise. This presents a problem for SMS in the following way. Since the laser beam must be focused to a small spot to maximize the optical transition probability, the power in the laser beam must be maintained below the value which would cause saturation broadening of the single-molecule lineshape. As a result, it is quite difficult to utilize laser powers in the milliWatt range for SMS of allowed transitions at low temperatures – in fact, powers below 100 nW are generally required with a corresponding increase in the relative size of the shot noise. This is one reason why the SNR of the original data on single molecules of pentacene in p-terphenyl in Fig. 2.3B was only on the order of 5. (The other reason was the use of relatively thick cleaved samples, which produced a population of weaker out-of-focus molecules in the probed volume. This problem was overcome with much thinner samples in later experiments.) In later work [28], frequency modulation of the absorption line itself (rather than the laser) was produced by an oscillating (Stark) electric field alone, and this method has also been used to detect the absorption from a single molecule at liquid helium temperatures. While successful, these methods are limited by the quantum shot noise of the laser beam, which is relatively large at the low laser intensity required to prevent saturation of the single-molecule absorption.

2.4 Selected Low-Temperature Milestones

The optical absorption experiments on pentacene in p-terphenyl showed that this material has sufficiently inefficient spectral hole-burning such that it should be a useful model system for SMS. In 1990, Orrit et al. demonstrated that sensing optical absorption by fluorescence excitation produces superior signal-to-noise if the emission is collected efficiently and the scattering sources are minimized [29]. Due to its relative simplicity, subsequent experiments have almost exclusively used this method. In fluorescence excitation, a tunable narrowband single-frequency laser is scanned over the absorption profile of the single molecule, and the presence of absorption is detected by measuring the fluorescence emitted (downward dashed arrows in Fig. 2.1C). A sketch of the optical components typically found in the cryostat in the author's laboratory is shown in Fig. 2.4A. Tunable pumping light is focused through lens L into sample S with a beam block B behind it. Emitted fluorescence is collected with high efficiency with paraboloid P and directed out of the cryostat. A long-pass filter is used to block any remaining pumping laser light, and the fluorescence shifted to long wavelengths from the tunable pump is detected with a photon-counting system. The detection is background-limited, and the shot noise of the probing laser is only important for the signal-to-noise and not the signal-to-background of the spectral feature. For this reason, it is critical to efficiently collect photons (as with the paraboloid or other high numerical aperture collection system), and to reject the pumping laser radiation. To illustrate, suppose a single molecule of pentacene in p-terphenyl is probed with $1\,\mathrm{mW/cm^2}$, near the onset of saturation of the absorption due to triplet level population. The resulting incident photon flux of 3×10^{15} photons/s-cm^2 will produce about 3×10^4 excitations per second. With a fluorescence quantum yield of 0.8 for pentacene, about 2.4×10^4 emitted photons/s can be expected. At the same time, 3×10^8 photons/s illuminate the focal spot $3\,\mu$m in diameter. Considering that the resonant 0–0 fluorescence from the molecule must be thrown away along with the pumping light, rejection of the pumping radiation by a factor greater than 10^6 is generally required, with minimal attenuation of the fluorescence. This is often accomplished by low-fluorescence thin-film interference filters or by holographic notch attenuation filters.

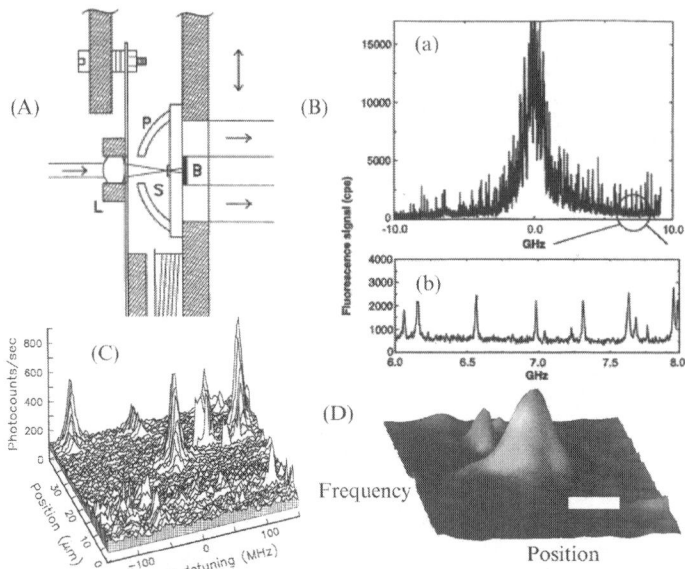

Fig. 2.4. (**A**) Sketch of the cryostat insert for single-molecule spectroscopy by fluorescence excitation. The focus of lens L is placed in the sample S by the magnet/coil pair M, C. (**B**) Scan over the inhomogeneous line (a) with a 2 GHz region expanded (b) to show isolated single-molecule absorption profiles. (**C**) Three-dimensional pseudo-image of single molecules of pentacene in *p*-terphenyl. The measured fluorescence signal (*z*-axis) is shown over a range of 300 MHz in excitation frequency (horizontal axis, center = 592.544 nm) and 40 μm in spatial position (axis into the page). (**D**) Rotation of the data in (**c**) to show that in the spatial domain, the single molecule maps out the shape of the laser focal spot. Bar, 5 μm. For details, see [33]

Table 2.1. Low-temperature single-molecule spectroscopy milestones from the Moerner Lab

Experiment	References
Observation of statistical fine structure scaling as $\sqrt{N}$	[21]
Optical detection and spectroscopy of a single molecule in a solid	[3, 26]
Optical temperature-dependent dephasing, nonlinear optical saturation	[33]
Imaging of a single molecule in space and frequency	[34]
Spectral diffusion and measurement of spectral trajectories	[30, 33, 34]
Photon antibunching for a single molecule	[53]
Photoswitching for single molecules in a polymer, Poisson kinetics	[36, 47]
Vibrational spectroscopy (resonance Raman)	[165, 166]
Magnetic resonance of a single molecular spin	[59]
Near-field optical spectroscopy of a single molecule in a solid	[63]
Pumping single molecules with morphology-dependent resonances of a microsphere	[167]

In the early 1990s, a number of first observations of physical phenomena for single molecules in solids were completed in the Moerner laboratory (Table 2.1). A selection of these advances will now be summarized.

2.4.1 Imaging Single Molecules in Frequency and Space

Figure 2.4B shows a scan over the inhomogeneously broadened optical absorption profile for pentacene in p-terphenyl. While molecules overlap near the center of the line at 592.321 nm (0 GHz), in the wings of the line, single, isolated Lorentzian profiles are observed as each molecule comes into resonance with the tunable laser. It is important to realize that due to inhomogeneous broadening, the laser frequency actually provides a tunable parameter that allows different single molecules to be selected. It is obvious that with a lower-concentration sample, single molecules at the center of the inhomogeneous line can also be studied. These beautiful spectra provided much of the basis for the early experiments. For example, at low pumping intensity, the lifetime-limited homogeneous linewidth of $7.8 + / -0.2$ MHz was directly observed [30]. This linewidth is the minimum value allowed by the lifetime of the S_1 excited state of 24 ns in agreement with previous photon echo measurements on large ensembles ([31, 32]. Such narrow single-molecule absorption lines are wonderful for the spectroscopist: many detailed studies of the local environment can be performed, because such narrow lines are much more sensitive to local perturbations than are broad spectral features.

Single-molecule spectra as a function of laser intensity provided details of the incoherent saturation behavior and the influence of the dark triplet state dynamics [33]. Clear heterogeneity in the observed saturation intensity was observed indicating that the individual molecules experience modifications in photophysical parameters due to differences in local environments. It was also possible to measure the linewidth of single pentacene molecules as a function of temperature in order to probe dephasing effects produced by coupling to a local phonon mode [33].

Going beyond spectral studies alone, a hybrid image of a single molecule can be obtained by acquiring excitation spectra as a function of the position of the laser focal spot in the sample. Spatial scanning was accomplished in a manner similar to confocal microscopy by tilting the incident laser beam in Fig. 2.4A to scan the focal spot across the sample in one spatial dimension. Figure 2.4C shows such a three-dimensional "pseudo-image" of single molecules of pentacene in p-terphenyl [34]. The z-axis of the image is the (red-shifted) emission signal, the horizontal axis is the laser frequency detuning (300 MHz range), and the axis going into the page is one transverse spatial dimension produced by scanning the laser focal spot (40 μm range). In the frequency domain, the spectral features are fully resolved because the laser linewidth of ~3 MHz is smaller than the molecular linewidth. However, as shown in Fig. 2.4D and first noted in [34], considering this image along the spatial dimension, the single molecule is actually serving as a highly localized nanoprobe of the laser beam diameter itself (here $\sim 5\,\mu$m, due to the poor quality of the focus produced by the lens in liquid helium). The molecule locally computes the transition probability proportional to the (squared) dot product of the absorption transition dipole and the local optical electric field, $\left|\vec{\mu}\cdot\vec{E}\right|^2$; hence, it may be regarded as a nanoscale probe of the focal spot, which is equivalent to the point-spread-function of the imaging system. This is the first example of a spatial image of a single molecule, and it illustrates the utility of the single molecule as a nanoscale probe, a property currently being utilized to provide localization of nanoscale objects far below the diffraction limit (see Sects. 2.5.3 and 2.5.4).

2.4.2 Observation of Spectral Diffusion of a Single Molecule

During the early SMS studies on pentacene in p-terphenyl in the Moerner group, an unexpected phenomenon appeared: resonance frequency shifts of individual pentacene molecules in a crystal at 1.5 K [30, 34]. This effect was called "spectral diffusion" in deference to similar spectral-shifting behavior long postulated for optical transitions of impurities in amorphous systems [35] (although this behavior is not diffusion in the strict sense, i.e., it is not governed by a spatial diffusion equation). Here, spectral diffusion means time-dependent changes in the center (resonance) frequency of a probe molecule due to configurational changes in the nearby host which affect the frequency of the electronic transition via guest–host coupling. For example, Fig. 2.5A shows a sequence of fluorescence excitation spectra of a single pentacene molecule in p-terphenyl taken as fast as allowed by the available SNR. The spectral shifting or hopping of this molecule from one resonance frequency to another from scan to scan is clearly evident.

Spectral shifting can be studied by (a) recording the observed lineshapes for many single molecules [36], (b) autocorrelation [37], and (c) measurement of the spectral trajectory $\omega_o(t)$ [34]. To record $\omega_o(t)$, one analyzes many sequentially acquired fluorescence excitation spectra like those in Fig. 2.5A and plots a trajectory or trend of the center frequency versus time as shown in Fig. 2.5B. The spectral trajectory shows that, for this molecule, the optical transition energy appears to have a preferred set of values and performs spectral jumps between these values that are discontinuous on the 2.5-s time resolution of the measurement. The behavior of other single molecules was qualitatively and quantitatively different, ranging from a wandering to creeping [33]. This is a direct demonstration of hidden heterogeneity from molecule to molecule that can only be observed by SMS – such spectral trajectories cannot be obtained when a large ensemble of molecules is in resonance. The individual jumps are generally uncorrelated; thus, the behavior of an ensemble-averaged quantity such as a spectral hole in this system would show only a broadening and smearing of the hole.

The first question which should be asked when such behavior is observed is this: is the effect spontaneous, occurring even in the absence of the probing

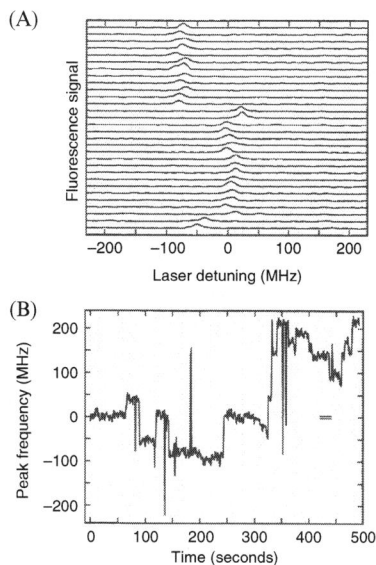

Fig. 2.5. Examples of single-molecule spectral diffusion for pentacene in p-terphenyl at 1.5 K. (**A**) A series of fluorescence excitation spectra each 2.5 s in duration spaced by 0.25 s showing discontinuous shifts in resonance frequency, with zero detuning = 592.546 nm. (**B**) Trend or trajectory of the resonance frequency over a long time scale for the molecule in (a). For details, see [34]

laser radiation, or is it light-driven, i.e., produced by the optical excitation itself? To distinguish spontaneous vs. light-driven behavior, the spectral diffusion was explored as a function of laser power, and for pentacene in p-terphenyl crystals, the effect was found to be predominantly spontaneous. Since the single-molecule absorption is extremely sensitive to the local strain field, it is reasonable to expect that the spectral jumps are due to internal dynamics of some configurational degrees of freedom in the surrounding lattice, driven by the phonons present at the experimental temperature. The situation is analogous to that for amorphous systems, where the local dynamics result from the two-level systems (TLSs) near the guest whose states change by phonon-assisted tunneling or thermally activated barrier crossing. One possible source for the tunneling states in this crystalline system could be discrete torsional librations of the central phenyl ring of the nearby p-terphenyl molecules about the molecular axis. The p-terphenyl molecules in a domain wall between two twins or near lattice defects may have lowered barriers to such central-ring tunneling motions. A theoretical study of the spectral diffusion trajectories ([38–40] has allowed determination of specific defects that can produce this behavior, attesting to the power of SMS in probing details of the local nanoenvironment.

Spectral shifts of single-molecule lineshapes were observed not only for certain crystalline hosts, but also essentially for all polymers studied [36], and even for polycrystalline Shpol'skii matrices [41]. What are the analogs of such effects at room temperature? Although molecular absorption lines are far broader in condensed phases due to dephasing and rovibrational effects, changes in the molecular configuration (twists, librations, etc.), energy transfer, or even photochemical changes can lead to changes in the absorption energies. Indeed, spectral shifting has been observed on larger frequency scales for single molecules on surfaces at room temperature in both near-field [42] and far-field studies [43]. When spectral shifts are large enough to move molecules into and out of resonance, blinking on and off of the emission is observed. Blinking behavior is turning out to be a common feature of many single-molecule experiments [44], giving us a new window into local dynamical behavior, and even leading to superresolution imaging in some cases (see Sects. 2.5.3 and 2.5.4 and [45]).

2.4.3 Optical Switching of Single Molecules

The experiments described so far were performed with single impurity molecules doped into a crystalline matrix. Amorphous systems such as glasses or polymers have a number of interesting physical properties at low temperatures which are quite different from those for crystalline materials, in particular a complex, multidimensional potential energy surface [46]. According to the TLS model, the material can be approximated by a distribution of asymmetric double-well potentials in which only the two lowest energy levels of each double-well are important. The effect of these TLSs on thermal and optical properties has been an area of extensive study. In 1992, perylene doped into a poly(ethylene) matrix (Fig. 2.6A) became the first polymeric system to allow SMS ([36,47]. Spectral diffusion manifested itself in this system in two different ways: as discontinuous jumps in frequency space on the scale of several hundred megahertz similar to pentacene in p-terphenyl discussed above, and in addition, the observed linewidths varied from molecule to molecule due to fast shifts of small amplitude.

Light-driven shifts in absorption frequency were also observed, in which the rate of the process clearly increased with increases in laser intensity [36]. This photoswitching effect may be called "spectral hole-burning" by analogy with the earlier hole-burning literature [20]; however, since here only one molecule is in resonance with the laser, the absorption line simply disappears. An example is presented in Fig. 2.6B. Traces (a), (b), and (c) show three successive scans of one perylene molecule. After trace (c) the laser was tuned into resonance with the molecule, and at this higher irradiation fluence, eventually the fluorescence signal dropped, that is, the molecule apparently switched off. Trace (d) was then acquired, which showed that the resonance frequency of the molecule apparently shifted by more than $+/-1.25\,\mathrm{GHz}$ as a result of the light-induced change in the nearby environment. Surprisingly, this effect was reversible for

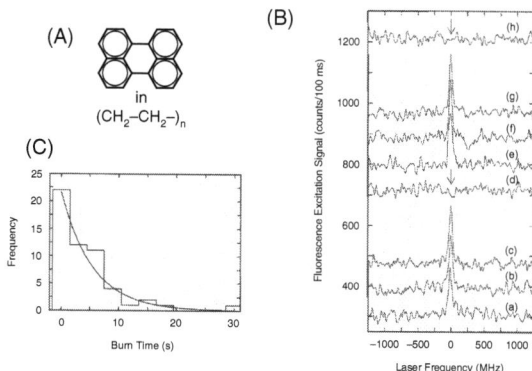

Fig. 2.6. (**A**) System for photoswitching: perylene impurity molecules in poly(ethylene). (**B**) Illustration of reversible frequency shifting for a single molecule at 1.4 K. (a–c) Single molecule scanned several times. After irradiating the molecule with higher fluence, the molecule shifts out of the scan range (d). After eventual reversal (e), the molecule may be scanned again (f, g) and photoswitched again (h). (**C**) Histogram of waiting times before the light-induced spectral shift (hole-burning), from 53 events for the same single perylene molecule. For details, see [36,47]

a good fraction of the molecules: a further scan some minutes later (trace (e)) showed that the molecule returned to the original absorption frequency. After trace (g) the molecule was photoswitched again and the whole sequence could be repeated many times.

The possibility to photoswitch one and the same molecule away from the laser wavelength multiple times was used to measure the kinetics of this process [47]. By measuring a large number of photoinduced spectral shifting events for one perylene molecule in the poly(ethylene) host, it was found that the various waiting times are exponentially distributed, suggesting that the underlying process obeys Poisson statistics and is Markovian (Fig. 2.6C). Due to the stable rigid structure of perylene, it is fairly clear that the spectral shift is caused by changes in the states of one or more nearby host TLSs coupled to the perylene electronic transition; however, the exact microscopic mechanism may be related to the generation of molecular internal vibrational modes during fluorescence emission or to nonradiative decay of the excited state. Several single-molecule systems have shown light-induced shifting behavior, for example, terrylene in poly(ethylene) [48], and terrylene in a Shpol'skii matrix [49], and it is hoped that future detailed study of this effect will shed light on the microscopic mechanisms of nonphotochemical hole-burning.

In principle, single-molecule light-induced spectral shifting (or single-molecule hole-burning) is a controllable process which could allow modification of the transition frequency of any arbitrary chosen molecule in the polymer host. This leads naturally to the possibility of optical storage at the single-molecule level. Such an idea would require more careful control of the spectral locations of the various molecules (formatting) as well as measures to deal with the exponential waiting times. In spite of these issues, optical modification of single-molecule spectra not only provides a unique window into the photophysics and low-temperature dynamics of the amorphous state, this effect also presages another area of current interest at room temperature: photoswitching of single molecules on and off, a powerful tool currently being used to achieve superresolution imaging of cellular structures (Sect. 2.5.3). Selected new molecules and mechanisms for photoswitching at room temperature are described in Sect. 2.5.4.

2.4.4 Photon Antibunching and Magnetic Resonance of Single Molecular Spins

One class of physical phenomena that have been demonstrated with single molecules concerns quantum optical effects which arise from the single (and therefore quantum mechanical) nature of the light emission. In particular, the stream of photons emitted by a single molecule contains information about the system encoded in the arrival times of the individual photons. On the timescale of the triplet lifetime, photons should be emitted in bunches, an effect first observed in the autocorrelation measurement by Orrit et al. [29], and this phenomenon has been used to measure the changes in the triplet yield and triplet lifetime from molecule to molecule which occur as a result of distortions of the molecule by the local nanoenvironment [50]. In contrast, in the nanosecond time regime within a single bunch, the emitted photons from a single quantum system are expected to show antibunching, which means that the photons "space themselves out in time," that is, the probability for two photons to arrive at the detector at the same time is small. This is a uniquely quantum-mechanical effect [51], which was first observed for single Na atoms in a low-density beam [52].

To observe antibunching correlations, the second-order correlation function $g^{(2)}(t)$ is generally measured by determining the distribution of time delays $N(\tau)$ between the arrival of successive photons in a dual-beam detector. Photon antibunching in single-molecule emission was first observed in the author's laboratory at IBM for the pentacene in p-terphenyl model system [53], demonstrating that quantum optics experiments can be performed in solids and on molecules for the first time (Fig. 2.7A). The high-contrast dip at $\tau = 0$ is strong proof that the spectral feature in resonance is indeed that of a single molecule. This observation has opened the door to a variety of other quantum-optical experiments with single molecules [54], such as measurement of the AC Stark shift [55]. The convenient "trap" that the solid forms for a single molecule has allowed multiple studies of the interactions of single molecules with the quantum radiation field, including single-photon sources [56–58].

Another class of physical phenomena demonstrated for single molecules involves magnetic resonance of a single molecular spin. Historically, the

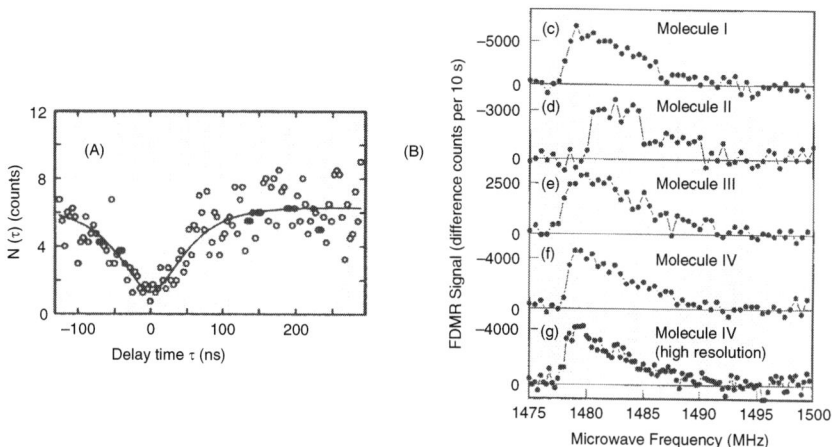

Fig. 2.7. (A) Measured distribution of time delays between successive detected fluorescence photons for a single molecule of pentacene in p-terphenyl showing antibunching at $\tau = 0$. For details, see [53]. (B) Magnetic resonance of a single molecular spin. Reductions in fluorescence as a function of microwave frequency for four different single molecules of pentacene in p-terphenyl. For details, see [59]

standard methods of electron paramagnetic resonance and nuclear magnetic resonance have been limited in sensitivity to about 10^8 electron spins and about 10^{15} nuclear spins, respectively, due chiefly to the weak interaction between the individual spins and the magnetic fields used to excite the transition. The power of SMS is that only one molecule is in resonance with the laser; hence, the detection of the effect of secondary perturbing fields on the optical emission can lead to observation of double resonances. For example, the magnetic resonance transition of a single molecular spin was observed at IBM [59] and at Bordeaux [60] using the pentacene in p-terphenyl model system and a combination of SMS and optically detected magnetic resonance (ODMR). The method involves selecting the optical absorption of a single molecule with the laser and monitoring the intensity of optical emission as the frequency of a microwave signal is scanned over the frequency range of the triplet spin sublevels. Since the emission rate is dependent upon the overall lifetime of the triplet (bottleneck) state, the emission rate is affected when the microwave frequency is resonant with transitions among the triplet spin sublevels. Figure 2.7B shows examples of the 1,480-MHz magnetic resonance transition among the T_x–T_z triplet spin sublevels at 1.5 K for a single molecule of pentacene in p-terphenyl, where the signal plotted is the change in the fluorescence emission rate as a function of the applied microwave frequency [59]. Traces (c)–(g) show the single-molecule lineshapes for four different molecules. Interestingly, the onset of the transition varies from molecule to molecule, and the lineshape of the microwave transition for a single spin is broadened by hyperfine interactions induced by the large number of different configurations possible for the nearby proton nuclear spins. These observations have opened the way for a variety of new studies of magnetic interactions in solids at the level of a single molecular spin, such as the use of external magnetic fields and deuteration to reduce proton spin flips, and hence the hyperfine broadening [61]. Roughly 10 years later, a single electron spin was detected by magnetic force microscopy, a complex new method that hopes to generally detect single spins by subtle shifts in a cantilever resonance [62].

2.4.5 Single Molecules Interacting with Novel Optical Fields

The nanometer-sized single molecule can be regarded as a local probe of electromagnetic and electrostatic interactions extremely close to the molecule, because the probability of excitation depends upon the local optical field as described above. In addition, under certain conditions, the molecule's emission can be influenced by the proximity of nanoscale metallic or dielectric structures.

One strategy for coupling a single molecule to a novel optical field involves pumping the molecule with the light from a near-field light source. To do this, a near-field probe composed of an aluminum-coated near-field fiber tip was used to excite single molecules of pentacene hidden below the surface of a p-terphenyl crystal as shown in Fig. 2.8A [63]. The near-field light beam resulted from light leaking out of the $\sim$ 70-nm hole in the metallic coating. In addition, the presence of the nearby metallic coating also allowed application

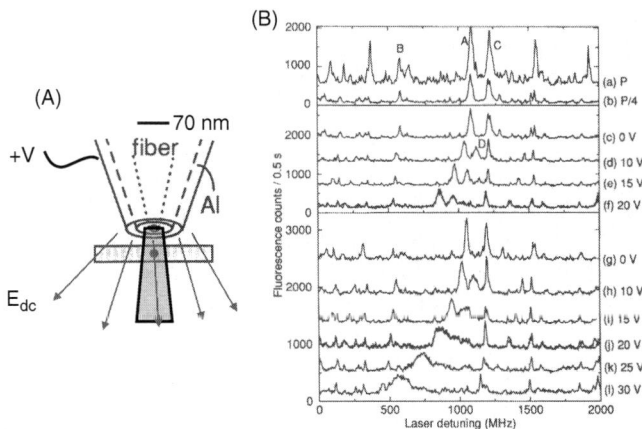

Fig. 2.8. Left: (**A**) Optical configuration for exciting a single molecule with a near-field light source, an Al-coated pulled optical fiber. Application of a potential V to the Al coating produces a highly anisotropic DC local electric field. (**B**) Spectra of pentacene in p-terphenyl molecules at various applied potentials. (a–b) saturation method to identify molecules close to the tip, (c–l) transverse dithering with Stark shift. For details, see [63]

of a dc voltage to the metal, with the effect that a highly anisotropic dc electric field could be generated close to the sample inside the cryostat (i.e., with the ground electrode effectively at infinity). Single molecules within a few hundred nanometers of this subwavelength light source were identified either by early saturation behavior or by analysis of Stark shifts of the absorption lines produced by a static electric potential applied to the Al coating as shown in Fig. 2.8B. Traces (a)–(f) show that some molecules show Stark shifts, while others do not because they are either in an unfavorable orientation or far away from the tip, deeper into the relatively thick crystalline sample. Traces (g)–(l) show both a shifting and broadening of a single-molecule absorption with transverse dithering. Given previous measurements of Stark shift coefficients, the observed shift of the single-molecule line can be viewed as a local sensor of the highly inhomogeneous electric field of the tip in order to estimate the distance between the molecule and the tip.

This early study was improved upon in the Sandoghdar laboratory by providing far more control of the molecule-tip distance and by providing very thin samples [64]. For example, pairs of nearby single molecules have been pushed into resonance using anisotropic electric fields from a nearby metal sphere [65], and the absorption and emission properties of single molecules have been modified by a nearby metallic sphere [66]. It remains an active research goal to couple the narrow optical absorption of a single molecule or ion trapped in a solid to a high-Q cavity in order to observe strong coupling of the molecule and the light field.

2.5 Selected Room Temperature Milestones

Single-molecule fluorescence detection was subsequently demonstrated at room temperature, first by detecting the burst of light as a molecule passes through the focus of a laser beam [67,68], but each molecule could be detected only once in this way. Correlation analysis of many such bursts provides a window into a variety of dynamical effects ranging from diffusion to intersystem crossing to rotational correlation [69], and this area termed "fluorescence correlation spectroscopy" (FCS, ([70–72]) has been reviewed in [73].

Of interest in this paper is the use of optical microscopy to observe the same single molecule for an extended period, measuring signal strength, lifetime, polarization, fluctuations, and so on, all as a function of time and with the express purpose of directly detecting any heterogeneity from molecule to molecule. First demonstrated at room temperature by near-field scanning optical microscopy (NSOM) [74–76], single molecules can now be detected using confocal microscopy [77], wide-field epifluorescence [78] and total internal reflection [79,80] methods.

These important steps are responsible for much of the current growth in the field, as a large array of molecules can now be studied with appropriate fluorescent labeling. In particular, a major driving force in single-molecule

Table 2.2. Room-temperature and single-biomolecule milestones from the Moerner Lab

Experiment	Refs.
Diffusion of single fluorophores in poly(acrylamide) gels	[80, 168]
Blinking and switching of single copies of green fluorescent protein	[82, 143]
Imaging z-oriented molecules with total internal reflection microscopy	[169]
Cameleon – single-pair FRET between two GFP mutants in a $[Ca^{++}]$-sensitive protein	[95]
Dynamics of a single-molecule pH sensor	[170]
Kinesin motor orientational dynamics driven by nucleotide state	[171, 172]
Single-molecule source of single photons on demand	[57]
Antibunching in single CdSe/ZnS quantum dot fluorescence	[173]
Single-molecule photophysics of DsRed, a red fluorescent protein	[119]
Diffusion in cell membranes, effects of cholesterol on membrane dynamics	[100–103]
New reporter fluorophores for SM studies	[174, 175]
Imaging of single histidine kinase proteins in bacteria	[114]
New single-molecule fluorophores for cellular imaging	[104]
Sagnac detection of ultrasmall phase shifts at room T ($\sim$19 molecules)	[176]
Single MreB cytoskeletal proteins treadmilling in a living bacterial cell	[115]
Anti-Brownian electrokinetic trap for single molecules in solution	[148]
Single-molecule nanoprobes explore defects in spin-grown crystals	[140]
Improved molecular beacon for DNA/RNA single-molecule detection	[177]
Motion of poly(arginine) cell-penetrating peptides interacting with cell membranes	[178]
Photoactivatable push–pull fluorophores for single-molecule cellular imaging	[159]
Cy3–Cy5 covalent heterodimers for single-molecule photoswitching	[135]

fluorescence studies is the study of biomolecules, in vitro and in vivo [81]. The Moerner laboratory has been involved in a variety of room-temperature single-molecule studies, many of which involve biomolecules both in and out of living cells (Table 2.2). Selected single-biomolecule experiments from the 1990s as well as several recent studies since 2000 will be summarized.

2.5.1 Blinking and Switching for Single Green Fluorescent Proteins

In 1996, partial immobilization of single molecules in an aqueous environment was achieved by using the water-filled pores of poly(acrylamide) gels, a technique that has been demonstrated for organic dye molecules [80] as well

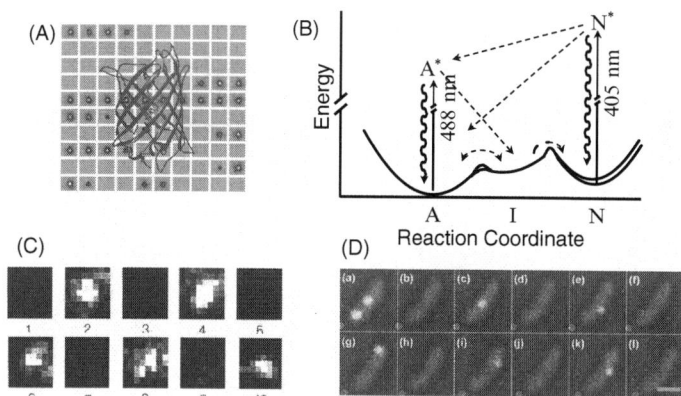

Fig. 2.9. (**A**) Structure of the green fluorescent protein [164] superimposed on a series of images of a single GFP trapped in a gel, 100 ms per image. (**B**) Schematic of energy-level structure consistent with the blinking and photoswitching effects [82]. (**C**) Images (600 nm × 600 nm) of 488-nm pumped emission from the long-lived dark state (odd panels) with the photoreactivated state (even panels) produced by 405 nm irradiation for the same single molecule of the T203F yellow mutant of GFP. Similar results occurred for the T203Y mutant [82]. (**D**) Reactivation of EYFP-MreB fusions in live *C. crescentus* cells. Fluorescence images show single EYFP-MreB molecules (white spots) overlaid on a reversed-contrast white-light image of the cell being examined. Only a few molecules are reactivated by 407 nm light in each image. Bar, 1 μm. For details, see [84]

as for green fluorescent proteins (GFPs) and other biomolecules [82]. GFP and its mutants are currently of great importance in molecular and cellular biology [83]. GFP is a small (238 amino acids), water-soluble protein isolated from the jellyfish *Aequoria victoria* which has a β-barrel structure as shown in Fig. 2.9A. The strongly absorbing and fluorescing chromophore of GFP located inside the barrel is formed spontaneously (in the presence of oxygen) from three amino acids in the native protein chain. No external cofactor (as in most other fluorescent proteins) is necessary, thus GFP is currently widely used as an indicator for gene expression or as a fluorescent fusion label for a large variety of proteins in cells. Since the quantum yield of the emission is fairly high and the absorption can be pumped by convenient Ar+ laser lines, single copies of GFP mutants were first observed using total-internal-reflection fluorescence microscopy [82].

These early experiments also yielded the first example of a room temperature single-molecule optical switch [82] and the first details of the photophysical character of GFP on the single-copy level. The experiments utilized two red-shifted GFP mutants (S65G/S72A/T203Y denoted "T203Y" and S65G/S72A/T203F denoted "T203F") which differ only by the presence of a hydroxyl group near the chromophore, both of which are quite similar to the currently widely used enhanced yellow fluorescent protein EYFP

(S65G/V68L/S72A/T203Y). In particular, a fascinating and unexpected blinking behavior appeared, discernable only on the single molecule level (see the background of Fig. 2.9A for a series of images on one molecule, for example). This blinking behavior likely results from transformations between at least two states of the chromophore (A and I, Fig. 2.9B), only one of which (A) is capable of being excited by the 488-nm pumping laser and producing fluorescence. Additionally, a much longer-lived dark state N is also accessible. Thermally stable in the dark for many minutes, this long-lived dark state is not permanently photobleached, but can be excited at 405 nm to regenerate the original fluorescent state as shown in the sequence of images in Fig. 2.9C. This means that the protein can be used as an emitting label until it enters the long-lived dark state, and it can be photo-reactivated back to the emissive form with the 405-nm light, a reversal of the apparent photobleaching. This can occur many times for the same single molecule.

In recent work, we have demonstrated that this reversible photoswitching effect also works for EYFP in living bacterial cells [84]. The sequence of images in Fig. 2.9D shows a single bacterial cell with many EYFP fusions to the protein MreB. After initially imaging all the molecules with 514 nm until they turn off, a few at a time are reactivated and bleached repetitively. In the years since this switching effect was first observed, much progress has been made in the development of improved photoactivatable (turn-on of emission by excitation at a control wavelength) and photoswitchable fluorescent proteins with colorful names (such as Kaede [85], PA-GFP [86], EosFP [87], and DRONPA [88]. All these molecules can be utilized for superresolution imaging [89, 90], but the widespread availability of fusions to EYFP should greatly enhance such applications.

2.5.2 FRET for a Dual-GFP Construct

SMS can be used to explore the dynamics of environment-dependent fluorescent molecules that are usually used as concentration reporters in biological media. One study concerns the "cameleon YC2.1" complex, whose structure is based on a cyan-emitting GFP (CFP) separated from a yellow-emitting GFP (YFP) by the calmodulin Ca^{2+}-binding protein and a calmodulin-binding peptide (M13) (see Fig. 2.10A). This complex was designed by A. Miyawaki and R. Y. Tsien to allow sensing of calcium concentration in cells by fluorescence resonant energy transfer (FRET). If Ca^{2+} ions are bound, the construct forms a more compact shape, leading to a higher efficiency of excitation transfer from the donor CFP to the acceptor YFP. The degree of FRET in cameleon is, therefore, a sensitive ratiometric reporter of the concentration of Ca^{2+} in solution and cells [91].

As is well-known, FRET is a powerful tool for sensing conformational changes in single biomolecules [92–94]. The first FRET study of a single dual-GFP construct was completed in the Moerner lab for the cameleon YC2.1 complex in 2000 [95]. Analysis of single-molecule signals from the

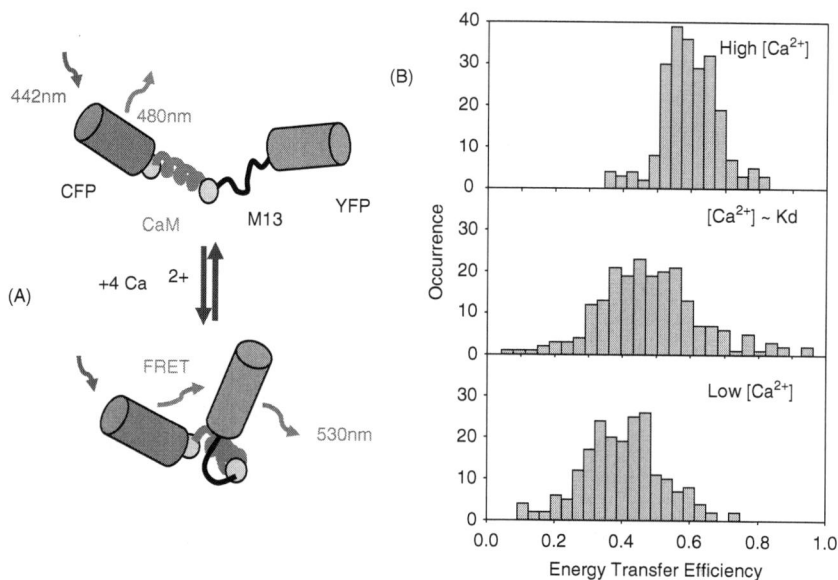

Fig. 2.10. (**A**) Schematic of the structure of cameleon containing CFP, YFP, CaM, and M13 showing the change in FRET upon binding and unbinding of Ca^{2+} ions. (**B**) Histograms of the energy transfer efficiency measured for single molecules of cameleon in agarose gels, at three different Ca^{2+} ion concentrations. For details, see [95]

proteins immobilized in aqueous agarose gels allowed retrieval of several interesting features of the energy transfer between the donor and acceptor mutants of the construct, as a function of the calcium concentration in the medium. The energy transfer efficiency distributions shift as expected with Ca^{2+} concentration, and show an increased width at the Ca^{2+} dissociation constant concentration (Fig. 2.10B, middle panel). This observation was consistent with the ligand binding kinetics, whose time scale at intermediate calcium concentration was close to the measurement time scale (20–200 ms). The complex dynamics of the fluctuations were examined using a combination of autocorrelation and cross-correlation in conjunction with polarization information.

2.5.3 imaging in cells

Diffusion

Another area of current interest involves further extensions of single-molecule studies to the surfaces or the interior of living cells. Tracking of moving single molecules on the plasma membrane or moving in the cytoplasm began some years ago [96] and was applied to several problems in cell biology starting around 2000 [97–99]. Around this time, the Moerner lab in collaboration with Harden McConnell began a single-molecule study of diffusion in the plasma membrane as modulated by cholesterol. Using a fluorescently-labeled small peptide antigen bound to major histocompatibility complexes of type II (MHCII), the diffusion of MHCII as a function of cholesterol concentration was explored to examine whether direct evidence could be obtained regarding the putative presence of membrane subdomains [100,101]. In this assay, only a tiny 1-nm Cy5 label was attached to the antigen to achieve labeling. Using epifluorescence imaging of a living Chinese hamster ovary (CHO) cell membrane, single molecules were observed performing a beautiful dance as thermal excitation drives their natural motion (Fig. 2.11A). From data like this, diffusion coefficients from single molecules were extracted, and distributions of measured values (Fig. 2.11B) were compared with the distribution which would be expected if the motion were Brownian but with a limited number of time steps (solid line). A key result of these studies was the observation of a large drop in diffusion coefficient upon the extraction of membrane cholesterol using β-cyclodextrin (Fig. 2.11B, (a) and (b)) [102–105]. Temperature-dependent studies confirmed that the drop in diffusion coefficient at ambient temper-

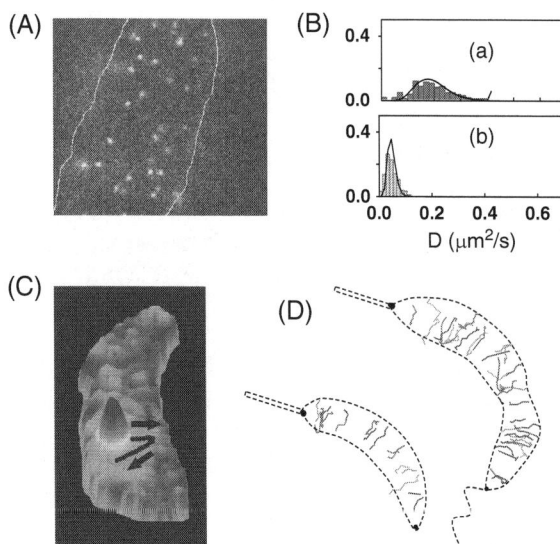

Fig. 2.11. (A) Example of the imaging of single MHCII proteins on the surface of a CHO cell by epifluorescence microscopy. The image represents 12 μm × 12 μm at the sample plane, with an integration time of 100 ms. (B) Probability distributions of single-molecule diffusion coefficients for (a) normal cholesterol concentration, and (b) after reduction of cholesterol concentration [102]. (C) Epifluorescence image of a *C. crescentus* cell with a single molecule of MreB-EYFP, and the motion observed during treadmilling. The cell width is ∼500 nm. (D) Representative trajectories of single MreB-EYFP treadmilling molecules showing filaments in a stalked (left) and in a predivisional (PD) cell. For details, see [115]

atures upon cholesterol extraction was likely due to a phase separation of cholesterol-rich domains which presented obstacles to the observed motion.

Many new investigators have been stepping up to the additional challenge of imaging in the higher background of the cell [106–109]. Naturally, autofluorescent proteins[110–112] are a powerful way to achieve genetically-directed labeling, and thus to follow intracellular events at the single-molecule level [113–118]. One continuing need is for mutants with absorption and emission at longer wavelengths; DsRed was thought to be particularly useful in this regard, but steps in single-molecule photobleaching traces showed that this protein retained its multimeric character even at very low concentrations [119]. Much progress has been made in mutation of this type of fluorescent protein to produce stable monomers.

Imaging of Single Molecules in Bacteria

In the early years of this century, the bright and red-shifted emission from single molecules of EYFP led to its use as a label for fusions to intracellular proteins in the Moerner lab in collaboration with the laboratory of Lucy Shapiro. The primary organism of interest has been *Caulobacter crescentus*, because cells of this bacterium display asymmetric division in the cell cycle: one daughter cell has a flagellum while the other has a stalk with a sticky end. This means that the cells have a genetic program that causes different groups of proteins to appear in the two different daughter cells, and understanding this process would contribute to the general problem of understanding developmental biology [120, 121]. The basic effect arises from spatial patterning of regulatory proteins, which leads to many interesting questions: how do the proteins actually produce patterns, how do these patterns lead to different phenotypes in the daughter cells, and so on.

The first study of this type in the Moerner lab explored the histidine kinase PleC, which localizes at one of the poles of the cell during the cell cycle. PleC fusions to EYFP were easily observed at the single-molecule level in *C. crescentus* cells, mostly diffusing via attachment to the inner membrane [114]. Because no directed transport or motional asymmetry was observed, the authors were able to conclude that a diffusion-to-capture model could explain the observed localization behavior.

A second localization study concentrated on the dynamics of the bacterial actin homolog, MreB [115], which had not been examined in vivo. The actin cytoskeleton represents a key regulator of multiple essential cellular functions in both eukaryotes and prokaryotes [122]. Using single-molecule epifluorescence microscopy, the motion of single fluorescent MreB-YFP fusions in living *C. crescentus* cells was observed in a background of unlabeled MreB. With time-lapse imaging, polymerized (filamentous or fMreB) MreB and unpolymerized (globular or gMreB) MreB monomers could be distinguished: gMreB showed fast motion characteristic of Brownian diffusion, while the labeled molecules in fMreB displayed slow, directed motion. This directional movement of labeled MreB in the polymer provided the first indication that like actin, MreB monomers treadmill through MreB filaments by preferential polymerization at one filament end and depolymerization at the other filament end. Figure 2.11C shows an image of a cell with the track of a single molecule indicated by the black arrow. The zig-zag motion represents a portion of a helix-like structure of the filament imbedded in the membrane. From this work, distributions of treadmilling speeds, lengths of filaments, and filament polarity could be extracted [115]. This is the first time that single-molecule imaging has been used to directly see treadmilling in any cellular system.

Superresolution Imaging by Treadmilling

As is well-known, biological fluorescence microscopy depends upon a variety of labeling techniques to light up different structures in cells, but the price often paid for using visible light is the relatively poor spatial resolution compared to X-ray or electron microscopy. The basic problem is that in conventional microscopes, fundamental diffraction effects limit the resolution to a dimension of roughly the optical wavelength λ divided by two times the numerical aperture (NA) of the imaging system, $\lambda/(2\text{xNA})$. Since the largest values of NA for state-of-the-art, highly corrected microscope objectives are in the range of about 1.3–1.6, the spatial resolution of optical imaging has been limited to about ∼180 nm for visible light of 500 nm wavelength.

In fact, the light from single fluorescent molecular labels about 1–2 nm in size provides a way around this problem, that is, a way to provide "superresolution," or resolution far better than the diffraction limit (reviewed in [45]). How can single molecules help? First, the observed "peak" from the single nanoscale source of light maps out the point-spread function (PSF) of the microscope, because the molecule is a nanoscale light absorber, far smaller than the size of the PSF. Actually, as described above, the molecule absorbs light with a probability proportional to the square of the dot product between the local optical electric field and the molecule's transition dipole moment. This point was realized at the very beginning of work in this field, where the fluorescence excitation signal from one molecule was used to map out the size of the focused pumping laser beam as in Fig. 2.4 CD [33]. Simply by measuring the shape of the PSF, the position of its center can be determined much more accurately than its width. This idea, digitizing the PSF to achieve "superlocalization," or position information far below the diffraction limit, is well-known, and was applied early-on to single nanoscale fluorescent beads with many emitters[123] and then to low-temperature single-molecule images [124, 125] where both spatial information and the secondary variable, laser wavelength, were used to separate molecules. The accuracy with which a single molecule can be located by digitizing the PSF depends fundamentally upon the Poisson process of photon detection, so the most important variable is the total number of photons detected above background, with a weaker dependence on the size of the detector pixels [126, 127]. In fact, Heisenberg noted in 1930 that the resolution improvement in localization improved by one over the square root of the number of photons detected [128].

To keep the PSFs from different molecules from overlapping, very low concentrations of the emitting molecule are usually required, although the early low temperature work in the field used spectral selection to identify the different individual molecules in the same laser focal volume [3, 29, 125]. In fact, the single MreB protein shown in Fig. 2.11C is part of an unlabeled filament of MreB molecules, and, over time, the molecule moves linearly through the filament by the treadmilling process described above [115]. By fitting the sequence of single PSFs to Gaussian profiles, an image of the shape of the filament can be obtained with 30 nm resolution as shown in Fig. 2.11D. In other words, here superresolution is obtained because the single molecule moves through the structure of interest. Placing many such single-molecule tracks on the shape of the bacterial cell provides new information about location, size, and length of the MreB filaments in a living cell, where in this case, superresolution arises from the treadmilling motion of the single label.

2.5.4 New Photoactivatable and Photoswitchable Single-Molecule Fluorophores

To achieve superresolution imaging in cells in a more general fashion, it is necessary to develop another way to be sure that only isolated single molecules are imaged at any given time [45, 129]. Very recently, researchers have shown that photoactivation or photoswitching of single molecules can directly be used to generate superresolution images. It is necessary to actively control the number of molecules emitted; hence, these methods might generally be termed single-molecule active-control microscopy (SMACM). Photoactivated GFP fusions were used in the method of Betzig et al. [89] termed PALM (PhotoActivated Localization Microscopy) and in the method of S. Hess et al. termed F-PALM (Fluorescence PhotoActivation Localization Microscopy) [90]. In fact, the photoswitching of EYFP originally reported in the Moerner lab in 1997 can also be used [84]. The basic idea of these methods involves only turning on a sparse subset of all the labels present. After imaging and localizing these single molecules and photobleaching them, a new subset is turned on and the process is repeated to build up a full image of the labeled structure. Final resolutions down to 10–20 nm have been achieved, and it is this impressive improvement of resolution that is causing obvious excitement in the field at the present time.

While fusions to fluorescent proteins are certainly powerful, a continuing challenge is to identify small-molecule labels for improved brightness and readout capability at the single-copy level [130]. Targeting of these molecules can be achieved by various strategies, but what about the important property of photoswitching needed for superresolution imaging? Recently, we have reengineered a red-emitting dicyanomethylenedihydrofuran (DCDHF) push–pull fluorophore so that it is dark until photoactivated with a short burst of low-intensity violet light (Fig. 2.12A). This molecule is one member of a broad class of push–pull single-molecule emitters with tunable wavelength, photostability, and sensitivity to local environmental rigidity [131–133]. Photoactivation of the dark fluorogen **1** leads to conversion of an azide to an amine, which shifts the absorption to long wavelengths as shown in Fig. 2.12B, into resonance with a pumping source at 594 nm. After photoactivation to produce the emission shown in Fig. 2.12B inset, the fluorophore **2** is bright and photostable enough to be imaged on the single-molecule level in living cells. This proof-of-principle demonstration provides a new class of bright photoactivatable fluorophores, as are needed for superresolution imaging schemes that require active control of single-molecule emission.

In another superresolution approach, Rust et al. have utilized controlled photoswitching of small molecule emitters for superresolution demonstrations [134] (STORM, for Stochastic Optical Reconstruction Microscopy). This method uses a Cy3–Cy5 emitter pair in close proximity in the presence of thiols that show a novel property: restoration of Cy5's photobleached emission can be achieved by brief pumping of the Cy3 molecule. In this way, the emission from a single Cy5 is turned on and off, again and again, to measure its position accurately multiple times. In the Moerner laboratory, a covalently bound Cy3–Cy5 dimer has been prepared as shown in Fig. 2.12C [135]. This molecule has the advantage that the need for random close associations of the Cy3 and

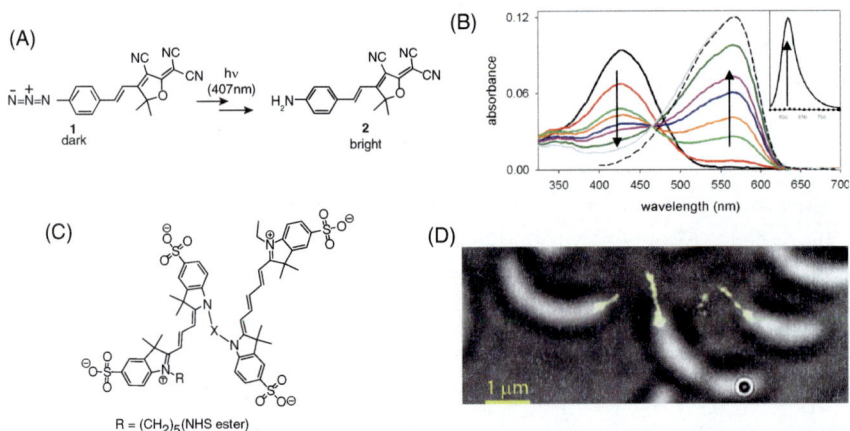

Fig. 2.12. (**A**) Scheme showing photoactivation of fluorescence for the DCDHF-azide **1** to produce the long-wavelength emissive molecule **2**. (**B**) Absorption spectra of **1** converting to **2** upon 407 nm irradiation. Inset: emission due to pumping at 594 nm. For details, see [159]. (**C**) A covalently bound dimer of Cy3 and Cy5 with a reactive functionality [135]. (**D**) Superresolution image of the stalks of living *C. crescentus* cells (yellow) acquired using the molecule in (**C**) bound to cell-surface amines. For details, see [135]

Cy5 is absent. Photoswitching of a succinimidyl ester derivative covalently attached to surface lysines has allowed generation of superresolution images of the stalks of live *C. crescentus* cells as shown in Fig. 2.12D.

In contrast to the previous approaches, it is important to note that there are also promising superresolution imaging methods [136] that do not specifically require single emitters, random photobleaching/blinking events, or photoswitching to be sure that only one emitter is present in a diffraction-limited volume. One key idea proposed by S. Hell et al. involves using optical saturation of the emission to provide a nonlinear response, which directly alters the shape of the PSF itself [137]. This approach has been termed RESOLFT, for REversible Saturable OpticaL Fluorescence Transitions[138], because reversible photoswitching into dark states may be regarded as a type of optical saturation which produces a nonlinear dependence upon the local pumping intensity, at much lower power levels. One earlier example of this has been termed STED (Stimulated Emission Depletion) microscopy [139], where pulsed stimulated emission in an annular region prevents emission of fluorophores not exactly at the center of a confocal pumping spot. All of these tantalizing new approaches to superresolution in biological imaging have advantages and disadvantages, and which method will achieve widespread use and useful time resolution for observing nanoscale cellular dynamics is still yet to be determined [45].

2.5.5 Trapping Single Molecules in Aqueous Solution: ABEL Trap

In single-molecule studies, one would like to observe each molecule for as long a time as possible in order to extract maximal information from the molecule, a requirement that is relatively easy to achieve for solid samples [140], but it is quite challenging for small biomolecules in their native aqueous environment. Brownian motion severe in solution at room temperature: a single 10 nm object in water diffuses through a diffraction-limited laser spot $\sim 0.3\,\mu$m in diameter in $\sim$1 ms. To address this, investigators have previously used a variety of strategies, such as immobilization on a transparent glass surface [141, 142], encapsulation within the water-filled pores of a gel [43, 80, 143], or enclosure in a vesicle [144].

One may ask, why not use an optical trap, such as the highly successful laser tweezer trap, first proposed in the mid-1980s by Ashkin and colleagues [2]? Indeed, this approach has led to major strides in biophysical understanding resulting from extremely precise biophysical force and sub-nanometer position measurements [145–147]. However, it is worth remembering that the object that is actually trapped in a laser tweezers system is a large dielectric bead, such as a poly(styrene) or glass sphere $\sim$1 μm in diameter. Laser tweezers cannot trap a single small biomolecule directly, because the polarizability falls as the volume d^3 (with d the diameter) and is simply too small.

Recently, a new trap was proposed and demonstrated in the Moerner lab, the Anti-Brownian ELectrokinetic trap (ABEL trap), which scales quite

favorably for trapping very small objects in solution [148]. The ABEL trap
uses low frequency electric fields and real-time feedback control [149–152]
to manipulate and trap individual nanoscale objects in solution at ambient temperature. Referring to Fig. 2.13A (side view section showing only two
of four electrodes) and 2.13B (top view of the microfluidic cell), the ABEL
trap works by (i) monitoring the Brownian motion of the particle by directly
measuring the particle's position with standard single-molecule fluorescence
imaging microscopy, and then (ii) applying a feedback voltage to the solution
so that the resulting electrokinetic drift (which may be either electrophoretic
or electro-osmotic in character) cancels the Brownian motion within the band-

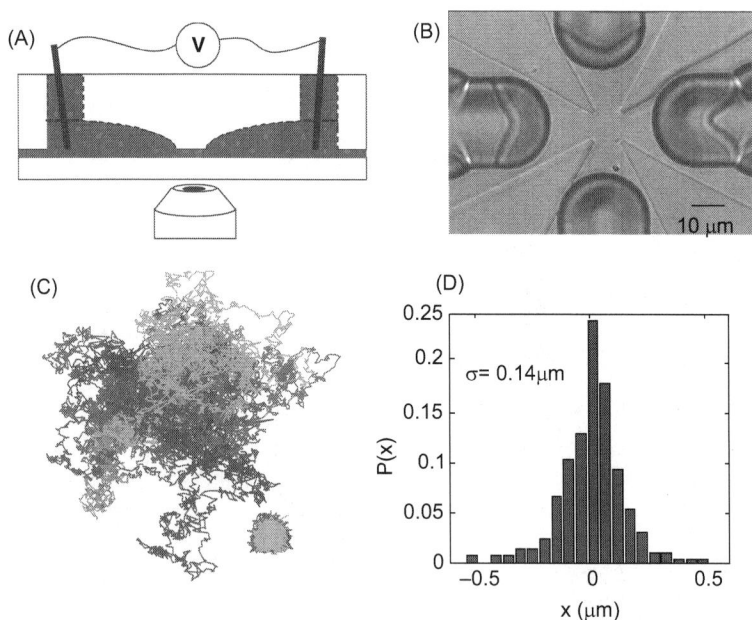

Fig. 2.13. Trapping single molecules in solution with the ABEL trap. (**A**) Schematic
side view of the trap showing that the microfluidic cell sits above the oil-immersion
objective of an inverted fluorescence microscope. Confinement in the z-direction
along the axis of the microscope is produced by the thin gap between the upper
transparent structure and a flat cover slip. Four electrodes are placed in the solution
far away from the central trapping region. (**B**) Top view of the microfluidic cell,
showing the trapping region about $10\,\mu m \times 10\,\mu m$ in size in the center. Four deep
milled channels extend out in the $+/-x$ and $+/-y$ directions. The four sharply
pointed raised regions serve to define the thickness of the trap in the z-direction
normal to the page. (**C**) Measured (lower right) and pseudo-free (central portion)
trajectories of 13 trapped particles of TMV. After [156], used by permission. (**D**)
Position probability distribution of a single fluorescently labeled molecule of the
chaperonin, GroEL, trapped in buffer. The standard deviation is shown. For details,
see [158]

width of the feedback system. The particle is confined in the z-direction by a thin channel on the order of 500 nm in thickness, and the microfluidic cell may be fabricated out of poly(dimethyl siloxane) [153], glass [154], or quartz [155] (shown in Fig. 2.13B). The control fields are applied in the x–y plane by four macroscopic electrodes placed in deep channels extending away from the trapping region in all four x–y directions.

The initial implementations of the ABEL trap used software-based feedback and electron-multiplying CCD imaging technology, in which the update time could be made as small as 4.5 ms [156]. In this configuration, the ABEL trap was used to trap a variety of small objects in solution, ranging from 20 nm fluorescent spheres to single fluorescently-labeled tobacco mosaic virus (TMV) particles [156]. Figure 2.13C (lower right) shows the x–y trajectory of single TMV particles, and one can see that a small residual motion of the particle naturally occurs due to the update time of the trap ($\sim$3.3 ms). Rather than being a nuisance, interestingly, this jiggling of the particle in the trap contains useful information [157]. A record of the actual positions of the particle and the applied electric fields at each update time can be used to remove the effects of the trapping and calculate a pseudofree-trajectory, which is statistically similar to the trajectory the particle would have followed if it had not been trapped (shown for the TMV particles in the main part of Fig. 2.13C) [156]. Under various assumptions, the pseudofree-trajectory can be used to extract information about both the mobility and the diffusion coefficient of the trapped particle [155].

To go faster, and therefore to trap single molecules without the need to artificially increase the solution viscosity, a hardware version of the ABEL trap has been constructed using rotation of the laser focus [158]. This method allows update times as small as $\sim$ 25 µs, and Fig. 2.13D shows the displacement histogram of a single fluorescently-labeled molecule of the chaperonin GroEL trapped in buffer [155]. Tracking and trapping single biomolecules in buffer is a fascinating new direction showing potential for improvements along many lines.

2.6 Summary

In this contribution, the early steps leading to the first single-molecule optical detection and spectroscopy [3,21] were described. The-low temperature experiments in the early 1990s yielded many novel physical effects which have reappeared in the later room temperature studies in different, but related forms. For example, the spectral diffusion process [34] can be thought of as an analogy to the blinking effects observed for many systems at room temperature [44]. Similarly, the light-induced spectral shifts at low temperature presaged the single-molecule photoswitching that was subsequently observed at room temperature for both single GFP proteins [82] and for single small molecules [134, 159–161]. The first single-molecule imaging [34] demonstrated that the nanoscale emitter allowed precise measurement of the size of a laser focal spot, leading to the use of single molecules to precisely locate the position of the PSF of a microscope [162,163]. These physical effects are currently being used for superresolution imaging, even in living cells; thus, the future of single-molecule fluorescence studies continues to be bright. It is to be expected that single molecules will continue to act as nanoscale probes of complex matter, providing an unprecedented view into heterogeneity, local dynamics, and the stochastic behavior of individuals.

Acknowledgements

The author warmly thanks many members of the Moerner single-molecule spectroscopy group for their crucial contributions to the work reported here, in particular: W. P. Ambrose, Th. Basché, T. P. Carter, L. Kador, J. Köhler, D. J. Norris, and T. Plakhotnik (low-temperature studies); J. Biteen, S. Brasselet, A. E. Cohen, N. R. Conley, J. Deich, R. M. Dickson, S.-Y. Kim, B. Lounis, S. Y. Nishimura, E. J. G. Peterman, M. Vrljic, and K. Willets (room-temperature studies). Stimulating collaborations with the groups of H. M. McConnell, M. Orrit, J. Schmidt, L. Shapiro, R. J. Twieg, and U. P. Wild are gratefully acknowledged. This work has been supported in part by prior grants from the U. S. Office of Naval Research and the National Science Foundation, and by current grants from the National Institutes of Health Nos. 1P20-HG003638, 1R21-RR023149, and 1R01-GM085437, from the Department of Energy Grant No. DE-FG02–07ER15892, and from the National Science Foundation Grant No. DMR-0507296.

References

1. Rigler, R., Orrit, M., Basche, T. (eds.), *Single Molecule Spectroscopy: Nobel Conference Lectures.* (Springer-Verlag, Berlin, 2001)
2. A. Ashkin, J.M. Dziedzic, J.E. Bjorkholm, S. Chu, Opt. Lett. **11**, 288–290 (1986)
3. W.E. Moerner, L. Kador, Phys. Rev. Lett. **62**, 2535–2538 (1989)
4. W.E. Moerner, T. Basché, Angew. Chem. **105**, 537 (1993)
5. W.E. Moerner, Science **265**, 46–53 (1994)
6. M. Orrit, J. Bernard, R. Brown, B. Lounis, in *Progress in Optics*, ed. by E. Wolf (Elsevier, Amsterdam, 1996), pp. 61–144
7. T. Plakhotnik, E.A. Donley, U.P. Wild, Annu. Rev. Phys. Chem. **48**, 181–212 (1996)
8. S. Nie, R.N. Zare, Annu. Rev. Biophys. Biomol. Struct. **26**, 567–596 (1997)
9. W.E. Moerner, M. Orrit, Science **283**, 1670–1676 (1999)
10. W.P. Ambrose, P.M. Goodwin, J.H. Jett, A. VanOrden, J.H. Werner, R.A. Keller, Chem. Revs. **99**(10), 2929–2956 (1999)
11. W.E. Moerner, J. Phys. Chem. B. **106**, 910–927 (2002a)
12. W.E. Moerner, D.P. Fromm, Rev. Sci. Instrum. **74**, 3597–3619 (2003)

13. P. Tinnefeld, M. Sauer, Angew. Chem. Int. Ed. **44**(18), 2642–2671 (2005)
14. P.V. Cornish, T. Ha, ACS Chem. Biol. **2**(1), 53 (2007)
15. W.E. Moerner, P.J. Schuck, D.P. Fromm, A. Kinkhabwala, S.J. Lord, S.Y. Nishimura, K.A. Willets, A. Sundaramurthy, G.S. Kino, M. He, Z. Lu, R.J. Twieg, in *Single Molecules and Nanotechnology*, eds. R. Rigler, H. Vogel (Springer-Verlag, Berlin, 2008), pp. 1–23
16. T. Basché, W.E. Moerner, M. Orrit, U.P. Wild, W.E. Moerner, M. Orrit, U.P. Wild, *Single Molecule Optical Detection, Imaging, and Spectroscopy.* (Verlag-Chemie, T, Munich, 1997)
17. Zander, C., Enderlein, J., Keller, R.A. (eds.), *Single-Molecule Detection in Solution: Methods and Applications*, (Wiley-VCH, Berlin, 2002)
18. C. Gell, D.J. Brockwell, A. Smith, *Handbook of Single Molecule Fluorescence Spectroscopy.* (Oxford Univ. Press, Oxford, 2006)
19. P.R. Selvin, T. Ha, *Single-Molecule Techniques: A Laboratory Manual.* (Cold Spring Harbor Laboratory Press, Cold Spring Harbor, NY, 2008)
20. W.E. Moerner Ed., *Topics in Current Physics 44; Persistent Spectral Hole-Burning: Science and Applications.* (Springer, Berlin, 1988)
21. W.E. Moerner, T.P. Carter, Phys. Rev. Lett. **59**, 2705 (1987)
22. T.P. Carter, M. Manavi, W.E. Moerner, J. Chem. Phys. **89**, 1768 (1988)
23. G.C. Bjorklund, Opt. Lett. **5**, 15 (1980)
24. T.P. Carter, D.E. Horne, W.E. Moerner, Chem. Phys. Lett. **151**, 102 (1988)
25. W.E. Moerner, in *Polymers for Microelectronics, Science, and Technology*, eds. Y. Tabata, I. Mita, S. Nonogaki, (Kodansha Scientific, Tokyo, 1990)
26. L. Kador, D.E. Horne, W.E. Moerner, J. Phys. Chem. **94**, 1237–1248 (1990)
27. G.C. Bjorklund, M.D. Levenson, W. Lenth, C. Ortiz, Appl. Phys. B. **32**, 145 (1983)
28. L. Kador, T. Latychevskaia, A. Renn, U.P. Wild, J. Chem. Phys. **111**, 8755–8758 (1999)
29. M. Orrit, J. Bernard, Phys. Rev. Lett. **65**, 2716–2719 (1990)
30. W.E. Moerner, W.P. Ambrose, Phys. Rev. Lett. **66**, 1376 (1991)
31. F.G. Patterson, H.W.H. Lee, W.L. Wilson, M.D. Fayer, Chem. Phys. **84**, 51 (1984)
32. H. de Vries, D.A. Wiersma, J. Chem. Phys. **70**, 5807 (1979)
33. W.P. Ambrose, T. Basché, W.E. Moerner, J. Chem. Phys. **95**, 7150 (1991)
34. W.P. Ambrose, W.E. Moerner, Nature **349**, 225–227 (1991)
35. J. Friedrich, D. Haarer, in *Optical Spectroscopy of Glasses*, ed. by I. Zschokke, (Reidel, Dordrecht, 1986), p. 149
36. T. Basché, W.E. Moerner, Nature **355**, 335–337 (1992)
37. A. Zumbusch, L. Fleury, R. Brown, J. Bernard, M. Orrit, Phys. Rev. Lett. **70**, 3584–3587 (1993)
38. P.D. Reilly, J.L. Skinner, Phys. Rev. Lett. **71**, 4257–4260 (1993)
39. P.D. Reilly, J.L. Skinner, J. Chem. Phys. **102**, 1540 (1995)
40. E. Geva, J.L. Skinner, J. Phys. Chem. B. **101**, 8920–8932 (1997)
41. T. Plakhotnik, W.E. Moerner, T. Irngartinger, U.P. Wild, Chimia **48**, 31 (1994)
42. J.K. Trautman, J.J. Macklin, L.E. Brus, E. Betzig, Nature **369**, 40–42 (1994)
43. X.S. Xie, Acc. Chem. Res. **29**(12), 598–606 (1996)
44. W.E. Moerner, Science **277**, 1059 (1997)
45. W.E. Moerner, Proc. Natl. Acad. Sci. USA. **104**, 12596–12602 (2007)
46. W.A. Phillips, *Topics in Current Physics 24; Amorphous Solids: Low-Temperature Properties.* (SpringerPhillips, W.A, Berlin, 1981)

47. T. Basché, W.P. Ambrose, W.E. Moerner, J. Opt. Soc. Am. B. **9**, 829 (1992)
48. P. Tchénio, A.B. Myers, W.E. Moerner, J. Lumin. **56**, 1 (1993b)
49. W.E. Moerner, T. Plakhotnik, T. Irngartinger, M. Croci, V. Palm, U.P. Wild, J. Phys. Chem. **98**, 7382–7389 (1994a)
50. J. Bernard, L. Fleury, H. Talon, M. Orrit, J. Chem. Phys. **98**, 850 (1993)
51. R. Loudon, *The Quantum Theory of Light.* (Clarendon, Oxford, 1983)
52. H.J. Kimble, M. Dagenais, L. Mandel, Phys. Rev. Lett. **39**, 691 (1977)
53. T. Basché, W.E. Moerner, M. Orrit, H. Talon, Phys. Rev. Lett. **69**, 1516–1519 (1992)
54. W.E. Moerner, R.M. Dickson, D.J. Norris, Adv. Atom. Mol. Opt. Phys. **38**, 193–236 (1997)
55. P. Tamarat, B. Lounis, J. Bernard, M. Orrit, S. Kummer, R. Kettner, S. Mais, T. Basché, Phys. Rev. Lett. **75**, 1514 (1995)
56. W.E. Moerner, New J. Phys. **6**, 88 (2004)
57. B. Lounis, W.E. Moerner, Nature **407**, 491–493 (2000)
58. B. Lounis, M. Orrit, Rep. Prog. Phys. **68**, 1129–1179 (2005)
59. J. Köhler, J.A.J.M. Disselhorst, M.C.J.M. Donckers, E.J.J. Groenen, J. Schmidt, W.E. Moerner, Nature **363**, 242–244 (1993)
60. J. Wrachtrup, C. von Borczyskowski, J. Bernard, M. Orrit, R. Brown, Nature **363**, 244 (1993)
61. J. Köhler, A.C.J. Brouwer, E.J.J. Groenen, J. Schmidt, Science **268**, 1457–1460 (1995)
62. D. Rugar, R. Budakian, H.J. Mamin, B.W. Chui, Nature **430**, 329–332 (2004)
63. W.E. Moerner, T. Plakhotnik, T. Irngartinger, U.P. Wild, D. Pohl, B. Hecht, Phys. Rev. Lett. **73**, 2764 (1994b)
64. R.J. Pfab, J. Zimmermann, C. Hettich, I. Gerhardt, A. Renn, V. Sandoghdar, Chem. Phys. Lett. **387**, 490–495 (2004)
65. C. Hettich, S. Schmitt, J. Zitzmann, S. Kuhn, I. Gerhardt, V. Sandoghdar, Science **298**, 385–389 (2003)
66. S. Kuhn, U. Hakanson, L. Rogobete, V. Sandoghdar, Phys. Rev. Lett. **97**(1), 017402 (2006)
67. E.B. Shera, N.K. Seitzinger, L.M. Davis, R.A. Keller, S.A. Soper, Chem. Phys. Lett. **174**, 553–557 (1990)
68. U. Mets, R. Rigler, J. Fluoresc. **4**, 259 (1994)
69. R. Rigler, U. Mets, J. Widengren, P. Kask, Eur. Biophys. J. **22**, 169–175 (1993)
70. E.L. Elson, D. Magde, Biopolymers **13**, 1–27 (1974)
71. D.L. Magde, E.L. Elson, W.W. Webb, Biopolymers **13**, 29–61 (1974)
72. M. Eigen, R. Rigler, Proc. Natl. Acad. Sci. **91**, 5740–5747 (1994)
73. R. Rigler, E. Elson, E. Elson, *Springer Series in Chemical Physics Vol. 65; Fluorescence Correlation Spectroscopy; Springer Series Chem. Phys.* (Springer-Rigler, Berlin, 2001)
74. E. Betzig, R.J. Chichester, Science **262**, 1422–1428 (1993)
75. W.P. Ambrose, P.M. Goodwin, J.C. Martin, R.A. Keller, Phys. Rev. Lett. **72**, 160–163 (1994)
76. X.S. Xie, R.C. Dunn, Science **265**, 361–364 (1994)
77. S. Nie, D.T. Chiu, R.N. Zare, Science **266**, 1018–1021 (1994)
78. J.K. Trautman, J.J. Macklin, Chem. Phys. **205**, 221–229 (1996)
79. T. Funatsu, Y. Harada, M. Tokunaga, K. Saito, T. Yanagida, Nature **374**, 555–559 (1995)

80. R.M. Dickson, D.J. Norris, Y. Tzeng, W.E. Moerner, Science **274**(5289), 966–969 (1996)
81. S. Weiss, Science **283**, 1676–1683 (1999)
82. R.M. Dickson, A.B. Cubitt, R.Y. Tsien, W.E. Moerner, Nature **388**(6640), 355–358 (1997)
83. R.Y. Tsien, Annu. Rev. Biochem. **67**, 509–544 (1998)
84. J.S. Biteen, M.A. Thompson, N.K. Tselentis, G.R. Bowman, L. Shapiro, W.E. Moerner, Proc. SPIE **7185**, 71850I (2009)
85. R. Ando, H. Hama, M. Yamamoto-Hino, H. Mizuno, A. Miyawaki, Proc. Natl. Acad. Sci. USA. **99**, 12651–12656 (2002)
86. G.H. Patterson, J. Lippincott-Schwartz, Science **297**, 1873–1877 (2002)
87. J. Wiedenmann, S. Ivanchenko, F. Oswald, F. Schmitt, C. Röcker, A. Salih, K. Spindler, G.U. Nienhaus, Proc. Natl. Acad. Sci. USA. **101**(45), 15905–15910 (2004)
88. R. Ando, H. Mizuno, A. Miyawaki, Science **306**, 1370–1373 (2004)
89. E. Betzig, G.H. Patterson, R. Sougrat, O.W. Lindwasser, S. Olenych, J.S. Bonifacino, M.W. Davidson, J. Lippincott-Schwartz, H.F. Hess, Science **313** (5793), 1642–1645 (2006)
90. S.T. Hess, T.P.K. Girirajan, M.D. Mason, Biophys. J. **91**, 4258–4272 (2006)
91. A. Miyawaki, J. Llopis, R. Heim, J.M. McCaffrey, J.A. Adams, M. Ikura, R.Y. Tsien, Nature **388**, 882 (1997)
92. T. Ha, T. Enderle, D.F. Ogletree, D.S. Chemla, P.R. Selvin, S. Weiss, Proc. Natl. Acad. Sci. **93**, 6264–6268 (1996)
93. T. Ha, Methods **25**, 78–86 (2001)
94. E. Nir, X. Michalet, K.M. Hamadani, T.A. Laurence, D. Neuhauser, Y. Kovchegov, S. Weiss, J. Phys. Chem. B. **110**, 22103–22124 (2006)
95. S. Brasselet, E.J.G. Peterman, A. Miyawaki, W.E. Moerner, J. Phys. Chem. B. **104**, 3676–3682 (2000)
96. T. Schmidt, G.J. Schutz, W. Baumgartner, H.J. Gruber, H. Schindler, Proc. Natl Acad. Sci. **93**, 2926–2929 (1996)
97. Y. Sako, S. Minoghchi, T. Yanagida, Nat. Cell Biol. **2**(3), 168–172 (2000)
98. T. Kues, R. Peters, U. Kubitscheck, Biophys. J. **80**, 2954–2967 (2001)
99. G.S. Harms, L. Cognet, P.H.M. Lommerse, G.A. Blab, H. Kahr, R. Gamsjaeger, H.P. Spaink, N.M. Soldatov, C. Romanin, T. Schmidt, Biophys. J. **81**, 2639–2646 (2001a)
100. M. Vrljic, S.Y. Nishimura, S. Brasselet, W.E. Moerner, H.M. McConnell, Biophys. J. **82**, 523A, (2002b)
101. M. Vrljic, S.Y. Nishimura, S. Brasselet, W.E. Moerner, H.M. McConnell, Biophys. J. **83**, 2681–2692 (2002a)
102. M. Vrljic, S.Y. Nishimura, W.E. Moerner, H.M. McConnell, Biophys. J. **88**, 334–347 (2005)
103. S. Nishimura, M. Vrljic, L.O. Klein, H.M. McConnell, W.E. Moerner, Biophys. J. **90**, 927–938 (2006a)
104. S.Y. Nishimura, S.J. Lord, L.O. Klein, K.A. Willets, M. He, Z.K. Lu, R.J. Twieg, W.E. Moerner, J. Phys. Chem. B. **110**(15), 8151–8157 (2006b)
105. M. Vrljic, S.Y. Nishimura, W.E. Moerner, H.M. McConnell, Biophys. J. **84**, 325A–325A (2003)
106. W.E. Moerner, Trends Analyt. Chem. **22**, 544–548 (2003)
107. M.C. Konopka, J.C. Weisshaar, J. Phys. Chem. A. **108**, 9814–9826 (2004)

108. J. Ichinose, S. Sako, Trends Analyt. Chem. **23**, 587–594 (2004)
109. X.S. Xie, J. Yu, W.Y. Yang, Science **312**(5771), 228–230 (2006)
110. J. Zhang, R.E. Campbell, A.Y. Ting, R.Y. Tsien, Nat. Rev. Mol. Cellular Biol. **3**, 906–918 (2002)
111. B.N.G. Giepmans, S.R. Adams, M.H. Ellisman, R.Y. Tsien, Science **312**, 217–224 (2006)
112. S.J. Remington, Curr. Opin. Struct. Biol. **16**, 714–721 (2006)
113. A.M. Femino, F.S. Fay, K. Fogarty, R.H. Singer, Science **280**(5363), 585–590 (1998)
114. J. Deich, E.M. Judd, H.H. McAdams, W.E. Moerner, Proc. Natl. Acad. Sci. USA. **101**, 15921–15926 (2004)
115. S.Y. Kim, Z. Gitai, A. Kinkhabwala, L. Shapiro, W.E. Moerner, Proc. Natl. Acad. Sci. U S A. **103**(29), 10929–10934 (2006)
116. G.S. Harms, L. Cognet, P.H.M. Lommerse, G.A. Blab, T. Schmidt, Biophys. J. **80**(5), 2396–2408 (2001b)
117. W.E. Moerner, J. Chem. Phys. **117**, 10925–10937 (2002b)
118. S. Courty, C. Luccardini, Y. Bellaiche, G. Cappello, M. Dahan, Nano Lett. **6**(7), 1491–1495 (2006)
119. B.L. Lounis, J. Deich, F.I. Rosell, S.G. Boxer, W.E. Moerner, J. Phys. Chem. B. **105**, 5048–5054 (2001)
120. L. Shapiro, H. McAdams, R. Losick, Science **298**, 1942–1946 (2002)
121. E.D. Goley, A.A. Iniesta, L. Shapiro, J. Cell Sci. **120**, 3501–3507 (2007)
122. I. Fujiwara, D. Vavylonis, T.D. Pollard, Proc. Natl Acad. Sci. **104**(21), 8827–8832 (2007)
123. J. Gelles, B.J. Schnapp, M.P. Sheetz, Nature **4**, 450–453 (1988)
124. F. Güttler, T. Irngartinger, T. Plakhotnik, A. Renn, U.P. Wild, Chem. Phys. Lett. **217**, 393 (1994)
125. A.M. van Oijen, J. Köhler, J. Schmidt, M. Müller, G.J. Brakenhoff, Chem. Phys. Lett. **292**, 183–187 (1998)
126. R.E. Thompson, D.R. Larson, W.W. Webb, Biophys. J. **82**, 2775–2783 (2002)
127. X. Michalet, S. Weiss, Proc. Natl. Acad. Sci. USA. **103**, 4797–4798 (2006)
128. W. Heisenberg, *The Physical Principles of Quantum Theory.* (University of Chicago Press, Chicago, 1930)
129. W.E. Moerner, Nat. Methods **3**, 781–782 (2006)
130. I. Chen, A. Ting, Curr. Opin. Biotechnol. **16**, 35–40 (2005)
131. K.A. Willets, S.Y. Nishimura, P.J. Schuck, R.J. Twieg, W.E. Moerner, Acc. Chem. Res. **38**(7), 549–556 (2005)
132. S.J. Lord, Z. Lu, H. Wang, K.A. Willets, P.J. Schuck, H.D. Lee, S.Y. Nishimura, R.J. Twieg, W.E. Moerner, J. Phys. Chem. A. **111**(37), 8934–8941 (2007)
133. H. Wang, Z. Lu, S.J. Lord, W.E. Moerner, R.J. Twieg, Tetrahedron Lett. **48**(19), 3471–3474 (2007)
134. M.J. Rust, M. Bates, X. Zhuang, Nat. Methods **3**, 793–795 (2006)
135. N.R. Conley, J.S. Biteen, W.E. Moerner, J. Phys. Chem. B. **112**, 11878–11880 (2008)
136. S.W. Hell, Science **316**, 1153–1158 (2006)
137. S.W. Hell, S. Jakobs, L. Kastrup, Appl. Phys. A. **77**, 859–860 (2003)
138. M. Hofmann, C. Eggeling, S. Jakobs, S.W. Hell, Proc. Natl. Acad. Sci. USA. **102**, 17565–17569 (2005)
139. S.W. Hell, J. Wichmann, Opt. Lett. **19**, 780–782 (1994)

140. C.A. Werley, W.E. Moerner, J. Phys. Chem. B. **110**, 18939–18944 (2006)
141. H. Yang, G. Luo, P. Karnchanaphanurach, T. Louie, I. Rech, S. Cova, L. Xun, X.S. Xie, Science **302**, 262–266 (2003)
142. I. Rasnik, S.A. McKinney, T. Ha, Acc. Chem. Res. **38**, 542–548 (2005)
143. E.J.G. Peterman, S. Brasselet, W.E. Moerner, J. Phys. Chem. A. **103**, 10553–10560 (1999)
144. E. Boukobza, A. Sonnenfeld, G. Haran, J. Phys. Chem. B. **105**, 12165–12170 (2001)
145. C. Bustamante, Z. Bryant, S.B. Smith, Nature **421**, 423–427 (2003)
146. K.C. Neuman, S.M. Block, Rev. Sci. Instrum. **75**, 2787–2809 (2004)
147. D.G. Grier, Nature **424**, 810–816 (2003)
148. A.E. Cohen, W.E. Moerner, Appl. Phys. Lett. **86**, 093109 (2005c)
149. M. Armani, S. Chaudhary, R. Probst, B. Shapiro, in *18th IEEE International Conference on Micro Electro Mechanical Systems, 2005*, p. 855 (2005)
150. J. Enderlein, Appl. Phys. B. **71**, 773–777 (2000)
151. A.J. Berglund, H. Mabuchi, Appl. Phys. B. **78**, 653–659 (2004)
152. S.B. Andersson, Appl. Phys. B. **80**(7), 809–816 (2005)
153. A.E. Cohen, W.E. Moerner, Proc. SPIE. **5699**, 296–305 (2005b)
154. A.E. Cohen, W.E. Moerner, Proc. SPIE. **5930**, 191–198 (2005a)
155. A.E. Cohen, *Trapping and manipulating single molecules in solution.* (Stanford University, Stanford, CA, 2006), Ph.D. dissertation
156. A.E. Cohen, W.E. Moerner, Proc. Nat. Acad. Sci. USA. **103**, 4362–4365 (2006)
157. A.E. Cohen, Phys. Rev. Lett. **94**, 118102 (2005)
158. A.E. Cohen, W.E. Moerner, Opt. Express **16**, 6941–6956 (2008)
159. S.J. Lord, N.R. Conley, H.D. Lee, R. Samuel, N. Liu, R.J. Twieg, W.E. Moerner, J. Am. Chem. Soc. **130**(29), 9204–9205 (2008)
160. M. Irie, T. Fukaminato, T. Sasaki, N. Tamai, T. Kawai, Nature **420**(6917), 759–760 (2002)
161. M. Heilemann, E. Margeat, R. Kasper, M. Sauer, P. Tinnefeld, J. Am. Chem. Soc. **127**(11), 3801–3806 (2005)
162. A.M. van Oijen, J. Köhler, J. Schmidt, M. Müller, G.J. Brakenhoff, J. Opt. Soc. Am. A. **16**, 909–915 (1999)
163. A. Yildiz, J.N. Forkey, S.A. McKinner, T. Ha, Y.E. Goldman, P.R. Selvin, Science **300**, 2061–2065 (2003)
164. M. Ormo, A.B. Cubitt, K. Kallio, L.A. Gross, R.Y. Tsien, S.J. Remington, Science **273**, 1392–1395 (1996)
165. P. Tchénio, A.B. Myers, W.E. Moerner, J. Phys. Chem. 97, 2491 (1993a)
166. A.B. Myers, P. Tchénio, M. Zgierski, W.E. Moerner, J. Phys. Chem. **98**, 10377 (1994)
167. D.J. Norris, M. Kuwata-Gonokami, W.E. Moerner, Appl. Phys. Lett. **71**, 297 (1997)
168. S. Kummer, R.M. Dickson, W.E. Moerner, Proc. Soc. Photo-Opt. Instrum. Engr. **3273**, 165–173 (1998)
169. R.M. Dickson, D.J. Norris, W.E. Moerner, Phys. Rev. Lett. **81**, 5322–5325 (1998)
170. S. Brasselet, W.E. Moerner, Single Mol. **1**, 15–21 (2000)
171. H. Sosa, E.J.G. Peterman, W.E. Moerner, L.S.B. Goldstein, Nat. Struct. Biol. **8**, 540–544 (2001)
172. E.J.G. Peterman, H. Sosa, L.S.B. Goldstein, W.E. Moerner, Biophys. J. **81**, 2851–2863 (2001)

173. B.L. Lounis, H.A. Bechtel, D. Gerion, P. Alivisatos, W.E. Moerner, Chem. Phys. Lett. **329**, 399–404 (2000)
174. K.A. Willets, O. Ostroverkhova, M. He, R.J. Twieg, W.E. Moerner, J. Am. Chem. Soc. **125**, 1174–1175 (2003)
175. K.A. Willets, P.R. Callis, W.E. Moerner, J. Phys. Chem. B. **108**, 10465–10473 (2004)
176. J. Hwang, M.M. Fejer, W.E. Moerner, Phys. Rev. A. **73**, 021802(R), (2006)
177. N.R. Conley, A.K. Pomerantz, H. Wang, R.J. Twieg, W.E. Moerner, J. Phys. Chem. B. **111**, 7929–7931 (2007)
178. H.D. Lee, E.A. Dubikovskaya, H. Hwang, A.N. Semyonov, H. Wang, L.R. Jones, R.J. Twieg, W.E. Moerner, P.A. Wender, J. Am. Chem. Soc. **130**, 9364–9370 (2008)

3

Single Molecules as Optical Probes for Structure and Dynamics

Michel Orrit

Summary. Single molecules and single nanoparticles are convenient links between the nanoscale world and the laboratory. We discuss the limits for their optical detection by three different methods: fluorescence, direct absorption, and photothermal detection. We briefly review some recent illustrations of qualitatively new information gathered from single-molecule signals: intermittency of the fluorescence intensity, acoustic vibrations of nanoparticles (1–100 GHz) or of extended defects in molecular crystals (0.1–1 MHz), and dynamical heterogeneity in glass-forming molecular liquids. We conclude with an outlook of future uses of single-molecule methods in physical chemistry, soft matter, and material science.

3.1 Introduction

3.1.1 Optical Signals from Single Nano-Objects in the Far Field

As a technique to study condensed matter, optical microscopy offers many advantages. It is noninvasive or only weakly invasive for the systems under study, it reaches deeply inside a sample (much beyond the first few molecular surface layers), it possesses high sensitivity thanks to single-photon detectors, and it is exceptionally versatile, because it builds up on a wealth of time-resolved and frequency-resolved laser techniques. However, in spite of recent progress in near-field optics and in nonlinear superresolution techniques, optics suffer from one great disadvantage, the low spatial resolution dictated by Abbe's diffraction. The diffraction-limited spot size in far-field optics is so much larger than a single molecule, that it took comparatively long to realize that far-field optical microscopy actually is sensitive enough to reach single-molecule detection [1,30]. Single-molecule fluorescence is currently used under many guises in various fields of scientific study, ranging from pure quantum optics to molecular cell biology. Beside the well-established fluorescence methods, new detection schemes have been proposed in the last years. One of the purposes of this chapter is to describe them briefly and to explore their limitations in delivering optical signals from single objects. In a second part, we present some uses of single molecules and nanoparticles to provide novel information on their environment and their interactions with it. The general directions are illustrated by means of recent work from the author's group, and a general outlook for future research is given.

3.1.2 Signal-to-Noise Ratio

Detecting single nano-objects in a far-field laser spot requires carefully optimized setups, discussed in several reviews and books [3, 19]. Hereafter, we assume ideal conditions, perfect optical elements (detectors, sources, filters, etc.), to concentrate on the fundamental limitations to signal-to-noise ratio in optical single-molecule studies. We consider three main detection techniques applied to individual small absorbers, emitting or not photoluminescence: direct absorption, fluorescence and photothermal contrast.

Direct Absorption

The simplest way to detect a small absorber is to look for the missing photons when we place it in a focused beam, possibly with lock-in detection methods to reduce noise sources. The latter point does not concern us here, as we suppose perfect conditions. This very experiment allowed Kador and Moerner to detect a single molecule optically for the first time [1]. With the use of intensity-squeezed light and of (yet hypothetical) perfect detectors, the sensitivity in this experiment could be pushed to arbitrary weak absorptions (Lounis 2005). However, practical measurements are done with regular laser (or coherent) light, in which irreducible intensity fluctuations are caused by photon noise. Let us place the absorber, with absorption cross-section σ at the waist of a tightly focused beam (beam sectional area A) and look at the missing photons. The signal in number of counts is thus:

$$S_{\mathrm{abs}} = \frac{\sigma}{A} \frac{P \cdot \Delta t}{h\nu},$$

where P is the power of the incident beam, Δt the integration time, and $h\nu$ the energy of the photons used in the experiment. This signal must be compared to the photon noise N of the beam in the same measurement time, so that:

$$(S/N)_{\mathrm{abs}} = \frac{\sigma}{A} \sqrt{\frac{P \cdot \Delta t}{h\nu}}.$$

To enhance the signal-to-noise ratio, one would like to increase the incident power P. Unfortunately, this power is limited to a saturation power P_{sat} by the object to be detected. For larger powers, saturation will significantly reduce the absorption cross-section. In an absorption measurement on a zero-phonon line at low temperature [1,6], the cross-section can reach a sizeable fraction of the squared light wavelength, $\sigma/A \approx 10^{-2}$ with a saturation intensity of about

$0.1 \, \text{W cm}^{-2}$, so that a single molecule can be detected with an integration time as short as some microseconds. At room temperature, the absorption cross-section of a single molecule is of the order of a fraction of square nanometers, $\sigma/A \approx 10^{-6}$, and the saturation intensity is about $1 \, \text{kW cm}^{-2}$. Therefore, the integration time for single-molecule detection becomes impractically long, of the order of 1 s. The situation is more favorable for metal nanoparticles, which have large absorption cross-sections ($5 \, \text{nm}^2$ for a gold nanosphere of $5 \, \text{nm}$ diameter at its plasmon resonance), and, more importantly, a saturation (or damage) intensity higher than $1 \, \text{MW cm}^{-2}$. The integration time for detection of such a particle by direct absorption becomes shorter than a microsecond. A major disadvantage of direct absorption, however, is that it does not distinguish true absorption losses from other loss processes; for example, scattering by specks of dust or by other inhomogeneities of the surrounding medium. Therefore, detection by direct absorption of small absorbing nanoparticles would only be possible in very pure and well-controlled environments, and it has not been demonstrated yet.

Note that the above arguments also apply to a broad range of scattering experiments. For an ideal emitter suspended in vacuum, dark-field scattering could provide the same signal-to-noise ratio as fluorescence. In practice, however, Rayleigh scattering from the matrix and from small inhomogeneities such as dust would completely overwhelm the weak scattering signal from a single molecule. The scattered wave contribution can be enhanced and made visible even for small objects by interference with a reference wave [8,10,26]. If this reference is the incident beam itself, the "optical theorem" of scattering theory tells us that the interference signal due to the scattered wave is none other than the absorption signal. One can easily see that the signal-to-noise ratio in these "interferential" scattering experiments can never exceed that of direct absorption [26].

Fluorescence

The crucial advantage of fluorescence for the detection of single molecules is the near absence of background under a broad range of usual conditions. In essence, each detected fluorescence photon indicates an absorption event (albeit with a low probability). By counting fluorescence photons directly, one removes the large statistical fluctuations inevitable in a direct absorption measurement. The number of counts detected within the integration time Δt is given by:

$$S_f = \eta \frac{\sigma}{A} \frac{P \cdot \Delta t}{h\nu},$$

where the fluorescence-detection yield η is the fraction of the absorbed photons that led to detection events. The noise now arises from number fluctuations in the background, which itself results from detector dark counts (rate D), stray light, fluorescence from impurities and out-of-focus molecules (cross-section σ_S), and from Raman scattering from matrix or solvent in the illuminated volume (cross-section σ_R). The signal-to-noise ratio in fluorescence is thus:

$$(S/N)_{\text{fluo}} = \eta \frac{\sigma}{A} \cdot \frac{\sqrt{P \cdot \Delta t / h\nu}}{\sqrt{D \cdot \frac{h\nu}{P} + \eta \left(\sigma_S + \sigma_R\right)/A}}.$$

Under ideal conditions, only Raman scattering is a fundamental limitation to the signal-to-noise ratio in single-molecule fluorescence. The ratio of fluorescence to Raman cross-sections is approximately given by:

$$\frac{\sigma_R}{\sigma_f} \sim \frac{\Gamma_{\text{rad}}\Gamma_{\text{fluo}}}{\Delta\omega^2} \sim 10^{-12},$$

where $\Gamma_{\text{rad}} \approx 10^{-3}\,\text{cm}^{-1}$ is a typical radiative rate, $\Gamma_{\text{fluo}} \approx 10^3\,\text{cm}^{-1}$ is the actual width of the fluorescing state at room temperature, and $\Delta\omega \sim 10^4\,\text{cm}^{-1}$ is the frequency difference between the exciting laser and the absorbing electronic states of the matrix or solvent. Because of the large detuning and of the weakness of spontaneous emission, fluorescence is at least a million of millions times more efficient than Raman scattering. This resonance enhancement causes the fluorescence signal of one resonant molecule to dominate the Raman signal of 10^{12} nonresonant solvent molecules [2]. It is the deep reason behind the enormous selectivity of fluorescence. It is interesting to remark that techniques lacking this resonant enhancement will never be able to reach the selectivity of one molecule in a sample volume of one cubic micrometer.

Comparing the signal-to-noise ratio of fluorescence to that of direct absorption, we see that the former ratio is larger by a factor of about $\sqrt{\eta A/(\sigma_S + \sigma_R)}$, which, for good emitters and favorable experimental conditions, is a very large number.

Photothermal Detection

For all its advantages, fluorescence does not apply to all objects. Metal nanoparticles, for example, have extremely weak photoluminescence quantum yields. It is, therefore, interesting to look for other optical contrast signals to detect such single nano-objects or possibly single nonfluorescent molecules. The photothermal method, in which a probe laser beam detects a weak change brought about by a strong pump beam, is very sensitive [23]. As an added bonus, its signal arises from the absorbed pump photons only, and not from the bulk of noninteracting pump photons. In this sense, photothermal detection is comparable to fluorescence, because it is a low-background method. It is, therefore, interesting to compare its sensitivity limit to those of direct absorption and fluorescence.

In current versions of the photothermal method adapted to single-object microscopy [12, 13], a first pump beam, modulated at high frequency, is absorbed by the object to be detected. As schematically shown in Fig. 3.1,

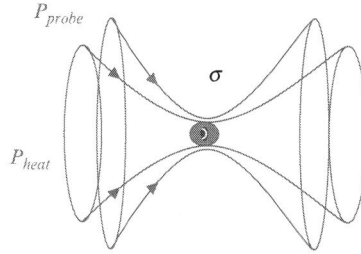

Fig. 3.1. Scheme of a photothermal experiment in which a small object (absorption cross-section σ) is heated by a pump beam (power P_{heat}). A probe beam (power P_{probe}) illuminates the object and its close surroundings. The wave scattered by the temperature inhomogeneity slightly changes the detected probe power

the absorbed heat diffuses into the object's surroundings, changing the local temperature and refractive index. A second beam at a different wavelength, the probe, illuminates the local volume around the object. By interfering with the probe beam itself (or with its reflection), the scattered probe wave slightly changes the transmitted (or reflected) probe intensity. This change is usually detected by a lock-in amplifier tuned to the modulation frequency of the pump beam. A detailed expression of the signal-to-noise ratio has been given in [13]. Here, we only give the main steps in the derivation, to obtain an estimate of this ratio.

Because it arises from interference, the probe intensity modulation is proportional to the scattered *field* $\overrightarrow{E}_{\text{scatt}}$, itself radiated by an equivalent dipole $\overrightarrow{p}$ arising from the change of dielectric permittivity $\Delta(n^2)$ of the probed volume V:

$$\left|\overrightarrow{E}_{\text{scatt}}\right| \approx \frac{1}{4\pi\varepsilon_0 w}\left(\frac{\omega}{c}\right)^2 \cdot |\overrightarrow{p}| \approx \frac{(\omega/c)^2}{2\pi w} n\Delta n \cdot V \cdot \left|\overrightarrow{E}_{\text{probe}}\right|,$$

where w and ω are the probe beam waist and angular frequency. As the heating is modulated at high frequency, the probed volume determined by heat diffusion is usually smaller than the focal spot. The change in probe intensity results from interference of the scattered field with the incident probe field and is therefore related to the energy $P_{\text{abs}} \cdot \tau$ absorbed during one modulation period τ, by:

$$\frac{\Delta P_{\text{probe}}}{P_{\text{probe}}} \approx \frac{1}{\pi w}\left(\frac{\omega}{c}\right)^2 n\frac{\partial n}{\partial T}\frac{P_{\text{abs}} \cdot \tau}{C_P},$$

where C_P is the volume-specific heat of the solvent, and $\partial n/\partial T$ the change of refractive index of the solvent with temperature. We note that the signal is proportional to both the pump and the probe intensities.

The noise, on the other hand, only arises from photon noise in the detected probe beam. Therefore,

$$(S/N)_{\text{photothermal}} = \sqrt{\frac{P_{\text{probe}} \cdot \Delta t}{h\nu}} \cdot \frac{(\omega/c)^2}{\pi w} \cdot \frac{\partial n}{\partial T}\frac{P_{\text{abs}} \cdot \tau}{C_P}.$$

Even though the absorbed power may be limited by saturation, the signal-to-noise ratio is not, because the probe intensity can be increased indefinitely, provided the probe wavelength is not absorbed by the object to be detected [12, 14]. In practice, the intensities cannot be arbitrarily increased because nonlinear and residual heating effects may take place, but the main conclusion is that photothermal detection can have a very different and much more favorable signal-to-noise ratio than direct absorption. For a gold nanoparticle of 5 nm diameter (cross-section 5 nm^2) and for pump and probe powers of 10 mW, the signal-to-noise ratio in a 1-ms integration time is larger than 300. Even for a single molecule at saturation, which absorbs a power of about 1 nW, the photothermal detection limit would be reached with a nonresonant probe power of less than 1 mW, within an integration time of 1 s. For molecules with very low fluorescence yields, the integration time could be even shorter.

Another original feature of the photothermal effect is the derivative $\partial n/\partial T$, typically on the order of $10^{-4}\,\mathrm{K}^{-1}$, but which can be enhanced dramatically close to phase transitions or to critical points of the matrix or solvent around the particle to be detected (provided the response is faster than the modulation time τ).

3.2 Examples of Nanoscale Probing

Hereafter, we give some examples of optical studies on single molecules which provide new information on condensed matter structure or dynamics. In these experiments, single molecules or single nanoparticles can be considered as probes of their environment.

3.2.1 Blinking

Even when a single molecule or a single semiconductor nanocrystal is illuminated by a continuous-wave source, its fluorescence intensity is usually not constant. Its brightness often changes suddenly – blinks – as a function of time, often between dark (off) and bright (on) states. Blinking was one of the first observations of single molecules [30], and it was soon found that many different mechanisms can cause it. Spontaneous or laser-induced spectral diffusion [15] can shift the single-molecule spectral line out of resonance with the exciting laser; photophysical transitions of the molecule to a dark state can interrupt the absorption and/or the fluorescence. The observation of individual blinking events, although related to quantum jumps and theoretically expected, was a great surprise because it revealed the complexity of photophysical processes at molecular scale and opened it to direct experimental study.

Blinking was soon observed in the photoluminescence of single semiconductor nanocrystals [17]. The distribution of off- and on-times was found to follow very remarkable power laws [18] from milliseconds to minutes (see Fig. 3.2). Later experiments [9, 21] extended the range to sub-millisecond times. More

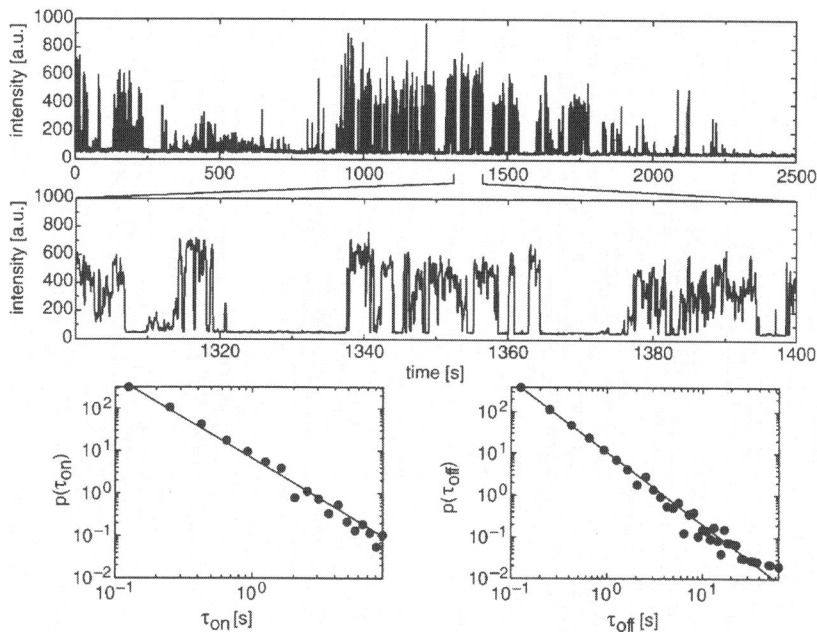

Fig. 3.2. Example of the blinking time trace of a single semiconductor nanocrystal. Magnification of the time-trace show the self-similar character of the on- and off-time pattern. The lower plots are probability distributions for on- and off-times fitted by power laws with exponents of about 1.5 (from [16])

recently, a similar power-law intermittency was also observed in fluorescence traces of single organic molecules [7, 22] and of other nanoparticles. These observations suggest that a single, robust mechanism is at work over this very broad range of times, spanning up to 9 orders of magnitude. A number of models have been proposed to explain power-law intermittency with such a broad dynamics. Most of them assume that blinking is related to the emitter's switching between a neutral bright state and a charge-separated dark state. Verberk et al. [9] have proposed that the broad range of times mainly reflects the distribution of tunneling distances between the emitter and charge acceptors (traps). The distribution of distances could be explained if charges can be self-trapped [7, 16] at different positions in the surrounding matrix. The combination of an exponentially small probability to be trapped at a large distance with an exponentially long recovery time leads to a power-law distribution for the distribution of off-times. A variation of this basic idea can also explain the power law distribution of on-times. The models of Marcus and colleagues assume the dynamics of the reaction coordinate in charge transfer, rather than the electronic matrix element, to govern the blinking times. Tang and Marcus [38] postulate a diffusive walk of the electron-transfer coordinate in a complex potential landscape. Other models [24] attribute the power law to spectral and/or spatial random walks. In the present author's view, however, these models face two big hurdles [16]:

(1) the spectral or spatial spaces in which the dynamics is supposed to take place is too limited to accommodate self-similarity over 9 orders of magnitude of times
(2) it is difficult to conciliate a leading role for nuclear coordinates with the weakness or total absence of a temperature effect on blinking

This is not saying, however, that the geometrical trapping model in its simplest version easily explains all the remarkable features of blinking. Many open questions remain:

(1) Why are there long on-times? Although the dot or molecule is excited continuously, electron transfer appears to remain largely ineffective during the long on-periods. Those are obviously the states one would like to stabilize to use the emitters as fluorescent probes
(2) What is the nature of the charge-separated states? Are those self-trapped states [24], are there multiple states stabilizing holes or electrons in different configurations?
(3) Why is power-law blinking observed in disordered systems only (polymers, glasses, surfaces, etc.), and not in well-ordered systems such as molecular crystals, self-assembled quantum dots, NV centers in diamond, or in colloidal quantum dots with thick ordered shells?

After they have unveiled the fascinating phenomenon of power-law intermittency, single-molecule studies are well placed to provide us with the key to its mechanism, thereby shedding light on complex and elusive charge-separated states in condensed matter.

3.2.2 Gold Nanoparticles

Gold nanoparticles are fascinating objects for basic studies and attractive labels because of their high chemical stability and photostability (no bleaching, no blinking), and because of their strong interaction with light. Because large nanoparticles have very low photoluminescence yields, they must be detected via their absorption or scattering. As we have seen, the photothermal method detects slight changes of index of refraction around the heated particle. An alternative way to study the nanoparticles is to probe their optical properties before heat has had time to diffuse into the environment, with a pump–probe technique. A short pump pulse excites the free electrons of the particle, which relax within a few picoseconds and transfer the absorbed energy to lattice vibrations. The change in optical properties of the particle can be monitored with a short probe pulse, as a function of the delay between pump and probe. The signal-to-noise ratio in this pump–probe experiment is similar to that in the photothermal method, with the differences that:

(1) the changing index of refraction is that of gold, instead of the environment's
(2) all the pump and probe photons are now concentrated in short pulses. Because the transient temperature of the gold particle is limited by melting or by other photo-damage processes, the average pump and probe intensities can be much higher in the cw photothermal experiment than in the pulsed pump–probe experiment

With a sensitive pump–probe technique, possibly within a common-path interferometer, one can detect the acoustic vibrations of an individual gold nanoparticle [36]. This measurement directly gives the vibration's damping time, a parameter inaccessible to measurements on ensembles of nanoparticles, because of the inhomogeneity in sizes and shapes of populations of nanoparticles. The damping of vibrations of a nanoparticle depends critically on the acoustic impedance mismatch between particle and substrate materials, as well as on the mechanical contact area between them. Acoustic damping is therefore a probe of this contact, which may often be limited to a few nanometers only in diameter.

More recently, we have studied the same nanoparticles by combining scanning electron microscopy, optical scattering in white light, and pump–probe spectroscopy of the acoustic modes. Whereas the radially symmetric breathing mode ($n = 0$, $\ell = 0$) is the main vibration excited in close-to-spherical nanoparticles, breaking of the spherical symmetry leads to excitation of the ellipsoidal elastic mode ($n = 0$, $\ell = 2$). This mode is significantly excited in ellipsoidal particles with aspect ratios larger than 1.2, and in dumbbells (contacting pairs) of spherical particles (see Fig. 3.3). In dumbbells, a new low-frequency mode consistently appears (at about 5 GHz for 80-nm diameter particles). This mode is a stretching vibration, where the centers of gravity of the two constituent particles are moving. The stretching frequency depends critically on the contact area between constituent particles, and could therefore be used as a probe of the contact.

3.2.3 Cryogenic Single-Molecule Spectroscopy

In carefully chosen host–guest systems, single guest molecules present extremely sharp zero-phonon lines at cryogenic temperatures, with widths of the order of 30 MHz [27]. As the quality factor of these resonances is larger than 10^7, these lines are extremely sensitive to perturbations and can be used as probes of various processes. A first field of study was the correlated tunneling movement of groups of atoms and molecules in polymers, glasses, and other disordered systems at low temperatures. Another exciting domain, for which optical exploration with single molecules has just started [28], is the motion of charge carriers giving rise to electrical conduction. To combine cryogenic single-molecule spectroscopy with electrical conduction, we looked for a host–guest system presenting sharp and intense zero-phonon lines in a host crystal where conduction is possible. High-purity anthracene crystals

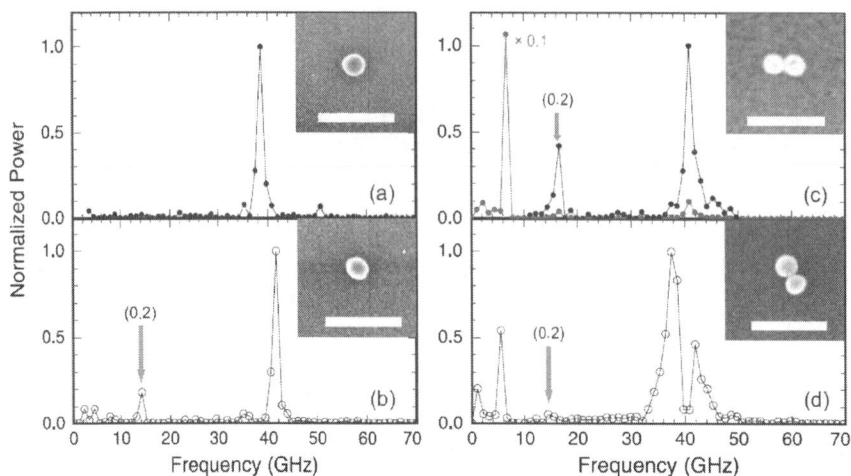

Fig. 3.3. Power spectra in the vibrational transients of single gold nanoparticles, shown together with their scanning electron micrographs. (**a**) Nanospheres (80 nm diameter) show only the radial breathing mode at ∼40 GHz. (**b**) Elongated particles show an ellipsoidal deformation mode at ∼14 GHz. (**c, d**) Dumbbells usually present the ellipsoidal mode and a stretching mode at ∼6 GHz

are good molecular conductors. While doping these crystals with one of the best single-molecule fluorophores, terrylene, we found that the anthracene matrix enhances the intersystem crossing of terrylene a thousand times [29]. Spectroscopy of single terrylene molecules in this matrix is thus impractical. We then shifted the guest's electronic transitions to the red by using dibenzoterrylene instead of terrylene. This new system is very convenient for single-molecule spectroscopy [30, 31]. We have started studies of single molecules in field-effect transistor structures. In the course of this study, we have discovered surprising low-frequency resonances, which we briefly discuss hereafter.

An oscillating electric field is applied to the sample by means of two gold electrodes on which the doped anthracene crystal is deposited. The frequency of the applied ac-voltage is varied at a constant amplitude, typically between 1 kHz and 1 MHz. We monitor a single-molecule line by repeatedly scanning the exciting laser in a range of a few gigahertz and measure its fluorescence intensity. Figure 3.4 shows an example of the time trace thus obtained, with the laser frequency as vertical axis and the ac-voltage frequency as the horizontal axis. In these scans, one often finds unexpected resonances in the amplitude of oscillation of the optical line of the single molecule. Similar resonances have previously been observed [28] upon application of an ac-voltage to a very different host–guest system, single terrylene molecules in an n-alkane matrix on an indium-tin-oxide thin conducting film.

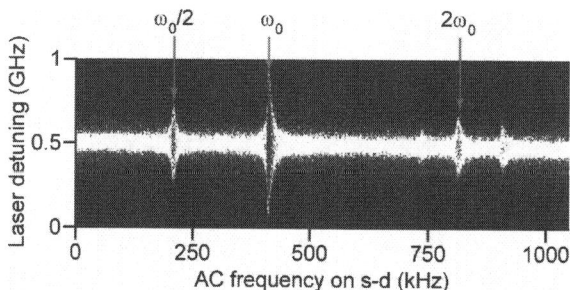

Fig. 3.4. Time-trace of a single molecule excitation spectrum as the frequency of an applied ac-voltage is varied. The vertical axis is the scanned laser frequency and the fluorescence intensity is color-coded. Note the main resonance at 350 kHz and the additional resonances at about twice and half the main frequency, which appear upon increasing the voltage amplitude

The main observations on the resonances may be summarized as follows:

(1) resonance frequencies are found in a range between 10 kHz and a few megahertz (higher frequencies are out of reach of our setup)
(2) there is a strong spatial correlation between the spectra of neighboring single molecules, and the spectra change across cracks in the crystals (thermal shrinkage upon cooling to helium temperature break the anthracene crystal into microcrystals, which present different resonance spectra). The correlation length appears to change with crystallinity. In the polycrystalline n-alkane matrix, it was smaller than the laser spot [28]
(3) the width of the resonances increases with the resonance frequency, and increases very steeply with temperature, as T^n with $n \approx 3-4$
(4) the resonances appear insensitive to the sample's contact with superfluid helium
(5) at high voltage values, multiples and sub-multiples of the resonances appear, which point to strong anharmonicities
(6) the resonances are not shifted by application of a gate voltage, only their amplitude increases when the gate voltage increases in absolute value

Observation (6) contradicts the interpretation of the resonances proposed in [28] as signatures of electric charge density waves. Observations (1)–(6) rather point to mechanical or acoustic oscillations of the whole microcrystal, which would be excited by the applied ac-field, for example, via trapped charge distributions. The oscillating mechanical strain would periodically shift the optical line of the single molecules. The damping of the resonance would occur mainly via Raman-like scattering of acoustic phonons, as was observed for quartz oscillators at low temperatures. In conclusion, the observation of these localized modes would be very difficult without a local probing by single-molecule lines. To our knowledge, such resonances have never been observed in ensemble measurements.

3.2.4 Supercooled Liquids

The Glass Transition

How does a liquid turn into a glass upon cooling? This is an old and still largely unsolved problem of condensed matter physics. A wide range of ensemble techniques has been applied to characterize glass-forming liquids. In spite of an enormous bulk of data, there is, to this day, no clear consensus on the length and time scales of dynamical inhomogeneities in glassy systems. Molecular-scale techniques such as dielectric relaxation or NMR, on the one hand, point to short length scales (1–2 nm in glycerol [32]) and to times comparable to the alpha-relaxation time [11]. On the other hand, many optical techniques, including light scattering, hole-burning, and single-molecule experiments [34, 39] indicate that at least some of the length scales are comparable to the wavelength of light (500 nm), and that relaxation can be very slow. Because they remove averaging, single-molecule measurements deliver molecular information at a truly local scale, and provide fresh and crucial insight into the dynamical heterogeneity of glassy matter.

Liquid-Like Pockets and Solid-Like Structures

Observations of condensed matter at nanometer scales with scanning probes often reveal a surprising time- and space-heterogeneity. Recent examples are the studies of single fluorescent molecules dissolved in supercooled molecular liquids and polymers [33, 34]. In our study of supercooled glycerol [39], we have followed the rotational diffusion of single perylene-diimide molecules by monitoring the polarization of their fluorescence (see Fig. 3.5). A correlation trace of a single molecule's linear dichroism provides the characteristic tumbling time, which can be related to the local viscosity by the Perrin equation. A first surprise is the large spread (about one decade) of the tumbling times measured for different molecules at different locations in the sample. An even bigger surprise was that the memory of the local viscosity at 205 K can exceed 1 day, i.e., a million times the molecular tumbling time. This demonstrates unambiguously the presence of a long-lived heterogeneity of this glassy system.

This observation points to the existence of long-lived, solid-like structures in a supercooled liquid, well above the glass transition. This conclusion is in stark contrast with the naïve picture of a glass former as a "liquid in slow motion," where all molecular-scale inhomogeneities would relax with the same characteristic molecular tumbling time. Instead, it rather agrees with the model of a mosaic-like landscape, with "lakes" filled with liquids having different viscosities and separated on the experimental timescale by long-lived (thus solid-like) dikes or walls, presumably made out of a denser phase. These structures would be responsible for the surprisingly long memory times.

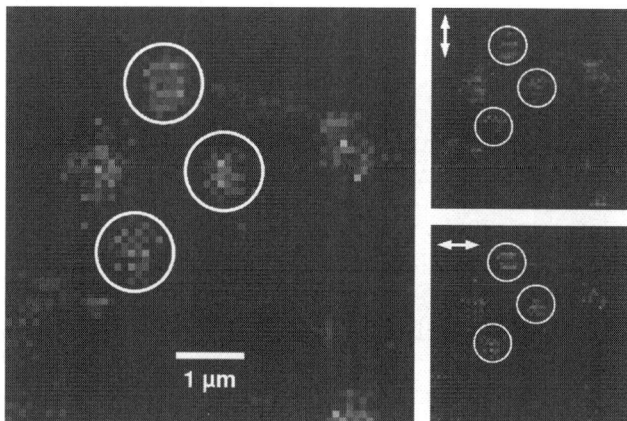

Fig. 3.5. Scanning confocal image of a glycerol thin film with a low concentration of fluorescent dye molecules at 204 K. The polarization of the fluorescence, recorded by two independent detectors, is rendered as the color of the pixels. The three circled single molecules demonstrate visible differences in their tumbling patterns. The upper one tumbles at the scanning rate (about 1 s per line scan), while the lower molecule keeps its orientation during a few successive scans. The middle molecule reorients faster

Rheology

With this mosaic model in mind, we have recently set up a macroscopic rheology experiment on molecular glass-forming liquids to confirm the presence of the hypothetical extended solid-like structures [37]. Because the solid-like structures might be minor components above the glass transition, and because we expected them to easily yield to applied stress, we designed a sensitive Couette rheometer for stresses as low as 100 Pa (corresponding to a hydrostatic pressure of only 10 mm of water). We indeed found the expected yield-stress behavior, but only after certain thermal histories had been applied to the sample. The rheological response of the supercooled liquid provided three parameters: the liquid's viscosity, a shear modulus corresponding to the elastic response of the solid-like structures, and an effective viscosity corresponding to their plastic deformation (or yield). Moreover, the latter two parameters were found to strongly increase with the aging time of the sample (over a period of 2 weeks at 205 K), which points to a slow continuous growth of the solid-like fraction.

The same type of thermal history caused solid-like behavior in the supercooled molecular liquid ortho-terphenyl [35], a fragile molecular glass-former, as well as in glycerol, a stronger, hydrogen-bonded one.

Open Questions

The rheological behavior of glycerol is thus fully consistent with single-molecule measurements, but it only proves that some solid-like structure connects the material at large distances. We still lack a direct confirmation and characterization of the mosaic picture, with the size and shape of the lakes, the dimensions and structure of the walls or separations between the lakes, the temperature and time dependence of the inhomogeneities. What is the *molecular* structure of the solid-like connections? How do dye molecules diffuse in this complex heterogeneous landscape? What happens when the solid structures yield to an external shear? All these questions can benefit from a single-molecule approach. Moreover, because of their chemical simplicity, molecular glass formers are excellent prototypes to understand such complex liquids as mixtures, polymers, colloidal suspensions, foams, or slurries [20].

3.3 Outlook and Conclusion

More than 15 years after the first demonstration experiments, single-molecule fluorescence has become a standard microscopy technique. The detectivity (number of counts, rate, background levels) already appears close to optimum for fluorescence experiments, although breathtaking progress is still made on the spatial resolution and in the tracking techniques. Significant improvements in the detection of fluorescence would probably require better dyes or new types of photoluminescent particles. However, other detection techniques may not have reached their full potential yet. Photothermal detection applies to nonfluorescent objects, with limits on the signal-to-noise ratio which are very different from those of direct absorption and of fluorescence, in principle extending down to the single-molecule level, although this still remains to be demonstrated. These methods could also be combined with near-field optics and with plasmonic local fields, which we did not discuss here.

What will be the use of single-molecule signals in science and applications? The investigation of complex processes, particularly in molecular cell biology, already benefits from the elimination of ensemble- and time-averaging by single-molecule fluorescence methods. At room temperature, however, the time resolution of fluorescence is limited. On the one hand, fluorescence rates are rarely higher than some tens of kilohertz, which imposes limitations on the fastest processes which can be followed at the single-molecule level (not enough photons are detected to fully reconstruct a dynamical trajectory). On the other hand, photobleaching limits the observation times of the same single molecule (even if the intensity is reduced, detector dark counts set an ultimate limit to acquisition times). It would be very appealing to combine the favorable features of low-temperature studies (photostability of the dyes and rigidity of the structures, which enable long integration times and statistically accurate measurements) with the relevant and interesting dynamics taking place at room temperature. We have recently proposed an all-optical method for this purpose [4]. Rapid optical heating and cooling could be realized by a modulated laser beam illuminating an absorbing film in a cryostat at low temperature. During the hot part of the temperature cycle, the molecule of interest could achieve dynamical transformations between substates, in the dark. The cold part of the cycle would be used for optical excitation and fluorescence detection, which would then provide structural information about the thermally trapped current substate of the molecule.

Although physical chemistry and soft matter science seem at first sight much more familiar and better explored than is molecular biology, the understanding of many chemical–physical problems is still fragmentary. Single molecules open unique windows on the molecular aspect of many problems in soft matter science. We have discussed the example of supercooled molecular liquids and of the approach of the glass transition in Sect. 3.2.4, but structural and dynamical inhomogeneity also plays important parts in many problems, for example, in the friction and adhesion between two solid surfaces. Optical observations of single molecules would complement truly molecular probing by atomic force microscopy by providing a more extended picture of the field of stresses and deformations in a real sample between mesoscopic and molecular scales. It is safe to predict that these and many more fields of physical chemistry will greatly benefit from future single-molecule observations.

Acknowledgements

Support of this work by the research program of "Stichting voor Fundamenteel Onderzoek der Materie" (FOM) supported by NWO (Nederlandse Organisatie voor Wetenschappelijk Onderzoek) is gratefully acknowledged.

References

1. L. Kador, W.E. Moerner, Phys. Rev. Lett. **62**, 2535 (1989)
2. M. Orrit, J. Bernard, Phys. Rev. Lett. **65**, 2716 (1990)
3. Basché T. et al., (eds.), *Single-Molecule Optical Detection, Imaging and Spectroscopy*, (VCH Weinheim Germany, 1997)
4. C. Zander et al., (eds.), *Single Molecule Detection in Solution* (Wiley-VCH, Berlin, 2002)
5. B. Lounis and M. Orrit, Rep. Prog. Phys. **68**, 1129–1179 (2005)
6. L. Kador, T. Latychevskaia, A. Renn, U.P. Wild, J. Chem. Phys. **111**, 8755 (1999)
7. T. Plakhotnik, V. Palm, Phys. Rev. Lett. **87**, 183602 (2001)
8. K. Lindfors et al., Phys. Rev. Lett. **93**, 127401 (2004)
9. M.A. van Dijk et al., Phys. Chem. Chem. Phys. **8**, 3486 (2006)
10. M. Orrit, in *Fluorescence of Supermolecules, polymers and Nanosystems*, ed. by O.S. Wolfbeis, M.N. Berberan-Santos (Springer, Berlin, 2008)
11. M. Tokeshi et al., Anal. Chem. **73**, 2112 (2001)
12. D. Boyer et al., Science **297**, 1160 (2002)

13. S. Berciaud et al., Phys. Rev. B **73**, 045424 (2006)
14. J. Hwang, W.E. Moerner, Phys. Rev. A **73**, 021802 (2006)
15. T. Basché, W.E. Moerner, Nature **355**, 335 (1992)
16. F. Cichos et al., Curr. Opin. Colloid Interf. Sci. **12**, 272 (2007)
17. M. Nirmal et al., Nature **383**, 802 (1996)
18. M. Kuno et al., J. Chem. Phys. **112**, 3117 (2000)
19. R. Verberk et al., Phys. Rev. B **66**, 233202 (2002)
20. P.H. Sher et al., Appl. Phys. Lett. **92**, 101111 (2008)
21. J. Schuster et al., Appl. Phys. Lett. **87**, 051915 (2005)
22. X. Hoogenboom et al., Chem. Phys. Chem. **8**, 823 (2007)
23. J. Tang, R.A. Marcus, Phys. Rev. Lett. **95**, 107401 (2005)
24. P. Frantsuzov, R.A. Marcus, Phys. Rev. B **72**, 155321 (2005)
25. A. Issac et al., Phys. Rev. B **71**, 161302 (2005)
26. M.A. van Dijk et al., Phys. Rev. Lett. **95**, 267406 (2005)
27. W.E. Moerner, M. Orrit, Science **283**, 1670 (1999)
28. J.M. Caruge, M. Orrit, Phys. Rev. B **64**, 205202 (2001)
29. A.A.L. Nicolet et al., J. Chem. Phys. **124**, 164711 (2006)
30. A.A.L. Nicolet et al., Chem. Phys. Chem. **8**, 1215 (2007a)
31. A.A.L. Nicolet et al., Chem. Phys. Chem. **8**, 1929 (2007b)
32. L. Berthier et al., Science **310**, 1797 (2005)
33. U. Tracht et al., Phys. Rev. Lett. **73**, 2727 (1998)
34. L.A. Deschenes, D.A. Vanden Bout, J. Phys. Chem. B **106**, 11438 (2002)
35. R. Zondervan et al., Proc. Nat. Acad. Sci. USA. **104**, 12628 (2007)
36. R.A.L. Vallee et al., J. Chem. Phys. **127**, (2007) 154903.
37. R. Zondervan et al., Proc. Nat. Acad. Sci. USA. **105**, 4993 (2008)
38. P. Sollich et al., Phys. Rev. Lett. **78**, 2020 (1997)
39. R. Zondervan et al., Biophys. J. **90**, 2958 (2006)

4

FCS and Single Molecule Spectroscopy

Rudolf Rigler

4.1 Introduction

The idea to develop Fluorescence Fluctuation spectroscopy started when working in Manfred Eigens' Laboratory in Göttingen in the Max Planck Institute for Physical Chemistry as a postdoctoral fellow. I had just finished the construction and testing of a fluorescence T-jump machine [1] when Jean Pierre Changeux from Institut Pasteur arrived in Göttingen with a bag of freshly isolated nicotinic acetyl choline receptor to use the new fluorescence T-jump apparatus for relaxation kinetic studies of the receptor. Due to the high concentration of detergents present in the preparation and limited conductivity of the solvent leading to strong cavitation in the T-jump cell we could not perform the temperature relaxation experiments. However, this experience raised the question whether the analysis of equilibrium fluctuations by observing changes in the quantum yield of fluorescence would not be an alternative way to follow kinetic processes. In this way all problems related to the instantaneous temperature change could be avoided.

Leo de Meyer a close collaborator of Manfred Eigen conducted at that time experiments to use dynamic laser light scattering for analyzing changes in the refraction index of solutions due to fluctuation in chemical equilibria as proposed by Yee and Keeler. I could convince Leo de Meyer and Klaus Gnädig his associate to change the setup for fluorescence observation. In order to test the system, we used a dye nucleic acid equilibrium, for which we determined the relaxation behavior [2, 3]. Notably, single photon detection at this time was in its infancy stage and Leo de Meyer was working on the first type of digital correlators. Since digital correlators were not available the fluctuation spectrum was analyzed by a spectrum analyzer.

Similar work was at this time conducted by Elliot Elson who had finished his PhD with Robert Baldwin in Stanford and was moving after a postdoctoral year with Bruno Zimm in San Diego to Cornell. Together with Watt W. Webb and Douglas Magde, Elliot Elson was able to show the first fluctuation data of the same dye–nucleic acid equilibrium we were working in Göttingen [4].

I had at this time moved back to the Karolinska Institute and started my own laboratory with a new Spectra Physics Model 164 Ar-ion laser which only exhibited "quantum" noise as compared to the heavy intensity fluctuation of the Argon laser in Göttingen which overshadowed our fluctuation spectra.

4.2 Fluorescence Correlation Spectroscopy

4.2.1 Kinetics of the Excited State and Rotational Correlations

Together with Måns Ehrenberg who joined me as my first doctoral student we had cleared up the theoretical background of rotational relaxation after anisotropic excitation of the groundstate with a pulse of polarized light [5], and we were involved in the first experiments using the distribution of photon arrival times after a pulsed excitation. The limitation of this approach in our eyes was the fact that the rotational anisotropy decay was linked directly to the life time of the excited state and no information on rotational motion was available after the excited state had vanished.

4.2.2 Rotational Diffusion

Given the successful experimental results of Elson and Webb [6] in using FCS (as it was called later on) to study chemical relaxations, our starting point in Göttingen, we concentrated our interest to analyze the first events related to photonic processes. The idea was to transpose the anisotropic relaxation of the excited state after a pulse of light into the observation of orientational fluctuations under continuous excitation [7]. The main result of our work was the demonstration that the strong coupling between the decay of the excited states and the orienational randomization process can be broken if the decay time is much faster than the rotational diffusion time $\tau << \tau_{\mathrm{rot}}$. This is due to the fact that in the pulsed excitation the anisotropic rotational decay of the excited state is observed while with stationary excitation the orientational fluctuation of the ground state is observed via its excited state. Thus, the time limitation inherent to pulsed decay (ps-ns) is absent and the time ranges between ns-s are accessible (Fig. 4.1). The usefulness of being able to measure rotational diffusion processes has been demonstrated [8, 9]) and recently by Zorilla et al. 2007 [10], Tsay et al. 2008 [11]. Rotational diffusion is principally more sensitive to conformational changes since D_{rot} scales with r^3 while $D_{\mathrm{transl.}}$ scales with r.

An important detail of the theory and its experimental verification is the fact that our setup uses isotropic detection (no polarizers) together with polarized excitation. With 2π or 4π detection a correct description of orientational correlation with three eigen- values of the diffusion tensor is obtained. They are sums of the rotational diffusion constants D_x, D_y, D_z around the individual axes of the rotational ellipsoid [5, 7] $6D_z$, $2D_z + 4D_{x,y}$, $5D_z + D_{x,y}$. For a spherical object with $D_x = D_y = D_z$ the rotational decay constants collapse to the well known result $1/\tau_{\mathrm{rot}} = 6D$ (Fig. 4.2).

Orientational fluctuation by groundstate excitation
with stationary illumination

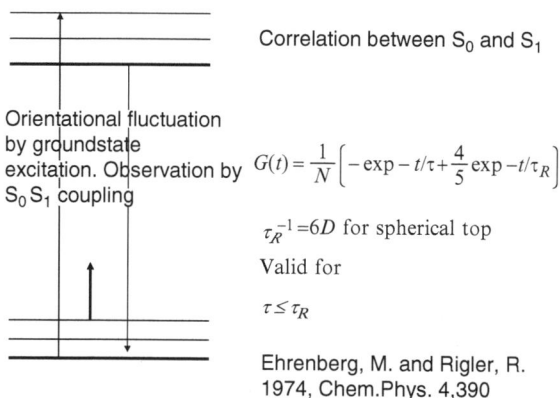

Correlation between S_0 and S_1

Orientational fluctuation
by groundstate
excitation. Observation by
$S_0 S_1$ coupling

$$G(t) = \frac{1}{N}\left[-\exp - t/\tau + \frac{4}{5}\exp - t/\tau_R\right]$$

$\tau_R^{-1} = 6D$ for spherical top

Valid for

$\tau \le \tau_R$

Ehrenberg, M. and Rigler, R.
1974, Chem.Phys. 4,390

Fig. 4.1. Correlation functions for excited states life time and rotational correlation time

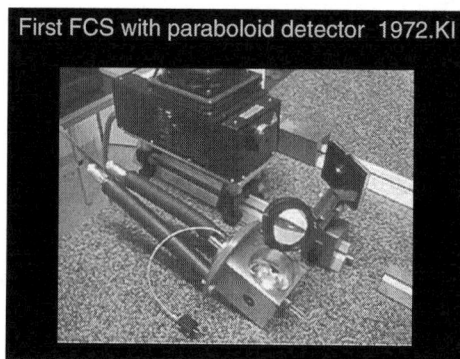

First FCS with paraboloid detector 1972.KI

Fig. 4.2. FCS setup from 1972. In front the 2π cavity with a parabolic reflector. In its focal point, a laminar cuvette with the sample was placed. For details see [12, 13]

An alternative setup is the classical one with two detectors and analyzers which have been used by Kask et al. 1987 and Tsai et al. 2008. Relating to the work of Aragon and Pecora (1975 [14]) the analysis with this setup leads to a complex solution with higher order eigenvalues. Our setup can easily be verified in the confocal setup [15], which started the new era of FCS.

4.3 Excited States Lifetime and Antibunching

The result of our analysis for stationary excitation showed the negative amplitude of the decay correlation function inherent to a system which can only be in one state: either the ground or the excited state. In contrast to excited

Antibunching and antibunching relaxation

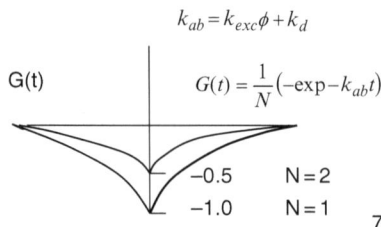

$$k_{ab} = k_{exc}\phi + k_d$$

$$G(t) = \frac{1}{N}(-\exp - k_{ab}t)$$

G(t)

−0.5   N = 2

−1.0   N = 1

7

Fig. 4.3. Excited states correlation function using stationary illumination (antibunching). Observe the amplitude is related to 1/N the number of emitted photons

states, decay after pulse excitation where only information on the decay constant (decay time) is available the excited states correlation time obtained with stationary illumination is the sum of excitation rate and decay rate $1/\tau_{corr} = (k_{exc} \cdot \phi + k_{decay})$. The correlation time is dependent on the flux of the excitation intensity ϕ, the excitation rate constant, and the decay constant ([16], Fig. 4.3). The realization that the statistics of single photon emission from the $S_1 - S_0$ transition has a negative correlation amplitude [7] has later been described as antibunching [17,66] and plays nowadays an important role in defining single photon emitters as sources for quantum cryptography.

For the analysis in which the correlation function has to be recorded in the ns regime a typical Hanbury Brown and Twiss [18] set up. Figure 4.4 shows a confocal setup (discussed later) in combination with a beam splitter and a time to amplitude converter (TAC). Since both pathways for the emitted photons acting as start or stop for the time amplitude converter (TAC) are equivalent the correlation curves registered in a multichannel analyzer (MCA) are symmetric relative to 0 time.

In the setup outlined in Fig. 4.4, which has an overall response time of 0.5 ns, both antibunching studies for performing excitation rates [16, 19] as well as rotational diffusion on polymers (Rhodamin labeled tRNA) were performed.

4.4 FCS in Confocal Volumes

FCS has for almost two decades suffered from the fact that single molecule detection was not feasible basically due to very unfavorable signal-to-background ratios. Usually, the correlation amplitudes in these times were of order 10^{-3} or 10^{-4} indicating that the background was equal to a few thousand fluorescing molecules. With the introduction of confocal illumination and the confocal volume in single molecule detection a real breakthrough was achieved for FCS [9, 15]. The reason was the fact that a single dye molecule like Rhodamine could be measured at a signal-to-background ratio of 1000:1

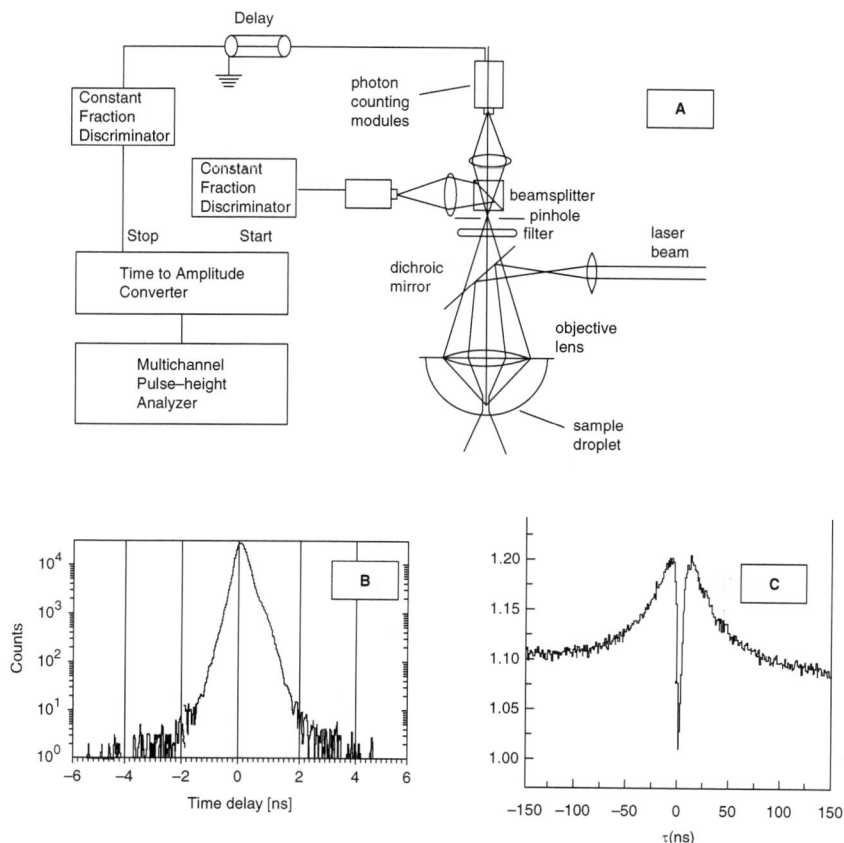

Fig. 4.4. Confocal setup for excited states correlation analysis using a TAC. (a) Time response (FWHM 0.5 ns) of the system using an Avalanche Photodiode (APD). (b) Rotational diffusion of Texas red labeled pancreatic porcine lipase and antibunching of Texas red [19] (c)

or 2000:1 and not with 10^{-3} or 10^{-4}:1 as in the previous years [21, 22]. Thus, the basis for FCS spectroscopy was changed by almost 6 orders of magnitude and it was the start for the rapid increase in FCS spectroscopy. An additional part was the development of Avalanche photodiodes originally by RCA and later by Perkin-Elmer. The high quantum yield in the visible spectrum, high time resolution, and very low background noise constitutes unique features.

4.5 Confocal Single Molecule Detection and Single Molecule Imaging

The proof that single molecules could be measured by confocal illumination was given in a study where fluorescence bursts of Rhodamine molecules passing through a sharply focused laser beam were analyzed. From the burst

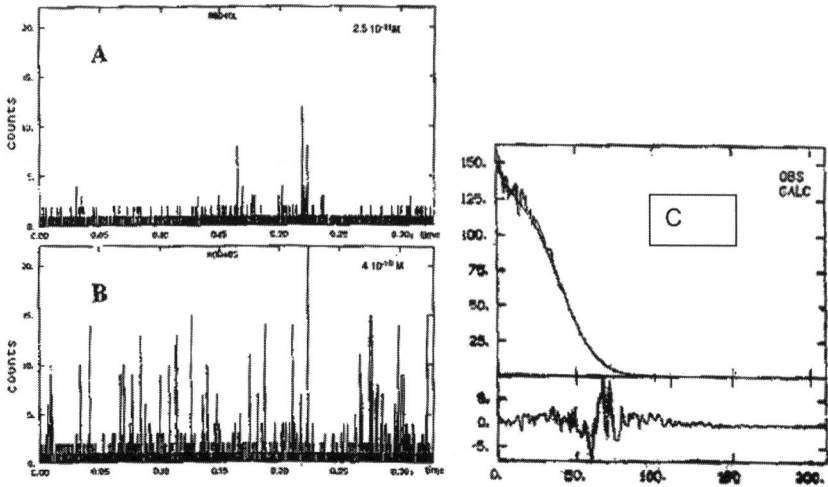

Fig. 4.5. Single molecule traces from Rhodamin 6G at different concentrations: 2.5×10^{-11}M (**a**) and 4×10^{-10}M (**b**). Autocorrelation function of intensity bursts: G0 = 150. $\tau_{\text{diff}} = 40\,\mu$s [21, 22] (**c**)

traces of Rhodamin molecules at different concentrations the autocorrelations curves were generated and analyzed. While the correlation times are due to the diffusion through the confocal volume ($\tau_{\text{diff}} = 40\,\mu$s), the amplitude of the correlation function G_0 at t_0 is equal to $1/N = 150$. This means that in 150 volume elements 1 Rh 6G molecule can be found. It is thus evident that the detected bursts are due to single molecules and the probability that an additional molecule may be involved is very rare (1:150) (Fig. 4.5).

Closely linked to the development of confocal single molecule detection was its use in single molecule imaging. The first demonstration of single molecule imaging in the far field was given by Dapprich et al. [23]. Here the positioning of dye labeled nucleic acid fragments attached to a streptavidinized glass cover slide by a biotin interaction was demonstrated (Fig. 4.6).

Confocal single molecule imaging today has different forms and the confocal volume elements can be generated in different ways: diffractive focusing of a pinhole on the objectplane by epi illumination [15, 24] or by total internal reflection optics [25]. Instead of a pinhole monomode or multimode glassfibers can be used (cf. Fig. 4.17) and is the standard with commercial FCS instruments like Confocor 1, 2, and 3 from Carl Zeiss, M10, M20 from Olympus, and the confocal imaging microscopes from Leica. Finally, the pixel of an optical detector like CCD cameras or matrix APDs can substitute the pinhole. The size of the confocal elements or of order 10^{-16} L and can be reduced by evanescent field excitation down to 10^{-18}L [25], W.W. Webb, this volume). An important development is single molecule imaging in single cells which will be treated later.

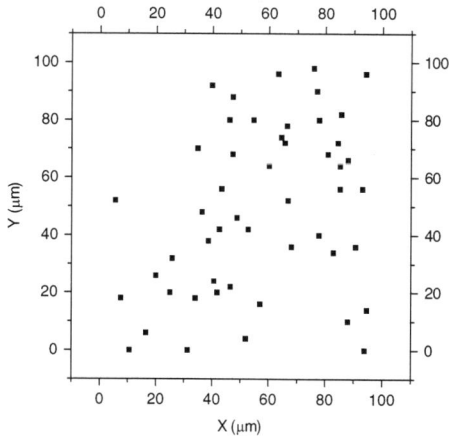

Fig. 4.6. Image of individual ss DNA fragments labeled at the 5' end with a tetramethyl Rhodamin molecule and positioned at the surface of a cover glass by biotin–streptavidin [23]

4.6 Single Molecule Dynamics

The first observation of the dynamic behavior of a single molecule was the observation that a tetra methyl rhodamin 217 bp tagged M13 DNA strand could achieve at least two different conformations as derived from the alternative appearance of two life times of the TMR tag for the same molecule [26, 27] (Fig. 4.7).

This behavior predicts that the exchange rate between these two conformations must be much smaller than the lifetime of the states. Indeed, we found the conformational relaxation to take place in the millisecond range (50 ms) by FCS [27]. Similar behavior has been reported by Hochstraser et al. for tRNA [28].

Instead of measuring the lifetime of a single molecule with pulsed excitation the time dependent intensity of a single molecule under stationary excitation can be measured. Fig. 4.8 shows the first fluctuation of the single DNA molecule. As evident from Fig. 4.8 the data are almost identical to a patch clamp signal from the experiments of Neher and Sackmann showing single ion channels and the fluctuation of the ion current.

The two intensity levels correspond to the two life times of the excited state of the TMR molecule which is due to electron transfer from neighboring Guanosin bases to TMR [27], [31]. We have chosen to calculate the correlation function (Fig. 4.8c) which shows the chemical relaxation rate of the conformational transition. The correlation function cannot be fitted with a single exponential as would be normal for first order kinetics. Instead, a stretched exponential of type $\exp\text{-}(kt)^{\beta}$ gave the best representation of the measured data. The stretched exponential can be envisaged as a distribution of relaxation rates with a mean relaxation rate k and a distribution defined by the

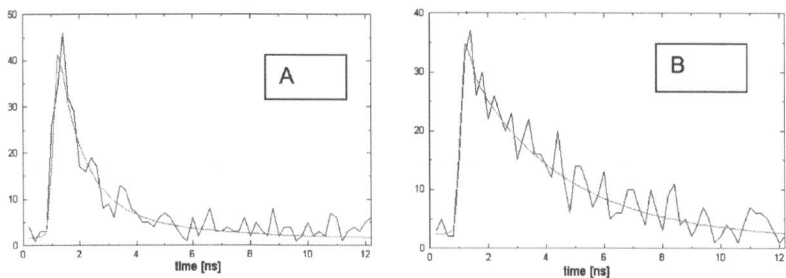

Fig. 4.7. Two conformations of single DNA strands labeled with TMR as evident from the lifetime of the TMR label ($\tau_1 = 0.85$ ns (a), $\tau_2 = 3.7$ ns (b)) [27]

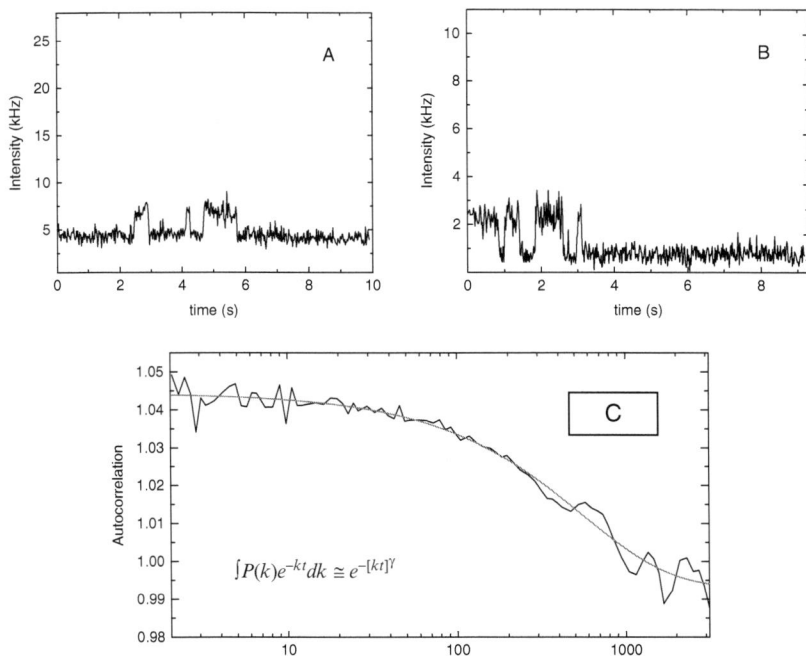

$$\int P(k)e^{-kt}dk \cong e^{-[kt]^\gamma}$$

Fig. 4.8. Conformational fluctuation of single DNA molecules (**a** and **b**). Correlation function of single molecule fluctuation fitted with stretched exponential. Time axis in milliseconds (**c**) [29]

stretch parameter β. $\beta = 1$ corresponds to single exponential state and $\beta < 1$ a broadened distribution in k.

4.7 Non-Ergodic Behavior

The ergodic theorem states the equivalence of the average of a molecular ensemble with the time average of a single molecule. We have, therefore, compared the values for the conformational transition of our standard molecule,

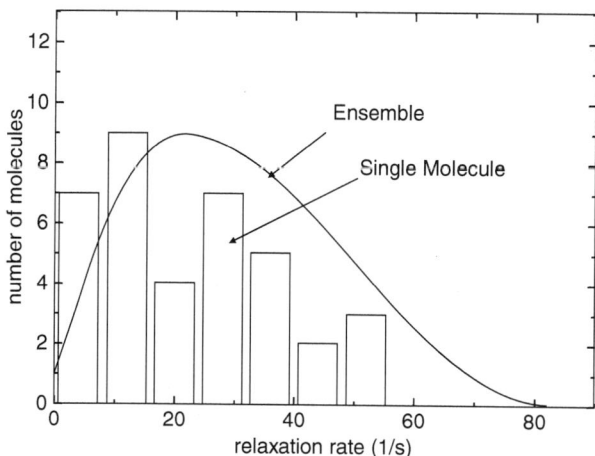

Fig. 4.9. Distribution of conformational rates from ensemble measurements by FCS [27] and comparison with the relaxation rates of 37 individual single molecules. The mean relaxation rates of the 37 molecules is $22.3\,\text{s}^{-1}$ which is close to the peak in the bulk measurement distribution [30]

the 217-bp M13 DNA piece labeled with a TMR molecule on a carbon linker at the 5′ end. From the ensemble analysis by FCS the relaxation curve was fitted with a stretched exponential of type $\exp\text{-}(\text{kt})^{\beta}$ with $k = 20\,\text{s}^{-1}$ and $\beta = 0.44$ [27] which demonstrates that the values of k are distributed between 1 and $80\,\text{s}^{-1}$ (Fig. 4.9).

A comparison with the data of individual single molecules was done by determination of on and off times between the two different conformation and calculating the individual relaxation rate $k = k_{\text{on}} + k_{\text{off}} = \tau_{\text{on}}^{-1} + \tau_{\text{off}}^{-1}$ [30]. Figure 4.8 shows the distribution of relaxation rates of 37 individual molecules which map under the ensemble distribution with a mean relaxation rate close to the peak of the ensemble distribution. The correct comparison would measure the correlation function of individual molecules which was done in Fig. 4.8. Stretched exponential analysis showed a β close to one, i.e., $\beta = 0.8$, indicating a very limited rate distribution compared to the ensemble with a broad rate distribution around two orders of magnitude.

The behavior is clearly non-ergodic in the sense that ensemble and time average differ considerably. One direct reason is the fact that the conformational transition can only be measured for a limited time until the molecule is bleached [30,31]. On the other hand, the relaxation rate of 37 classes of single DNA molecules cover relaxation rates between 1 and $55\,\text{s}^{-1}$. This means that under the assumption that all TMR tagged DNA molecules are identical they can switch from one relaxation regime to another which stays constant during the survival time of photobleaching [31].

4.8 Triplet Kinetics

An important development for confocal FCS analysis was the identification of the influence of excited states triplet kinetics on the FCS correlation time spectrum, particularly on chemical relaxation as well as on diffusion. The first theoretical and experimental analysis was done by Widengren et al. 1994 [32], Widengren et al. 1995 [33].

For this purpose, the first microscope-based single colour cross correlation FCS instrument was developed which still is in use 20 years after its start (Fig. 4.10).

Figure 4.11 shows a summary of the triplet states kinetics with the intersystem crossing between excited singlet state and the triplet state and the triplet decay rate.

A detailed summary of the triplet kinetics and its relation to photobleaching is given by J. Widengren in this volume [34] (Fig. 4.12).

4.9 Relaxation Kinetics

The first use of FCS in chemical relaxation was demonstrated by Elliott Elson, Douglas Magde, and Watt Webb [4], [6]. The system analyzed was the fluctuation due to association and dissociation between an intercalating fluorescent dye and double stranded DNA. Since the diffusion terms constitute own eigenvalues of the diffusion reaction equation, differences in the diffusion of free and bound dye constitute an obstacle in the analysis of the relaxation terms by FCS. We have for this reason focused our interest on cases with little or no change in translational diffusion when using the new FCS in its confocal form.

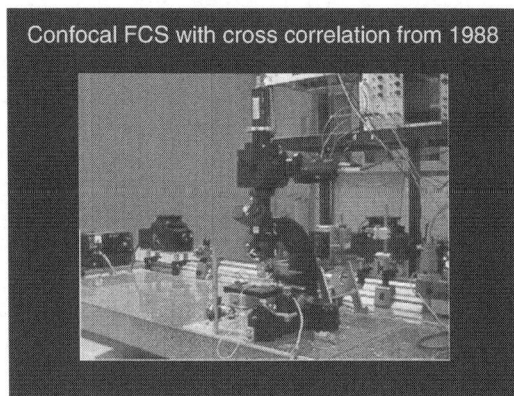

Fig. 4.10. First confocal epi-illumination microscope with dichroic mirror and beam splitter with two RCA avalange photodiodes and identical cutoff filters. Building year 1988 and still in use

Triplet relaxation

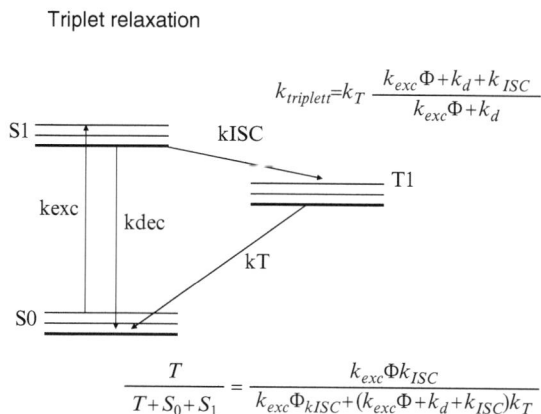

$$k_{triplett}=k_T \frac{k_{exc}\Phi+k_d+k_{ISC}}{k_{exc}\Phi+k_d}$$

$$\frac{T}{T+S_0+S_1}=\frac{k_{exc}\Phi k_{ISC}}{k_{exc}\Phi k_{ISC}+(k_{exc}\Phi+k_d+k_{ISC})k_T}$$

Fig. 4.11. Triplet scheme with S_0, S_1, and T_1 manifolds. Φ excitation flux (photons/cm^2)

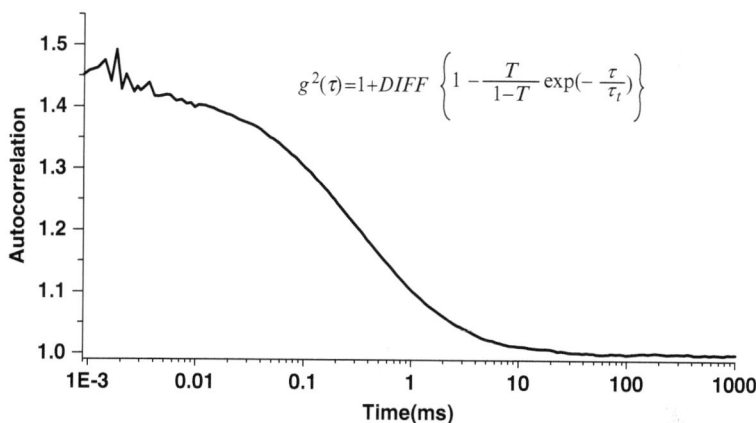

$$g^2(\tau)=1+DIFF\left\{1-\frac{T}{1-T}\exp(-\frac{\tau}{\tau_t})\right\}$$

Fig. 4.12. Triplet autocorrelation function of Rodamin 6G $\tau_{trip}=3\,\mu s$, $T=0.1$, $\tau_{diff}=40\,\mu s$

The first system was the protonization kinetics of Fluorescein which can exist in single and double ionized form with large difference in their quantum yield [35] (Fig. 4.13).

In Fig. 4.14, the concentration dependent chemical relaxation term of Fluorescein isothiocyanate as a function of pH is shown with faster relaxation times at higher proton concentration. From the plot of proton concentration versus the inverse relaxation time (relaxation rate) recombination rate constant k_{rec} of the protonization reaction can be calculated. k_{rec} amounts to $4\times10^{11}\,M^{1-}\,s^{-1}$. This value belongs to the highest measured e.g. for protonisation in ice lattice (Eigen & De Meyer).

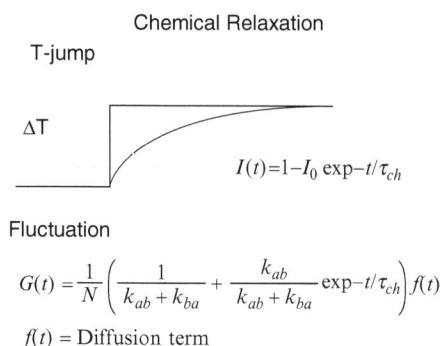

Chemical Relaxation

T-jump

$$I(t)=1-I_0 \exp-t/\tau_{ch}$$

Fluctuation

$$G(t) = \frac{1}{N}\left(\frac{1}{k_{ab}+k_{ba}} + \frac{k_{ab}}{k_{ab}+k_{ba}}\exp-t/\tau_{ch}\right)f(t)$$

$f(t)$ = Diffusion term

Fig. 4.13. Chemical relaxation with a T-jump experiment and by fluctuation spectroscopy (FCS) with the reaction diffusion term

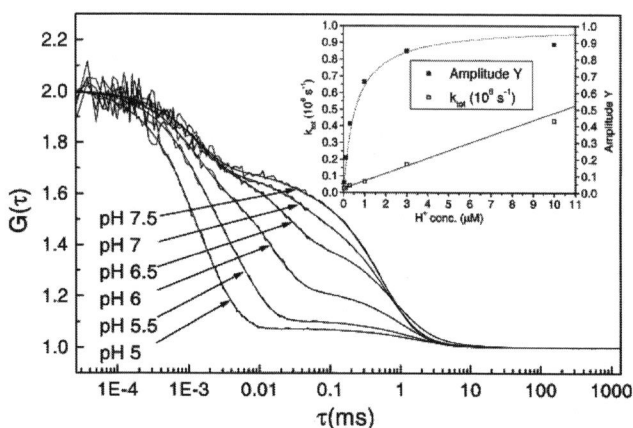

Fig. 4.14. Autocorrelation function of FITC as a function of the proton concentration. Insert: relaxation rate as a function of the proton concentration [35,36]

As a related case, the protonization of different forms of the green fluorescent protein GFP were analyzed [36,37].

4.10 Single Enzyme Molecule Kinetics

The idea to analyze the catalysis by a single enzyme molecule was an obvious step. This was first demonstrated by Peter Lu and Sunney Xie [38]. They were able to show intensity fluctuation of the flavin cofactor of cholesterol oxidase using confocal single molecule detection. Our idea of single enzyme catalysis was to follow the turnover of a substrate into the product by catalysis of a single enzyme molecule. We chose horse radish peroxidase and a non-fluorescence substrate dihydro-rhodamine which is turned over in the fluorescent product

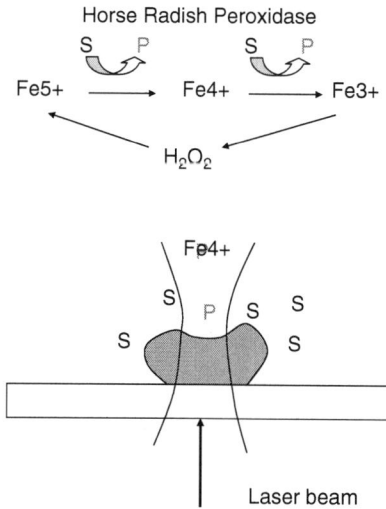

Fig. 4.15. Setup for measuring the turnover of dihydro-rhodamin by horse radish peroxidase bound to a coverslip by biotin–streptavidin interaction

rhodamin. We were able to show the turnover of single biotinyated horseradish peroxidase molecules bound to a coverslip with streptavidin [39] (Figs. 4.15 and 4.16). In this reaction, the enzyme's hemin ring containing a Fe^{5+} is reduced to Fe^{4+} and Fe^{3+}, respectively, with concomitant oxidation of the substrate. Reoxidation of the hemin is performed by H_2O_2. The substrate specificity of the enzyme is low and a variety of substrates like rhodamines and others can be used.

4.11 Stretched Exponentials

The first analysis was done with classical confocal epi-illumination [20] in a hanging drop and showed the fluctuation in the turnover of dihydro-rhodamin [39]. The traces show periods with high turnover followed by periods with sporadic enzyme activity (Fig. 4.16a). This behavior is also evident from the complex correlation function obtained from the turnover fluctuation. In contrast to the expected single exponential behavior we found here again the stretched exponential to be the best choice (Fig. 4.16b). Models describing a manifold of conformational states with different turnover rates have been proposed [39] and are signified by the β parameter of the stretched exponential $\exp-(kt)^\beta$. β equal or close to 1 indicates the presence of a single state. Values between 0.5 and 0.2 as have been found are characteristic for a multitude of states. The model using β as a description for conformational distributions is supported by its temperature dependence: increasing temperatures increase

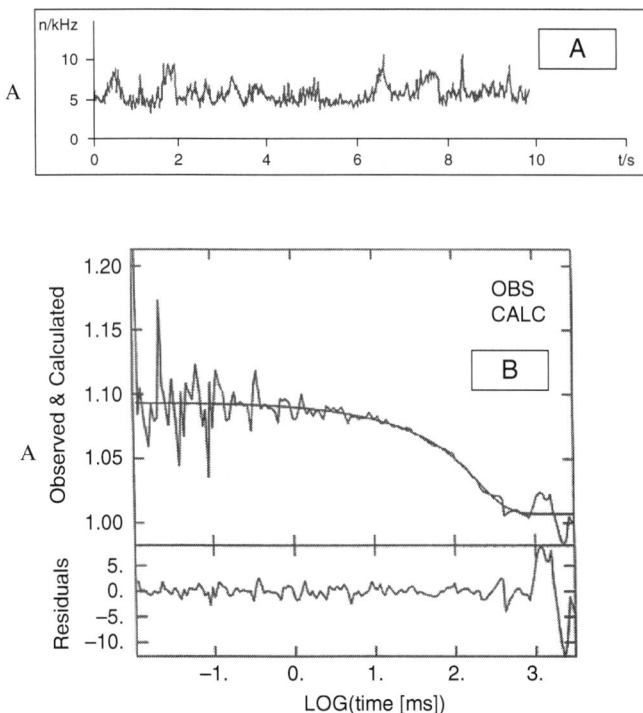

Fig. 4.16. Fluctuation of product formation from dehydro-rhodamin [130 nM] by a single horse radish peroxidase molecule in the presence hydrogen peroxide [10 μM]. Observation by confocal epi-illumination (**a**) Autocorrelation function of product fluctuation (**b**) [39]

the mean turnover rate k but also the β parameter indicating a less distributed conformational landscape (Fig. 4.17).

A drawback of the excitation with epi-illumination lies in the back scattering of the excitation light close to the surface. We have, therefore, changed to an inverse epi-illumination rendering the excitation light into a well controlled beam by total internal reflection (TIRF) and using the evanescent wave transmitted by the reflecting surface for excitation [25,42]. By this change we created an excitation volume of at least one order of magnitude smaller then the typical 0.2 femtoliter as well as a high reduction of the background scattering. Figure 4.18 shows the improvement of the signal-to-background ratio for excitation [41,42].

The development of the epi-TIRF FCS between 2002 and 2007 was an important step to improve FCS on surface bound objects and had lead to a significant reduction of the background radiation, as well as of the confocal volume from 0.2 fL by a factor 20 by the reduction of the extension in the z-axis from 2000 to 100 nm. Due to increased collection efficiency the cpm (counts per molecule and sec) increased from 100 kHz for a typical dye molecule to

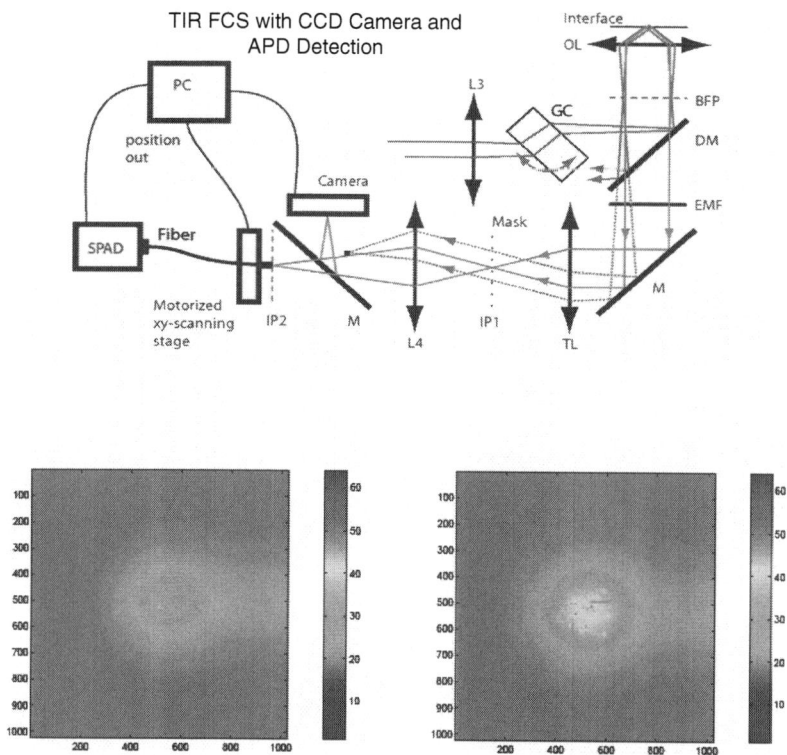

Fig. 4.17. TIRF setup with inverse epi-illumination for FCS analysis. The excitation beam is focused in the back focal plane of the objective and totally reflected at the glass surface carrying the sample. The evanescent component on the other side of the glass surface is exiting the sample. The emission is collected by a high NA objective and focused in the glass fiber cable serving as confocal pinhole. For fast overall detection of fluorescent spots generated by the enzyme catalysis a CCD camera is used. For single molecule detection and FCS the confocal fiber is moved to the position indicated by the CCD camera with a motorized $x-y$ stage. Below: turnover of horse radish peroxidase in the absence (*left image*) and presence of cosubstrate (H_2O_2) (*right image*) [41]

about 2 MHz. The classical HRP experiment from 1999 (Fig. 4.18) has been repeated in the new setup (Fig. 4.17) and the data show the typical stretched behavior with a turnover rate of $144\,s^{-1}$ and a β of 0.34 for Rh6G and of $1729\,s^{-1}$ and 0.23 for Rh 123, respectively (Fig. 4.18). We have introduced the idea of the conformational manifold as a basis of catalysis. This view has been adopted for all the enzymes which have been investigated (lipase B, [43,44]; β-galactosidase [45]).

The HRP experiments have been performed in three laboratories with classical FCS (Rigler Karolinska) and TIRF FCS (Rigler/Lasser EPFL, Widengren KTH) with the same results for different substrates which are turned over with different rates (Fig. 4.18 and Table) [39,41,42].

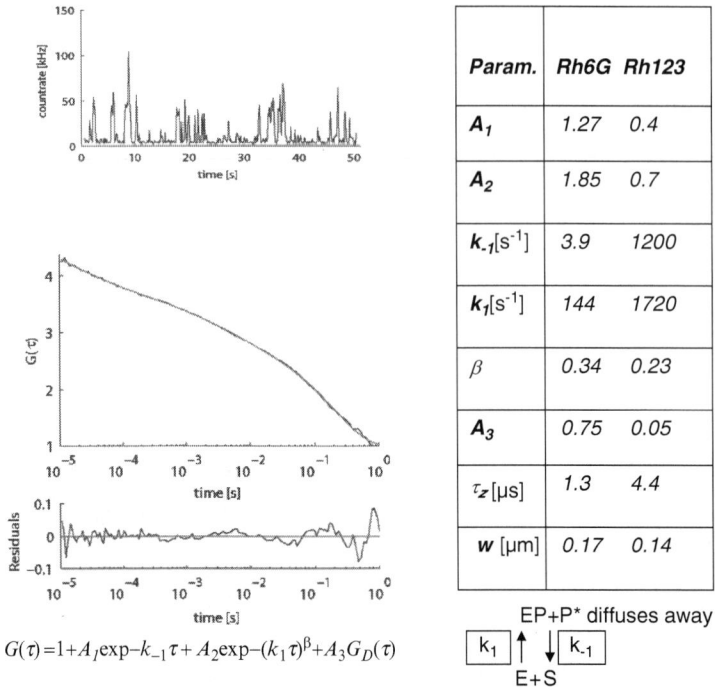

Param.	Rh6G	Rh123
A_1	1.27	0.4
A_2	1.85	0.7
$k_{-1}[s^{-1}]$	3.9	1200
$k_1[s^{-1}]$	144	1720
β	0.34	0.23
A_3	0.75	0.05
τ_z [µs]	1.3	4.4
w [µm]	0.17	0.14

EP+P* diffuses away

$k_1 \uparrow \quad \downarrow k_{-1}$

E+S

$$G(\tau)=1+A_1\exp-k_{-1}\tau+A_2\exp-(k_1\tau)^\beta+A_3G_D(\tau)$$

Fig. 4.18. Measurements on a coverslip with immobilized horse radish peroxidase. Substrate: dihydrorhodamine 6G Cosubstrate: H_2O_2 in Epi-TIRF FCS setup. Evaluation according to Edman et al. 1999 [39, 41]. Product fluctuations (*upper graph*). Autocorrelation function of the product fluctuation (*lower graph*). G_D is the diffusion term. Table of turnover rates for two rhodamine substrates (Rh6G and Rh 123) [42]

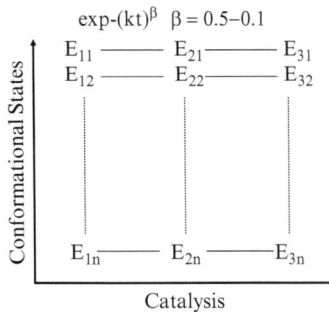

$\exp-(kt)^\beta \quad \beta = 0.5-0.1$

Conformational States

| E_{11} —— E_{21} —— E_{31} |
| E_{12} —— E_{22} —— E_{32} |

E_{1n} —— E_{2n} —— E_{3n}

Catalysis

Fig. 4.19. Conformational Landscapes in Horse radish peroxidase from [39]

We have interpreted the observation of the stretched exponential behavior as being caused by the manifold of transition rates involved in the substrate turnover. This has been observed for all cases where the single enzyme catalysis has been studied by the product turnover [43–45] (Fig. 4.19).

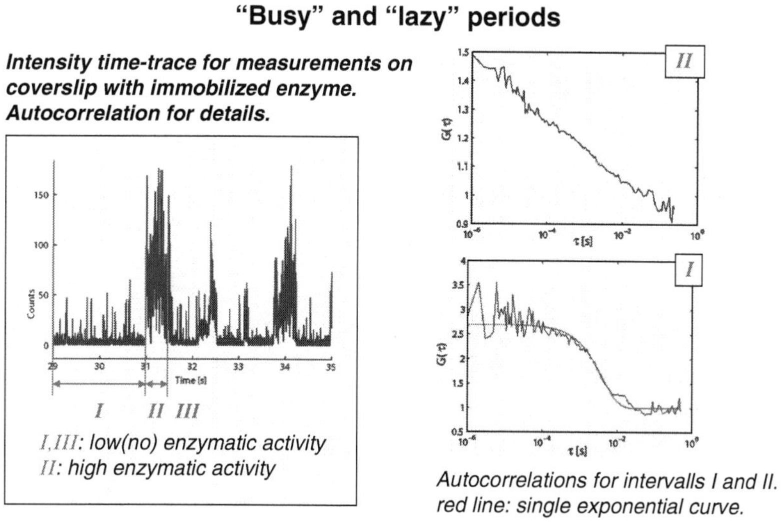

"Busy" and "lazy" periods

Intensity time-trace for measurements on coverslip with immobilized enzyme. Autocorrelation for details.

I,III: low(no) enzymatic activity
II: high enzymatic activity

Autocorrelations for intervals I and II.
red line: single exponential curve.

Fig. 4.20. Busy and lazy periods of horse radish peroxidase. In the busy periods stretched exponential kinetics are observed while in the lazy periods single exponetial kinetics are found [40]

Hassler et al. [40, 41] observed that the stretched exponential behavior was related to the active phase of the enzyme (busy phase) while the less active phase (lazy phase) exhibited a single exponential phase (Fig. 4.20). The behavior which was originally observed by Edman et al. 1999 [39] at a much lower sensitivity had been observed again, however, with a much better intensity and time resolution. As can be seen from the trace, the turnover frequency is more than 4-fold higher in the busy periods. It is tempting to relate this to the memory effect decribed in the next chapter (Fig. 4.21).

4.12 Memory Landscapes

A first attempt to analyze memory effects in catalytic activity was suggested by Lu et al. [38] comparing the relation of enzymatic processes occurring close and distant in time. As an alternative strategy we proposed the non-Markovian function NMF [46].

The NMF is described as difference between higher (three time) and a lower order (two time) correlation functions obtained from the same data set.

$$\text{NMF} = G(\tau_1, \tau_2)/G(\tau_2) - G(\tau_1)$$

NMF has two time axes τ_1 and τ_2 and the value zero for Markov processes. We have been able to show for the HRP system that the NMF decays in milliseconds from a positive value to the zero (Markov) plane. Simulations

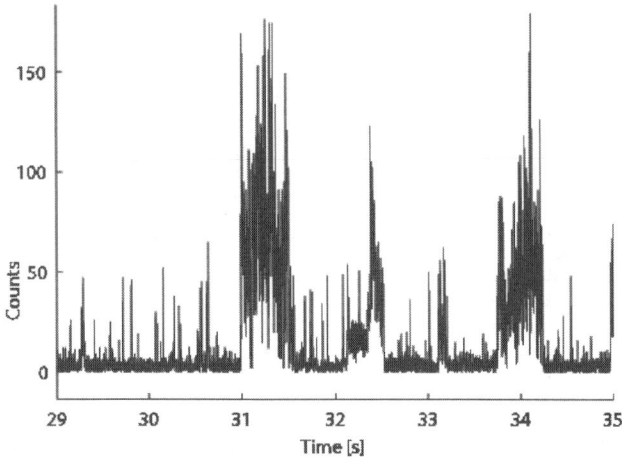

Fig. 4.21. Enlargement of trace from Fig. 4.16. In the lazy period about 10 turnover s^{-1} are seen while in the busy period more than 40 turnovers s^{-1} are seen

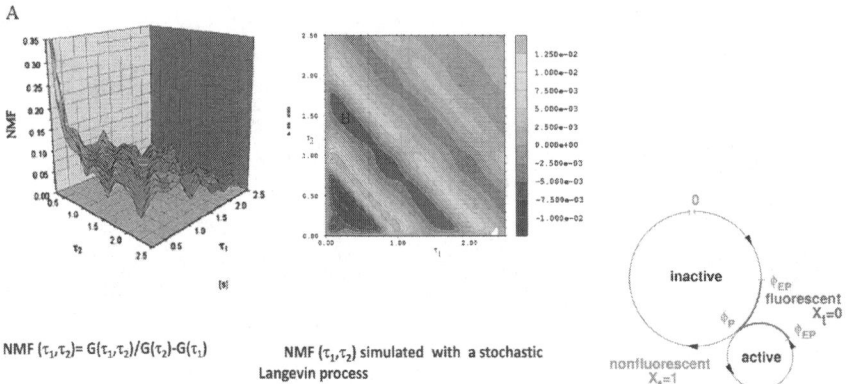

NMF $(\tau_1,\tau_2)= G(\tau_1,\tau_2)/G(\tau_2)-G(\tau_1)$ NMF (τ_1,τ_2) simulated with a stochastic
                                                            Langevin process

Fig. 4.22. Comparison between NMF as measured by FCS [46] and simulation using Langevin description [67] with a 2-loop cycle

using a Langevin description of stochastic and deterministic processes generate the experimental data from HRP.

The periodicity in the NMF indicates the presence of non-equilibrium processes with periodic solutions and is related to the number of states and the transition rates [46, 67]. It is important to mention that the Langevin simulations are very sensitive to the model used and the identity with the experimental data set is indicative for a driven conformational loop which is coupled to the catalytic step generating the substrate product conversion with photon emission (Fig. 4.22).

The Langevin approach was also used for the oxidase system [38] and showed that in addition to an equilibrium distribution of on times also a

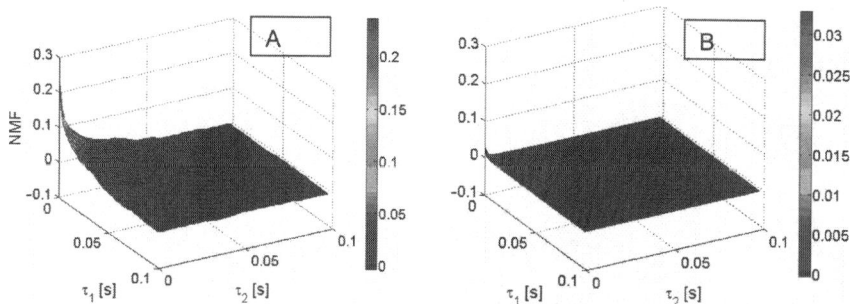

Fig. 4.23. NMF for surface bound horse radish peroxidase in the presence of dihydro-rhodamin (100 nM) as substrate (**a**). Control without enzyme (**b**)

non-equilibrium distribution of on times was found indicating the presence of memory echo [47, 48].

Cao has described the "event echo" [49] as difference between a joint (correlated) probability to detect photons in time $p(\tau_1, \tau_2)$ and a disjoint (uncorrelated) propability $p(\tau_1)p(\tau_2)$. The difference distribution function $\delta(\tau_1, \tau_2)$ is given by

$$\delta(\tau_1, \tau_2) = p(\tau_1, \tau_2) - p(\tau_1)p(\tau_2)$$

When rearranging the

$$\text{NMF} = [G(\tau_1, \tau_2) - G(\tau_1)G(\tau_2)]/G(\tau_2)$$

the identity relating to the difference of joint and disjoint correlation functions is evident. It seems that this memory effect is general for enzymatic catalysis and has also been observed for the turnover of β-galactosidase [45].

It can be seen from Fig. 4.23 that the catalysis is causing a clear NMF effect which relaxes below 100 ms to its groundstate. It is near at hand to assume that the relaxation of the NMF is related to a delay in the hysteretic reorganization of the activated (equilibrium) state. The conformations of the transition state supporting a rapid turnover are still present during a "memory" period. With an encounter probability of substrate and enzyme, which is in the time range of the memory delay, the turnover is accelerated and perpetuated for some time until the system falls again into its catalytic "groundstate".

4.13 Single Molecules in the Electric Field

The motion of single molecules in electric fields has been investigated using oscillating fields between two path clamp tips (Fig. 4.24a).

From the velocity v in the oscillating field dV/dx $v = Dq/kT$ dV/dx the charge q can be determined. For Rh-dUTP a effective charge of 1.55×10^{-19} C

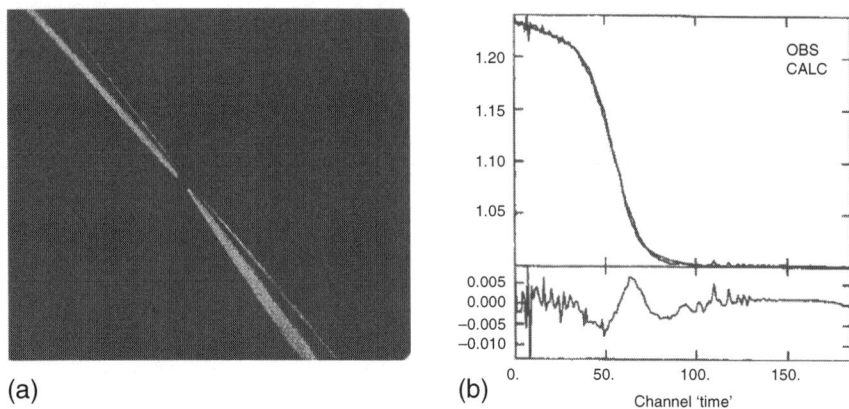

Fig. 4.24. Gold plated patch clamp tips (corteousy E. Neher). An oscillating field of 12 V at 100 Hz was applied (**a**) and the fluctuation of TMR labeled DNA molecules and/or nucleotide was measured (**b**) [50]

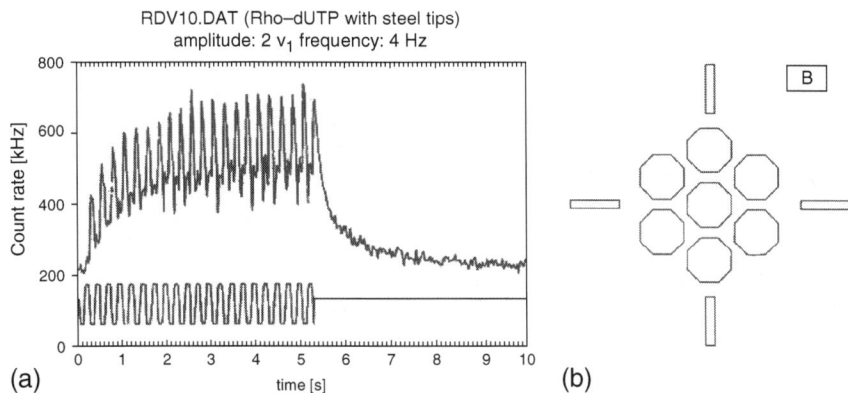

Fig. 4.25. Increase of the density of Rh-dUTP in the oscillating field (4 kV cm^{-1} at 4 Hz) observed in the confocal element between the tips and decay of the concentration as the field is switched off [50]. (**a**) Concept of the quadruple trap with a seven-element detector projected into the cavity (diameter 6 μm) surrounded by four electrodes (metal tips or microcapillaries). The observed particle undergoing diffusion will be repositioned by the electric field whose magnitude and direction is computed from the optical feedback (**b**) [51]

was determined at high oscillating field. Further, it was found that the concentration of labeled charged nucleotide could be increased by a factor 2 in the oscillating field (Fig. 4.25a).

In the active electric field trap, as outlined above, charged particles (nucleotide) can be moved at a field strength of 10 kV cm^{-1} by 10 μm ms^{-1} equivalent to 50 μs per confocal volume element. This is the range of the diffusion time of a dye labeled charged nucleotide (cf. also Fig. 4.24b). For larger charged particles with slower diffusion the situation for trapping is even more favorable. This has recently been demonstrated by Cohen and Moerner [52] in their ABEL trap [53].

The usefulness of electric field for concentrating very dilute samples (viral particles) is evident and an increase in concentration by 4 orders of magnitude was observed for labeled DNA molecules [50]. The importance for medical diagnostics of viral diseases is at hand.

4.14 FCS Cross Correlation and Applications

The potential of fluorescence cross correlation (FXS) has been outlined by Eigen and Rigler [54]. Compared to FRET which gives information about the vicinity of fluorescent centers FXS gives information about their connectivity in the dynamic sense independent of how close or far away the fluorescent centers are positioned. A detailed theoretical description has been provided [55,56] and FXS has been used successfully in a variety of applications both in solution as well as at the cellular level. C. Zeiss with his Confocor 2 (for solution measurements) and Confocor 3 (for cellular measurements) has provided instruments of highest sensitivity.

Examples are the use in gene expression analysis where the gene sequence is detected by the simultaneous presence of two complementary gene probes with two different fluorescence probes [57,58] which is used for gene expression profiling. Recently also the DGA (direct gene analysis) has been performed in individual cells for visualizing individual genes. Other examples are the analysis of RNAi and related enzymes at the cellular level [59] as well as the use of copolymeric nanocontainers for transporting cargo into the cell [60].

FCS has become due to its sensitivity, detection speed and the need for minute volumes (1 µl) the prime optical screening technology for drug screening with a throughput up to 300,000 compounds per day. It is used by all major Pharma companies. With the need for more sophistication in drug screening solutions assays are being replaced by image based assays.

4.15 Single Molecule Detection in Single Cells

The development of single molecule imaging has been based on the properties of avalanche photodiodes. We have for this reason introduced this technology into the imaging of single molecules in the cellular domain. For this reason we have (in collaboration with C. Zeiss) incorporated two APD detectors into the pathway of collecting photons in a scanning confocal microscope. An image can be obtained in the scanning mode by storing the detected photons pixel by pixel. Since the image is generated by a moving mirror (scanning mirror) the detected image area will contain individual photons obtained during the time interval of the scanning process. It will constitute a Poisson process with

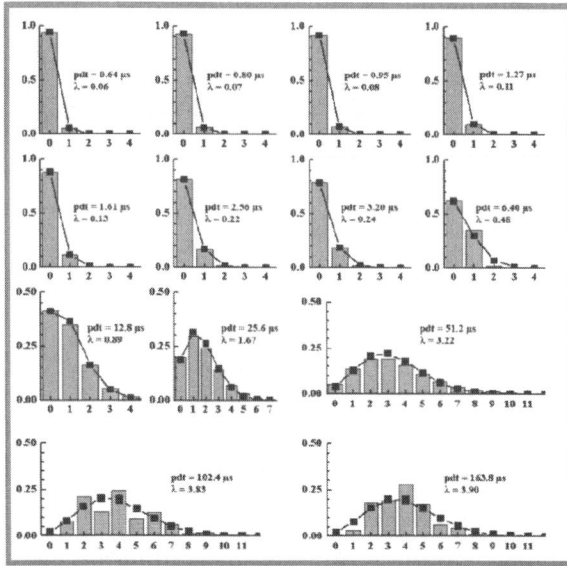

Fig. 4.26. Photon distribution as a function of the scanning speed. Dependent on the scanning speed the photons are squeezed to the left side of the distribution (fast scan) or well resolved (slow speed). From the dependence of λ both number as well as diffusion rate can be determined

a mean value $\lambda = kt$ where k is defined as arrival rate. When determining the count distribution of the image with different scanning rates, the number of photons and molecules as well as their dynamic properties such as diffusion time, flow speed can be obtained [61] (Fig. 4.26).

When the mean of the Poisson distribution λ is plotted against the scan time θ both the number of molecules N as well as their mobility (diffusion time) can be determined. Examples on the mobility of Green fluorescent protein expressed in SK-NM-C cells has been obtained and compared with classical confocal FCS.

$$\lambda = kt = N \left[\frac{\theta}{\tau_{\text{diff}}} + \frac{\text{Pa}}{\text{Ca}} \right]$$

θ = scan time

τ_{diff} = diffusion time

Pa = pixel area

Ca = confocal area

The advantage of APD imaging is the virtual absence of background noise and the high quantum yield of detection in the visible spectrum of light. The sensitivity is orders of magnitude higher than that with standard detectors (PMT) and has led to the situation that only low expression levels are needed.

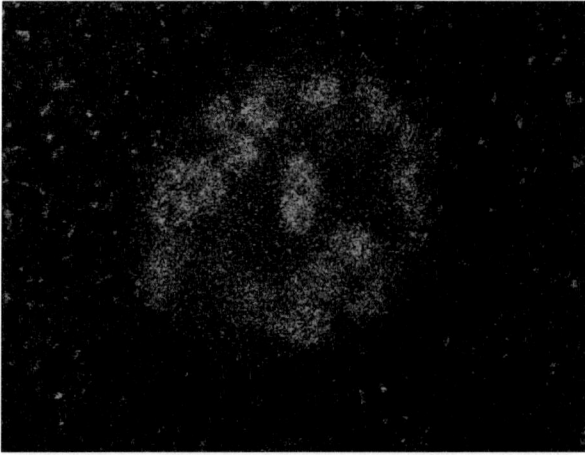

Fig. 4.27. Expression of GFP labeled transcription factor in the salivary gland cells of *Drosophila* [62]. The generation of transcription factor molecules can be seen in the cytoplasmic reticulum as well as the distribution on the polytenic chromosomes in the nucleus. Image taken in a confocal laser scan microscope from *C. Zeiss*, modified for APD Imaging [61]

The image below (Fig. 4.27) showing the presence of transcription factors interacting with *Drosophila* chromosomes would have been impossible to obtain since at higher expression levels the nucleus is filled with GFP-transcription factor which overshadows the chromosomal bands.

4.16 Outlook

It is evident that a new area of Single Molecule Analysis has started and the interest will be concentrated on the cellular level. The development of genetic markers like GFP and others will make molecular details and their connection visible and accessible for detailed studies. FCS as a powerful tool of dynamic analysis has already substituted classical relaxation kinetics in equilibrium but is also able to work at non-equilibrium conditions as found in a living cell. It will play a very important role and we are seeing just the beginning.

The sensitivity of photon detectors and the level of their background noise are paramount for dynamic analysis of single molecules. We have analyzed a new development of APDs produced in the CMOS form [63]. For a 2×2 APD array, the detected photons have been measured in all four detectors using illumination by diffractive optics ([64]) and simultaneous analysis with four correlators [65] (Fig. 4.28).

Recently, a 128×128 CMOS APD detector has been produced at the ETH Lausanne (EPFL). This development will provide a 10^4 pixel area for simultaneous kinetic and dynamic analysis with a time resolution of ns and

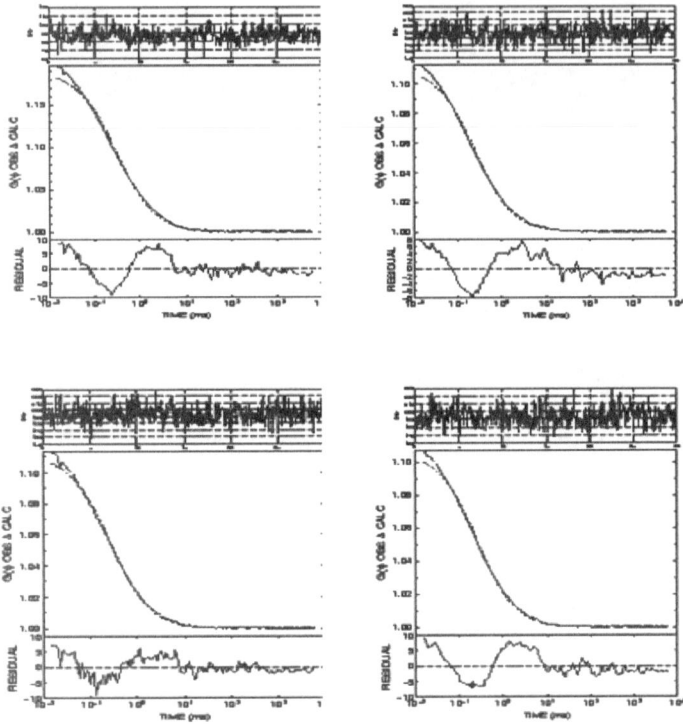

Fig. 4.28. Autocorrelation function of diffusing dye molecules in 2×2 volume elements excited with four laser beams and measured with the 2×2 CMOS APD

single photon sensitivity. We believe that this development will add a new dimension to confocal image analysis.

Acknowledgement

I want to thank all my collaborators which had been involved in the endeavor of FCS and Single Molecule Spectroscopy starting in Manfred Eigen and Leo de Meyer's Laboratory in 1969: they are Manfred Eigen, Leo de Meyer, Klaus Gnädig, Måns Ehrenberg, Pietro Graselli, Jerker Widengren, Michael Brinkmeier, Petra Schwille, Lars Edman, Stefan Wennmalm, Per Thyberg, Ûlo Mets, Peet Kask, Zeno Földes Papp, Aladin Pramanik, Hans Blom, Michael Gösch, Kai Hassler, Per Rigler, Thiemo Anhut, Marcel Leutenegger, Theo Lasser, Vladana Vukojevic, and Lars Terenius. They have with their enthusiasm and dedication contributed to the creative atmosphere which was necessary to develop FCS and single molecule detection.

The work has been supported during the years by continuous support from the Knut and Alice Wallenberg foundation, the National Science Research

Council (NFR), and the Technical Science Research Council (TFR) now VR (vetenskapsrådet).

References

1. R. Rigler, R. Rabl, T. Jovin, Rev. Sci. Instrum. **45**, 580 (1974)
2. R. Rigler, M. Ehrenberg, Quart. Rev. Biophys.. **6**, 13 (1973)
3. J. Ramstein, M. Ehrenberg, R. Rigler, Biochemistry **19**, 3938 (1980)
4. D. Magde, E. Elson, W.W. Webb, Phys. Rev. Lett. **29**, 705 (1972)
5. M. Ehrenberg, R. Rigler, Chem. Phys. Lett. **14**, 53 (1972)
6. E. Elson, D. Magde, Biopolymers **14**, 13, 1–27 (1974)
7. M. Ehrenberg, R. Rigler, Chem. Phys. **4**, 390 (1974)
8. P. Kask, P. Piksarv, Ü. Mets, M. Pooga, E. Lippmaa, Eur. Biophys. J. **14**, 257–261 (1987)
9. R. Rigler, J. Widengren, U. Mets, in *Fluorescence Spectroscopy*, ed. by O.S. Wolfbeis, (Springer, Berlin, 1992) pp. 13–24
10. S. Zorilla, M.A. Hink, A.J.W.G. Visser, M. Pillar Lillo, Biophys. Chem. **125**, 298–305 (2007)
11. J.M. Tsay, S. Doose, S. Weiss, J. Am. Chem. Soc. **128**, 1639–1547 (2008)
12. R. Rigler, M. Ehrenberg, Quart. Rev. Biophys. **9**, 1 (1976)
13. R. Rigler, M. Ehrenberg, Quart. Rev. Biophys. **9**, 69 (1976)
14. S.R. Aragon, R. Pecora, J. Chem. Phys. **64**, 1791–1803 (1976)
15. R. Rigler, J. Widengren, BioScience **3**, 180–183 (1990)
16. Ü. Mets, J. Widengren, R. Rigler, Chem. Phys. **218**, 191–198 (1997)
17. H.J. Carmichael, D.F. Walls, J. Phys. **B9**, 1199–1219 (1976)
18. R. Hanbury Brown, R.R.Q. Twiss, Nature **177**, 27–29 (1956)
19. Ü. Mets, in *Fluorescence Correlation spectroscopy. Theory and Application*, ed. by R. Rigler, E. Elson (Springer, New York, 2000) pp. 346–359
20. R. Rigler, J. Widengren, U. Mets, in *Fluorescence spectroscopy*, ed. by O.S. Wolfbeis (Springer, 1992), pp. 13–24
21. Rigler, R., Mets, Ü., Soc. Photo-Opt. Instrum. Eng. **1921**, 239–248 (1993)
22. Mets, Ü., Rigler, R., J. Fluorescence **4**(3) 259–264 (1994)
23. J. Dapprich, Ü. Mets, W. Simm, M. Eigen, R. Rigler, Exp. Tech. Phys. **41**(2), 259–264 (1995)
24. Rigler, R., Mets, U., Widengren, J., Kask, P., Eur. Biophys. J. **22**, 169–175 (1993)
25. Hassler Anhut, T., Rigler, R., Gösch, M., Lasser, Th., Biophys. J. Biophys. Lett. (2005) doi:10.1529/biophysj.104.053884. L01–L03
26. L. Edman, Ü. Mets, R. Rigler, Exp. Tech. Phys. **41**(2), 157–163 (1995)
27. L. Edman, Ü. Mets, R. Rigler, Proc. Natl. Acad. Sci. USA. **93**, 6710–6715 (1996)
28. Y. JIa, A. Sytnik, L. Li, S. Vladimirov, B.S. Cooperman, R.M. Hochstrasser, Proc. Natl. Acad. Sci. USA. **94**, 7932 (1997)
29. S. Wennmalm, L. Edman, R. Rigler, Proc. Natl. Acad. Sci. USA. **94**, 10641–10646 (1997)
30. S. Wennmalm, L. Edman, R. Rigler, Chem. Phys. **247**, 61–67 (1999)
31. S. Wennmalm, R. Rigler, J. Phys. Chem. B. **103**(13), 2516–2519 (1998)
32. J. Widengren, R. Rigler, Ü. Mets, J. Fluorescence. **4**(3), 255–258 (1994)
33. J. Widengren, Ü. Mets, R. Rigler, J. Phys. Chem. **99**, 13368–13379 (1995)

34. J. Widengren, R. Rigler, Bioimaging **4**, 149–157 (1996)
35. J. Widengren, R. Rigler, J. Fluorescence. **7**(1), 211–213 (1996)
36. J. Widengren, T. Terry, R. Rigler, Chem. Phys. **249**, 259–271 (1999)
37. J. Widengren, Ü. Mets, R. Rigler, Chem. Phys. **250**, 171–186 (1999)
38. P. Lu, L. Xun, S.X. Xie, Science **282**, 1877–1882 (1998)
39. L. Edman, Z. Földes-Papp, S. Wennmalm, R. Rigler, Chem. Phys. **247**, 11–22 (1999)
40. K. Hassler, Thesis EPFL, (2005)
41. K. Hassler, P. Rigler, M. Leutenegger, R. Rao, R. Rigler, M. Gösch, T. Lasser, Opt. Express **13**, 7415–7423 (2005)
42. K. Hassler, P. Rigler, H. Blom, R. RiglerWidengren, J., T. Lasser, Opt. Express **15**, 5366–5375 (2007)
43. K. Velonia, O. Flomenborn, D. Loos, S. Masuo, M. Cotlet, Y. Engelborghs, J. Hofkens, A.E. Rowan, J. Klafter, R.J.M. Nolte, Van der Auweraer, F.C. De Schryver, Ang. Chemie. Int. Ed. **44**, 560–564 (2005)
44. O. Flomenbom, K. Velonia, D. Loos, S. Masuo, M. Cotlet, Y. Engelborghs, J. Hofkens, A.E. Rowan, R.J.M. Nolte, Van der Auweraer, F.C. De Schryver, J. Klafter, **102**, 2368–2372 (2005)
45. B. English, W. Min, A. van Oijen, T.G. Kang, L. Guobin, S. Hongye, B.J. Cherayil, S.C. Kou, S.X. Xie, Nat. Chem. Biol. **2**, 87–94 (2006)
46. L. Edman, R. Rigler, Proc. Natl. Acad. Sci. USA. **97**, 8266–8271 (2000)
47. H.P.h. Lerch, R. Rigler, A.S. Mikhailov, Proc. Natl. Acad. U S A. **102**, 10807–10812 (2005)
48. H.P.h. Lerch, A.S. Mikhailov, R. Rigler, Chem. Phys. **331**, 304–308 (2007)
49. J. Cao, Chem. Phys. Lett. **327**, 38–44. (2000)
50. R. Rigler, Electrophoresis **23**, 605–608 (2002)
51. R. Rigler, J. Biotechnol. **41**, 179–186 (1995)
52. A.E. Cohen, W.E. Moerner, Appl. Phys. Lett. **86**, 093109 (2005)
53. W.E. Moerner, Proc. Natl. Acad. Sci. USA. **104**, 12596–12603 (2007)
54. M. Eigen, R. Rigler, Proc. Natl. Acad. Sci. USA. **91**, 5740–5745 (1994)
55. F.J. Schwille Meyer-Almes, R. Rigler, Biophys. J. **72**, 1878–1886 (1997)
56. R. Rigler, Z. Földes-Papp, F.J. Meyer-Almes, C. Sammet, M. Völcker, A. Schnetz, J. Biotechnol. **63**, 97–10 (1998)
57. K. Korn, P. Gardellin, B. Liao, M. Amacker, A. Bergström, H. Björkman, A. Camacho, S. Dörhöfer, K. Dörre, J. Enström, T.h. Ericson, T. Favez, M. Gösch, A. Honegger, S. Jaccoud, M. Lapczyna, E. Litborn, P. Thyberg, H. Winter, R. Rigler, Nucl. Acid Res. **31**(e 89), 1–8 (2003)
58. K. Korn, M. Damond, J.F. Cajot, E. Litborn, B. Liao, P. Thyberg, H. Winter, A. Honegger, P. Gardellin, R. Rigler, J. Biotechnol. **107**, 107–114 (2004)
59. T. Ohrt, D. Merkle, K. Birkenfeld, C.J. Echeverri, P. Schwille, Nucl. Acid Res. **34**, 1369–1380 (2000)**
60. P. Rigler, W. Meyer, J. Am. Chem. Soc. **121**, 367–373 (2006)
61. V. Vukojevic, M. Heidkamp, Y. Ming, B. Johansson, L. Terenius, R. Rigler, Proc. Natl. Acad. Sci. USA. **105**, 18176–18181 (2008)
62. K.D. Papadopoulos, W. Gehring, V. Vukojevic, R. Rigler, (2008), June 1–6
63. A. Rochas, M. Gösch, A. Serov, P.A. Besse, R.S. Popovic, T. Lasser, R. Rigler, IEEE Photonics Technol. Lett. **15**, 963–965 (2003)

64. H. Blom, M. Johansson, A.S. Hedman,Lundberg, S., A. Hanning, S. Hard, R. Rigler, Appl. Opt. **41**, 3336–3342 (2002)
65. M. Gösch, A.S. Serov, T. Anhut, Th. Lasser, A. Rochas, H. Blom, R. Rigler, J. BioMed Optics **9**, 913–921 (2004)
66. H.J. Kimble, M. Deganais, L. Mandel Phys. Rev. Lett. **39**, 691–695 (1977)
67. H. Ph. Lerch, A.S. Michailov, B. Hess, Proc. Natl. Acad. Sci US, **99**, 15410–15415 (2002)
68. S. Yang, J. Cao, J. Phys. Chem. B **105**, 6536–6549 (2001)

Part III

Fluorescence-Correlation Spectroscopy

Single-Molecule Spectroscopy Illuminating the Molecular Dynamics of Life

Watt W. Webb

Summary. This chapter summarizes a series of new single-molecule spectroscopy investigations in the life sciences at Cornell University that began with our invention of Fluorescence Correlation Spectroscopy (FCS) about 1970. Our invention of FCS became my first focus on the "Molecular Dynamics of Life." It motivated my transition from research on quantum fluctuations and transport in condensed matter physics including superconductivity and in the molecular dynamics of coherent fluctuations and nano-transport in inanimate physical and chemical systems subject to the nonlinear dynamics of continuous phase transitions. These interdisciplinary transitions exemplify the productivity of such interdisciplinary interactions in science.

5.1 Fluorescence Correlation Spectroscopy in Its Historical Context

"The brightest flashes in the world of thought are incomplete until they have been proven to have their counterparts in the world of fact," wrote John Tyndall more than a century ago. We conceived Fluorescence Correlation Spectroscopy (FCS) [1–7] to explore experimentally one such "flash" concept about what might enable DNA transcription. In the 1960s and 1970s, it was not known how the double helix of DNA was separated to accommodate what we now know as transcription enabled by DNA polymerase and molecular cofactors. Elliot Elson hypothesized that the DNA double helix might partition via spontaneous structural fluctuations, so he proposed that we apply Quasi-elastic Light Scattering (QELS), which I had been optimizing to measure concentration fluctuations in and in between fluids displaying "critical phenomena" as they approached conditions for a continuous phase transitions. However, I calculated that QELS would not work well for chemical reactions, but Elliot resolved the problem by finding a report that the drug ethidium can intercalate in DNA, distorting both the DNA chain and the ethidium to produce a fluorescent ethidium signal on binding. We conceived the theory for the fluorescence correlation assays of the coupled diffusion, chemical reactions, and fluorescence switching which we incorporated into tidy dynamic matrix equations clearly enough that we could even discover an unexpected diffusive coupling interaction, as reported in two of our first FCS publications [2,8].

In one of our earlier applications, FCS diagnosed unanticipated micelle formation and led to the first development of confocal image microscopy for smaller focal volumes [3]. Recognizing the effective applications of fluorescent marker dynamics to understand cell membrane dynamics, we applied FCS to molecular diffusion on cell membranes, entering thereby into a long series of studies of the dynamics of membrane processes in life, which was at that time a quagmire of conflicting ideas [4]. Later, we also extended FCS theory to fluid flow analysis [9]. It has proven useful for a diversity of ultrafast chemical kinetics as well, c.f. [10–13].

The scientific background in which FCS was created does illuminate the processes of creative discovery. The chemical physics challenges of the 1960s were to measure and understand the statistical fluctuations of fluid transitions from liquid to gas, etc., and phase separation and solubilization in mixtures energizing bulk and interfacial fluctuations, which were frequently measured by QELS and revealed fluctuating optical properties of the fluctuating fluids. We had been focusing on the fluctuations of interfaces between superimposed fluids as their mutual solubility increased to complete solubility. Our attraction to these continuous phase transitions had originated in quantum superconductivity, where we analyzed and measured quantum superconducting fluctuations associated with magnetic flux quanta and/or quantum fluctuations that dominated the behavior of superconducting magnets and quantum counting Josephson junctions.

Thus, we saw precursors of FCS as early experimental indicators of the broad range of dynamic life processes that are now frequently susceptible to optical experiments.

5.2 Multiphoton Microscope and Photophysical Technologies Probing Molecular Dynamics of Life

Although FCS has now been invoked in about 3,000 scientific publications, now at ∼400 per year, its use before about 1990 was limited by severe technological barriers involving instability of laser light sources, poor sensitivity of photon detectors, noisy electronics, and insufficient computer capacities for the correlation computations. These problems inhibited application of FCS, until suddenly about 1990 the electro-optical and computational technologies advanced so that it became feasible. These advances occurred in synchrony with our creation of Multiphoton Laser Scanning Microscopy, which has enabled effective research on the molecular dynamics of life in living tissues and animals [14].

5.3 Single-Molecule Diffusion Anomalies (and Membrane Structures)

Not only has fluorescence provided excellent indicators of dynamics in the bio-physical dynamics of life, but also it usefully detects charge transfers through cell membranes, where individual molecular channels regenerate life processes in modes detected by quantum charge transfers and other optical indicators of the electrodynamics of life, particularly in neural signaling.

The early years of measuring fluorescent labeled molecular diffusion in membranes had shown diffusion of proteins on cell membranes to be orders of magnitude slower than diffusion of lipids on cell membranes; this led us to focus on the dynamics of molecular interactions and transport on cell membranes, which were quite mysterious in 1972 (and in some ways still are!). Many of our early experiments measured molecular diffusion on membranes [4] as well as chemical interactions in solution [3]. In fact, for our early micelle formation reactions in solution, our creation of the first confocal microscopy instrument [3] was so far ahead of its time that its report has generally been ignored. Our research group also collaboratively developed "Fluorescence Photobleaching Recovery," now called Fluorescence Recovery After Photobleaching (FRAP), to measure diffusion on membranes rapidly, thus avoiding the technical restrictions that were limiting FCS before 1990 [15]. FRAP measures protein diffusion many orders of magnitude slower than lipids on cell membranes, while on quasi-pure lipid membranes the diffusing coefficients were generally within a factor of 10 of the lipids. Later, we extended FRAP analysis to three-dimensional diffusion using MPM to define a three-dimensional localized focal volume [16]. We explored these effects on membranes earlier by electrophoretic diffusion and aggregation of proteins [17,18]. This analysis eventually led us to understand the diffusion anomalies on cells called "anomalous subdiffusion," understandable by a simple theory quantitating the idea that the energy barrier profiles in which diffusion takes place in inhomogeneous solutions can be described by expressing the energy landscape dependent pattern of energy states for the diffusing molecules.

This led to a series of experimental measurements tracking the time-dependent trajectories of individual protein molecules on living cell surfaces. An important target was a slow transport of the low-density lipoprotein receptor (LDR-R), which had very low mobility on the cell surfaces, as had been pointed out to us by Brown and Goldstein as a quandary in their research on inhibition of hypercholesterolemia [19]. In 1982, we learned how to label LDL so brightly that we could track its diffusion for the first time [20], at first by imaging it with a film-based movie camera, played back on our football coach's stopped flow projector. Twelve years later, high-precision tracking over normal cell surfaces revealed the theoretically predicted phenomenon of anomalous sub-diffusion based on local energy states for a mobile molecule varying in space, time, and energy in a way that slowed molecular displacements from $(\Delta r)^2 \propto Dt^1$ to a slower rate of displacement $\propto Dt^\alpha \leqslant 1$, where the exponent α could have any values in the range $0 \leqslant \alpha \leqslant 1$ [21]. Applying this type of measurements explained the puzzling requirement to fit photobleaching recovery experiments by this inhibited spatiotemporal inhomogeneity of cell membranes [22]. This diagnosis appropriately replaces in most membranes the concept of immobile fraction of molecules that had been applied in FRAP measurements. More recently, we developed a simple algorithm for highly precise analysis of molecular tracking experiments that is widely used for molecular motor step analysis [23]. This technique has become so popular that some of its users have renamed it after their girlfriends.

FCS has provided a powerful tool since the 1990s for the investigation of protein structure and molecular fluctuations. One of our most interesting targets for application of FCS was the Green Fluorescent Protein family, which has now been expanded to an elegant collection of switchable fluorescent molecules. The elegant colors of the GFP family are epitomized by Roger Tsien's Crayola GFP family [24]. Now the family of switchable GFP labeling molecules includes light-controlled modulators of gene expression, calcium release, etc. [25]. One of our first studies of GFP relied on FCS to reveal a very complex physical chemistry due to strong pH sensitivity associated with protein binding to the chromophore and related structural transformations. The dynamics and thermodynamics of these reactions were resolved by FCS [26]. Many other features of photophysics of the GFP family have been analyzed in a series of our publications led by Dr. Ahmed Heikal, c.f. [27–29].

5.4 Single-Molecule Spectroscopy in Neuroscience

Single-molecule spectroscopy of the molecular conductance of ion charge transfer through individual transmembrane protein ion channels is such an established procedure in neuroscience today that its discovery by Neher and Sakmann in 1976 [30] is not often considered as a single-molecule spectroscopy concept. But one may define the spectrum of conductance fluctuations through ion channels. Their original report aroused some skepticism amongst many biologists who conjectured that the charge conductance bursts were not due to individual transmembrane channel conformational switching but were simply fluctuating membrane structural leaks. However, we were able to resolve this controversy. Cornell's Professor Ephraim Racker had succeeded in isolating the ion-channel molecule activated by binding of acetylcholine from membranes, and with the help of his former student we reconstituted it inside pure lipid membranes and demonstrated that it accounted for the known properties of the acetylcholine activated channel. Its specificity in these clean systems confirmed the specificity of the original patch clamping. It also confirmed the suspected existence of two different conductance states and a set of the molecular conformations and two binding states for acetylcholine [31].

5.5 FCS Protein Structure Fluctuations in Neurodegenerative Disease

Our most recent studies of molecular fluctuations have focused on the conformational fluctuations of molecular structures and processes involved in the neurodegenerative diseases Alzheimer's Disease, Prefrontal Temporal Dementia, and Parkinson's Disease. Our thinking with our collaborator at Weill Medical College, Professor David Eliezer, that binding of α-synuclein (α-syn) to certain membranes might be a nucleating aggregation in Parkinson's Disease led us to an FCS analysis of the molecular dynamics of α-syn binding to lipid vesicle membranes which showed strong binding of α-syn to phosphatidyl serine (negatively charged) membranes to the supersaturation limit, on which further research is needed [32].

We have also used FCS to analyze the structural fluctuations of apoMyoglobin with a biochemical collaborator, Professor Stewart Loh of SUNY Upstate Medical College. We discovered that denaturation of this set of α-helices by low pH could be measured using FCS with a terminal fluorophore Alexa 488 attached. We discovered that this fluorescence is quenched by contact with four different amino acids Tyr, Trip, Met, and His, which are conveniently dispersed along this molecular sequence, providing good indicators of structural fluctuation dynamics vs. pH [33].

The apo-Mb molecule also has provided a research model for the Amyloid-β to β-Amyloid transition of Alzheimer's Disease, having been shown in the horse Myoglobin sequence to undergo aggregation as an Amyloid under aggressive conditions [34, 35]. Recently, we have also found that whale apo-Mb also undergoes an Amyloid transition, but only under an even more aggressive environment. Our Transmission Electron Microscopy (TEM) imaging of β-Amyloid formation indicates that the aggregate structures, as well as the dynamics of aggregation depend sensitively on the solution concentration of A-β, cofactors, temperature, and stirring.

Although our exploration of the interactions of β-Amyloid plaques with neurons in transgenic Alzheimer's mice has tentatively indicated that neuronal dendrites could pass through the mature Amyloid aggregate, the contrast of recent imaging of the process of fast Amyloid aggregate formation and prompt interaction with its glial and neuronal environment [36] implies that the earlier stages of Amyloid formation may be more destructive than we had observed previously [37]. We have a manuscript in preparation on our MPM imaging of neurons intersecting β-Amyloid plaques without signs of damage that supports this view.

Our recent neuronal imaging in transgenic mice has yielded an unexpectedly important discovery [38]. We found that Second Harmonic Generation (SHG) imaging of neurons by SHG radiation by the microtubule assemblies in dendrites showed parallel polarity orientation in the microtubule orientations in dendrites. Microtubule polarity dictates the selective directions of motion of the molecular motors on microtubules serving as the neuron's messengers for cargo transport. Previous measurements on the short dendrites in hip-pocampal cultures had shown mixed polarities. However, in the CA1 and CA3 bands in maturing mouse brain slices, the polarities were generally par-allel and increasing as the dendrites grew longer as the mice matured with greater age.

5.6 Molecular Transcription Factor Activation in the Cell Nucleus

The next two topics of this chapter describe our recent discoveries that return to the motivation for our original development of FCS, which is the enabling of gene expression. We have discovered and demonstrated that Multiphoton Microscopy could provide the three-dimensional resolution fluorophore imag-ing necessary to detect transcription factors in action as they stimulate gene expression of crucial proteins such as hsp70, which serve as protein folding chaperones needed for cells to respond to stress. Our experiments record acti-vation of expression of the appropriate genes for chaperones like hsp70 by activation of the heat-shock factor (HSF) in the living polytene cells of the *Drosophila* salivary gland. As the HSF which is initially dispersed in the nucle-oplasm associates with native gene loci for chaperone molecules like hsp70 on heating the tissues to 37 °C, the RNA polymerase also accumulates and recy-cles in the active volume to express the triggered genes in the same vicinity instead of drifting away, c.f. [39–42]. These techniques have enabled imaging of the dynamics of gene expression in intact living cells. Our molecular genet-ics colleague Professor John Lis is continuing this research, stimulated by our initial success, to apply it in diploid cells.

Figure 5.1 illustrates the process of assembly of the Heat Shock Factors labeled green in the free nucleoplasm of the nucleus of the polytene Drosophila nucleus as the temperature is rapidly raised from 23 to 37 °C. Its segregation to the genes it activates in the red-labeled chromatin ensemble is seen as the short stripes of yellowish green on the chromatin.

5.7 Efficient DNA Sequencing by Single-Molecule Spectroscopy

But a second outcome of our interest in research on gene expression, which began with the motivation for FCS in the 1960s, was our development of an effective, potentially inexpensive method for DNA sequencing that we reported in 2003 [43]. Isolating a DNA polymerase molecule attached to the illuminated bottom of a zero-mode waveguide of $\sim$50 nm diameter, which limits the depth of this illumination volume to $\sim$10–100 zeptoliters even in an open channel, as shown by FCS measurements, while leaving the solu-tion volume open at the top for the fluorescence labeled bases to circulate

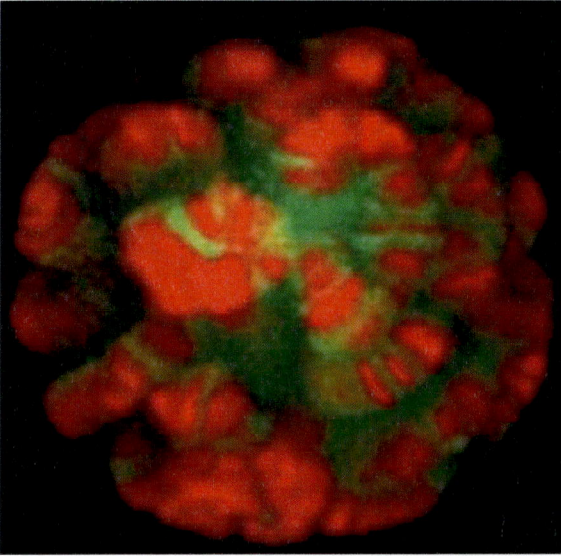

Fig. 5.1. 3D reconstructed polytene nucleus [39]

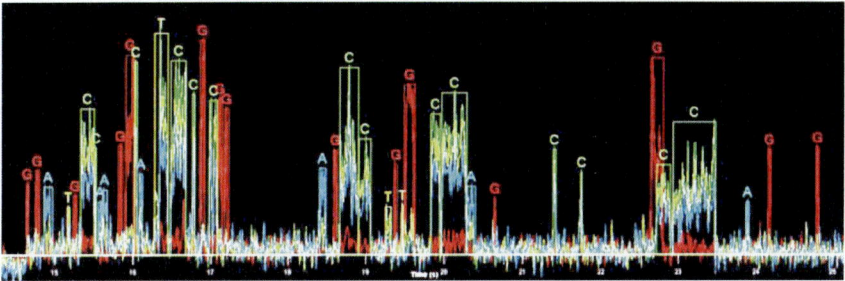

Fig. 5.2. DNA sequencing by recording fluorescence burst sequence from the zero-mode waveguide as complementary DNA strand is formed

at appropriate concentration. Illumination of the four different fluorophore-labeled nucleotides through the transparent bottom of the waveguide excites each fluorophore bound to the sequentially appropriate base for ∼1 ms during the time the polymerase binds the new base, shifts to the next sequential site, and separates the γ-phosphate to which the fluorophore is bound. This ultimate result in the data for rapid, accurate DNA sequencing is now being developed for worldwide application by Pacific Biosciences in Menlo Park, CA.

Figure 5.2 illustrates the fluorescence light bursts that provide the four colors encoding the four nucleotide species: A, G, C, and T as they are added to the complementary strand by the DNA polymerase. The flashes illustrate signals that are resolved into the rapid sequence determination illustrated by the punctate signal with each burst well above the actual noise background in the Pacific Biosciences sequencing apparatus model system (Copyright 2008, Pacific Biosciences, Inc.)

5.8 A Medical Future for Our "Single-Molecule Spectroscopy Illuminating the Molecular Dynamics of Life"

Currently integrating many consequences of our research history is our new program to create a future for MPM molecular micro-spectroscopy by creating a medically diagnostic "Medical Multiphoton Microscopic Endoscopy." The concept is to add the capability of MPM microscopy imaging of the intrinsic fluorescence and SHG of tissues into conventional medical endoscopy apparatus. These devices now rely only on scattered light images, to select sites for tissue biopsies to be removed surgically to be fixed and stained for subsequent pathologists' diagnostics. The objective of the MPM imaging inside body tissues is to provide microscopy imaging to guide the best selection of appropriate locations for biopsy locations and in some cases to provide preliminary in situ, in real-time diagnostic imaging to guide surgery.

Until recently, the MPM imaging experience on living tissue had largely been based on small animals, including transgenic mice subject to various disease states, c.f. [37, 42–51]. Now, motivated by observations of intrinsic tissue fluorescence and SHG in these model systems, we aim to enable diagnostic MPM imaging within the living human patient by adapting and miniaturizing multiphoton microscopy apparatus for incorporation in medical endoscopes for various specialties in human medical care.

The first specialization has been to human bladder cancer, where the potential for major human benefit in cancer treatment would be enabled by reducing the need for numerous traumatic biopsies to monitor the progress of otherwise very efficient pharmacological treatment. After 2 years of development, our colleagues at Cornell's Weill Medical College in New York City have demonstrated diagnostic effectiveness for this purpose with MPM imaging of fresh biopsy specimens that are subsequently fixed and stained for the pathologist usual preparations. It now appears that the MPM diagnostics can work for the purpose. This result has attracted the attention of MDs specializing in many additional fields for initial trials.

Design and construction of the MPM Endoscopes is also required, of course. Our first Endoscope prototypes are designed and are under construction for initial trial this fall or winter at the Cornell Veterinary College facilities.

Thus, our initial objectives for FCS are being realized with continuing surprises!

References

1. D. Magde, E. Elson, W.W. Webb, Thermodynamic fluctuations in a reacting system – measurement by fluorescence correlation spectroscopy. Phys. Rev. Lett. 29(11), 705–708 (1972)
2. D. Magde, E.L. Elson, W.W. Webb, Fluorescence correlation spectroscopy 2 – experimental realization. Biopolymers 13(1), 29–61 (1974)
3. D.E. Koppel, D. Axelrod, J. Schlessinger, E.L. Elson, W.W. Webb, Dynamics of fluorescence marker concentration as a probe of mobility. Biophys. J. 16(2), 1315–1329 (1976)
4. P.F. Fahey, D.E. Koppel, L.S. Barak, D.E. Wolf, E.L. Elson, W.W. Webb, Lateral diffusion in planar lipid bilayers. Science 195, 305–306 (1977)
5. S. Maiti, U. Haupts, W.W. Webb, Fluorescence correlation spectroscopy: Diagnostics for sparse molecules. PNAS 94(22), 11753–11757 (1997)
6. W.W. Webb, Fluorescence correlation spectroscopy: genesis, evolution, maturation and prognosis, in Fluorescence Correlation Spectroscopy Theory and Applications, ed. by R. Rigler, E.S. Elson (Springer, Berlin, 2001)
7. W.W. Webb, Fluorescence correlation spectroscopy: inception, biophysical experimentations and prospectus. Appl. Opt. 40(24), 3969–3983 (2001)
8. E.L. Elson, D. Magde, Fluorescence correlation spectroscopy.1. Conceptual Basis and Theory. Biopolymers 13(1), 1–27 (1974)
9. D. Magde, W.W. Webb, E.L. Elson, Fluorescence correlation spectroscopy. III. Uniform Translation and Laminar Flow. Biopolymers 17, 361–376 (1978)
10. M. Foquet, J. Korlach, W.R. Zipfel, W.W. Webb, H.G. Craighead, Focal volume confinement by submicrometer-sized fluidic channels. Anal. Chem. 76(6), 1618–1626, (2004)
11. D.R. Larson, J.A. Gosse, D. Holowka, B. Baird, W.W. Webb, Temporally resolved interactions between antigen-stimulated IgE receptors and Lyn Kinase on living cells. J. Cell Biol. 171(3), 527–536 (2005)
12. H.Y. Park, X. Qiu, E. Rhoades, J. Korlach, L.W. Kwok, W.R. Zipfel, W.W. Webb, L. Pollack, Achieving uniform mixing in a microfluidic device: hydrodynamic focusing prior to mixing. Anal. Chem. 78(13), 4465–4473 (2006)
13. H.Y. Park, S.A. Kim, J. Korlach, E. Rhoades, L.W. Kwok, W.R. Zipfel, M.N. Waxham, W.W. Webb, L. Pollack, Conformational changes of calmodulin upon Ca2 + -binding studied with a microfluidic mixer. PNAS 105(2), 542–547 (2008)
14. W. Denk, J.H. Strickler, W.W. Webb, Two-photon laser scanning fluorescence microscopy. Science 248, 73–76 (1990)
15. D. Axelrod, D.E. Koppel, J. Schlessinger, E. Elson, W.W. Webb, Mobility measurement by analysis of fluorescence photobleaching recovery kinetics. Biophys. J. 16(9), 1055–1069 (1976)
16. J.B. Shear, C. Xu, W.W. Webb, Multiphoton-excited visible emission by serotonin solutions. Photochem. Photobiol. 65(6), 931–936 (1997)
17. D.W. Tank, W.J. Fredericks, L.S. Barak, W.W. Webb, Electric-field induced redistribution and post-field relaxation of low-density lipoprotein receptors on cultured human-fibroblasts. J. Cell Biol 101(1), 148–157 (1985)
18. T.A. Ryan, J. Myers, D. Holowka, B. Baird, W.W. Webb, Molecular crowding on the cell-surface. Science 239(4835), 61–64 (1988)
19. M.S. Brown, J.L. Goldstein, A receptor-mediated pathway for cholesterol homeostasis. Science 232(4746), 34–47 (1986)

20. L.S. Barak, W.W. Webb, Diffusion of low-density lipoprotein-receptor complex on human- fibroblasts. J. Cell Biol. 95(3), 846–852 (1982)

21. R.N. Ghosh, W.W. Webb, Automated detection and tracking of individual and clustered cell-surface low-density-lipoprotein receptor molecules. Biophys. J. 66(5), 1301–1318 (1994)

22. T.J. Feder, I. Brust-Mascher, J.P. Slattery, B. Baird, W.W. Webb, Constrained diffusion or immobile fraction on cell surfaces: A new interpretation. Biophys. J. 70(6), 2767–2773 (1996)

23. R.E. Thompson, D.R. Larson, W.W. Webb, Precise nanometer localization analysis for individual fluorescent probes. Biophys J. 82(5), 2775–2783 (2002)

24. R. Heim, A.B. Cubitt, R.Y. Tsien, Improved green fluorescence. Nature 373(6516), 663–664 (1995)

25. G. Mayer, A. Heckel, Biologically active molecules with a "Light Switch". Angewandte Chemie Int. Ed. 45(30), 4900–4921 (2006)

26. U. Haupts, S. Maiti, P. Schwille, W.W. Webb, Dynamics of fluorescence fluctuations in green fluorescent protein observed by fluorescence correlation spectroscopy. PNAS 95(23), 13573–13578 (1998)

27. S.T. Hess, A.A. Heikal, W.W. Webb, Fluorescence photoconversion kinetics in novel green fluorescent protein pH-sensors (pHluorins). J. Phys. Chem. B 108, 10138–10148 (2004)

28. S.T. Hess, S. Huang, A.A. Heikal, W.W. Webb, Biological and chemical applications of fluorescence correlation spectroscopy: a review. Biochemistry 41(3), 697–705 (2002)

29. A.A. Heikal, S.T. Hess, G.S. Baird, R.Y. Tsien, W.W. Webb, Molecular spectroscopy and dynamics of intrinsically fluorescent proteins: Coral red (dsRed) and yellow (Citrine). PNAS 97(22), 11996–12001 (2000)

30. E. Neher, B. Sakmann, Single-channel currents recorded from membrane of denervated frog muscle fibers. Nature (Lond.) 260, 119–802 (1976)

31. D.W. Tank, R.L. Huganir, P. Greengard, W.W. Webb, Patch-recorded single-channel currents of the purified and reconstituted torpedo acetylcholine-receptor. PNAS 80(16), 5129–5133 (1983)

32. E. Rhoades, T.F. Ramlall, W.W. Webb, D. Eliezer, Quantitation of alpha-synuclein binding to lipid vesicles using fluorescence correlation spectroscopy. Biophys. J. 90(12), 4692–4700 (2006)

33. H. Chen, E. Rhoades, J.S. Butler, S.N. Loh, W.W. Webb, Dynamics of equilibrium folding fluctuations of apomyoglobin measured by fluorescence correlation spectroscopy. PNAS 104(25), 10459–10464 (2007)

34. F. Chiti, C.M. Dobson, Protein misfolding, functional amyloid, and human disease. Ann. Rev. Biochem 75, 333–366 (2006)

35. M. Fandrich, M.A. Fletcher, C.M. Dobson, Amyloid fibrils from muscle myoglobin – Even an ordinary globular protein can assume a rogue guise if conditions are right. Nature 410(6825), 165–166 (2001)

36. M. Meyer-Luehmann, T.L. Spires-Jones, C. Claudia Prada, M. Garcia-Alloza, A. de Calignon, A. Rozkalne, J. Koenigsknecht-Talboo, D.M. Holtzman, B.J. Bacskai, B.T. Hyman, Rapid appearance and local toxicity of amyloid-beta plaques in a mouse model of Alzheimer's disease. Nature 451, 720–724 (2008)

37. R.H. Christie, B.J. Bacskai, W.R. Zipfel, R.M. Williams, S.T. Kajdasz, W.W. Webb, B.T. Hyman, Growth arrest of individual senile plaques in a model of Alzheimer's disease observed by *in vivo* multiphoton microscopy. J. Neurosci. 21(3), 858–864 (2001)

38. A.C. Kwan, D.A. Dombeck, W.W. Webb, Proximal apical dendrites have uniform polarity microtubules. PNAS 105(32), 11370–11375 (2008)
39. J. Yao, K. Munson, W.W. Webb, J.T. Lis, Dynamics of heat shock factor association with native gene loci in living cells. Nature 442(7106), 1050–1053 (2006)
40. J. Yao, C.J. Fecko, J.T. Lis, W.W. Webb, Intranuclear distribution and local dynamics of RNA polymerase II during transcription activation. Mol. Cell 28, 978–990 (2007)
41. Z. Ni, A. Saunders, N.J. Fuda, J. Yao, J.-R. Suarez, W.W. Webb, J.T. Lis, P-TEFb is critical for the maturation of RNA polymerase II into productive elongation in vivo. Mol. Cellular Biol. 28(3), 1161–1170 (2008)
42. J. Yao, K.L. Zobeck, W.W. Webb, J.T. Lis, Imaging transcription dynamics at endogenous genes in living Drosophila tissues. Methods 45(3), 233–241 (2008)
43. M.J. Levene, J. Korlach, S.W. Turner, M. Foquet, H.G. Craighead, W.W. Webb, Zero-mode waveguides for single molecule analysis at high concentrations. Science 299, 682–686 (2003)
44. R.M. Williams, W.R. Zipfel, W.W. Webb, Multiphoton microscopy in biological research. Current Opin. Chem. Biol. 5, 603–608 (2001)
45. D.A. Dombeck, K.A. Kasischke, H.D. Vishwasrao, M. Ingelsson, B.T. Hyman, W.W. Webb, Uniform polarity microtubule assemblies imaged in native brain tissue by second-harmonic generation microscopy. PNAS 100(12), 7081–7086 (2003)
46. W.R. Zipfel, R.M. Williams, R.H. Christie, A.Y. Nikitin, B.T. Hyman, W.W. Webb, Live tissue intrinsic emission microscopy using multiphoton excited intrinsic fluorescence and second harmonic generation. PNAS 100(12), 7075–7080 (2003)
47. W.R. Zipfel, R.M. Williams, W.W. Webb, Nonlinear magic: multiphoton microscopy in the biosciences. Nature Biotechnol. 21(11), 1369–1377 (2003)
48. M.J. Levene, D.A. Dombeck, R.P. Molloy, K.A. Kasischke, R.M. Williams, W.R. Zipfel, W.W. Webb, In vivo multiphoton microscopy of deep brain tissue. J. Neurophysiol. 91, 1908–1912 (2004)
49. K.A. Kasischke, H.D. Vishwasrao, P.J. Fisher, W.R. Zipfel, W.W. Webb, Neural activity triggers neuronal oxidative metabolism followed by astrocytic glycolysis. Science 305(5680), 99–103 (2004)
50. R.M. Williams, W.R. Zipfel, W.W. Webb, Interpreting second harmonic generation images of collagen I fibrils. Biophys. J. 88, 1377–1386 (2005)
51. H.D. Vishwasrao, A.A. Heikal, K.A. Kasischke, W.W. Webb, Conformational dependence of intracellular NADH on metabolic state revealed by associated fluorescence anisotropy. J. Biol. Chem. 280(26), 25119–25126 (2005)

6

Chemical Fluxes in Cellular Steady States Measured by Fluorescence-Correlation Spectroscopy

Hong Qian and Elliot L. Elson

6.1 Introduction

Genetically, identical cells adopt phenotypes that have different structures, functions, and metabolic properties. In multi-cellular organisms, for example, tissue-specific phenotypes distinguish muscle cells, liver cells, fibroblasts, and blood cells that differ in biochemical functions, geometric forms, and interactions with extracellular environments. Tissue-specific cells usually have different metabolic functions such as synthesis of distinct spectra of secreted proteins, e.g., by liver or pancreatic cells, or of structural proteins, e.g., muscle vs. epithelial cells. But more importantly, a phenotype should include a dynamic aspect: different phenotypes can have distinctly different dynamic functions such as contraction of muscle cells and locomotion of leukocytes. The phenotypes of differentiated tissue cells are typically stable, but they can respond to changes in external conditions, e.g., as in the hypertrophy of muscle cells in response to extra load [1] or the phenotypic shift of fibroblasts to myofibroblasts as part of the wound healing response [2]. Cells pass through sequences of phenotypes during development and also undergo malignant phenotypic transformations as occur in cancer and heart disease.

How can we define the molecular basis of specific phenotypic states? Cells such as muscle or secretory cells must have characteristic patterns of biochemical reactions that support their structural properties and metabolic functions. The persistence of a cell in a steady phenotypic state suggests that the underlying network of metabolic reactions also operates in a steady state.[1] The metabolic network is controlled by signaling molecules that promote or

---

[1] We intend the term "steady state" to be understood in the statistical thermodynamic sense of allowing stationary concentration fluctuations around a steady mean concentration This mean concentration could be established either in equilibrium or in a nonequilibrium steady state. If several steady states are compatible with a given set of concentration constraints (as is discussed later), there will be concentration fluctuations centered around each of the steady states.

inhibit cellular processes. These include autocrine and paracrine hormones and growth factors as well as mechanical interactions with the extracellular matrix and other cells. One presumes that maintenance of a metabolic steady state requires that these signals are also in a steady state [3].

When a macroscopic chemical reaction system is in a steady state, the concentrations of the reactants are constant in time. One type of steady state is chemical equilibrium, a state in which there is a detailed balance across each and every chemical reaction in the system. In a reversible chemical reaction the flux, i.e., the number of molecules produced by the reaction in a given time, in the forward direction is balanced by an equal flux in the backward direction. For example, in the simple isomerization, A $\leftrightarrow$ B, with forward and backward rate constants, k_+ and k_-, respectively the fluxes in the forward and backward directions are k_+C_A and k_-C_B, where C_A and C_B are the concentrations of A and B. In equilibrium detailed balance ensures that $k_+C_A = k_-C_B$, yielding the equilibrium constant, $K = C_A/C_B = k_+/k_-$. In general, however, biochemical networks cannot be in overall equilibrium in living cells. Equilibrium systems are dead. Rather, cells are continually transforming chemical energy into biochemical synthesis and degradation as well as dynamic functions. Hence, cells operate in nonequilibrium steady states (NESSs) [4]. Although the concentrations of the reactants are constant, i.e., constant in a statistical sense with stationary fluctuations, detailed balance no longer holds. There is a continuous and steady conversion of chemical species that enter the NESS into products that are withdrawn. For the above example in an NESS $k_+C_A \neq k_-C_B$ and so the fluxes in the forward and reverse directions are different. To maintain this state, cells expend chemical energy typically in the form of ATP. Chemical energy is used to hold some reactant concentrations in the reaction network constant, e.g., by continuously supplying reactants and withdrawing products. The system is driven by these nonequilibrium conditions. These chemical constraints on the reaction participants determine the values of the other reactant concentrations and fluxes in the NESS system [5].

In a linear chemical reaction system, there is a unique steady state determined by the chemical constraints that establish the NESS. For nonlinear reactions, however, there can be multiple steady states [6]. A network comprised of many nonlinear reactions can have many steady states consistent with a given set of chemical constraints. This fact leads to the suggestion that a specific stable cellular phenotypic state can result from a specific NESS in which the steady operation of metabolic reactions maintains a balance of cellular components and products with the expenditure of biochemical energy [4]. Similarly, the network of chemical and mechanical signals that regulate the metabolic network must also be in a steady state. Important problems, then, are to determine the variety of steady states available to a system under a given set of chemical constraints and the mechanisms by which cells undergo transitions among steady states as their phenotypic properties change.[2] As long as the chemical participants, e.g., enzymes or signaling molecules, are present in high concentrations, these cellular NESSs behave as deterministic systems. Chemical fluctuations are insignificant. Suppose, however, that a metabolic enzyme or a signaling intermediate is present in only a few copies. Then, fluctuations in the number or function of this cellular component could lead to stochastic behavior of the biochemical reaction network. As we shall see below, there is ample evidence for this stochastic behavior in simple prokaryotic systems. In multi-cellular organisms, the fluctuations in the behavior of a chemical reaction in a signaling network could lead to stochastic behavior of the network. These considerations might help to explain phenomena such as incomplete penetrance of mutations among organisms with identical genetic makeup and the diversion of cells from normal to pathological phenotypic states, e.g., [7–9].

Elaborate models have been developed to account for the behavior of cellular biochemical networks. Boolean network models use a set of logical rules to illustrate the progress of the network reactions [10]. These models do not take into explicit account the participation of specific biochemical reactions. Models that account for the details of biochemical reactions have been proposed [11,12]. The behavior of these models depends on the rate constants of the chemical reactions and the concentrations of the reactants. Measurements like those described below of reaction fluxes and reactant concentrations will be able to test such network models. In the following sections, we will use simple examples to illustrate the characteristic steady-state behavior and propose an approach to measure fluxes and concentrations.

6.2 Steady States of Nonequilibrium Chemical Reaction Networks

The life of a cell is maintained by the continuous activity of a myriad of biochemical reactions that provide metabolic energy, synthesize (and degrade) structural and functional molecules such as proteins, nucleic acids and lipids, and drive cellular dynamic functions such as contraction, locomotion, and cytokinesis. These reactions are organized into networks or modules that have specific functions such as protein synthesis or production of energy by oxidative phosphorylation. Then, when a cell is in a stable state, these reaction networks must also operate stably, i.e., energy is continuously generated and

---

[2] There is a formal similarity between the possibilities of multiple steady states for a biochemical reaction system and of multiple conformational states available to a protein. Ultimately, if concentration constraints are removed from the reaction system, it will relax to equilibrium. Similarly, proteins in nonequilibrium conformational states, i.e., states that do not have the global minimum free energy, will eventually relax to the equilibrium state, although this may take a long time if there are substantial activation energy barriers among the metastable and equilibrium conformational states.

proteins are continuously synthesized, both at stable levels. Thus, the reaction networks operate as NESS systems: the rates of energy production and dissipation and the rates of protein production and degradation are balanced. Reaction intermediates are at constant concentration. For example, suppose one or a sequence of reversible chemical reactions transforms input reagents into output products, e.g., amino acids into proteins. At some cost of chemical energy, typically in the form of ATP, specific values of the input and output concentrations are held constant. Depending on the values of these concentration constraints, the system will eventually find one or more states in which all the chemical reaction rates and the concentrations of the reaction intermediates remain constant in an NESS. As input and output constraints are changed, the reaction rates and intermediate concentrations change correspondingly. Therefore, a cell could control its phenotypic state by regulating the biochemical concentrations that maintain its underlying reaction network steady states. An important task of systems biology is to understand the factors that control cellular biochemical steady states.

For each set of constant input and output concentration constraints a system of linear chemical reactions has a unique steady state. For a network of nonlinear biochemical reactions, however, there could be several steady states compatible with a given set of constraints. The number and character of these steady states are determined by the structure of the network including the extent of nonlinearity, the number and connectivity of the individual chemical reactions and the values of the reaction rate constants and the concentrations of the reactants. The higher the order of a chemical reaction, the more steady states may be compatible with a given set of chemical constraints. The simple trimolecular reaction system of Schlogl [13] illustrates how a third-order chemical reaction can have two stable steady states compatible with a single set of chemical constraints:

$$A + 2X \overset{k_1}{\underset{k_2}{\rightleftharpoons}} 3X, \tag{6.1}$$

$$B \overset{k_3}{\underset{k_4}{\rightleftharpoons}} X. \tag{6.2}$$

Steady states can be established by fixing the concentrations of A and B (at values a and b). The reaction kinetics are described by the following:

$$dx/dt = k_1 a x^2 - k_2 x^3 - k_4 x + k_3 b. \tag{6.3}$$

The properties of this system have been discussed in detail in a recent publication [6]. In equilibrium, the conditions of detailed balance hold

$$k_1 a x_{eq}^2 = k_2 x_{eq}^3 \text{ and } k_3 b = k_4 x_{eq}.$$

These lead to the condition that $R = k_1 k_4 a / k_2 k_3 b = 1$. If, however, the values of a and b are maintained such that this condition does not hold, the system is no longer in equilibrium but can, nevertheless, be in an NESS. If $R > 1$, there is constant and continuous transformation of A into B; if $R < 1$, B is transformed into A. In the former case, molecules of A are continuously supplied and B molecules are continuously withdrawn to hold a and b constant. These steady states can be identified from the condition $dx/dt = 0 = k_1 ax^2 - k_2 x^3 - k_4 x + k_3 b$. This cubic reaction can have as many as three positive real solutions, depending on the values of the rate constants and a and b. For example, if $k_1 = 3$, $k_2 = 0.6$, $k_3 = 0.25$, and $k_4 = 2.95$, and setting $a = 1$ and $b = 0.9$, solution of the cubic equation yields steady states at $x = 0.0832$, 1.219, and 3.6978. From a plot of dx/dt vs. x, one sees that the slope is negative at values of x before the first and third roots, which are therefore stable steady states, while the positive slope before the middle root indicates that it is unstable [6]. Which of the two stable steady states the system reaches, depends only on the initial value of x. If the initial value of $x < 1.22$, the system goes to the steady-state value of 0.083, for an initial value $x > 1.22$, the system settles at $x = 3.70$ (Fig. 6.1). The net fluxes in the "forward" (A → B) and in the "backward" (B → A) directions are different for the two steady states. The forward and

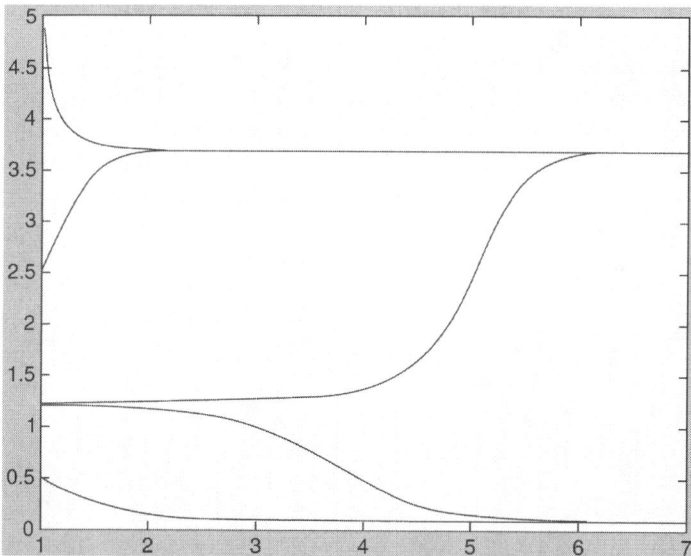

Fig. 6.1. Approach to steady states from different initial states. The nonequilibrium steady state that the system reaches depends on the system's initial state. Under the conditions listed below there are two stable steady states, at $x = 0.08$ and 3.70. There is an unstable state at 1.22. If the initial concentration of x is greater than 1.22, the system goes to the higher steady state. If the initial concentration is less than 1.22, the lower steady state is occupied
Parameters: $k_1 = 3.0$, $k_2 = 0.6$, $k_3 = 0.25$, $k_4 = 2.95$, $a = 1.0$, $b = 0.9$

backward fluxes for reaction 1 are $f_{1f} = k_1 ax^2$ and $f_{1b} = k_2 x^3$ and the net flux is $f_{1f} - f_{1b}$. In a steady state, this must equal the net flux through reaction 2, $f_{2f^-} - f_{2b} = k_4 x - k_3 b$. For the steady state with $x_{ss} = 3.6978$, the net flux is 10.6836 [concentration/time], while for the steady state with $x = 0.0832$, the net flux is 0.0204 [concentration/time]. For a metabolic or signaling network, this kind of difference in flux could be an important functional difference between the two steady states. Although the Schlogl reaction is a simplified and rather artificial example, it provides a useful illustration of the behavior of steady states for higher order chemical reactions. Evidently, the higher the order of the reaction, the greater the number of potential steady states.

For an extensive network composed of many biochemical reactions, the analysis becomes correspondingly more complex. One simplified type of analysis makes use of Boolean network models [10] in which each protein or gene in the network can be only either active or inactive and processes advance through discrete time steps. It has been argued that the state of the system, is determined by the activity state of each of the N genes or enzymes of the network. Hence, there are 2^N states available. The chemical details of the processes that lead to enzyme activation or gene transcription are omitted from this type of model. The binary on–off character of the decision steps enforces a strong nonlinearity in a Boolean model, cf. [14]. As a result, these models can lead to multiple steady states for a reaction system. An interesting and extensively studied example is the control of the yeast cell cycle. A simplified version takes account of 11 genes/proteins that control and drive replication. These include cyclins, inhibitors of cyclin-kinase complexes, transcription factors, and check points, e.g., of cell size [15]. The state of a cell is determined by which of these are activated. For example, in the stationary state, G_1, the inhibitors Cdh1 and Sic1 are active, and all other relevant proteins are inactive. A given activated protein may transmit either an activation or inhibition signal to one or more other proteins. Activation or inhibition of a protein results from the weighted input of all the other active proteins. When a start signal is given by activating the cyclin Cln3, the cell is driven through a series of states by the sequential activation or inhibition of various proteins. Remarkably, activation of the system from each of the $2^{11} = 2,048$ possible initial states leads eventually to the population of only seven stationary states of which one state, corresponding to G_1, contained 86% of the final states [15]. Hence, G_1 represents a strong attractor of the system, corresponding to its relatively high stability. In this case, therefore, of the seven steady states revealed, one is much more stable than all the others. Additionally, this approach was able to demonstrate that activating the G_1 stationary state caused the system to pass in sequence through the S-, G_2-, and M-phases as it eventually returned to the stable G_1-state, thereby recapitulating the normal cell cycle pathway [15]. Although this elegant and relatively simple model can reveal important properties of a gene system, it cannot be used to probe the mechanisms of processes at the molecular level of the chemical reactions that drive the network. One strategy for introducing chemical reactions into the Boolean network approach is to represent complex chemical processes in terms of "effective" or "virtual" reactions [16]. The former represent complex processes that transform input to output molecules in a single step without taking into account the multi-step intermediate reactions. For example, the single effective reaction, mRNA → protein, includes in a single step the many reactions required for both transcription and translation. (Virtual reactions do not correspond to specific biochemical reactions, but are built from mathematical functions that model biochemical processes.) The finite time required for complex effective reactions is accounted for by introducing explicit time delays for specific processes. The time delays are unimportant in studies of steady states in which the transients seen as the system approaches the steady state have died out. For systems with long transient times, however, the approach to some steady states may be incomplete during normal cell function. Then, the inclusion of the time lags is necessary for an adequate description of the operation of the network. Although using effective reactions with time lags explicitly considers the overall chemical processes that drive network behavior, a thorough analysis of NESS systems, e.g., to determine the fluxes through the individual biochemical reactions, requires a more detailed treatment of the individual reactions at the molecular level.

The yeast cell cycle has also been analyzed at this high level of chemical detail [17]. The molecular mechanism of the cycle in the form of a series of chemical equations was described by a set of ten nonlinear ordinary differential kinetic rate equations for the concentrations of the cyclins and associated proteins and the cell mass, derived using the standard principles of biochemical kinetics. Numerical solution of these equations yielded the concentrations of molecules such as the cyclin, Cln2, which is required to activate the cell cycle, or the inhibitor, Sic1, which helps to retain the cell in the "resting" G_1 phase. The rate constants and concentrations (~50 parameters) were estimated from published measurements and adjusted so that the solutions of the equations yielded appropriate variations, i.e., similar to those experimentally measured, of the concentrations of the constituents of the system and the cell mass. The model also provides a rationalization of the behavior of cells with mutant forms of various system constituents.

There are two main steady states in the cycle. One is G_1 in which the cell is maintained by inhibitors such as Sic1. The other is at the S/M phase boundary to which the cell migrates due to an increase in cyclins, e.g., Cln2, and cyclin-dependent kinase activity. (These two states are controlled by the antagonism between the kinases and inhibitors.) Because of the complexity of the model, however, it is difficult to obtain an overall view of the location and character of all the steady states that might be accessible to the system. This approach has also been used to characterize the behavior of the morphogenesis checkpoint in budding yeast and also mutants [18] as well as to provide a more complete integrative analysis of the control of the yeast cell cycle [17].

Cells can change their steady-state functions either by changing the constraints that govern a specific steady state, thereby changing the fluxes of the reactions in that steady state or, if multiple steady states are consistent with a given set of constraint conditions, by jumping among the various steady states available. For the latter to happen there must be molecular fluctuations to drive the state changes. As we have seen in the simple Schlogl reaction illustration, for deterministic systems the steady state into which a system settles is determined only by the system's initial state. Hence, deterministic systems will not undergo transitions among multiple steady states available under a single set of concentration constraints. As we shall see, composition fluctuations allow the system to fluctuate among steady states that are comparably stable. The greater the disparity in stability among the steady states, the smaller the likelihood of jumping from the more to the less stable state. For this among other reasons it is, therefore, useful to discuss fluctuations or noise in biological reaction networks.

6.3 Noise and Fluctuations in Biological Reaction Networks

A wide variety of evidence supports the notion that cellular biochemical processes are subject to fluctuations that influence cellular functions. This has been well characterized in studies of protein production in prokaryotic systems. It has been proposed that large numbers of protein molecules are produced in bursts from individual mRNA molecules. Hence, protein translation amplifies fluctuations in the number of mRNA molecules [9,19]. This hypothesis has been confirmed experimentally in a study of the effect of varying independently the rates of transcription and translation of a single fluorescent reporter gene in *B. subtillis* [20]. A single copy of the gene (gfp) for the green fluorescent protein (GFP) was incorporated into the bacterial chromosome under the control of an inducible promoter. Transcriptional efficiency was controlled by varying the inducer concentration. Translational efficiency was varied by using bacterial strains with different mutations in the ribosomal binding site and initiation codon of gfp. Expression of GFP within individual cells of a population was measured by flow cytometry to determine the level of "phenotypic noise," i.e., the variation of GFP expression from cell to cell within the population. This revealed a roughly Gaussian distribution of expression levels. The "phenotypic noise strength" is the variance of this distribution divided by its mean. Consistent with the idea that phenotypic noise is amplified at the level of translation, the measured noise strength increased with the mean level of protein expression as the translational efficiency was increased, but was relatively insensitive to variation of transcriptional efficiency. Further insight into this behavior has been supplied by observations of protein production in single living cells in real time [21]. Measuring with single molecule sensitivity, the appearance of the yellow fluorescent protein

(YFP) fused to a membrane targeting peptide showed that the YFP molecules appear in bursts originating from individual stochastically transcribed mRNA molecules.

A similar approach was used to discriminate between intrinsic noise, stochastic fluctuations that arise from fluctuations of the molecules directly involved in gene expression, and extrinsic contributions that arise form fluctuations in other molecules that regulate or otherwise contribute to the process [22]. The latter are uniform within a cell but vary from cell to cell while the former produce fluctuations over time within an individual cell. The two forms of noise were distinguished by determining the extent of correlation between the expression of cyan and yellow fluorescent proteins under the control of identical promoters in the same cell. The intrinsic noise was measured as the difference in fluorescence intensity of the YFP and CFP. The extrinsic noise was determined by using the fact that the square of the total noise equals the sum of the squares of the intrinsic and extrinsic noises. The results show that both types of noise contribute significantly.

One approach to including noise (fluctuations) in the analysis of the yeast cell cycle reaction network is an extension of the Boolean network approach discussed earlier. The time evolution of the probability of a state is computed from a master equation for each of the 2^{11} states The transition rules defined by Li et al. [15] are embodied in transition probabilities used in the master equations [14]. These equations provide the steady-state probabilities for each of the states, P_i^{ss}. In analogy with the Boltzmann distribution, a generalized energy, U_i, is defined as $P_i^{ss} = \exp[-U_i]$. Fluctuation probabilities could be calculated from these energies and an energy landscape can be developed that gives a global picture of the relative stabilities of the various states of the system. In this representation, both intrinsic and extrinsic noises are lumped together. In the following, we intend to focus on the intrinsic noise of the system that results from statistical fluctuations of the numbers of specific molecules that participate in individual reactions in a network and that are present in small numbers.

A further discussion of the Schlogl reaction provides an illustration of an important difference in the behavior of steady states in deterministic systems and in systems subject to stochastic fluctuations. In contrast to the deterministic Schlogl system analyzed earlier in terms of conventional chemical concentrations, the analysis of the stochastic system is carried out in terms of the probability $p_n(t)$ that there are n $x(t)$ molecules in the system at time t in addition to $n_A = a$ and $n_B = b$ molecules of A and B that are held constant. Figure 6.2 illustrates the individual pathways and rates by which a system with n molecules could undergo a transition to a system containing either $n + 1$ or $n - 1$ molecules. This leads to the master equation for each of the $p_n(t)$ [6]:

$$\frac{dp_n(t)}{dt} = \lambda_{n-1}p_{n-1} + \mu_{n+1}p_{n+1} - (\lambda_n + \mu_n)p_n, \quad \text{for } n = 0 \cdots \infty,$$

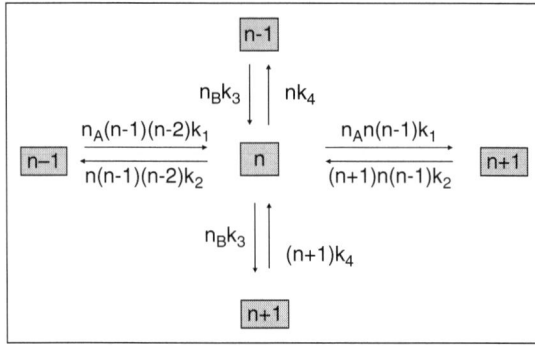

Fig. 6.2. Kinetic pathways for the Schlogl reaction. There are four pathways by which the system can either increase or decrease the number of x molecules. The rates depend on the numbers of the interacting molecules and the rate constants as shown

where
$$\lambda_n = \frac{ak_1 n(n-1)}{V} + bk_3 V$$

and
$$\mu_n = nk_4 + \frac{k_2 n(n-1)(n-2)}{V^2}.$$

The rate constants used in the deterministic model, expressed in terms of molecular concentrations, are related to the constants for the stochastic model, expressed in terms of molecule numbers by taking into account the volume, V, of the system as in the earlier expressions for λ_n and μ_n [23].

The steady-state distribution, $p_n{}^{ss}$, is determined from the condition [6] $\lambda_{n-1} p_{n-1}{}^{ss} = \mu_n p_n{}^{ss}$, and so:

$$\frac{p_n^{ss}}{p_o^{ss}} = \prod_{i=0}^{n-1} \frac{\lambda_i}{\mu_{i+1}} \tag{6.4a}$$

with
$$p_0^{ss} = 1 - \sum_{j=1}^{\infty} p_j^{ss}. \tag{6.4b}$$

For an illustrative selection of parameters, $k_1 = 2.7$, $k_2 = 0.6$, $k_3 = 0.25$, $k_4 = 2.95$ with $a = 1.0$ and $b = 2.0$, these equations yield the steady-state distribution shown in Fig. 6.3a in which the probability of being in the state with fewer x molecules ($x^{ss} \sim 3$) is slightly greater than being in the state with more x molecules ($x^{ss} \sim 40$). There are two peaks representing the two steady states. The time behavior of the system can be determined from a Monte Carlo simulation based on the transition probabilities shown in Fig. 6.2, and illustrated in Fig. 6.3b. The system frequently passes between the two steady states, but spends most of its time in the more probable state. This unstable behavior would be unsuitable for a typical reaction network steady state. As illustrated in Fig. 6.3c d, if the relative probabilities of the two states are reversed, again the system spends proportionally longer in the more stable state ($x^{ss} \sim 40$).

An important contrast illustrated by Figs. 6.1 and 6.3 has previously been discussed [6]: the state in which a deterministic system settles is governed entirely by its initial state. Once it has reached its stable steady state, it does not undergo transitions to another steady state. In contrast, the initial state of the stochastic system has no effect on its ultimate long-time behavior. Rather the system passes between the available steady states dwelling in each steady state for times related to the overall stability of the state.

Although this simple example is far from capturing the complexity of a cellular biochemical network, stochastic networks would also be able to fluctuate among steady states. Any biochemical network that supported a stable phenotypic state would have to be far more stable than the example of the Schlogl reaction shown here, cf. [14]. Nevertheless, even systems in which one (normal) network steady state was far more stable than all the others compatible with the chemical constraints could undergo rare transitions to other available, less probable (abnormal) states. Rare events like this could have significance for the initiation of pathological conditions such as cancer [8] even as rare protein conformational fluctuations might be important for initiating prion diseases.

The proposed relationship discussed here between stable cellular phenotypic states and stable nonequilibrium steady states of metabolic or signaling biochemical networks suggests that the latter should be stable over the lifetime of the phenotypic state. The extent to which there are multiple steady states compatible with the conditions (reactant concentrations and chemical rate constants) of specific phenotypic states is unknown. For example, depending on conditions, there might be only one stable steady state for the Schlogl reaction. Similarly, in a cellular biochemical network not only the order of the reactions but also the large collection of rate constants and the range of reactant concentrations would determine the number of stable steady states. This number is not obtainable from a Boolean network analysis. Rather, it must be obtained by analysis of a biochemical network at the molecular level of its individual reactions (Chen et al., 2000; [17]). At present it is difficult to explore completely the huge parameter space of such large and complex networks to discover all of the steady states that might be available. Even over a range limited to realistic reactant concentrations, a thorough exploration would be difficult.

An alternative approach is to characterize *experimentally* the steady-state fluxes of individual reactions in the network. The pattern of fluxes of all of the reactions in the system provides a comprehensive definition of an NESS system. As we have seen, different steady states have different flux magnitudes and these quantities can provide an operational definition of these states.

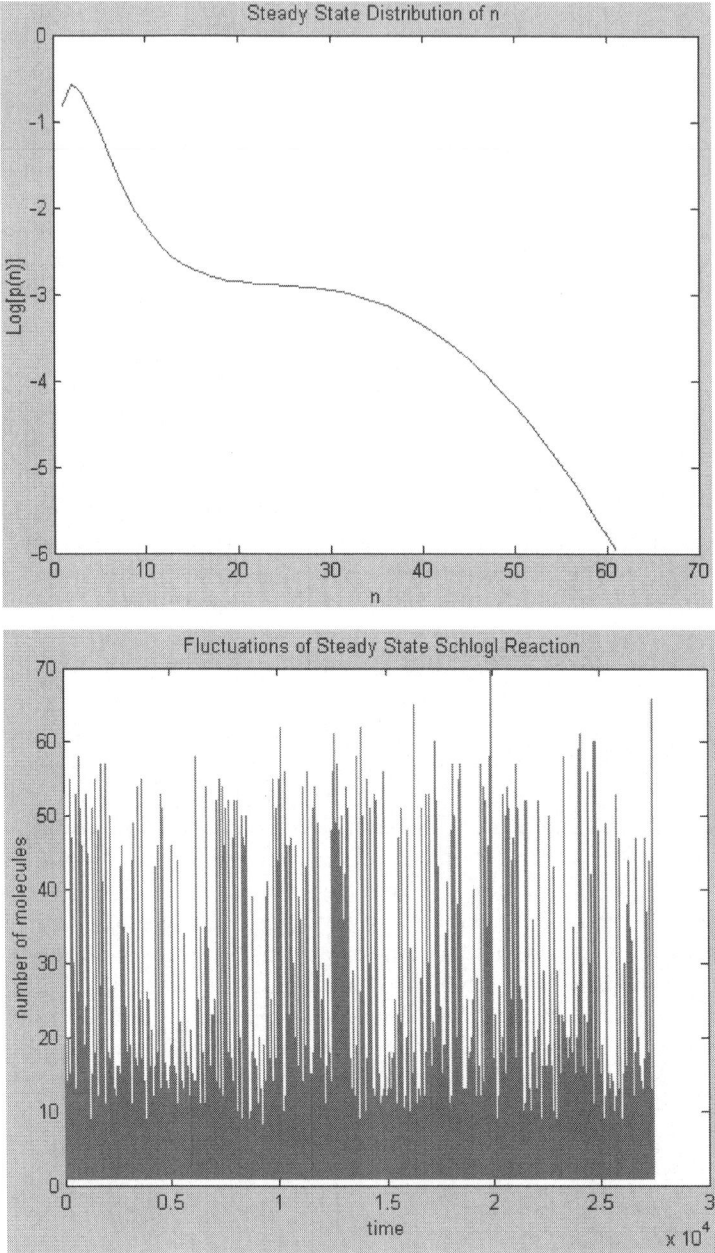

Fig. 6.3. Distributions of x in between two steady states (A, C), and dynamic fluctuations between these two states (B, D). The steady-state distributions (A and C) were calculated using (6.4) in the text. The fluctuations in x were calculated using a Gillespie-type Monte Carlo algorithm to the chemical master equation (Beard and Qian, 2008). Parameters: Panels (**a**) and (**b**), $k_1 = 2.7$, $k_2 = 0.6$, $k_3 = 0.25$, $k_4 = 2.95$, $a = 1$, $b = 2$. In panels (**c**) and (**d**), $k_1 = 3.3$; others are the same as in (**a**) and (**b**). In panels (**b**) and (**d**), the volume of the system is set to be 10

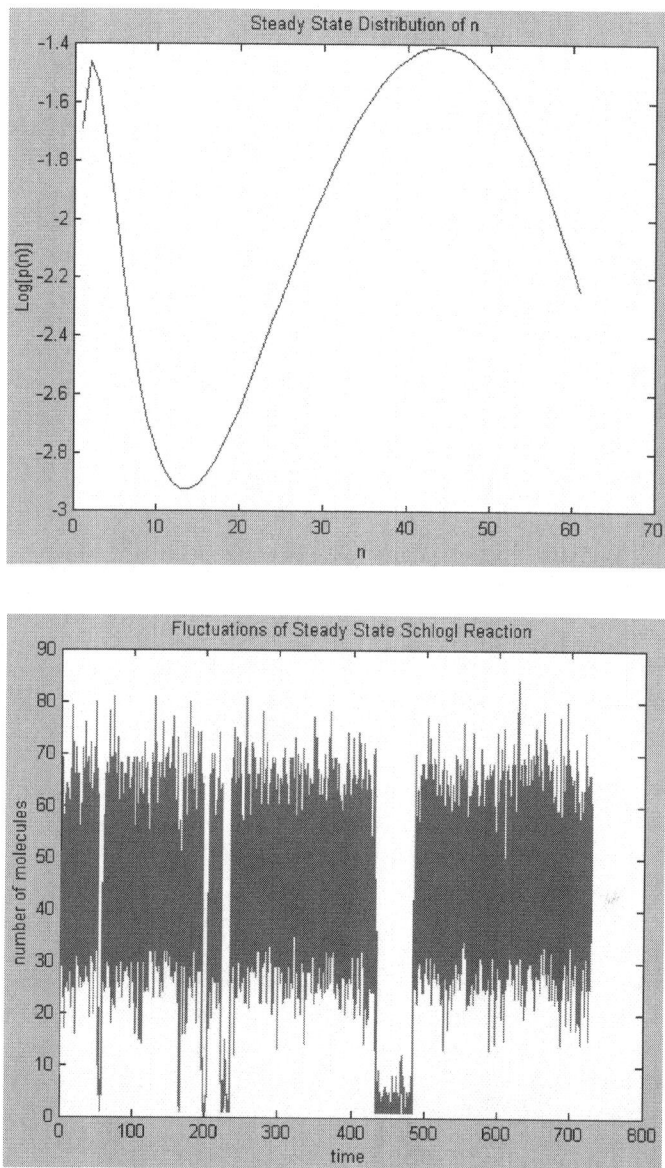

Fig. 6.3. (continued)

An extended Boolean network analysis of the yeast cell cycle, suggests that there are large "energy barriers," i.e., the system must pass through states of low probability to reach relatively higher probability steady states that coexist under one set of constraint conditions, cf. [14]. In analogy with chemical systems that have high activation energy barriers, this suggests that transitions among these states should be abrupt. Based on the analysis of individual network biochemical reactions, however, neither the number of steady states that exist under one set of realistic chemical constraint conditions nor whether transitions among steady states are abrupt or continuous is clear. If the constraint conditions change as the cell undergoes a change of phenotype, would the biochemical networks undergo abrupt changes to different steady states or rather remain in steady states that changed their character continuously during the phenotypic transformation? Experimental measurements even of only a few reaction fluxes in steady state could help to answer these questions and test important hypotheses proposed in systems biology.

6.4 Experimental Characterization of Steady State Fluxes

Fluorescence correlation spectroscopy (FCS) provides an approach to measure reaction fluxes of NESS systems both in living cells and in vitro. Consider the reaction A $\leftrightarrow$ B with rate constants k_{ab} and k_{ba}. The flux in the directions A $\rightarrow$ B and B $\rightarrow$ A are $k_{ab}C_a^{ss}$ and $k_{ba}C_b^{ss}$, respectively, where C_a^{ss} and C_b^{ss} are the steady-state concentrations of A and B. The possibility of directly measuring these fluxes and the relationship between the fluxes and the two-color FCS cross-correlation function has been previously demonstrated [24]. We provide a different version more closely tied to the experimental measurement. The FCS cross-correlation function measures the correlation of the fluctuations of the concentrations of A and B in a small laser-illuminated sample region of a solution [25–27]. The cross-correlation functions can be defined as $G_{AB}(\tau) = <\delta F_A(0)\ \delta F_B(\tau)>$ and $G_{BA}(\tau) = <\delta F_B(0)\ \delta F_A(\tau)>$, where $\delta F_j(t) = F_j(t) - F_j^{ss}$ is the departure of the fluorescence of the jth component from its steady-state mean value at time t and $< \cdots >$ indicates either an ensemble or time average. The two-color cross-correlation function for the reaction between A and B has been developed in earlier work ([25], Appendix 3). For simplicity, we suppose that the diffusion coefficients of A and B have the same value, D. (This will be a good approximation for most cases of interest.) We note that the derivation of the FCS correlation functions for equilibrium systems will be the same as for NESS systems except that the correlation function amplitudes will be proportional to C_a^{ss} or C_b^{ss} in the latter rather than to C_a^{eq} and C_b^{eq} as in the former. This is because the amplitudes are determined by the magnitude of the fluctuations of the number of molecules in the laser-illuminated observation region, which are governed by a Poisson distribution both for equilibrium and NESS systems

(Beard, 2008). Hence, for equilibrium systems $<(\delta C_A{}^{eq})^2> = C_A{}^{eq}$, where $\delta C_A{}^{eq}$ is the fluctuation amplitude $C_A - C_A{}^{eq}$ and for first-order NESS systems $<(\delta C_A{}^{ss})^2> = C_A{}^{ss}$ with $\delta C_A{}^{ss} = C_A - C_A{}^{ss}$. Then, treating a two-dimensional (planar) system for simplicity we have [25],

$$G_{BA}^{SS}(\tau) = \left[\frac{Q_A Q_B C_B^{SS}}{S(1+K)}\right]\left[\frac{1 - \exp(-R\tau)}{1 + \tau/\tau_D}\right]$$

and

$$G_{AB}^{SS}(\tau) = \left[\frac{Q_A Q_B K C_A^{SS}}{S(1+K)}\right]\left[\frac{1 - \exp(-R\tau)}{1 + \tau/\tau_D}\right]$$

Here, Q_A and Q_B account for the different optical properties of the fluorophores that distinguish A and B as well as the laser intensity and other instrumental factors. $K = k_{ab}/k_{ba}$, and $\tau_D = w^2/4D$ is the characteristic diffusion time for a Gaussian excitation intensity profile with $\exp(-2)$ radius w, and S is the area of the laser spot. It is readily shown that for equilibrium systems $G_{AB}{}^{eq}(\tau) = G_{BA}{}^{eq}(\tau)$ due to the fact that $k_{ab}C_B{}^{eq} = k_{ba}C_A{}^{eq}$ [25]. The NESS fluxes can be obtained from the initial slope of the correlation functions,

$$\lim_{\tau \to 0}\left[\frac{G_{BA}{}^{ss}(\tau)}{\tau}\right] = \frac{Q_A\ Q_B\ k_{ba}\ C_b{}^{ss}}{S}\ \text{and}$$

$$\lim_{\tau \to 0}\left[\frac{G_{AB}{}^{ss}(\tau)}{\tau}\right] = \frac{Q_A\ Q_B\ k_{ab}\ C_a{}^{ss}}{S}.$$

The factors Q_A and Q_B can be determined from the zero-time amplitudes of FCS autocorrelation functions: $G_{AA}{}^{ss}(0) = Q_A{}^2 C_A{}^{ss}/S$ and $G_{BB}{}^{ss}(0) = Q_B{}^2 C_B{}^{ss}/S$ and from the mean fluorescence of A and B in the steady state, $<F_A> = Q_A C_A{}^{ss} S$ and $<F_B> = Q_B C_B{}^{ss} S$. Hence, the fluxes are

$$\text{flux}_{A \to B} = k_{ab}\ C_A{}^{ss} S = Z_{AB} \lim_{\tau \to 0}\left[\frac{G_{AB}{}^{ss}(\tau)}{\tau}\right];$$

$$\text{the flux}_{B \to A} = k_{ba}\ C_B{}^{ss} S = Z_{AB} \lim_{\tau \to 0}\left[\frac{G_{BA}{}^{ss}(\tau)}{\tau}\right]$$

with $Z_{AB} = <F_A><F_B>/(G_{AA}(0)\ G_{BB}(0))$. Of course, the net flux in the direction A → B is $\text{flux}_{net} = \text{flux}_{A \to B} - \text{flux}_{B \to A}$. It is also important to realize that, as indicated above, the steady-state concentrations of A and B can be directly obtained from the mean fluorescence and $G_{ii}{}^{ss}(0)$, e.g., $G_{AA}{}^{ss}(0)/(<F_A>)^2 = (SC_A{}^{ss})^{-1}$. Knowing $C_A{}^{ss}$ and S, one can determine k_{ab} directly from the appropriate flux; $k_{ab} = (\text{flux}_{A \to B})/C_A{}^{ss}$ and likewise for k_{ba}.

Using this approach, one could determine the NESS flux for any reaction A ↔ B that caused a large enough change in the color of the measured fluorescence to permit an accurate measurement of the two-color cross-correlation function. One attractive strategy would use Forster resonance energy transfer (FRET). If the reaction caused a large enough conformation change that the efficiency of FRET were large for A and small for

B, then the fluorescence of A would be dominated by the acceptor emission spectrum and that of B by the donor spectrum. Thus, the reaction would cause a blue shift, which, if it were big enough, would enable a cross-correlation measurement. The reverse process with FRET large for B and small for A, would work equally well. For the bimolecular reaction $A + L \leftrightarrow B$, the flux could be measured if the binding caused a large enough shift in the fluorescence spectrum of A. The analysis for the bimolecular reaction is the same as given earlier; in a steady state, the concentration of L is constant and so can be absorbed into a pseudo-first-order rate constant.

The required reaction-dependent change in FRET is illustrated by several molecules that have been developed as FRET biosensors to detect specific biochemical reactions. One is a version of myosin light chain kinase (MLCK) that allows the determination of its location as well as its activation state in living cells [28, 29]. MLCK is activated by calcium-bound calmodulin. The biosensor molecule is constructed by fusing to MLCK an indicator protein that links blue and green fluorescence proteins (BFP and GFP, respectively) by a $[Ca^{2+}]_4$/calmodulin binding site. In the absence of $[Ca^{2+}]_4$/calmodulin the flexible binding domain allows FRET between BFP and GFP. When present, $[Ca^{2+}]_4$/calmodulin binds to its normal site, activating MLCK, and also binds to a site on the indicator protein between BFP and GFP, disrupting FRET. Hence, activation of MLCK is associated with a substantial change of fluorescence from green, the acceptor, to blue, the donor. This appears to be a favorable possibility for cross-correlation measurement. The binding of $[Ca^{2+}]_4$/calmodulin to MLCK is, however, likely to be in equilibrium in cells, and therefore is not a good prospect for measuring NESS fluxes. A similar approach has been used to construct calcium biosensors called "chameleons" based on calmodulin [30]. Enzymes that are activated by phosphorylation in NESS states are continually being phosphorylated by ATP catalyzed by kinases and being dephosphorylated by phosphatases. The activity of such an enzyme is determined by the level of ATP and the activity of the kinases as well as that of the phosphatases in an energy-consuming NESS. MAP kinase cascades in which a sequence of kinases are successively phosphorylated and activated by other enzymes of the cascades are important examples of this sort of NESS. A FRET-based derivative of the MAP kinase ERK2 is an example of an enzyme that responds to phosphorylation by changing fluorescence color [31] and so could, in principle, be susceptible to the sort of FCS-based flux measurements described earlier. This particular example, however, may not have a sufficiently large FRET change to allow accurate measurements.

6.5 Summary and Conclusions

Metabolic and signaling networks of biochemical reactions provide the molecular bases of cellular phenotypes. Many of the reactions in these networks are maintained in NESS states by chemical constraints that require the expenditure of chemical free energy. Nonlinear reaction networks can have more than one steady state compatible with a given set of chemical constraints. For reactions that have small numbers of reactant molecules, the system can fluctuate among the available steady states. To understand the behavior of a system, it is important to know the states that are available to it and also to understand how the cell undergoes transitions among these available steady states. The number of states that are available under a given set of chemical constraints depends on the concentrations of the reactants and the rate constants of the reactions. These are difficult to determine experimentally. Also, it is difficult for computational analyses of networks to sample a large portion of the parameter space available to a system. Therefore, it would be helpful to have a direct experimental approach to characterizing NESS states of reaction networks. FCS provides one potentially useful approach. Two-color cross correlation can yield a direct measurement of reaction fluxes. FCS measurements can be carried out in the cytoplasm of both prokaryotic and eukaryotic cells. It is crucial, however, to have fluorescent reactant molecules that sensitively indicate the reaction progress by a change in fluorescence color. One promising approach would be to develop FRET probes for this purpose. The greatest challenge for the implementation of this approach is the selection or construction of these probes.

References

1. P.H. Sugden, Mechanotransduction in cardiomyocyte hypertrophy. Circulation 103(10), 1375–1377 (2001)
2. J.J. Tomasek, G. Gabbiani, B. Hinz, C. Chaponnier, R.A. Brown, Myofibroblasts and mechano-regulation of connective tissue remodelling. Nat. Rev. Mol. Cell Biol. 3(5), 349–363 (2002)
3. J.B. Bassingthwaighte, The modeling for a primitivesustainablecell. Philos. Trans. R. Soc. Lond. A 359, 1055–1072 (2001)
4. H. Qian, Phosphorylation energy hypothesis: open chemical systems and their biological functions. Annu. Rev. Phys. Chem. 58, 113–142 (2007)
5. H. Qian, S. Saffarian, E.L. Elson, Concentration fluctuations in a mesoscopic oscillating chemical reaction system. Proc. Natl. Acad. Sci. U S A 99(16), 10376–10381 (2002)
6. M. Vellela, H. Qian, Stochastic dynamics and non-equilibrium thermodynamics of a bistable chemical system: the Schlogl model revisited. J. R Soc. Interface. (doi:10.1098/rsif.2008.0476) (2008)
7. J. Ansel, H. Bottin, C. Rodriguez-Beltran, C. Damon, M. Nagarajan, S. Fehrmann, J. Francois, G. Yvert, Cell-to-cell stochastic variation in gene expression is a complex genetic trait. PLoS Genet 4(4), e1000049 (2008)

8. P. Ao, D. Galas, L. Hood, X. Zhu, Cancer as robust intrinsic state of endogenous molecular-cellular network shaped by evolution. Med. Hypotheses. 70(3), 678–684 (2008)

9. C.V. Rao, D.M. Wolf, A.P. Arkin, Control, exploitation and tolerance of intracellular noise. Nature 420(6912), 231–237 (2002)

10. S.A. Kauffman, Metabolic stability and epigenesis in randomly constructed genetic nets. J. Theor. Biol. 22(3), 437–467 (1969)

11. J.J. Tyson, K. Chen, B. Novak, Network dynamics and cell physiology. Nat. Rev. Mol. Cell Biol. 2(12), 908–916 (2001)

12. J.J. Tyson, K.C. Chen, B. Novak, Sniffers, buzzers, toggles and blinkers: dynamics of regulatory and signaling pathways in the cell. Curr. Opin. Cell Biol. 15(2), 221–231 (2003)

13. F. Schlogl, Chemical reaction models for non-equilibrium phase transition. Z. Physik. 253, 147–161 (1972)

14. B. Han, J. Wang, Quantifying robustness and dissipation cost of yeast cell cycle network: the funneled energy landscape perspectives. Biophys. J. 92(11), 3755–3763 (2007)

15. F. Li, T. Long, Y. Lu, Q. Ouyang, C. Tang, The yeast cell-cycle network is robustly designed. Proc. Natl. Acad. Sci. U S A 101(14), 4781–4786 (2004)

16. R. Zhu, A.S. Ribeiro, D. Salahub, S.A. Kauffman, Studying genetic regulatory networks at the molecular level: delayed reaction stochastic models. J. Theor. Biol. 246(4), 725–745 (2007)

17. K.C. Chen, L. Calzone, A. Csikasz-Nagy, F.R. Cross, B. Novak, J.J. Tyson,Integrative analysis of cell cycle control in budding yeast. Mol. Biol. Cell 15(8), 3841–3862 (2004)

18. A. Ciliberto, B. Novak, J.J. Tyson,Mathematical model of the morphogenesis checkpoint in budding yeast. J. Cell Biol. 163(6), 1243–1254 (2003)

19. H.H. McAdams, A. Arkin, Stochastic mechanisms in gene expression. Proc. Natl. Acad. Sci. U S A 94(3), 814–819 (1997)

20. E.M. Ozbudak, M. Thattai, I. Kurtser, A.D. Grossman, A. van Oudenaarden, Regulation of noise in the expression of a single gene. Nat. Genet. 31(1), 69–73 (2002)

21. J. Yu, J. Xiao, X. Ren, K. Lao, X.S. Xie, Probing gene expression in live cells, one protein molecule at a time. Science 311(5767), 1600–1603 (2006)

22. M.B. Elowitz, A.J. Levine, E.D. Siggia, P.S. Swain, Stochastic gene expression in a single cell. Science 297(5584), 1183–1186 (2002)

23. D.A. Beard, H. Qian, Chemical Biophysics: Quantitative Analysis of Cellular Systems (Cambridge University Press, New York, 2008)

24. H. Qian, E.L. Elson, Fluorescence correlation spectroscopy with high-order and dual-color correlation to probe nonequilibrium steady states. Proc. Natl. Acad. Sci. U S A 101(9), 2828–2833 (2004)

25. E. Elson, D. Magde, Fluorescence correlation spectroscopy. I. conceptual basis and theory. Biopolymers 13, 1–27 (1974)

26. D. Magde, E.L. Elson, W.W. Webb, Thermodynamic fluctuations in a reacting system – measurement by fluorescence correlation spectroscopy. Phys. Rev. Lett. 29, 705–708 (1972)

27. D. Magde, E.L. Elson, W.W. Webb, Fluorescence correlation spectroscopy. II. An experimental realization. Biopolymers 13(1), 29–61 (1974)

28. T.L. Chew, W.A. Wolf, P.J. Gallagher, F. Matsumura, R.L. Chisholm,A fluo-
    rescent resonant energy transfer-based biosensor reveals transient and regional
    myosin light chain kinase activation in lamella and cleavage furrows. J. Cell Biol.
    156(3), 543–553 (2002)
29. E. Isotani, G. Zhi, K.S. Lau, J. Huang, Y. Mizuno, A. Persechini, R. Geguchadze,
    K.E. Kamm, J.T. Stull, Real-time evaluation of myosin light chain kinase acti-
    vation in smooth muscle tissues from a transgenic calmodulin-biosensor mouse.
    Proc. Natl. Acad. Sci. U S A 101(16), 6279–6284 (2004)
30. A. Miyawaki, J. Llopis, R. Heim, J.M. McCaffery, J.A. Adams, M. Ikura, R.Y.
    Tsien, Fluorescent indicators for Ca2+ based on green fluorescent proteins and
    calmodulin. Nature 388(6645), 882–887 (1997)
31. A. Fujioka, K. Terai, R.E. Itoh, K. Aoki, T. Nakamura, S. Kuroda, E. Nishida,
    M. Matsuda, Dynamics of the Ras/ERK MAPK cascade as monitored by
    fluorescent probes. J. Biol. Chem. 281(13), 8917–8926 (2006)
32. K.C. Chen, A. Csikasz-Nagy, B. Gyorffy, J. Val, B. Novak, J.J. Tyson. Kinetic
    analysis of a molecular model of the budding yeast cell cycle. Mol Biol Cell
    11(1), 369–391 (2000)

In Vivo Fluorescence Correlation
and Cross-Correlation Spectroscopy

Jörg Mütze, Thomas Ohrt, Zdeněk Petrášek, and Petra Schwille

Summary. In this manuscript, we describe the application of Fluorescence Correlation Spectroscopy (FCS), Fluorescence Cross-Correlation Spectroscopy (FCCS), and scanning FCS (sFCS) to two in vivo systems. In the first part, we describe the application of two-photon standard and scanning FCS in Caenorhabditis elegans embryos. The differentiation of a single fertilized egg into a complex organism in C. elegans is regulated by a number of protein-dependent processes. The oocyte divides asymmetrically into two daughter cells of different developmental fate. Two of the involved proteins, PAR-2 and NMY-2, are studied. The second investigated system is the mechanism of RNA interference in human cells. An EGFP based cell line that allows to study the dynamics and localization of the RNA-induced silencing complex (RISC) with FCS in vivo is created, which has so far been inaccessible with other experimental methods. Furthermore, Fluorescence Cross-Correlation Spectroscopy is employed to highlight the asymmetric incorporation of labeled siRNAs into RISC.

Fluorescence Correlation Spectroscopy is not a single molecule method in strict sense, because it relies on drawing statistically relevant data on molecular systems from analyzing fluctuations around a mean (of molecular concentration or brightness) at thermodynamic equilibrium [1, 2]. This can only be achieved by averaging over many hundreds or thousands of molecular events per measurement. Thus, it distinguishes itself from techniques such as single molecule imaging or manipulation discussed elsewhere in this volume. However, the optical setup that was employed in the early 1990s for FCS, confocal illumination, and detection of fluorophores in a less than femtoliter volume laid the ground for many single molecule applications to come [3, 4], particularly the analysis of intramolecular fluctuations by FRET [5] or the identification of FRET-active subpopulations on the basis of fluorescence burst analysis [6, 7]. Therefore, it is fair to say that FCS is a method with single molecule sensitivity and potential, and instrumentally supports the other optical single molecule techniques. The true virtue of FCS is the comparative ease of carrying out measurements in any environment, compared with tedious individual-molecule tracking methods or data processing-intensive burst analysis, for which reason it belongs to the single molecule-like methods that have nowadays been used most frequently in biology labs. In particular, in conjunction with dual-color cross-correlation FCCS, [8] which allows to quantitatively analyze molecular interactions, FCS gives direct access to parameters of fundamental biological relevance, such as molecular concentrations, mobility coefficients, and binding rates. As the life sciences advance to ever higher levels of complexity, from cells to organisms and from molecules to molecular networks, there is a large need for a relatively simple method that nevertheless allows to draw quantitative information about molecular dynamics and interactions. Our hypothesis, to be supported by examples of ongoing work in our laboratory, is that FCS/FCCS is a method with enormous potential for cell and developmental biology, which should definitely complement the future imaging facilities of biological centers, because it delivers information not directly accessible to any of the established imaging techniques. A field that will, to our belief, benefit significantly from FCS is the field of developmental biology, i.e., the characterization of cell and tissue polarization and morphogenesis, because in addition to structural features, it is primarily molecular concentration gradients and distribution characteristics that need to be quantified. For this reason, one of the examples of ongoing work in our lab will focus on the measurement of polarization-controlling proteins in early C. elegans embryos.

After fertilization of the oocyte, the single cell embryo divides asymmetrically, resulting in two daughter cells of different sizes and different developmental potential. The symmetry of the oocyte is broken by the sperm-derived centrosome, and results in a polarization of the embryo along the anterior–posterior axis. Among the factors regulating this division are PAR proteins, whose asymmetric distribution is supposed to depend on the contraction of the actomyosin network. FCS applications in C. elegans are relatively straightforward due to the transparency of the animal; however, in deeper cell layers, the employment of two-photon FCS is beneficial [9]. In addition, characterization of very slowly moving particles, as in our case, the PAR-2 protein on the cortex of the oocyte, and, later, the embryo, requires additional instrumental features such as a moving (or scanning) illumination beam in order to prevent or limit photobleaching [10,11]. For this reason, FCS applications in C. elegans will be carried out with a circularly scanned laser beam. We will demonstrate the ability to quantify PAR-2 mobility and clearly distinguish it from cortical activity, displayed by myosin NMY-2, showing how factors other than cortical flow regulate the polarization of cells.

As a very recent example for the virtue of cross-correlation (FCCS) in determining molecular interactions, we give some insight in our ongoing work on elucidating the mechanisms or RNA interference on a single molecule level. RNA interference (RNAi) is an evolutionary conserved posttranscriptional gene-silencing mechanism by which short double-stranded RNAs inhibit the expression of genes by the specific degradation of RNA molecules of complementary sequence. This highly potent and specific phenomenon possesses great potential not only as widely used lab tool to study the function of specific genes, but also as a novel approach for medical therapies. The short double-stranded RNAs are of either exogenous (siRNA) or endogenous (miRNA) origin. To act as triggers for mRNA degradation, they are incorporated into a ribonucleoprotein effector complex, known as the RNA-induced silencing complex (RISC). Only one of the two strands, termed guide strand, is assembled into the endonucleolytically active component of RISC, the protein Argonaute2 (Ago2). The guide strand is defined as the strand with the lower thermodynamic stability at the 5-end of the duplex. The other strand, termed passenger strand, is released from the complex and gets degraded. RISC identifies target mRNAs based on complementary base pairing between the guide strand and the mRNA. The target mRNA is bound by hybridization, and gene expression is silenced by either endonucleolytic cleavage of the mRNA or translational repression.

7.1 Fluorescence Correlation and Cross-Correlation Spectroscopy

Setup and principle of FCS and FCCS have been reviewed extensively previously [12, 13]. The technique is based on the statistical analysis of equilibrium fluorescence fluctuations induced by, e.g., the dynamics of fluorescent molecules in a tiny observation volume. By correlating these fluctuations with itself at a later time τ, an autocorrelation curve is obtained, which can be fitted to an appropriate model function to extract the characteristic time scales of the system. The two basic parameters of a FCS autocorrelation curve are the decay time, reflecting time scales of molecules dynamics, and the amplitude, indicating the average number of particles in the detection volume.

A basic FCS/FCCS setup is depicted in Fig. 7.1a. A laser beam is guided by a dichroic mirror onto the back aperture of a high numerical aperture objective, creating an optically defined, diffraction-limited open observation volume in the sample. The emitted fluorescence is collected by the same objective and is separated from the excitation light by a dichroic mirror and an additional emission filter. In case of two-photon excitation, an axial confinement is inherent, while for excitation with continuous wave lasers, a pinhole in the image plane is necessary to reach a small and restraint detection volume. The signal is recorded by detectors with single-photon sensitivity, usually avalanche photodiodes (APDs), and correlated with a hard- or software correlator. The recorded fluorescence signal is analyzed according to the autocorrelation function illustrated in Fig. 7.1b and fitted with the appropriate model. For dual color FCCS, a second laser beam of different wavelength is coupled into the microscope via a second dichroic beamsplitter. The two excitation lines are focused on the same diffraction limited spot, with the longer wavelength excitation volume being larger due to the wavelength dependent diffraction limit. The emitted fluorescence of the two spectrally different dyes

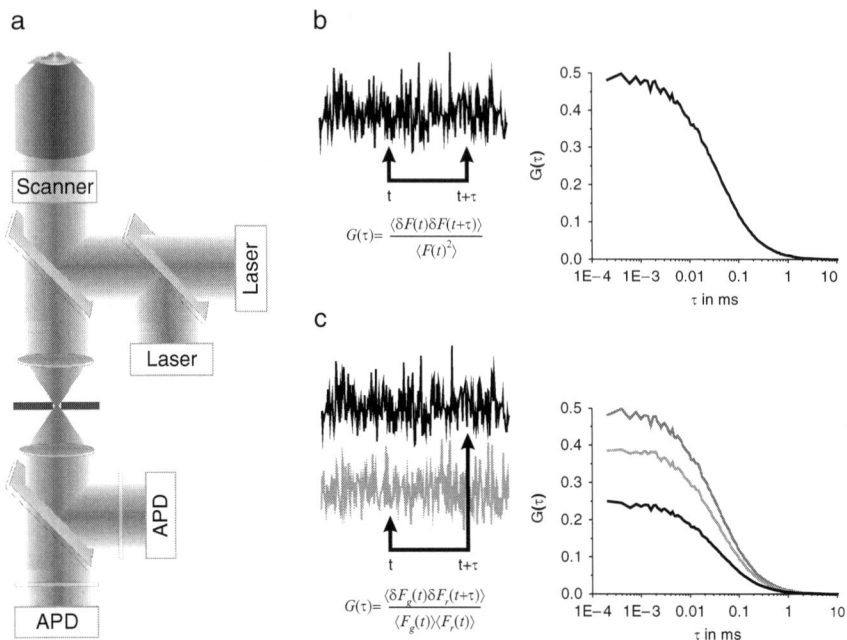

Fig. 7.1. Setup and principle of FCS. (a) Typical setup for dual-color FCCS. Two laser beams of different wavelength are merged via a dichroic mirror. A second dichroic directs the light via a scanning module onto the back aperture of a high N.A. objective. The laser light is focused inside the sample, and the emitted fluorescence light is collected by the same objective. The longer wavelength fluorescence light passes the dichroic mirror and is focused onto a confocal pinhole. Another dichroic beamsplitter splits the fluorescence in two spectral channels. Further emission filters in front of the avalanche photodiode detectors eliminate any residual laser light. (b) For FCS, the fluorescence fluctuations from one channel are temporally autocorrelated giving rise to a correlation function, which can be fitted to a model function. (c) In FCCS experiments, two spectrally separated fluorescence signals are recorded and their auto- and cross-correlation result in three curves, one autocorrelation curve for each spectral channel and a cross-correlation curve

is divided by an additional dichroic mirror. The two channels are correlated with each other, resulting in a cross-correlation curve. Only when a dual-labeled molecule translocates through the detection volume, it contributes to the cross-correlation curve. Hence, the amplitude of the cross-correlation curve is directly proportional to the concentration of the double labeled species and allows a quantitative analysis of interactions in vivo. Therefore, these techniques allow to determine the size of analyzed molecules as well as the interaction between two labeled molecules.

7.2 Two-Photon Scanning FCS in C. elegans embryos

The Caenorhabditis elegans embryo divides from a single cell asymmetrically into two blastomeres, each with a different developmental fate. The sperm-derived centrosome initiates an axis of polarity inside the embryo, followed by movements of the actomyosin cortex. Anterior and posterior cortical domains are established defining the polarity of the embryo. These cortical domains are associated with the asymmetric localization of PAR proteins [14, 15]. From a uniform distribution, these proteins rearrange dramatically, localizing to either the anterior or posterior half of the cortex [16,17]. Then, the first division of the oocyte occurs along the anterior–posterior axis. This division is asymmetric, producing two daughter cells which undergo further repeated asymmetric cell divisions [18].

The localization and dynamics of the involved proteins can be studied with fluorescence microscopy on a sub-second time scale, with faster dynamics, however, remaining unresolved. Fluorescence correlation spectroscopy can access molecular dynamics on a much faster timescale, providing information on diffusion properties and absolute concentrations. Fluorophore excitation by the simultaneous absorption of two photons, termed two-photon excitation, provides numerous advantages over conventional one-photon excitation. The illumination is restricted to the focal vicinity, reducing background and confining photobleaching to the focal region. This is crucial, especially when performing FCS measurements in spatially confined volumes such as a single cell, wherein with one photon excitation the reservoir of dye molecules can be depleted during the experiment. Furthermore, the distribution of potentially toxic excitation light over a large part of the embryo in one-photon FCS can be disadvantageous for the further development of the embryo [19].

So far, it is not fully understood how the asymmetric PAR protein distribution is connected with the reorganization of the actomyosin cortex. Here, we have analyzed the polarity protein PAR-2 and a component of the actomyosin cortex, the non-muscle myosin NMY-2, in vivo with standard and scanning FCS inside the cytosol and on the cortex of C. elegans embryos.

7.2.1 Scanning FCS

Fluorescence correlation spectroscopy with a stationary measurement volume can be prone to artifacts such as photobleaching of slowly moving or immobile molecules. Slow molecular motion also requires long measurement times in order to increase the statistical accuracy. These drawbacks can be circumvented by moving the excitation volume relative to the sample along a defined path. The total laser light dose in a single measurement point is reduced, and the depleted concentration due to bleaching can recover before reexcitation in the following scan cycle. Several types of scanning have been implemented, such as scanning in a line, raster or circular fashion [20–22]. Here, we employ scanning FCS in a circle of different radii on the cortex of a C. elegans embryo.

When scanning on a two-dimensional surface with radius R and angular frequency ω, the model autocorrelation function for free diffusion can be written as:

$$G(\tau) = \frac{G_0}{1 + \frac{\tau}{\tau_D}} e^{-\frac{R^2 \sin^2(\omega\tau/2)}{\omega_0^2(1+\tau/\tau_D)}} \tag{7.1}$$

with the amplitude of the correlation function G_0, the diffusion time τ_D, and the size of the detection volume ω_0. The exponential term expresses the periodic motion of the scanning beam. Peaks in the resulting oscillating auto-correlation curve at integer multiples of the scan period T relate to the usual autocorrelation curve.

The setup is very similar to that described in Fig. 7.1a, with the exception that here a tunable Ti:Sapphire laser (Mira 900-F, Coherent) was coupled via two galvanometer scanners into a Olympus IX71 fluorescence microscope equipped with a home built detection unit [23]. An Olympus 60x W/IR objective focused the light into the sample and collected the fluorescence light. After passing a dichroic mirror it was detected by an avalanche photodiode (SPCM-CD2901, PerkinElmer). Average laser power was 5 mW before entering the objective at a wavelength of 920 nm. Scan radii ranged in between 2 and 9 μm at a scan frequency of 300 Hz. The collected autocorrelation curves were fitted to the model in (7.1) for free diffusion on a two-dimensional surface.

Preparation of the worms and transfection of GFP::PAR-2 and NMY-2::GFP are described elsewhere [24]. Fertilized eggs were placed in agarose gel in between two coverslides. Experiments were performed either with the objective focus on the equatorial plane to measure in the cytoplasm or with the focus near the lower coverslip to measure on the cortex.

7.2.2 FCS in the Cytoplasm

Single cell C. elegans embryos transfected with GFP::PAR-2 and NMY-2::GFP were imaged using two-photon excitation (Fig. 7.2a, b). Both proteins were present in the cytoplasm and on the cortex, exhibiting characteristic distributions during different stages of development as reported in the literature [25]. In the presented case, PAR-2 localizes to the posterior cortex after the establishment of anterior–posterior polarity. Standard two-photon FCS in a single location inside the cytoplasm of the embryo was performed to investigate the two proteins. The obtained autocorrelation functions were fitted to a model for free diffusion in three dimensions. Histograms of the measured diffusion coefficients are depicted in Fig. 7.2c, d. Since the autocorrelation curves can be fitted to a model for free diffusion in three dimensions, the motion of the two proteins seems to be mainly governed by free diffusion. The differences in the diffusion coefficients indicate that PAR-2 and NMY-2 do not diffuse together in one complex in the cytoplasm.

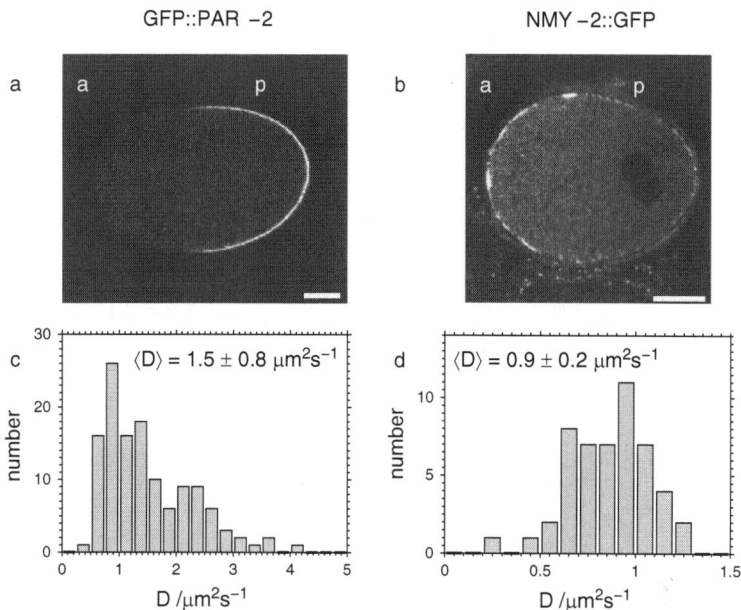

Fig. 7.2. GFP::PAR-2 (**a**) and NMY-2::GFP (**b**) imaged before the first cell division. Histograms of measured diffusion coefficients of (**c**) GFP::PAR-2 and (**d**) NMY-2::GFP with FCS in the cytoplasm of C. elegans embryos. Scale bar represents 10 μm. a: anterior, p: posterior

7.2.3 sFCS in the Cortex

Measurements of GFP::PAR-2 and NMY-2::GFP on the cortex using conventional FCS failed because of motion of the living embryo. The developing embryo causes motions that cannot be separated from the slow motion of proteins on the cortex. Shorter measurement times and the low mobility of proteins on the cortex result in a statistical accuracy insufficient for FCS. Scanning FCS probes many different measurement volumes and can therefore increase the number of probed molecules, increasing the statistical accuracy. Photobleaching is also minimized since the scanning beam is spending only a fraction of the measurement time in one location. Experiments were performed on the flattened bottom part of the embryo by scanning in a circle at the largest possible radius. Larger radii correspond to an increased number of independent probed volumes.

Autocorrelation curves for GFP::PAR-2 and NMY-2::GFP on the cortex are depicted in Fig. 7.3. The peaks in the autocorrelation curve correspond to the scan beam returning to the same measurement volume after $T = 1/300$s and form the autocorrelation at the same location after each cycle. The diffusion behavior of GFP::PAR-2 differs from that of NMY-2::GFP on the cortex. Free two-dimensional diffusion cannot describe the behavior of GFP::PAR-2

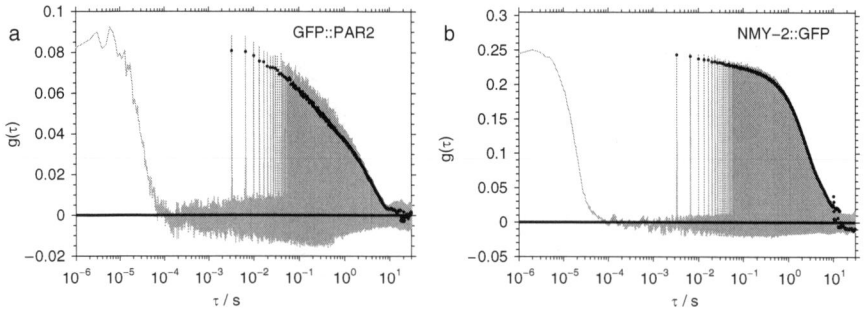

Fig. 7.3. Autocorrelation curves from sFCS measurements of GFP::PAR-2 (**a**) and NMY-2::GFP (**b**) on the cortex. Peak amplitudes are denoted by *black dots*

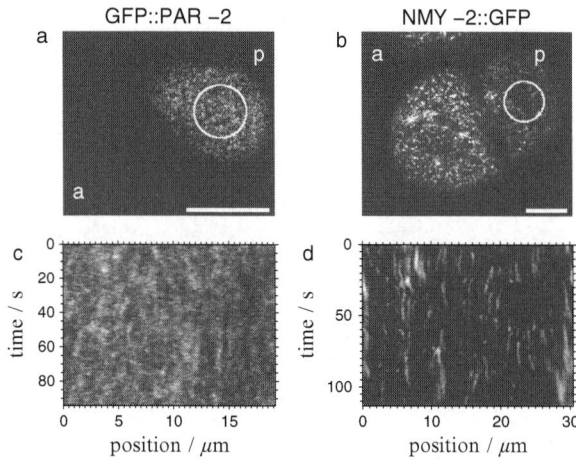

Fig. 7.4. GFP::PAR-2 (**a**, **c**) and NMY-2::GFP (**b**, **d**) on the cortex. The scan path of sFCS measurements is indicated by a *white circle*. (**c**) and (**d**) indicate the non-uniform fluorescence pattern of GFP::PAR-2 and NMY-2::GFP. Fluorescence intensity is plotted as a 2D-plot, where one horizontal line represents one full circular scan with each subsequent cycle displayed on the vertical axis from *top* to *bottom*. Scale bar represents 10 μm. a: anterior, p: posterior

on the cortex but rather resembles that of anomalous diffusion or a multi-component model. The decay of the autocorrelation function occurs faster for NMY-2::GFP than for GFP::PAR-2, indicating a faster movement of the proteins. The decay is steeper than for two-dimensional diffusion, possibly due to the contribution of translational motion.

Figure 7.4 depicts the nonuniform distribution of GFP::PAR-2 and NMY-2::GFP on the cortex. To illustrate the movements of the proteins, the fluorescence intensity was plotted in a two-dimensional graph with one horizontal line corresponding to one revolution of the scanner, with each subsequent scan plotted below the previous (Fig. 7.4c, d). For NMY-2::GFP, coordinate motion is visible in features moving parallel to the vertical axis. GFP::PAR-2 exhibits less discrete spots, and, therefore, less of this coordinate motion is visible.

7.2.4 Discussion

Two photon FCS and sFCS have been employed to analyze the polarity protein PAR-2 and the actomyosin cortex protein NMY-2 in the optically challenging system of a developing C. elegans embryo. Drawbacks of conventional FCS, such as photobleaching and low statistical accuracy, were overcome by scanning the beam, making it possible to analyze the dynamics of slowly moving molecules on the cortex of the embryo. Since each position along the scan path is only excited for a small fraction of the scan period, photobleaching is minimized and could not be observed during the measurement time (Fig. 7.4c, d). By probing numerous measurement volumes along the scan path, the statistical accuracy for slowly moving molecules, such as proteins on a membrane, is significantly increased.

Imaging of the labeled embryos displayed the known distribution of the polarity involved protein PAR-2 and the actomyosin cortex protein NMY-2 during different stages of development. FCS measurements in the cytoplasm yielded autocorrelation curves for GFP::PAR-2 with a diffusive or subdiffusive behavior. The diffusion coefficient for NMY-2::GFP is smaller, indicating much slower diffusion. The autocorrelation functions of NMY-2::GFP exhibit a sharp decay of the autocorrelation function, suggesting contributions of directed flow. Comparison of the diffusion coefficient of PAR-2 and NMY-2 with a reference protein of similar size in the cytoplasm indicates that both proteins are part of a larger complex or multimerize (data not shown).

PAR-2 seems to be uncoupled from movement of NMY-2, a member of the actomyosin cortex, as observed in sFCS measurements on the cortex. Further investigations into the different cortical elements and polarity proteins are necessary to elude the mechanisms of asymmetric cell division in developing C. elegans embryos. FCS and sFCS are two techniques capable of investigating the dynamics of the involved proteins, providing a complementary approach to existing techniques.

7.3 FCS and FCCS Elucidate the RNA Interference Pathway

RNA interference (RNAi) is a physiological mechanism that inhibits gene expression by the specific degradation or repression of messenger RNA in virtue of the RNA-induced silencing complex (RISC). Argonaute proteins are the catalytic components of RISC. A single strand of a small interfering RNA is assembled into Argonaute2 (Ago2), providing the complementary nucleotide sequence to a specific messenger RNA (mRNA). Ago2 is then silencing gene expression by either endonucleolytic cleavage of the mRNA or translational repression.

Recent studies have shown that next to the well-characterized cytoplasmic mechanisms also nuclear functions of Argonaute proteins exist [26]. So far, the investigation into the nuclear RISC complexes by biochemical approaches was difficult due to the contamination of nuclear extracts with cytoplasmic RISC, caused by their association with the nuclear envelope. Furthermore, the very low endogenous expression levels in the nucleus, compared to the cytoplasm, do not allow for specific nuclear localization studies by immunofluorescence approaches. Altogether, the concomitant analysis of the nuclear and cytoplasmic localized RISC was so far not possible [27].

On this basis, we developed an FCS/FCCS platform to study nuclear and cytoplasmic RNAi pathways to further characterize the relationship, if it actually exists, of the nuclear and cytoplasmic RISC. To study the mechanism of RNA Interference in human cells in vivo, we created stable EGFP–Ago2 expressing cell line, based on ER293 cells, in which the effector protein of RNAi, Ago2, was tagged with the fluorescent protein EGFP. It has been shown that labeling Ago2 on the N-terminus with EGFP has no effect on its silencing efficiency [28, 29]. The cell line exhibited EGFP–Ago2 expression levels similar to the endogenous expression of Ago2, a concentration level suitable for FCS measurements.

As previously reported, confocal imaging showed the primarily localization of Ago2 (Fig. 7.5a) to the cytoplasm [29]. The biochemical characterization showed that the EGFP–Ago2 complex contained endogenous miRNAs,

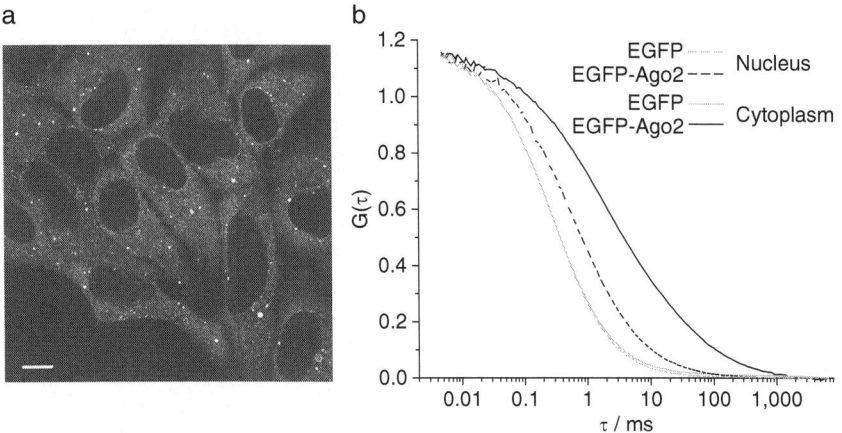

Fig. 7.5. EGFP–Ago2 cell line. (**a**) Image of several 10G cells. Scale bar: 10 μm. (**b**) FCS curves of EGFP–Ago2 taken in the cytoplasm (*black line*) and nucleus (*black dashed line*) of 10G cells. Curves are normalized and the average of at least ten measurements. *Grey line* and *grey dashed line* indicate measurements of EGFP in ER293 cells as control

as observed by immunoprecipitation studies and a coimmunoprecipitation revealed that EGFP–Ago2 is associated with Dicer, a member of the RISC loading complex. Compared to the endogenous Ago2 level, 10G cells showed a 2–2.5 higher expression level of EGFP–Ago2 (data not shown). These results demonstrate that EGFP–Ago2, expressed in the generated cell line 10G, mimics the natural behavior of endogenous Ago2 with the possibility to analyze the localization, concentration, and diffusion properties in the living cell with FCS and FCCS.

7.3.1 FCS and FCCS Methods

All measurements were performed on a commercial LSM510 and a Confo-Cor3 (Zeiss, Jena, Germany). Before entering the back aperture of a Zeiss C-Apochromat 40x, N.A. = 1.2, water immersion objective, the laser lines were set to a power of $3.5\,kW\,cm^{-2}$ (488 nm) and $1.05\,kW\,cm^{-2}$ (633 nm). The confocal pinhole was adjusted to 70 μm. The main beam splitter (488/633 dichroic mirror) and a 505 nm longpass emission filter (FCS) or a second dichroic beamsplitter (LP635) and a 505–610 nm bandpass and 655 nm longpass (FCCS) removed any residual excitation light. A calibration with the known diffusion coefficient of Alexa Fluor 488 (Invitrogen, Karlsruhe, Germany) and Atto 655 (ATTO-TEC GmbH, Siegen, Germany) resulted in a $1/e^2$ lateral radius of the detection volume of ∼0.19 μm for 488 nm excitaion [30] and ∼0.24 μm for 633 nm excitaion [31]. A maximum cross-correlation of $(80.04 \pm 0.13)\%$ of the setup was determined with a reference sample made of a Rhodamine 6G and Cy5 labeled 30nt 2-O-Methyl modified dsRNA (IBA GmbH, Germany).

ER293 and 10G cells were cultured at 37°C in DMEM (high glucose, Sigma) with 10% fetal calf serum, 2 mM glutamine, $0.3\,mg\,ml^{-1}$ G418, and $0.4\,mg\,ml^{-1}$ Hydromycin B (only 10G cells). Cell measurements were performed at room temperature in air-buffer [32]. Labeled siRNAs were delivered into 10G cells via microinjection, providing a defined starting point of the experiment, ensuring a known concentration and a homogenous distribution within the cell. Cells were imaged by laser scanning microscopy, and the laser beam was parked 1 μm above the lower membrane in either the nucleus or the cytoplasm of a individual cell for FCS. Each measurement lasted eight times 30 s, at an average molecular brightness of ∼4 kHz in each channel. Individual runs that differed significantly from the average due to a strong change in fluorescence were discarded from analysis. The obtained autocorrelation curves were fitted to a two component FCS model. The structural parameter and the blinking dynamics of EGFP and Cy5 were determined in calibration measurements to 130 μs for EGFP and 70 μs for Cy5. Results were corrected for background, caused by autofluorescence, by measuring the count rate in ER293 cells lacking EGFP [9]. All mentioned errors are the standard error of the mean.

7.3.2 FCS Results

So far it has not been possible to determine the size and mass of the human RISC complex in vivo. The literature states molecular weights from 160 kDa to ~2 MDa, depending on the experimental procedure applied in vitro [33,34]. Fluorescence correlation spectroscopy offers the possibility to investigate the diffusion properties of fluorescing molecules, and therefore to deduce their molecular weight via the Stokes-Einstein Relation [35].

To rule out effects of different viscosity in the cytoplasm and the nucleus, ER293 cells were transfected with EGFP and analyzed via FCS. Autocorrelation curves obtained from the cytoplasm yielded a diffusion coefficient of $D_{EGFP(Cyt)} = (25.5 \pm 0.9) \, \mu m^2 \, s^{-1}$, similar to the diffusion coefficient of EGFP in the nucleus $D_{EGFP(Nuc)} = (24.5 \pm 0.5) \, \mu m^2 \, s^{-1}$, as reported in the literature [36]. Measurements on EGFP–Ago2 in 10G cells resulted in two different diffusion coefficients of $D_{EGFP-Ago2(Cyt)} = (5.4 \pm 0.2) \, \mu m^2 \, s^{-1}$ in the cytoplasm and $D_{EGFP-Ago2(Nuc)} = (13.7 \pm 0.5) \, \mu m^2 \, s^{-1}$ in the nucleus. Using the EGFP measurements as a reference, the molecular weights can be calculated from the diffusion coefficients. The obtained cytoplasmic molecular weight of RISC was (3.0 ± 0.6) MDa and (158 ± 26) kDa for RISC in the nucleus. The calculated value for RISC in the nucleus is within the margin of error to resemble the EGFP–Ago2 protein alone, while the molecular weight of RISC in the cytoplasm is ~20 times larger, representing a much larger and therefore different protein complex. So far the RNA-induced silencing complex was only accessed in size by complicated biochemical methods in vitro with a huge spread of determined molecular weights. This is the first time that RISC could be probed in its native environment and at equilibrium, in different compartments of the cell.

FCS also determines the particle number in the detection volume and, therefore, the absolute concentration of fluorescent molecules. The FCS measurements yielded a 4.2 ± 0.5 times higher abundance of EGFP-Ago2 in the cytoplasm than in the nucleus. A similar result was obtained by the quantification via imaging.

7.3.3 FCCS Results

The asymmetric incorporation of only one strand of the double stranded siRNA or miRNA can be studied employing dual color FCCS. It is known that the strand with the less tightly paired 5'-end is transferred into the core protein of RISC, Ago2, while the other strand is released from the complex [37,38]. This thermodynamic asymmetry of the siRNA or miRNA defines the guide strand, which later acts as a template for RISC to identify its homologous target. By delivering Cy5-labeled siRNAs via microinjection into 10G cells, it is possible to monitor the incorporation into EGFP–Ago2 of either the labeled guide or the labeled passenger strand in vivo. The siRNA TK3, targeting the mRNA of Renilla luciferase, was labeled covalently on the 3'-end of the

guide or passenger strand with Cy5. It has been shown that the labeling on the
3'-end has no effect on the siRNAs silencing efficiency [32]. Cross-correlation
measurements of the labeled passenger strand yielded cross-correlation amplitudes below 5%, in the cytoplasm and nucleus for various time points up to
12 h after microinjection (Fig. 7.6a). The passenger strand is excluded from
RISC, and therefore no significant cross-correlation is present. In contrast, the
labeled guide strand exhibited a steady increase in cross-correlation after up
to 6 h of incubation (Fig. 7.6b). The labeled passenger strand could not be
detected for incubation times longer than 12 h, because of the higher degradation of the strand, whereas the labeled guide strand could be detected up
to 48 h. This specific incorporation of the guide strand was observed for several different siRNAs with a similar time dependence (data not shown). The
labeled guide strand also displayed cross-correlation amplitudes of up to 10%
in the nucleus, corroborating the findings of a functional nuclear RISC.

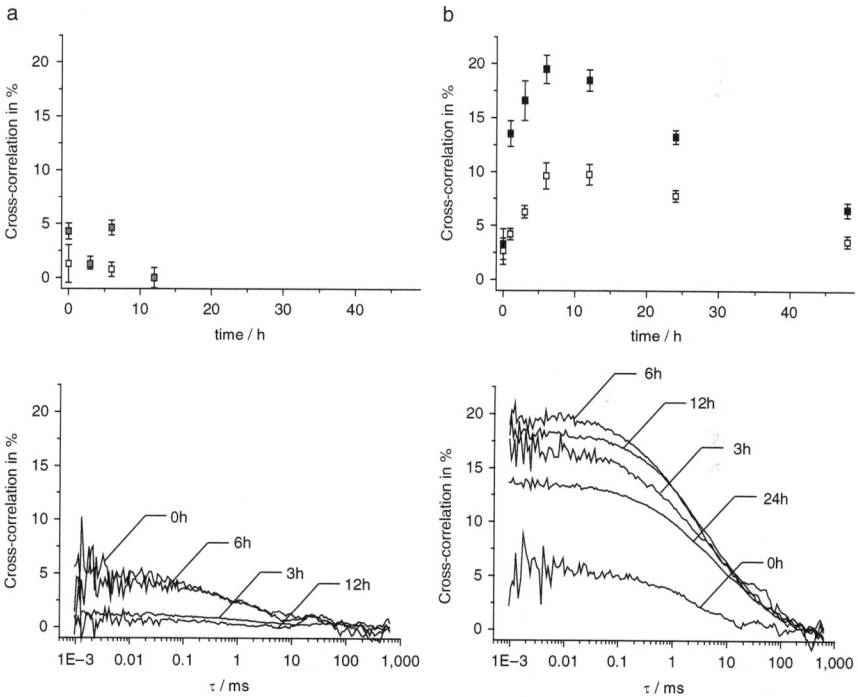

Fig. 7.6. Guide and passenger strand loading into RISC in the cytoplasm and
nucleus. Normalized cross-correlation in the cytoplasm (*filled boxes*) and nucleus
(*open boxes*) of EGFP–Ago2 and Cy5 labeled passenger (**a**) or guide strand (**b**) of
siTK3 for several time points after injection. The *bottom* graphs display the
corresponding cross-correlation curves

7.3.4 Discussion

Several features of FCS and FCCS proved to be advantageous when studying the dynamics of a protein and its interaction with other molecules of interest in their native environment. The noninvasiveness of these methods allow to access the diffusion characteristics and therefore the molecular weight of a complex in vivo, in different compartments of the cell. The single molecule sensitivity is favorable when studying proteins at expression levels similar to the endogenous level. Once the confocal volume of the FCS setup is calibrated with a reference dye of known diffusion coefficient, the absolute concentration of biomolecules can be determined. Recent implementation into commercial laser scanning microscopes provides extraordinary stability, crucial for experiments that rely on many repetitive measurements. For example, the time course for the labeled guide strand presented in Fig. 7.5b is the result of seven experiments, each consisting of at least 10 cells with 8 min acquisition time per cell. Advanced optics provides superior background suppression and a nearly perfect overlap of the excitation volumes for dual color experiments. Therefore, FCCS can be used to examine the binding of two spectrally different labeled species.

To summarize our findings, by fusing EGFP to the endonuclease Ago2, a stable and functional cell line was created, with expression levels suitable for FCS and FCCS. We could study the dynamics of the RNA-induced silencing complex in the cytoplasm and nucleus of human cells in vivo for the first time with FCS. Our data show that two different RISC are present in the cell, one large complex of $\sim$3 MDa in the cytoplasm and one smaller of $\sim$160 kDa in the nucleus, which most likely represents the Ago2 protein alone. It has been shown previously that RNAi also functions in the nucleus of human cells, by knocking down RNAs, only localized to the nucleus [26]. Our cross-correlation experiments showed that EGFP–Ago2 contained the labeled guide strand of an siRNA, thereby forming activated RISC in the nucleus. With cross-correlation amplitudes of up to 20% in the cytoplasm after 6 h of incubation, the amount of cross-correlation decreased for later times after microinjection. We were able to show in vivo that the RLC is able to detect the siRNA asymmetry and load only the guide strand into Ago2.

References

1. M. Ehrenberg, R. Rigler, Chem. Phys. **4**, 390–401 (1974)
2. D. Magde, W.W. Webb, E. Elson, Phys. Rev. Lett. **29**, 705 (1972)
3. M. Eigen, R. Rigler, Proc. Natl. Acad. Sci. U S A **91**(13), 5740–5747 (1994)
4. R. Rigler, U. Mets, J. Widengren, P. Kask, Eur. Biophys. J. Biophys. Lett. **22**, 169–175 (1993)
5. T. Ha, T. Enderle, D.F. Ogletree, D.S. Chemla, P.R. Selvin, S. Weiss, Proc. Natl. Acad. Sci. U S A **93**(13), 6264–6268 (1996)

6. A.A. Deniz, M. Dahan, J.R. Grunwell, T. Ha, A.E. Faulhaber, D.S. Chemla, S. Weiss, P.G. Schultz, Proc. Natl. Acad. Sci. U S A **96**(7), 3670–3675 (1999)

7. M. Margittai, J. Widengren, E. Schweinberger, G.F. Schröder, S. Felekyan, E. Haustein, M. König, D. Fasshauer, H. Grubmüller, R. Jahn, C.A.M. Seidel, Proc. Natl. Acad. Sci. U S A **100**(26), 15516–15521 (2003)

8. P. Schwille, F.J. Meyer-Almes, R. Rigler, Biophys. J. **72**(4), 1878–1886 (1997)

9. P. Schwille, U. Haupts, S. Maiti, W.W. Webb, Biophys. J. **77**(4), 2251–2265 (1999)

10. Z. Petrášek, P. Schwille, Biophys. J. **94**(4), 1437–1448 (2008)

11. J. Ries, P. Schwille, Biophys. J. **91**(5), 1915–1924 (2006)

12. K. Bacia, S.A. Kim, P. Schwille, Nat. Meth. **3**(2), 83–89 (2006)

13. O. Krichevsky, G. Bonnet, Rep. Prog. Phys. **65**, 251–297 (2002)

14. C.R. Cowan, A.A. Hyman, Annu. Rev. Cell. Dev. Biol. **20**, 427–453 (2004)

15. M. Schaefer, J.A. Knoblich, Exp. Cell. Res. **271**(1), 66–74 (2001)

16. R.J. Cheeks, J.C. Canman, W.N. Gabriel, N. Meyer, S. Strome, B. Goldstein, Curr. Biol. **14**(10), 851–862 (2004)

17. E.M. Munro, Curr. Opin. Cell. Biol. **18**(1), 86–94 (2006)

18. S. Guo, K.J. Kemphues, Curr. Opin. Genet. Dev. **6**(4), 408–415 (1996)

19. P.T. So, C.Y. Dong, B.R. Masters, K.M. Berland, Annu. Rev. Biomed. Eng. **2**, 399–429 (2000)

20. K.M. Berland, P.T. So, Y. Chen, W.W. Mantulin, E. Gratton, Biophys. J. **71**(1), 410–420 (1996)

21. M.A. Digman, C.M. Brown, P. Sengupta, P.W. Wiseman, A.R. Horwitz, E. Gratton, Biophys. J. **89**(2), 1317–1327 (2005)

22. J.P. Skinner, Y. Chen, J.D. Müller, Biophys. J. **89**(2), 1288–1301 (2005)

23. Z. Petrášek, M. Krishnan, I. Mönch, P. Schwille, Microsc. Res. Tech. **70**(5), 459–466 (2007)

24. Z. Petrášek, C. Hoege, A. Mashaghi, T. Ohrt, A.A. Hyman, P. Schwille, Biophys. J. **95**(11), 5476–5486 (2008)

25. C.R. Cowan, A.A. Hyman, Development **134**(6), 1035–1043 (2007)

26. G.B. Robb, K.M. Brown, J. Khurana, T.M. Rana, Nat. Struct. Mol. Biol. **12**(2), 133–137 (2005)

27. L. Peters, G. Meister, Mol. Cell. **26**(5), 611–623 (2007)

28. A.K.L. Leung, J.M. Calabrese, P.A. Sharp, Proc. Natl. Acad. Sci. U S A **103**(48), 18125–18130 (2006)

29. G.L. Sen, H.M. Blau, Nat. Cell. Biol. **7**(6), 633–636 (2005)

30. E. Petrov, P. Schwille, in *State of the Art and Novel Trends in fluorescence Correlation Spectroscopy*. Standardization and Quality Assurance in Fluorescence Measurements II: Bioanalytical and Biomedical Applications (Springer, Heidelberg, 2007)

31. T. Dertinger, V. Pacheco, I. von der Hocht, R. Hartmann, I. Gregor, J. Enderlein, Chemphyschem **8**(3), 433–443 (2007)

32. T. Ohrt, D. Merkle, K. Birkenfeld, C.J. Echeverri, P. Schwille, Nucleic Acid Res. **34**(5), 1369–1380 (2006)

33. T.P. Chendrimada, K.J. Finn, X. Ji, D. Baillat, R.I. Gregory, S.A. Liebhaber, A.E. Pasquinelli, R. Shiekhattar, Nature **447**(7146), 823–828 (2007)

34. J. Martinez, T. Tuschl, Genes. Dev. **18**(9), 975–980 (2004)

35. J. Lippincott-Schwartz, E. Snapp, A. Kenworthy, Nat. Rev. Mol. Cell. Biol. **2**(6), 444–456 (2001)

36. Y. Chen, J.D. Müller, Q. Ruan, E. Gratton, Biophys. J. **82**(1 Pt 1), 133–144 (2002)
37. A. Khvorova, A. Reynolds, S.D. Jayasena, Cell **115**(2), 209–216 (2003)
38. D.S. Schwarz, G. Hutvgner, T. Du, Z. Xu, N. Aronin, P.D. Zamore, Cell **115**(2), 199–208 (2003)

8

Fluorescence Flicker as a Read-Out in FCS: Principles, Applications, and Further Developments

Jerker Widengren

8.1 Introduction

Fluorescence Correlation Spectroscopy (FCS) exploits fluorescence intensity fluctuations brought about by the behavior of individual molecules, which are typically subject to continuous excitation in a defined observation volume. This behavior means either molecules moving in and out of the observation volume due to translational diffusion or flow, or undergoing other processes, which affect the detected fluorescence brightness of the molecules themselves. The FCS technique has developed very strongly, in parallel with the strong development of fluorescence-based single-molecule spectroscopy, and is now established as a standard technique used in many laboratories all over the world. However, the concept of number fluctuation analysis, where the fluctuating number of particles within a fixed volume is analyzed in terms of its time dependence, was introduced a long time ago by Svedberg [1], Smoluchowski [2], Chandrasekhar [3], and others. Several other techniques have also been developed that provide indications of number fluctuations, i.e., the dynamic light scattering technique, exploiting the light scattering intensity from the particles of interest [4,5], and the voltage clamp approach [6], where fluctuation analysis of electrical currents over sections of cellular or artificial membranes is performed (with the later developed patch clamp technique [7] as its single-molecule counterpart).

FCS and the utilization of fluorescence intensity as the fluctuating quantity was originally introduced in the 1970s with a series of papers presenting the theory and the first experimental realization of the technique [8–11]. In the first FCS experiments, the average number of fluorescent molecules in the observation volume was about 10,000. However, it was still possible to extract the motion of individual molecules from the large uncorrelated bulk fluorescence. In this particular respect, FCS is perhaps the first realization of fluorescence-based single-molecule spectroscopy.

Although FCS showed a great potential at this time, its applicability was strongly reduced due to high-background light levels and low-detection quantum yields. The situation was dramatically changed by the introduction of extremely small observation volumes in FCS measurements. This, in combination with confocal epi-illumination, highly sensitive avalanche photodiodes for fluorescence detection and very selective band-pass filters to discriminate the fluorescence from the background, made it possible to improve signal-to-background ratios in FCS-measurements by several orders of magnitude [12–14]. In a dynamic system, the relative fluctuations increase as the concentration of molecules decreases. Therefore, in order to obtain high enough relative fluctuations in fluorescence, the number of molecules residing in the detection volume at the same time should be kept low. By the improvements mentioned earlier, this could be realized, without compromising the signal level, and measurement times could be shortened drastically, enabling a more widespread use of the FCS technique.

In a confocal epi-illuminated FCS arrangement, the total detected fluorescence intensity is given from:

$$F(t) = Q \int_{-\infty}^{\infty} CEF(\bar{r}) I_{\text{exc}}(\bar{r}) C(\bar{r}, t) \mathrm{d}V. \tag{8.1}$$

Here, $C(\bar{r}, t)$ is the concentration of fluorescent molecules at position $\bar{r}$ and time t. $CEF(\bar{r})$ is the collection efficiency function of the confocal microscope setup and $I_{\text{exc}}(\bar{r})$ denotes the excitation intensity. $Q = q\sigma_{\text{exc}} \Phi_f$, is a brightness coefficient, where σ_{exc} is the excitation cross-section of the fluorescent molecules under study, Φ_f is their fluorescence quantum yield, and q signifies the efficiency of detection of fluorescence which is emitted from the center of the laser focus. The parameter q includes the solid angle of light collection, the transmission of the microscope optics and the spectral filters, as well as the detection quantum yield of the detector.

In a standard FCS measurement, fluorescence fluctuations arise from translational diffusion, as the fluorescent molecules are diffusing into and out of the confocal observation volume, providing information about the translational diffusion coefficients and the average number of molecules residing simultaneously in the observation volume. In the absence of any other kinetic process affecting the fluorescent molecules the time-dependent normalized intensity autocorrelation function (ACF) can be written as:

$$G(\tau) = \lim_{T \to \infty} \frac{1}{T} \int_0^T \frac{F(t+\tau)F(t)}{<F>^2} dt$$

$$= \frac{1}{N} \left(\frac{1}{1 + 4D\tau/\omega_1^2} \right) \left(\frac{1}{1 + 4D\tau/\omega_2^2} \right)^{1/2} + 1. \tag{8.2}$$

Here, N is the mean number of molecules within the sample volume element, and D is the translational diffusion coefficient of the fluorescent molecules.

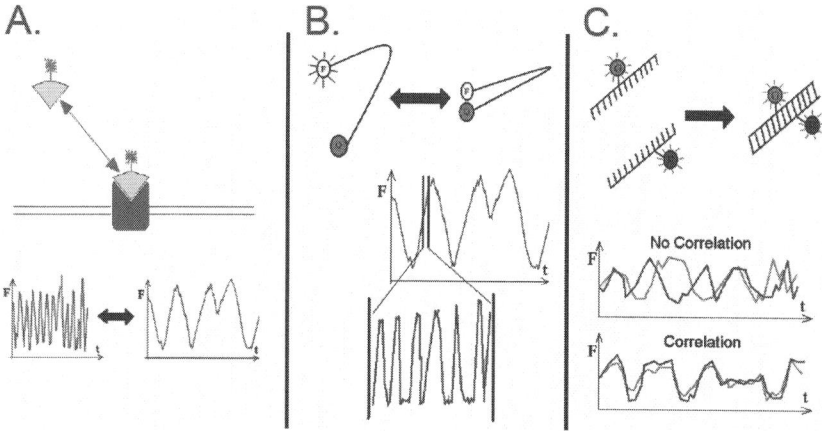

Fig. 8.1. Three major categories of molecular dynamic processes or reactions that can be analyzed by FCS. (a) Reactions that lead to a significant [15] change in the translational diffusion coefficient of the fluorescent reaction partner. (b) Molecular dynamic processes or reactions that change the fluorescence brightness of the studied molecules. (c) Spectral cross-correlation [16], e.g., of reaction partners labeled with *green* (G) and *red* (R) emitting fluorophores that upon association move in concert and generate correlated fluctuations in the green and red emission range

Brackets denote the time average. ω_1 and ω_2 denote the distances in the radial and axial dimensions, respectively, at which the detected fluorescence per unit volume has decreased by a factor of e^2.

However, the available information is not limited to translational diffusion coefficients and concentrations. By FCS, a wide range of processes can be studied, spanning a time range from nanoseconds to seconds. In principle, any process at equilibrium conditions, which reflects itself in terms of a change of the detected fluorescence $F(t)$, can be measured. Three of the most widely applied FCS-based approaches to extract information from molecules undergoing or participating in dynamic processes or reactions are outlined in Fig. 8.1.

In a standard FCS experiment, exploiting the fluctuations in $F(t)$ within one emission wavelength range, there are two criteria, of which at least one needs to be fulfilled in order to gain information about a reaction process by FCS:

1. The reaction must lead to a change of the diffusion properties of the fluorescent species under investigation (Fig. 8.1a). Thereby, the typical duration of the fluorescence bursts due to the passage of fluorescent molecules through the excitation volume is changed. This has been exploited in particular for ligand-receptor interaction studies [17].
2. The reaction needs to change the fluorescence quantum yield or the excitation cross-section of the reactants (Fig. 8.1b). Thereby, the magnitude of the fluorescence emission per molecule is changed, reflected by the fluorescence brightness coefficient Q of (8.1).

Alternatively, in the absence of a significant change in neither diffusion properties nor fluorescence brightness as a consequence of the reaction taking
place, various cross-correlation approaches can be applied (Fig. 8.1c). For
this purpose, cross-correlation FCS has been introduced correlating different
detection channels, separated with respect to emission wavelength range [16]
or spatial localization of the observation volume [18].

In this chapter, the focus will be on how information can be extracted, utilizing the second category described earlier (Fig. 8.1b). In its general form, the
normalized autocorrelation function of the detected fluorescence fluctuations
will show a complex dependence on the reaction rates and the coefficients of
the translational diffusion, and cannot be expressed in an analytical form. For-
tunately, for a rather broad range of molecular reactions the reaction-induced
fluorescence fluctuations can be treated separately from those due to translational diffusion [19]. If diffusion is much slower than the chemical relaxation
time(s) and/or the diffusion coefficients of all fluorescent species are equal,
then the time-dependent fluorescence correlation function can be separated
into two factors. The first factor, $G_D(\tau)$, depends on transport properties
(diffusion or flow) and the second, $R(\tau)$, depends only on the reaction rate
constants:

$$G(\tau) = G_D(\tau)R(\tau) + 1, \tag{8.3}$$

where

$$R(\tau) = \frac{\sum\limits_{i,j=1}^{M} Q_i Q_j X_{ij}(\tau)}{\sum\limits_{i=1}^{M} Q_i^2 \bar{C}_i}. \tag{8.4}$$

Here, Q_i is the fluorescence brightness coefficient of state i and $X_{ij}(\tau)$ is the
solution to the following set of differential equations and initial conditions:

$$\begin{aligned} dX_{ik}(\tau)/d\tau &= \sum\limits_{j=1}^{M} T_{ij} X_{jk}(\tau), \\ X_{ik}(0) &= \bar{C}_i \delta_{ik}. \end{aligned} \tag{8.5}$$

$X_{ij}(\tau)$ describes the probability of finding a molecule in state j at time τ,
given that it was in state i at time 0. M is the number of species participating
in the chemical reaction, and T_{ij} represents the corresponding matrix of the
kinetic rate coefficients. C, denotes the average concentration of state i.

It is, when possible, very convenient to be able to treat the kinetics of the
chemical reaction separately from the translational diffusion in the fluctuation
analysis. As mentioned, a rather broad range of chemical reactions fulfills
the criteria for (8.3)–(8.5). In addition, for a reaction which under standard
conditions does not fulfill these criteria, it is sometimes possible to modify the
conditions. For instance, the dwell times can be retarded with respect to the
chemical relaxation times by expanding the observation volume or by speeding
up the reactions under study, for instance by using higher concentrations of
unlabelled reactants.

In this chapter, two different realizations will be discussed in which FCS is used to monitor reactions that generate changes in Q of the fluorescent species. First, it will be shown how FCS by this approach provides an alternative way of monitoring ion concentration exchange and how the approach can be used to investigate protonation kinetics at biological membranes. Thereafter, an overview will be given of how FCS can be used to monitor a range of photo-induced transitions in fluorophores. It will also be shown how these photodynamic processes can be investigated and manipulated by the use of modulated excitation, in the context of FCS measurements, and as a means to allow monitoring of photo-induced transient states on a more general and massive parallel basis.

8.2 Monitoring of Ion Concentration and Exchange

Fluorescent indicators are since long well established and useful tools to monitor concentrations of physiologically important ions within living cells (see [20] for a review). These indicators typically respond with changes in fluorescence intensities and/or lifetimes, or with shifts in emission or excitation spectra. Such shifts allow ratioing between different excitation or emission wavelengths [21], whereby differences in dye concentrations, optical pathlengths, or other parameters that affect the fluorescence output are cancelled out. FCS can provide an alternative method to these approaches to monitor ion concentrations, offering some specific advantages [22, 23]. The FCS approach exploits the strong contrast in the fluorescence brightness coefficient Q that follows from ions binding to or dissociating from a range of ion sensitive fluorophores and yields also temporal information about the binding-process. For the case of pH-sensitive dyes, there are several different dyes commercially available of which the most commonly used are derivatives of fluorescein. Taking the pH-sensitive dye Fluorescein Isothiocyanate (FITC) in a buffered aqueous solution as an example, the following three equilibria need to be considered:

$$Fl^{2-} + H^+ \underset{k_-}{\overset{k_+}{\rightleftharpoons}} HFl^- \tag{8.6}$$

$$B^- + H^+ \underset{k_{diss}}{\overset{k_{ass}}{\rightleftharpoons}} BH \tag{8.7}$$

$$BH + Fl^{2-} \underset{k_{-1}}{\overset{k_1}{\rightleftharpoons}} B^- + HFl^- \tag{8.8}$$

Here, HFl^- and Fl^{2-} denote the protonated anion and the dianion forms of the FITC. BH and B^- denote the protonated and unprotonated forms of the buffer that are active in the proton exchange with FITC at a certain pH.

The ratios HFl^-/Fl^{2-} and BH/B^- for a given pH are determined by the pK_a of FITC and the buffer, respectively. For FITC, the proton exchange will not notably affect the diffusion properties of the dye. Typically, the proton exchange rate will also take place on a time scale at least an order of magnitude faster than that of translational diffusion of the dye molecules through the observation volume in an FCS experiment. Consequently, in the context of an FCS experiment one or both of the prerequisites are fulfilled to separate the contribution from proton exchange into a separate factor in the correlation function (8.3). This protonation-dependent factor can be described by:

$$R(\tau) = \frac{1}{1-P}\left(1 - P + P\exp(-k_{\text{prot}}\tau)\right). \qquad (8.9)$$

Assuming that the protonated form of FITC is non-fluorescent, i.e., that $Q_{HFl^-} = 0$, P corresponds to the fraction of protonated FITC, i.e., $P = [HFl^-]/([Fl^{2-}] + [HFl^-])$. k_{prot} is the protonation rate constant given by:

$$k_{\text{prot}} = k_+\left[H^+\right] + k_- + \left(\left[B^-\right] + [BH]\right)\frac{k_1\left[H^+\right]/K_a + k_{-1}}{\left[H^+\right]/K_a + 1} \qquad (8.10)$$

Here, K_a is the acidity constant of the buffer.

The basic features of this protonation monitoring approach are illustrated in Fig. 8.2a–c. In Fig. 8.2a, a series of FCS curves are shown, recorded from FITC in aqueous solution at different pH, with no added buffer. The curves were fitted to (8.3) using (8.9) (also adding a second factor to the correlation function describing singlet–triplet transitions, which is described further in Sect. 8.3 of this chapter).

Figure 8.2b demonstrates the effect of buffer strength on the correlation curves of FITC at a fixed pH. In Fig. 8.2b (inset), the relaxation rate of ion exchange, k_{prot}, as measured by FCS, can be seen to increase linearly with the concentration of an added phosphate buffer. For comparison, graphs of HEPES, and citric acid buffers at the same pH have been added to illustrate the general influence on k_{prot} by an added buffer and the difference observed from one buffer to another. Also included in the inset graph of Fig. 8.2 is the relaxation rates measured at different concentrations of sodium chloride in the absence of a buffer at pH 6.5, which had no effect on k_{prot} in the FCS curves. In Fig. 8.2c, a set of correlation curves of the Green Fluorescent Protein (GFP) mutant (F64L, S65T) is shown. For this protein, a pH-dependence similar to that for FITC was found. However, much higher buffer concentrations were required in order to significantly increase the relaxation rate of the proton transfer kinetics, and the relation between relaxation rate of GFP and buffer concentration was nonlinear.

The X-ray structure of GFP (Fig. 8.2c, inset) shows that the fluorescently active part (residues 65–67) of the GFP molecule is surrounded by a very tight barrel, formed by the 11-stranded β-barrel. A likely explanation to the FCS measurements is that this barrel not only slows the exchange of protons

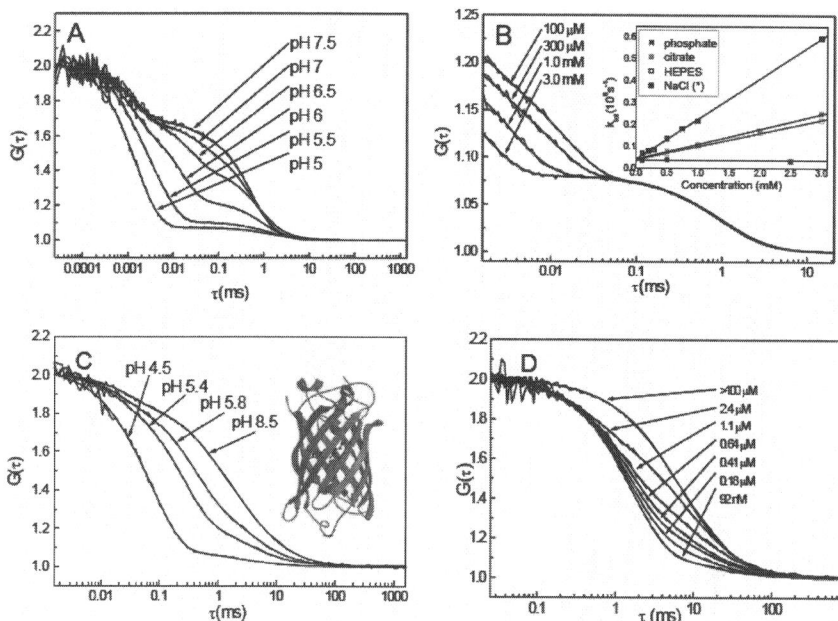

Fig. 8.2. Examples of ion exchange, monitored by FCS. (**a**) FITC measured at different pH. Decays of the correlation curves in the 1 ms, 2–80 µs, and 1 µs time range are attributed to translational diffusion, proton exchange and single–triplet state transitions of the fluorophores, respectively. (**b**) FITC at pH6, with different concentrations of phosphate buffer. Inset: Measured k_{prot} vs. concentration of buffer/salt at pH6. (**c**) FCS curves of GFP (S65T) at different pH. (**d**) The calcium sensitive dye Rh-II, measured at different concentrations of free calcium

within the microenvironment of the GFP molecule but also physically prevents buffer molecules and protons from directly reaching the fluorescently active residues in the interior of the barrel. Hence, the active residues are only affected indirectly, where changes in fluorescence are a secondary effect mediated intramolecularly, following a proton exchange at some exterior part of the molecule. The examples in Fig. 8.2a–c illustrate that FCS gives information about both the fraction of protonated dyes as well as the rate of their proton exchange with their environment. Thereby, in addition to the local pH, also the local buffering properties can be monitored. The protonation rates also reflect to what extent the fluorophore units are shielded from proton exchange with the bulk solution. The fluorophore unit of GFP can serve as a clear example of this phenomenon. In this sense, the exchange rate of hydrogen ions can give sensitive and direct information about the extent of shielding and thus provide a read-out of the conformational states of biomolecules. The concept is also applicable to other ions than hydrogen. Figure 8.2d illustrates the use of a calcium sensitive fluorophore, Rhod-II (Invitrogen). For this dye, in contrast to FITC and many other pH sensitive fluorophores, the fluorescence emission is quenched in the absence of calcium.

Although FCS does not provide any significant advantages for monitoring ion concentrations over extended three-dimensional volumes it can offer selective advantages over other techniques for measuring local ion exchange. In contrast to the laser-induced proton pulse approach [24] and other relaxation techniques frequently used for molecular proton exchange studies, FCS measurements are performed at equilibrium conditions, not requiring any perturbation into some, often strongly un-physiological, initial condition. Moreover, while most other fluorescence-based techniques for ion concentration monitoring require dye concentrations of the order μM or more, FCS is typically performed at about 1,000-fold lower concentrations. This minimizes the influence of buffering effects due to ions binding to the fluorophores themselves, or to any light absorbing proton emitter. Finally, the proton exchange seen in the FCS measurements reflects the immediate environment around the fluorescent probes and is not influenced by the conditions in the media the protons pass on their way from a light absorbing proton emitter.

Recently, we exploited these selective advantages of the FCS approach to investigate the principal role of biological membranes for proton uptake of membrane incorporated proteins [25]. Transport of ions across biological membranes is of fundamental importance for a range of cellular processes such as nerve conduction, energy metabolism, and import of nutrients into cells. Proton transport at and across membranes plays a fundamental role, e.g., in oxidative phosphorylation and involves a series of membrane-spanning proteins in the inner membranes of the mitochondriae. Observations have been made that certain membrane incorporated proteins can take up protons from bulk water at a rate faster than that limited by proton diffusion in the bulk. However, the occurrence and mechanism of such proton-collecting antennae or localized proton circuits at the surface of biological membranes have not been directly verified experimentally. To better understand these phenomena, it is an advantage to study the protonation kinetics at the level of individual surface proton acceptors/donors, at physiologically relevant conditions, and at thermodynamic equilibrium.

On this basis, we used FCS to investigate the particular role of the biological membrane for proton uptake and transport by monitoring the protonation dynamics at the surface of liposomes with well-defined compositions. In each liposome (approximately 30 nm in diameter), only one of the lipid head groups was covalently labeled with FITC (Fig. 8.3a). In Fig. 8.3b, a set of FCS curves are shown, measured from liposomes, composed of the lipid DOPG. The inset shows the corresponding protonation rates, k_{prot}, obtained by fitting the curves to (8.3), (8.9), and (8.10). At low proton concentrations, from the intercept and the slope of k_{prot} vs. the proton concentration (inset, Fig. 8.3b), the protonation association, and dissociation rates of liposome-associated FITC, k_+ and k_- could be determined. While the proton dissociation rate of FITC in the liposomes was found to be indistinguishable from that of free FITC in aqueous solution, the association rate was measured to be two orders of magnitude faster. Changing the lipids of the liposomes from DOPG

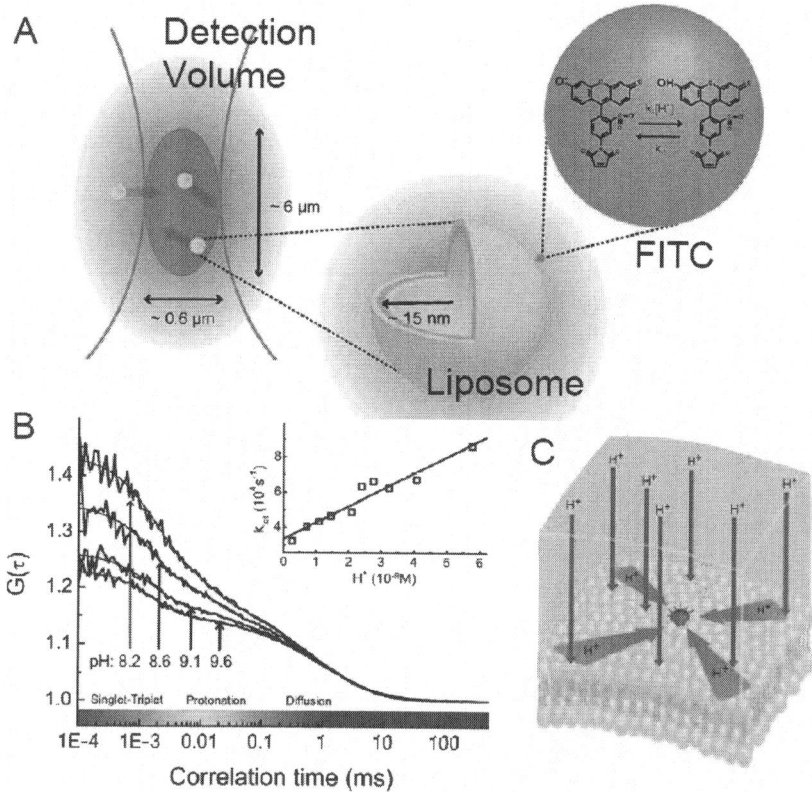

Fig. 8.3. FCS Proton exchange kinetics measurements at biological membranes. (a) Principal design of experiment. Liposomes were labeled with one FITC fluorophore undergoing fluorescence fluctuations due to protonation/deprotonation. (b) Collection of FCS curves of the vesicles at different pH. The FCS curves reflect singlet–triplet transitions in the microsecond time range, protonation kinetics in the 10–100 µs time range and translational diffusion in the milliseconds time range. *Inset*: measured protonation relaxation rates vs. proton concentration. (c) Principle of the proton collecting antenna effect

(negatively charged head group) to DOPC (zwitterionic head group) slowed down the association rate k_+, but it remained much larger than that of FITC free in solution. The increased association rates thus cannot be attributed to electrostatic effects. Instead, the lipid head groups seem to collectively act as a proton-collecting antenna, dramatically accelerating proton uptake from water to a membrane-anchored proton acceptor. This implies that proton migration along the surface can be significantly faster than that between the lipid head groups and the surrounding water phase. If a specific proton acceptor (in this case a dye molecule) at a surface is surrounded by protonatable groups (lipid heads groups), its proton collision cross-section, and therefore its proton uptake rate, can increase dramatically (Fig. 8.3c). Both DOPG and DOPC are protonatable, and can thus provide that migration medium. However, their pK_a values are in the range of 2, and the lipids are therefore hardly protonated themselves. On the other hand, addition of the lipid DOPA, with a pK_a close to that of FITC (around 7), into the DOPC/DOPG liposomes results in a further significant increase of the protonation relaxation rate, k_{prot}. The increase in k_{prot} is linearly dependent on the added DOPA concentration and can be described by (8.10). In this sense, DOPA acts as a two-dimensional buffer. Taken together, the study of [25] shows that ion translocation across membranes and between the different membrane protein components is a complex interplay between the proteins and the membrane itself. The membrane, or the ordered water layer close to the membrane, acts as a proton-conducting link between membrane-spanning proton transporters, and the interplay is modulated by the lipid composition of the membrane. FCS is well suited to investigate this interplay and can be applied further to a broad range of investigations of membrane-associated ion-exchange, also involving membrane proteins as well as other ions than hydrogen.

8.3 Photo-Induced Transient States

Photophysical properties of the fluorescently labeled molecules set the fundamental limits for the sensitivity and overall performance of virtually all forms of fluorescence-based single-molecule measurements. Important figures of merit are the fluorescence flux, and the total number of photons emitted per molecule, which are also highly relevant for all applications of fluorescence spectroscopy and imaging where a high sensitivity or a fast readout is important. The flux of the emitted fluorescence photons is mainly limited by the finite de-excitation rate of the excited fluorophores, given by the fluorescence lifetime and by the extent to which the fluorophores are transformed into transient non-fluorescent states, such as the triplet state. The total number of photons emitted by a fluorophore molecule, n_f, is in the end limited by the photochemical lifetime. In the absence of nonlinear effects, it is given by $n_f = \Phi_f/\Phi_D$, where Φ_f is the fluorescence quantum yield and Φ_D is the photodestruction quantum efficiency. Φ_f and Φ_D can be defined as the fractions of excitations to the excited singlet state, S_1, that lead to fluorescence emission and photodegradation, respectively.

Photo-induced states, such as triplet, photo-isomerized, and photo-oxidized states reduce Φ_f. Also, since several of these states act as precursor states for photobleaching, generation of photo-induced states can reduce n_f. Finally, the blinking caused by transitions to and from these photo-induced states may cause problems in single-molecule experiments, in that this blinking may shadow other molecular processes of interest, taking place in the same time range.

However, in addition to their role as limiting factors for fluorescence signal strength and for generating blinking artifacts, photo-induced transient states of fluorophores and fluorescent proteins have also in the last few years attracted a large interest due to their key role in a range of applications. Photo-switching into long-lived transient states can be exploited for protein transport and localization studies in cells [26]. Photo-switching in general also provides a core mechanism in practically all recently developed approaches for fluorescence-based ultra-high resolution microscopy [27–30]. Moreover, it also deserves to be emphasized that the population kinetics of long-lived photo-induced transient states can contribute with additional dimensions of fluorescence information. Multiplexing by recording two or more of the traditional fluorescence parameters: intensity, emission wavelength, lifetime, and polarization are widely used as a means to increase the information content in fluorescence spectroscopy and imaging. In contrast, the information contained in the population dynamics of photo-induced, long-lived, non- or weakly fluorescent transient states, e.g., states generated by *trans-cis* isomerization, intersystem crossing, or photo-induced charge transfer within fluorescent marker molecules seem not to have been exploited to their full potential. An attractive feature of these states is their long lifetimes, rendering them highly sensitive to the immediate environment of their fluorescent host molecules. While the fluorescence lifetime of a singlet excited state of a fluorophore is $\sim 10^{-9}$ s, the lifetimes of these transient states are $\sim 10^{-6}$–10^{-3} s. Consequently, these states have $\sim 10^3$–10^6 more time to interact with the immediate environment of the fluorophore. Their kinetics can thus change considerably, also due to small changes in, e.g., accessibility of quencher molecules or microviscosities, caused by a biomolecular interaction.

Over the years, a range of techniques have been used to characterize these transient states and their population kinetics. Transient absorption spectroscopy is a well-established technique that has been extensively used to characterize fluorophore photodynamics [31,32]. However, the technique is relatively technically complicated, lacks the sensitivity for measurements at low ($<\mu$M) concentrations and is mainly restricted to cuvette experiments. The emission originating either directly (phosphorescence) or indirectly (delayed fluorescence) from a long-lived first excited triplet state can also be used to characterize its population [33]. This has also been implemented for microscopic imaging [34]. However, coupled to the long-lived emission is also the susceptibility of the triplet state to dynamic quenching by oxygen and trace impurities, which can be circumvented only after elaborate and careful sample preparation. This artifactual quenching not only shortens the triplet lifetime, but practically makes the luminescence un-detectable. Biomolecular monitoring by this readout is thus largely restricted to deoxygenized carefully prepared samples, which restricts its applicability to biological specimen.

As an alternative read-out to the transient state absorption, or the phosphorescence of the triplet state, the population and kinetics of transient states of fluorophores can be followed by FCS, via the fluorescence fluctuations generated as individual fluorophores transit to and from the (non-fluorescent) transient state [35–37]. FCS experiments reflect the same blinking behavior as can also be observed from individual fluorescent molecules, but without the need to look at truly individual molecules. In the FCS experiment, these fluctuations are superimposed on those due to translational diffusion, and their kinetics can be analyzed by the approach of (8.3)–(8.5). As for the protonation kinetics, the transitions to and from the transient states typically take place on a time scale at least an order of magnitude faster than that of translational diffusion of the dye molecules through the observation volume in an FCS experiment and will not influence the diffusion properties of the fluorescent molecules. For FCS experiments, and in analogy to (8.9), transient state kinetics will contribute to the correlation curves by a factor:

$$R(\tau) = \frac{1}{1-T}\left(1 - T + T\exp(-k_T\tau)\right) + 1 \qquad (8.11)$$

Here, T signifies the steady-state population of the non-fluorescent transient state of the otherwise fluorescent molecules in the detection volume, and k_T is the relaxation rate for the transitions to and from this transient state.

FCS measurements have proven to be useful for the monitoring of different photo-induced transient states, including triplet states [35], isomerized states [36], and states generated by photo-induced charge transfer [37]. These states in turn reflect a range of environmental properties, including oxygen and other quencher molecule concentrations, viscosity, and local temperature. Transient state parameters, monitored by FCS, can also be used as a measure of the extent to which Fluorescence (or Förster) Resonance Energy Transfer (FRET) occurs between two fluorophores, reflecting intra- or intermolecular distances [38].

For triplet state monitoring by FCS, the highly sensitive fluorescence readout is used for recording, rather than the faint, easily quenched, phosphorescence signal from the triplet state itself. Thereby, a favorable combination of a high signal level (given by the readout of fluorescence photons) and an outstanding environmental sensitivity (given by the long lifetimes of the transient states) can be obtained. The experimental realization is relatively simple and quenching of the triplet states of the fluorophores by oxygen or other compounds will not ruin the read-out signal. The FCS concept for triplet state monitoring is therefore applicable to a wide range of samples.

One limitation, however, is that only a limited number of spots can be measured simultaneously. A compromise between the temporal analysis of FCS and fluorescence fluctuation analysis in the spatial domain [39] can be obtained by exploiting the time structure of sample/laser scanning confocal microscope images [40,41]. Thereby, spatial correlation analysis of the emitted fluorescence is combined with temporal characterization of the fluorescence emission from the serial data stream of subsequently scanned pixels. This means, however, that the temporal characterization refers to the average of a sample distributed over a relatively large part of the image recorded, rather than to individual pixels. Moreover, all forms of FCS rely on spontaneous fluorescence fluctuations. Thus, only very few molecules ($<$ $\sim$200) can be detected at a time. A low fluorescence signal can for that reason not be compensated by addition of more fluorescent material, and these approaches are therefore dependent on a high fluorescent brightness of the molecules investigated [42]. Moreover, since the fluctuations of the transient states to be investigated typically take place in the microsecond time range, a relatively high time resolution of the detection is required. This limits FCS-based approaches to dilute samples, puts high demands on molecular brightness of the sample, and requires a combination of high sensitivity, time-resolution, and noise suppression of the detection.

Recently, we presented a concept and its experimental realization which circumvents the above limitations, yet maintains the favorable combination of a high detection sensitivity (given by the readout of fluorescence photons) and a strong environmental responsiveness (given by the long lifetimes of the transient states) [43]. By modulation of the excitation source, the kinetics of photo-induced transient states of fluorescent molecules can be determined from the plain time-averaged fluorescence. When the excitation source is modulated in the range of the relaxation time of the transient state of a fluorescent dye, and has an intensity that is high enough to drive the fluorophores into this transient state to a significant degree under continuous wave (CW) excitation, this state can also be populated to significantly different degrees depending on the repetition rate and duration of the pulses. From the systematic variation of the time-averaged fluorescence following from these changes in the pulse characteristics, reflecting the corresponding population variations of the electronic states under modulated excitation, the transient state parameters can be extracted. The procedure is illustrated in Fig. 8.4, for the case the kinetics of singlet–triplet state transitions are to be determined.

Apart from triplet state readout, the approach is applicable to several other photo-induced, transient states, including photo-isomerized states and states generated by photo-induced charge transfer. The modulation approach can analyze samples at higher concentrations, where the spontaneous fluorescence fluctuations would not be detectable with FCS. In contrast to FCS, the modulated excitation concept does not rely on a strong fluorescence brightness of the molecules investigated and is fully compatible with low time-resolution detection, for instance by a CCD camera. The time information is entirely kept in the modulation of the excitation, and the process of acquisition and analysis can easily be automated, and possibly included as an add-on feature on various microscopes and plate-readers, for e.g., automated transient state monitoring or imaging on a massively parallel scale, for HTS as well as for fundamental biomolecular studies. As long as sufficient transient state populations can be induced, and proper corrections for possible fluorescence decay due to photodegradation can be asserted, the proposed approach is

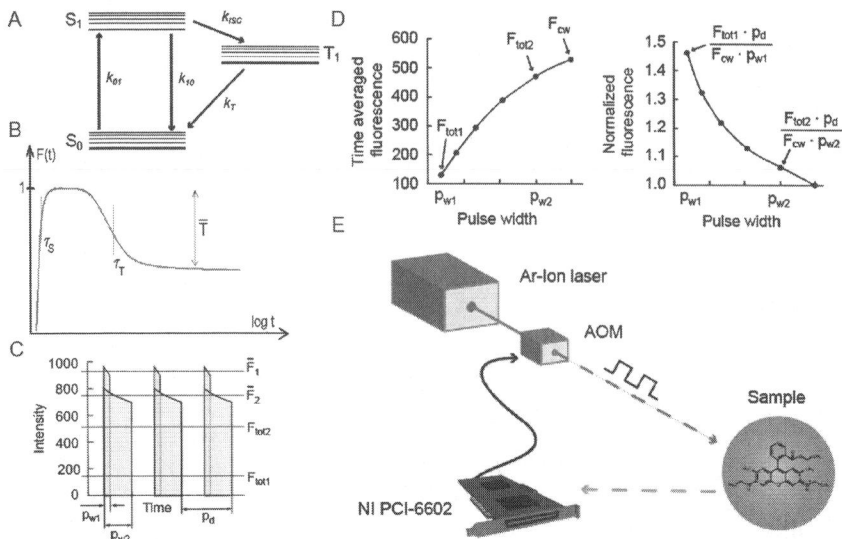

Fig. 8.4. Modulated excitation approach for transient state imaging. (**a**) Jablonski diagram, including the ground singlet (S_0), excited singlet (S_1) and the lowest triplet (T_1) state. (**b**) Principal time development of the fluorescence intensity $F(t)$ as a function of time after onset of excitation at time $t = 0$. Typically, the populations of S_0 and S_1 equilibrate in the nanosecond time range (τ_S, anti-bunching relaxation time), while equilibration between the singlet states and the triplet state (τ_T, bunching time) occurs in the time range of microseconds after onset of excitation. Finally, a steady-state is established with a constant population of T_1. (**c**) Variation of average fluorescence intensity with pulse duration: For excitation pulses with a duration $p_{w1} < \tau_T$ the recorded fluorescence intensity during excitation can be significantly higher than that recorded during a longer pulse with duration $p_{w2} > \tau_T$. F_{tot1} and F_{tot2} denote the corresponding fluorescence intensity averaged over the total duration of the excitation pulse train. (**d**) Typical variation of the time-averaged fluorescence intensity (F_{tot}, *left*) and the fluorescence normalized by the excitation pulse train duty cycle (*right*). (**e**) Example of an experimental realization of the excitation modulation by use of a CW laser and an acousto-optical modulator

compatible with a range of modalities for excitation time-modulation (including time-modulated wide-field excitation, moving arrays of laser foci or laser excitation fringe patterns), as well as various spatial confinement strategies of the excitation and/or the detection (including evanescent-field excitation via total-internal-reflection, two-photon-excitation, or simply by use of thin, or otherwise physically confined, samples). Evidently, the time-modulated excitation experienced by a stationary sample can also be generated by translation of the sample with respect to the excitation, or vice versa. In this way, the excitation need not be idle, and in particular for low duty cycle excitation a larger sample volume can be interrogated within the same period of time. Based on this notion, we recently established the transient state imaging concept by use of a laser scanning confocal microscope (LSCM) [44] (Fig. 8.5).

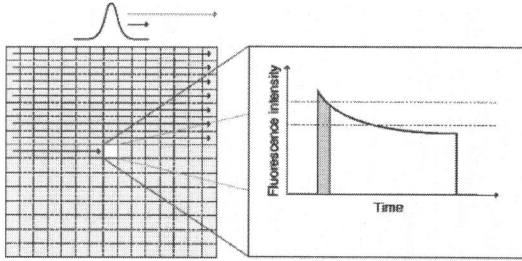

Fig. 8.5. Principle of the modulated excitation approach in a LSCM. In an LSCM, variation of scanning speed/pixel dwell time is from the sample point of view equivalent to varying the duration of the excitation pulse of a stationary excitation field

From the above, one may conclude that a systematic variation of the modulation characteristics of the excitation intensity provides a means to circumvent the limitations of FCS as a tool for monitoring photo-induced long-lived transient states. However, it can also be interesting to consider intensity-modulated excitation for FCS measurements themselves. In FCS measurements, it is typically crucial to find appropriate excitation intensities which on the one hand do not generate strong photobleaching, on the other hand render the studied fluorescent molecules sufficiently bright. In order to find an optimal combination of photobleaching and molecular brightness, it may for some FCS applications be of interest not only to adjust the excitation intensity levels but also the excitation intensity distribution over time. In general, modulated excitation in FCS experiments opens for combinations of FCS with various photoswitching procedures, with the purpose to get ultra-small detection volumes [27] or simply as an additional means to control and manipulate photo-induced states. Recently, we demonstrated the combination of FCS with full correlation time range information with time-modulated excitation. The approach is based on a combination of excitation pulse-train design, covering all possible time-intervals between excitation periods (fluorescence photons), and a proper normalization of the fluorescence ACF (8.2). The normalization is either performed with respect to the ACF of the excitation light, or the ACF of the time-trace of the fluorescence signal itself, time-shifted by an integer number of pulse periods. Analogous to the transient-state imaging described earlier, the excitation modulation for FCS measurements can equally well be transformed into the spatial domain, as long as a spatio-temporal modulation scheme of the excitation can be applied that provides excitation time periods covering all possible time-intervals between excitation periods for all sample locations (Fig. 8.5).

8.4 Conclusions

The FCS technique and its applications have developed very strongly over the last 15 years. Still, more than 30 years after its introduction and almost 20 years after its breakthrough in the early 1990s, there is a strong expansion of new applications and an active method development. Obviously, the recording of spontaneous molecular fluctuations by fluorescence still remains to be exploited for a range of applications. Also, the continuous strong development of lasers, detectors, and data processing tools spur the generation of yet improved versions of FCS in terms of sensitivity, specificity, temporal and spatial resolution, and read-out speed. In this presentation, the use of fluorescence flicker superimposed onto the fluctuations generated by translational diffusion of the studied molecules is highlighted. This flicker can be generated from and thus reflect many different processes, including conformational transitions, molecular interactions, and micro-environmental conditions. Here, it is specifically reviewed how FCS provides a useful tool for monitoring kinetics of ion exchange, and more specifically how basic mechanisms for proton exchange at and along biological membranes can be studied, offering new angles of view compared to present methods of choice, in particular proton pulse relaxation methods. It is also discussed how FCS offers a simple, yet powerful means to characterize photo-induced long-lived transient states of fluorophores. Knowledge of these states is highly relevant for single-molecule spectroscopy, providing limiting factors for fluorescence signal strength and for generating blinking artifacts. Moreover, the interest to characterize photo-induced transient states has strongly increased in the last few years due to applications of photoswitching, for protein transport and localization studies in cells, and as a core mechanism to increase resolution in fluorescence-based light microscopy. What is emphasized here is that the population kinetics of long-lived photo-induced transient states also can contribute with additional dimensions of fluorescence information. By modulating the excitation intensity, it is possible to extend the transient state characterization beyond FCS measurements, thereby circumventing the need for time-resolution in the detection, and not relying on spontaneous fluctuations from low numbers of molecules. Excitation modulation opens for massive parallel imaging of transient states via a sensitive fluorescence signal, possibly yielding new dimensions of fluorescence-based information.

Acknowledgements

Several people have contributed to the work presented in this chapter. I acknowledge all of them in my research group at Exp. Biomol. Physics, KTH, in particular Tor Sandén and Gustav Persson for their work on the modulated excitation approaches [43–45], and the proton exchange study [25]. The latter study is part of a collaboration with the group of Professor Peter Brzezinski,

Stockholm University. Finally, I acknowledge Professor Rudolf Rigler, for the collaboration and his inspiring enthusiasm over the years.

References

1. T. Svedberg, Z Phys. Chem. 77, 147 (1911)
2. M. von Smoluchowski, Wien Berichte 123, 2381 (1914)
3. S. Chandrasekhar, Rev. Mod. Phys., 15, 1 (1943)
4. D.W. Schaefer, Science 180, 1293 (1973)
5. B.J. Berne, R. Peccora Dynamic Light Scattering, with Applications to Chemistry, Biology and Physics (Wiley, New York, 1975)
6. A.L. Hodgkin, A.F. Huxley, B. Katz, Arch. Sci. Physiol. (Paris), 3, 129–150 (1949)
7. E. Neher, B. Sakmann, Nature, 260, 779–802 (1976)
8. D. Madge, E.L. Elson, W.W. Webb, Phys. Rev. Lett. 29, 705–711 (1972)
9. E.L. Elson, D. Magde, Biopolymers 13, 1–27 (1974)
10. D. Magde, E.L. Elson, W.W. Webb, Biopolymers 13, 29–61 (1974)
11. M. Ehrenberg, R. Rigler, Chem. Phys. 4, 390–401 (1974)
12. R. Rigler, J. Widengren, Ultrasensitive detection of single molecules by Fluorescence Correlation Spectroscopy, in Bioscience, ed. by B. Klinge, C. Owman (Lund University Press, Lund, 1990), pp. 180–183
13. R. Rigler, J. Widengren, U. Mets, in Fluorescence Spectroscopy, ed. by O.S. Wolfbeis (Springer, Berlin, 1992), pp. 13–24
14. R. Rigler, U. Mets, J. Widengren, P. Kask, Eur. Biophys. J. 22, 169–175 (1993)
15. U. Meseth, T. Wohland, R. Rigler, H. Vogel, Biophys. J., 76, 1619–1631 (1999)
16. P. Schwille, F.J. Meyer-Almes, R. Rigler, Biophys. J. 72, 1878–1886 (1997)
17. B. Rauer, E. Neumann, J. Widengren, R. Rigler Biophys. Chem. 58, 3–12 (1996)
18. M. Brinkmeier, K. Dörre, J. Stephan, M. Eigen Anal. Chem. 71, 609–616 (1999)
19. A.G. Palmer, N.L. Thompson, Biophys. J. 51, 339–343 (1987)
20. R.Y. Tsien, Meth. Cell Biol. 30, 127–156 (1989)
21. R.B. Silver, Meth. Cell Biol. 56, 237–251 (1998)
22. J. Widengren, R. Rigler, J. Fluorescence 7(1), 211–213 (1997)
23. J. Widengren, B. Terry, R. Rigler, Chem. Phys. 249, 259–271 (1999)
24. M. Gutman, Meth. Enzymol. 127, 522–538 (1986)
25. M. Brändén, T. Sandén, P. Brzezinski, J. Widengren, Proc. Natl. Acad. Sci. USA 103, 19766–19770 (2006)
26. J. Lippincott-Schwartz, N. Altan-Bonnet, G.H. Patterson, Nature Cell Biol. S7–S14 (2003)
27. S.W. Hell, Science, 316, 1153–1158 (2007)
28. S.W. Hell, J. Wichmann, Opt. Lett. 19, 780–782 (1994)
29. E. Betzig, G.H. Patterson, R. Sougrat, O.W. Lindwasser, S. Olenych, J.S. Bonifacino, M.W. Davidson, J. Lippincott-Schwartz, H.F. Hess, Science, 313, 1642–1645 (2006)
30. M.J. Rust, B. Bates, X.W. Zhuang, Nat. Meth. 3, 793–795 (2006)
31. H. van Amerongen, R. van Grondelle, Meth. Enzymol. 246, 201–226 (1995)
32. V.E. Korobov, A.K. Chibisov, Russ. Chem. Rev. 52, 27–42 (1983)
33. P. Cioni, G.B. Strambini, Biochim. Biophys. Acta. 1595, 116–130 (2002)
34. G. Marriott, R.M. Clegg, D.J. Arndt-Jovin, T.M. Jovin, Biophys. J. 60, 1374–1387 (1991)

35. J. Widengren, U. Mets, R. Rigler, J. Phys. Chem. 99, 13368–13379 (1995)
36. J. Widengren, P. Schwille, J. Phys. Chem. A, 104, 6416–6428 (2000)
37. J. Widengren, J. Dapprich, R. Rigler, Chem. Phys. 216, 417–426 (1997)
38. J. Widengren, E. Schweinberger, S. Berger, C.A.M. Seidel, J. Phys. Chem. A 105, 6851–6866 (2001)
39. N.O. Petersen, P.L. Höddelius, P.W. Wiseman, O. Seger, K.E. Magnusson, Biophys. J. 65, 1135–1146 (1993)
40. Y. Xiao, V. Buschmann, K. Weston, Anal. Chem. 77, 36–46 (2005)
41. M.A. Digman, P. Sengupta, P.W. Wiseman, C.M. Brown, A.R. Horwitz, E. Gratton, Biophys. J. 88, L33–36 (2005)
42. D.E. Koppel, Phys. Rev. A 10, 1938–1945 (1974)
43. T. Sandén, G. Persson, P. Thyberg, H. Blom, J. Widengren, Anal. Chem. 79(9), 3330–3341 (2007)
44. T. Sandén, G. Persson, J. Widengren, Anal. Chem. 80, 9589–9596 (2008)
45. G. Persson, P. Thyberg, J. Widengren, Biophys. J. 94, 977–985 (2008)

Part IV

Quantum Dots and Single Molecule Behaviour

Development of Nanocrystal Molecules for Plasmon Rulers and Single Molecule Biological Imaging

A.P. Alivisatos

Summary. Nanocrystals play an increasingly important role as labels in biological imaging. This chapter considers distinct groupings of nanocrystals, termed nanocrystal molecules, in which the individual nanocrystals are plasmonically coupled. These new structures can be used to investigate biomolecular dynamics.

Nanocrystals can play an extremely important role as labels and contrast enhancers in biological imaging. The most famous example of this is the use of colloidal quantum dots, which were introduced as biological labels just 10 years ago in simultaneous publications by Alivisatos and Weiss [1] (Fig. 9.1), and by Nie [2]. In the intervening decade, these new labels have become commercially available, and very widely used because they do not bleach readily and they offer the possibility of multiple colors of emission with a single excitation source [3]. Today, there are many new developments in nanomaterial design and synthesis, and these newer materials will form the basis of yet another wave of new biological labels with enhanced functionalities, enabling new modes of sensing and detection at the single molecule level in biology. A goal of this paper is to show an example of nanomaterial design that enables such new labels to be created.

This paper introduces the reader to a new class of biological label, nanocrystal molecules. This term can readily be understood when we consider that a single isolated colloidal quantum dot, a nanocrystal of a semiconductor, can be thought of as an artificial atom. This designation arises because the fundamental electronic and optical properties are controlled by the designer. The density of electronic states, the wavelength of light emission, the fluorescence radiative rate, and other properties are all adjusted by synthetically controlling nanocrystal parameters such as size, shape, and choice of core and shell material [4]. As the ability to make such artificial atoms has increased in sophistication, it becomes natural to consider whether such artificial molecules could be created by combining the nanocrystals into well-defined groupings. A key feature of an artificial molecule is that it should exhibit collective behavior; that is, the properties should be clearly distinct from those of

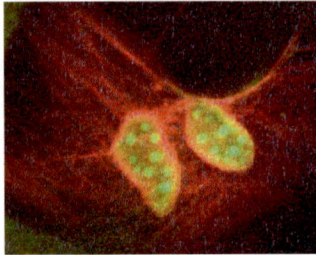

Fig. 9.1. The first biological labeling experiment with colloidal semiconductor nanocrystal quantum dots, reproduced from the 1998 paper by Alivisatos and Weiss [1]. A larger size of dot, red emitting, has been used to label the actin fibers of the fibroblast cells, while a smaller, green-emitting set of dots is used to label the histone proteins in the nuclei. Today colloidal quantum dots are widely used in biological imaging, demonstrating the important role of nanoparticles in this field

the component nanoparticles. In the biological imaging context, this distinguishes the work here from many current and important efforts in which, for instance, quantum dots are combined with magnetic nanoparticles to provide for multi-modal imaging [5]. Here, we seek distinct groups of nanoparticles with strong coupling between the particles.

A further important consideration for the work described here will be the means by which the nanoparticles are brought into proximity with each other. In principle, there are several approaches for achieving this goal. In this chapter, we focus on only one such approach, the use of biological macromolecules to effect the nanocrystal assembly (note: a second approach, based on inorganic interconnections, is also being explored, but will not be covered here). There are two reasons for using biological macromolecules as the glue. First, biological macromolecules are capable of chemical recognition and assembly to a degree of sophistication that substantially exceeds what is possible otherwise. Thus, a wide range of nanocrystal molecules, with a variety of symmetries and degrees of interconnection, can be achieved readily using oligonucleotides or peptides to arrange the nanoparticle assembly. Second, by using biological macromolecules as the glue that holds the nanoparticles together, we automatically have built biological sensing into the resulting nanocrystal molecules. Any conformational change in the linking biological molecules will automatically result in a change in the distance between the nanoparticles, and if the nanoparticles are coupled to each other strongly, it should produce a signal that can be detected.

The next question to consider is what physical mechanism of nanoparticle coupling we could rely upon in order to have strong coupling between two nanocrystals that are separated by a few nanometers of, for instance, DNA. Recall that DNA requires about 12 base pairs for a thermally stable duplex, and that this corresponds to a length of about 4 nm, which is also the same size as many proteins that could be used as linkers. Thus, we can expect the biological glue that could hold nanoparticle together into a nanocrystal molecule to be a few nanometers in size. Further, the biological environment can be thought of as an electrically insulating, optically transparent, dielectric medium. In such a context, the most natural interparticle coupling we can look at is the through-space electromagnetic interaction of the nanoparticles. Such a coupling can readily be provided by the plasmons of noble metal nanoparticles [6].

A plasmon is a collective oscillation of the valence electrons in a metallic particle. The resonance frequency of a metal plasmon depends critically on the material composition, especially the electron density in the metal. Most metals exhibit plasmon resonances in the far ultraviolet, but the noble metal nanoparticles are different. In the noble metals, Cu, Ag, and Au, there is a d to s interband transition that mixes with the plasmon resonance in such a way as to shift the plasmon resonance to lower energy, to the visible region of the electromagnetic spectrum. The plasmon resonance of spherically shaped Ag nanoparticles produces a very strong light scattering resonance in the 420 nm wavelength region, while for Au nanoparticles, it is in the 550 nm region. In nanoparticles, the plasmons couple very strongly with optical fields, producing very intense light scattering spectra. The intensity of the light scattering from a single 20 nm diameter Au nanoparticle is readily detected in a dark field microscope, using just the white light illumination from a simple tungsten lamp. The light scattering from the noble metal nanoparticles depends strongly on size; the intensity varies as the sixth power of the radius [7]. Thus, a 40 nm Au nanoparticle is readily seen in this way, but a 5 nm nanoparticle is difficult to detect in the same apparatus.

Now imagine that two noble metal nanoparticles are brought into close proximity to each other, with a separation that is less than their diameter. The charge oscillations in the two particles will strongly couple with each other. The single plasmon resonance of one particle will now split into two resonances, a longitudinal mode, in which the charge oscillations are aligned along the interparticle axis, and a second transverse mode, in which the charge oscillations are perpendicular to the interparticle axis. The strength of this coupling will depend in detail on the dielectric constant of the surrounding medium, but will be most sensitive to the interparticle separation. As the two particles get closer together, the longitudinal mode will shift to lower energy, and transverse to higher energy than the isolated particle resonance (Fig. 9.2). Further, the light scattering intensity will also change. The longitudinal mode will be substantially more intense than the single particle resonance. The longitudinal plasmon resonance form a pair of strongly coupled particles and will be four times the scattering from one particle. Thus, plasmon coupling readily provides a system for looking at interparticle separation, with shifts in the plasmon resonance and strong changes in the light scattering intensity as a function of the interparticle distance.

One feature of plasmon scattering cannot be overemphasized when considering the use of plasmons in biological single molecule spectroscopy. The light scattering from a single noble metal nanoparticle does not change with

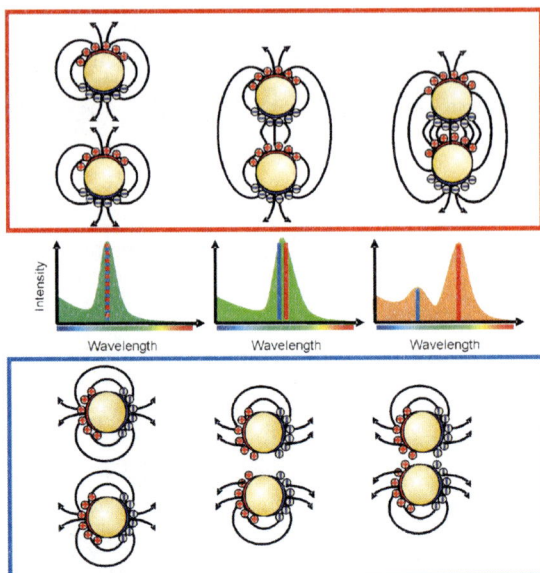

Fig. 9.2. Cartoon illustrating changes in the light scattering spectra when two noble metal nanoparticles are brought into close proximity to each other. The plasmon resonance splits into longitudinal and transverse modes, with scattering from the longitudinal mode more intense. This can be used as the basis for a plasmon ruler with application in single molecule biophysics

time at all. It is just constant. In a noble-metal nanocrystal molecule, the only reason why the light scattering would change is that the distances between the nanoparticle changes (small changes due to variation of the dielectric constant of the surrounding medium can be detected, but they are an order of magnitude smaller than those due to particle proximity) [8]. Otherwise, the spectra are invariant. This is not true of any other type of label. Dyes show intersystem crossing and bleach. Under typical single molecule conditions, they may only last for a few tens of seconds as a label [9]. Colloidal dots last much longer, but they also show blinking, and the timescale for the blinking is stochastic, with a wide range of on and off times [10]. As a consequence, most optical single molecule probes exhibit time-dependent characteristics, which frequently interfere with their use in biological detection. Indeed, much of the art of single molecule biophysics today involves the development of tricks and tools for disentangling the variations in probe properties from those of the biological phenomena under study. Plasmon resonances do not suffer from these drawbacks. Indeed, the plasmon rulers, which we will discuss later provide faithful trajectories for the analysis of single molecule biophysical phenomena.

Our goal then is clear, to construct nanocrystal molecules in which noble metal nanoparticles of 10–40 nm diameter are joined together using oligonucleotides or peptides, and to observe the light scattering from single molecules

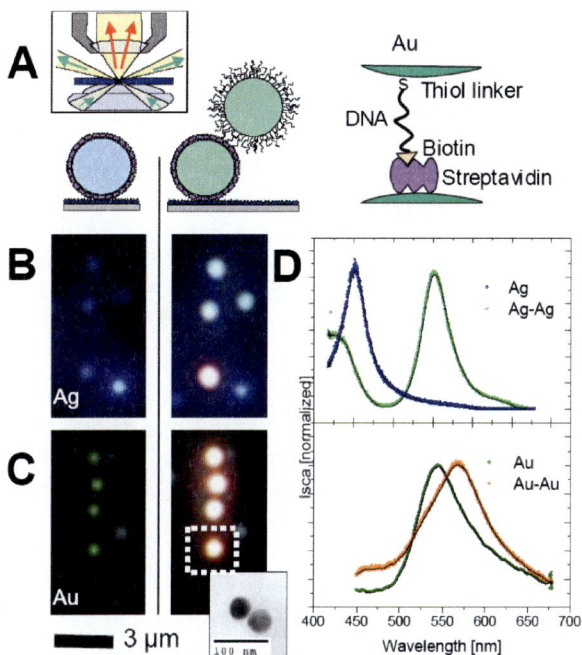

Fig. 9.3. Dark field light scattering spectra from single particles and pairs of particles illustrate plasmon coupling (see text) [11]

of such constructs. Further, we would like to see how these light scattering spectra change when the biological macromolecules joining them together change their conformation. The starting point for such studies is the simplest nanocrystal molecules, homo-dimers in which two nanoparticles of the same size are joined together. A first experiment of this type can be seen in Fig. 9.3 [11]. The left panel of sections B and C shows the light scattering from single Ag and Au nanoparticles immobilized on a surface. The right panel shows the light scattering when a second particle binds to the first. The light scattering intensity increases markedly, and the spectrum changes as expected when the second particle binds. Indeed, the dark field single molecule light scattering spectra in panel D show all the expected features for plasmon coupling.

Over the last decade, we have shown that DNA can be used to construct a wide range of such nanocrystal molecules. The first such example of using DNA to organize nanocrystal molecules was shown in 1996 by Alivisatos and Schultz [12] (Fig. 9.4). This work appeared simultaneously with the work by Mirkin and Letsinger, who used DNA to assemble arrays of nanoparticles [13]. Indeed, if the nanocrystal is thought of as an artificial atom, then the construction of artificial solids is a complementary activity to the construction of artificial molecules, and this has been investigated extensively as well [14].

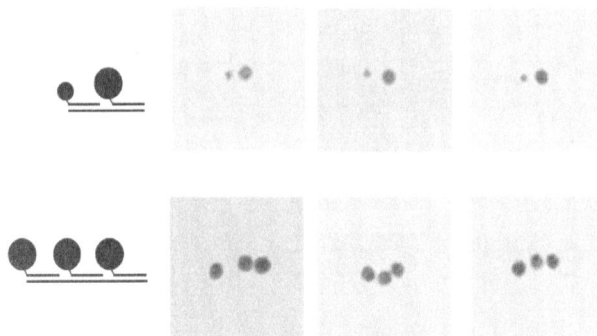

Fig. 9.4. DNA directed assembly of 5 and 10 nm Au nanocrystals using DNA. From the 1996 paper by Alivisatos and Schultz [12]

In the intervening years, we have developed the ability to systematically prepare designed nanocrystal molecules. In this work, we first isolate nanoparticle bearing an exact, discrete number of single stranded oligonucleotides with chosen sequences, using electrophoresis or HPLC to separate particles containing none, one, two, or more oligonucleotides each [15, 16]. These groupings can then be assembled together by adding the nanoparticle-DNA conjugates together along with the appropriate complementary DNA strands to join them. Many symmetries of nanocrystal molecule can be created in this way.

We have studied the plasmon light scattering from individual nanoparticle pairs as a function of the length of the DNA joining them together [11]. Using this method, we can calibrate the so-called plasmon ruler, so that we can develop a clear understanding of how the light scattering spectra depend upon the distance between the particles [17]. There are substantial shifts in the light scattering spectra when the particles are within one diameter from each other (Fig. 9.5). Thus a 40 nm nanoparticle pair can be used to measure distance changes of significantly less than 1 nm, and can measure distances as long as 60 nm. It is important to note that the DNA joining the nanocrystals together is somewhat floppy. DNA is thought to have a persistence length of about 50 nm [18]. Thus, the DNA in use, here, as a linker can be just a bit shorter or longer than a persistence length, and we can expect that the DNA will flex and fluctuate, altering the distance between the particles as a function of the time. Indeed, the time-dependent light scattering spectra can be used to probe these motions.

To illustrate the power of these nanocrystal molecules for single particle biophysics, we have investigated the cutting of DNA by the Eco RV restriction enzyme [19] (Fig. 9.6). We chose this system because the kinetics of the enzyme has been studied previously with a wide range of tools, ranging from ensemble biochemistry methods to dye-labeled single molecule techniques. We have deposited dimers of nanoparticles on a substrate, joined together by oligonucleotides that contain a sequence, which is cut by the Eco RV restriction

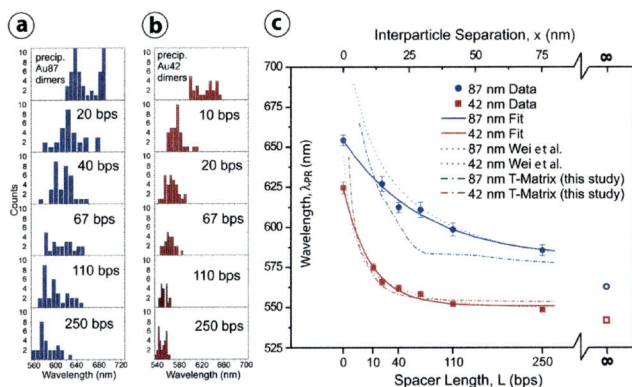

Fig. 9.5. Calibration of the plasmon ruler by single molecule light scattering. The peak of the light scattering spectrum is observed for different lengths of DNA. Collaborative work with Prof. Jan Liphardt [17]

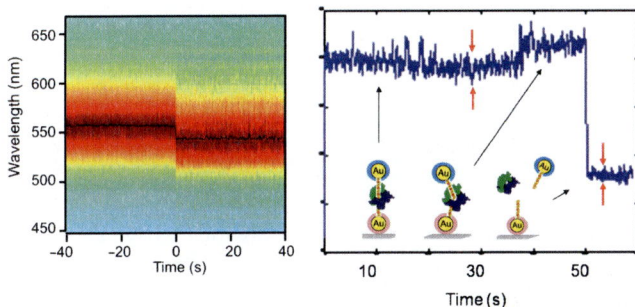

Fig. 9.6. Plasmon ruler study of DNA cleavage by the Eco RV restriction enzyme. Upon cleavage, the light scattering spectrum and intensity change abruptly. The single particle trajectories can be used to observe the sequence of events prior to cutting, including bending of the DNA. Collaborative work with Prof. Jan Liphardt [19]

enzyme. This field of dimers is imaged in a dark field microscope such that a wide field of individual dimers is observed simultaneously. When the enzyme is flowed into the experimental chamber, we see the DNA cutting events clearly as abrupt changes in the light scattering from any given dimer (the second particle flows away as soon as it is cut). The intensity of the light scattering drops abruptly when the cut takes place, and the weaker scattering spectrum of the single particle left behind is blue shifted from that of the dimer. When the cutting times of a group of such particles are tabulated, we are able to recover the known ensemble rate constant for the cutting of the DNA by the Eco RV restriction enzyme. The presence of the Au particles does not perturb the DNA cutting by the enzyme.

It is instructive to look in detail at single nanocrystal molecule plasmon ruler trajectory of the restriction enzyme cutting DNA. Each trajectory shows a cleavage event, a point in time at which the light scattering intensity abruptly drops. One can also see that the noise in the light scattering spectra after the cut is clearly less than before. This is because the interparticle separation is fluctuating before the cut, modulating slightly the scattering power. We can focus our attention now on the time interval just before the cut takes place. We see that in the time before the cut, there is typically a brief interval when the light scattering intensity actually goes up. This can be understood as arising from the two particles getting closer together, so that their coupling increases, and their pair polarizability goes up. This decrease in interparticle separation can be understood as arising from the bending of the DNA by the Eco RV restriction enzyme, a step that again is known to precede the cutting from ensemble studies. Here, however, the time intervals for the bending can be clearly discerned for each case. Further, we have shown that it is possible to see how fluctional the DNA is during this time interval when it is bent (the DNA is actually a little bit more floppy when it is bent by the enzyme.) This study clearly establishes that nanocrystal molecules and the plasmon ruler can be powerful tools for observing in vitro single molecule biophysical phenomena.

The plasmon rulers shown here can be extended to work in vivo as well. We have recently developed nanocrystal molecules joined together by peptides, and we have shown that their light scattering can be detected when the particles are introduced into the cytoplasm of living cells. Further, we have shown that we can detect the cleavage of the peptides when specific byproteases are activated [20].

So far, we have only considered the simplest of nanocrystal molecules, the homo-dimer. Many more complex compositions and symmetries of nanocrystal molecule can be created. Two of them will be described here briefly: one in which a central colloidal semiconductor quantum dot is surrounded by a specific number of Au nanocrystals and another in which nanoparticles are assembled into chains. In the central symmetry case, colloidal quantum dots with many oligonucleotides per particle are mixed with Au nanocrystals, each attached to just a single oligonucleotide that is complementary to the sequence on the DNA on the quantum dot. The resulting nanocrystal molecules can now be run through a gel, in which they separate into discrete bands. Each band consists of colloidal Qdots joined with a specific number of Au particles. $QD(Au)_n$ molecules, where n = 1 to 8 have been isolated in this way [21] (Fig. 9.7). In the case of chains, we start with particles that contain two oligonucleotides per particle. These can be joined together using the ligase chain reaction [22]. This is an example wherein enzymatic operation can be used in the construction of nanocrystal molecules. Further, such strings with complementary sequences can be joined, providing for chains of nanocrystal dimers. An example of tetramers and hexamers is shown in Figure 9.8. In the future, combinations of these two arrangements, central branch points

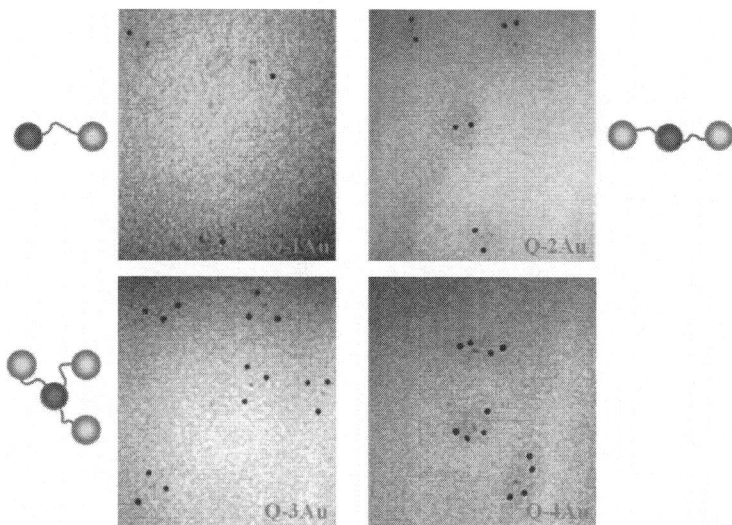

Fig. 9.7. QD(Au)$_n$, with n=1 to 4 assembled for a colloidal quantum dot and 1 to 4 10 nm diameter Au nanocrystals [21]

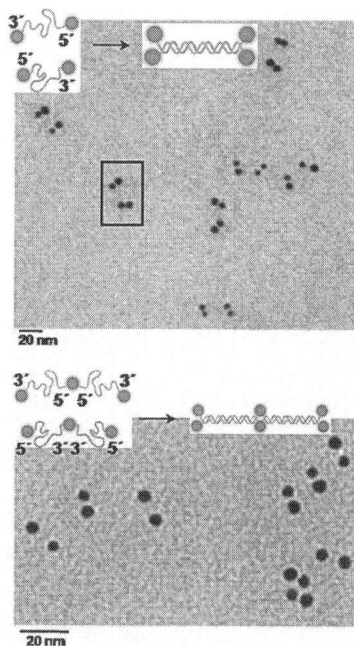

Fig. 9.8. Tetramers and hexamers of 10 nm Au nanocrystals assembled as chains of pairs of particles using DNA and enzymatic ligation. Collaborative work with Prof. Jean Frechet [22]

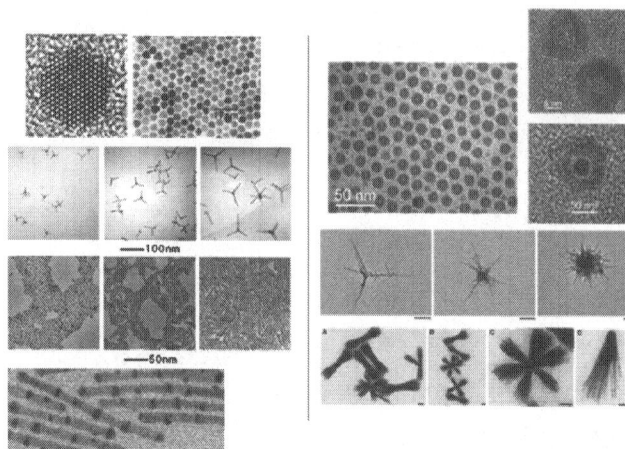

Fig. 9.9. Gallery of nanocrystal molecules prepared by inorganic routes, rather than by biological assembly. These nanocrystal molecules can be branched or chained, as well as be hollow and nested. They can be coupled together electronically, as opposed to through space by plasmon coupling

and linear extensions, can be used to create a huge variety of nanocrystal molecules. Such assemblies will exhibit collective behaviors, which can be interrogated in future nanocrystal molecule and single molecule biophysics studies.

To illustrate just how far the nanocrystal molecule concept looks like it can go, consider a last case, chiral nanocrystal molecules. Over the last 25 years, Nadrian Seeman has shown that it is possible to make DNA nanostructures in which multiple DNA strands are used to weave a more rigid pattern [23]. He has made DNA tiles and DNA cubes and pyramids. We have worked with him to integrate nanoparticles into some of these DNA structures. In one recent study, as yet unpublished, we have made DNA pyramids using a variant of the Seeman designs and then attached a different nanoparticle to each vertex of the pyramid [24]. In this way, we have made left- and right-handed four particle nanocrystal molecules. Such constructs are of particular interest because they offer the possibility of examining enzyme cutting of DNA in more detail, as the change in each nanoparticle separation can be seen independently. In such assemblies, we should be able to use plasmon rulers in the future to see more than simple distance changes; for instance, twists and torsions could also be detected.

In the work described here, we see one specific class of new nanostructure, nanocrystal molecules joined by DNA or peptides. The coupling between the particles is due to plasmon coupling. We can envision other types of nanocrystal molecule, in which the coupling between the particles arises, for instance, by direct electronic coupling. To achieve such goals, we need to make nanocrystal molecules joined together by inorganic linkers. In recent years, we have created branched, striped, and nested nanocrystals connected inorganically (Fig. 9.9), showing many of the same arrangements as the nanocrystal molecules assembled by DNA [25–27]. It is beyond the scope of the present paper to describe the properties of these inorganically connected nanocrystal molecules. However, the strong analogy between biologically assembled and inorganically grown nanocrystal molecules further helps us to develop the concept of artificial molecule and provides a further illustration of how rich the field of nanoscience continues to be.

References

1. M. Bruchez Jr., M. Moronne, P. Gin, S. Weiss, A.P. Alivisatos, Science **281**, 2013–2016 (1998)
2. W.C.W. Chan, S. Nie, Science **281**, 2016–2018 (1998)
3. A.P. Alivisatos, W.W. Gu, C. Larabell, Annu. Rev. Biomed. Eng. **7**, 55–76 (2005)
4. A.P. Alivisatos, Science **271**, 933–937 (1996)
5. J.H. Lee, Y.W. Jun, S.I. Yeon, J.S. Shin, J. Cheon, Angew. Chem. Int. Ed. **45**, 8160–8162 (2006)
6. H. Wang, D.W. Brandl, P. Nordlander, N.J. Halas, Acc. Chem. Res. **40**, 53–62 (2007)
7. U. Kreiberg, M. Vollmer, *Optical Properties of Metal Clusters* (Springer, New York, 1995)
8. P.K. Jain, W. Huang, M.A. El-Sayed, Nano Lett. **7**, 2080–2088 (2007)
9. U. Resch-Genger, M. Grabolle, S. Cavaliere-Jaricot, R. Nitschke, T. Nann, Nat. Meth. **5**, 763–775 (2008)
10. M. Kuno, D.P. Fromm, H.F. Hamann, A. Gallagher, D.J. Nesbitt, J. Chem. Phys. **112**, 3117–3120 (2000)
11. C.S. Snnichsen, B.M. Reinhard, J. Liphard, A.P. Alivisatos, Nat. Biotechnol. **23**, 741–745 (2005)
12. A.P. Alivisatos, K.P. Johnsson, X.G. Peng, T.E. Wilson, C.J. Loweth, M.P. Bruchez, P.G. Schultz, Nature **382**, 609–611 (1996)
13. C.A. Mirkin, R.L. Letsinger, R.C. Mucic, J.J. Storhoff, Nature **382**, 607–609 (1996)
14. A.P. Alivisatos, Cytometry **59A**, 29 (2004)
15. D. Zanchet, C.M. Micheel, W.J. Parak, D. Gerion, A.P. Alivisatos, Nano. Lett. **1**, 32–35 (2001)
16. S.A. Claridge, H.W. Liang, S.R. Basu, J.M.J. Frchet, A.P. Alivisatos, Nano. Lett. **8**, 1202–1206 (2008)
17. B.M. Reinhard, M. Siu, H. Agarwal, A.P. Alivisatos, J. Liphardt, Nano. Lett. **5**, 2246–2252 (2005)
18. C. Bustamante, S.B. Smith, J. Liphardt, D. Smith, Curr. Opin. Struct. Biol. **10**, 279–285 (2000)
19. B.M. Reinhard, S. Sheikholeslami, A. Mastroianni, A.P. Alivisatos, J. Liphardt, Proc. Nat. Acad. Sci. **104**, 2667–2672 (2007)

20. Y.W. Jun, S. Sheikholeslami, D. Hotstetter, C. Tajon, C. Craik, A.P. Alivisatos, Proc. Natl. Acad. Sci. in press (2009)
21. A.H. Fu, C.M. Micheel, J. Cha, H. Chang, H. Yang, A.P. Alivisatos, J. Am. Chem. Soc. **126**, 10832–10833 (2004)
22. S.A. Claridge, A.J. Mastroianni, Y.B. Au, H.W. Liang, C.M. Micheel, J.M.J. Frechet, A.P. Alivisatos, J. Am. Chem. Soc. **130**, 9598–9605 (2008)
23. N.C. Seeman, Annu. Rev. Biophys. Biomol. Struct. **27**, 225–248 (1998)
24. A. Mastroianni, S.A. Claridge, A.P. Alivisatos, J. Am. Chem. Soc. 131, 8455-8459 (2009)
25. L. Manna, D.J. Milliron, A. Meisel, E.C. Scher, A.P. Alivisatos, Nat. Mater. **2**, 382–385 (2003)
26. R.D. Robinson, B. Sadtler, D.O. Demchenko, C.K. Erdonmez, L.W. Wang, A.P. Alivisatos, Science **317**, 355–358 (2007)
27. Y.D. Yin, R.M. Rioux, C.K. Erdonmez, S. Hughes, G.A. Somorjai, A.P. Alivisatos, Science **304**, 711–713 (2004)

10

Size-Minimized Quantum Dots for Molecular and Cellular Imaging

Andrew M. Smith, Mary M. Wen, May D. Wang, and Shuming Nie

Summary. Semiconductor quantum dots, tiny light-emitting particles on the nanometer scale, are emerging as a new class of fluorescent labels for a broad range of molecular and cellular applications. In comparison with organic dyes and fluorescent proteins, they have unique optical and electronic properties such as size-tunable light emission, intense signal brightness, resistance to photobleaching, and broadband absorption for simultaneous excitation of multiple fluorescence colors. Here we report new advances in minimizing the hydrodynamic sizes of quantum dots using multidentate and multifunctional polymer coatings. A key finding is that a linear polymer containing grafted amine and thiol coordinating groups can coat nanocrystals and lead to a highly compact size, exceptional colloidal stability, strong resistance to photobleaching, and high fluorescence quantum yields. This has allowed a new generation of bright and stable quantum dots with small hydrodynamic diameters between 5.6 and 9.7 nm with tunable fluorescence emission from the visible (515 nm) to the near infrared (720 nm). These quantum dots are well suited for molecular and cellular imaging applications in which the nanoparticle hydrodynamic size needs to be minimized. Together with the novel properties of new strain-tunable quantum dots, these findings will be especially useful for multicolor and super-resolution imaging at the single-molecule level.

10.1 Introduction

Semiconductor quantum dots (QDs) are currently under intense research and development for use in molecular and cellular imaging, but a major limiting factor is still their relatively large sizes [1–13]. This bulkiness is not an intrinsic problem of QD nanocrystals, but arises mainly from organic surface coatings used for encapsulation and stabilization. In fact, small 4–7-nm QDs have been shown to have hydrodynamic sizes (diameters) of 20–40 nm when coated with amphiphilic polymers [5–7]. Small hydrophilic and cross-linked ligands such as thioglycerol and dihydrolipoic acid have been used to reduce the coating layer thickness [12, 14–21], but the resulting dots often suffer from low colloidal stability, photobleaching, or low quantum yields. Consequently, these size-reduced QDs have found only limited utility in live cell and in vivo applications [8, 9].

Here we report a new class of multifunctional multidentate polymer ligands not only for minimizing the hydrodynamic size of QDs, but also for overcoming the colloidal stability and photobleaching/signal brightness problems encountered in previous research. A major finding is that a mixed composition of thiol ($-SH$) and amine ($-NH_2$) groups grafted to a linear polymer chain can lead to a highly compact QD with long-term colloidal stability, strong resistance to photobleaching, and high fluorescence quantum yield. In contrast to the standing brush-like conformations of pegylated dihydrolipoic acid ligands and monovalent thiols, we believe that these multidentate polymer ligands can wrap around the QD in a closed "loops-and-trains" conformation [22–24]. This structure is highly stable from a thermodynamic perspective and is responsible for the excellent colloidal and optical properties observed. As a result, we have prepared a new generation of bright and stable CdTe QDs with small hydrodynamic diameters between 5.6 and 9.7 nm, with fluorescence emission tunable from the visible (515 nm) to the near infrared (720 nm) spectra. In addition to CdTe nanocrystals, we find that this new class of multidentate polymers is applicable to a broad range of core nanocrystals as well as core/shell nanostructures including CdS, ZnSe, CdSe/ZnS, and CdTe/CdS. These coatings have been adapted for minimizing the hydrodynamic size of core/shell particles with their optical properties tuned by lattice strain, which may open doors to improved super-resolution imaging of individual molecules.

10.2 Results and Discussion

Figure 10.1 shows the synthesis of multidentate polymer ligands as well as the methods for encapsulating CdTe quantum dots. Roughly 35% of the carboxylic acids of polyacrylic acid (PAA, MW $\sim 1, 800$) were covalently modified with cysteamine and N-Fmoc-ethylenediamine using diisopropylcarbodiimide (DIC) and N-hydroxysuccinimide (NHS). After deprotection of the amine with piperidine and purification, each polymer molecule contained approximately 3.5 active thiols and 3.0 active amines, as determined via Ellman's reagent and fluorescamine assays. For coating quantum dots, this balanced composition of amines and thiols was found to provide superior monodispersity, photostability, and fluorescence quantum yield compared to either amines or thiols alone. Further studies are still needed to understand the underlying mechanisms for this effect. The CdTe quantum dots were prepared in a high temperature organic solvent using hydrophobic ligands (e.g., alkylamines), and it is necessary to first exchange the native ligands with hydrophilic thioglycerol due to the insolubility of the polymer in nonpolar solvents. These polar monovalent ligands are then replaced with the multidentate ligand.

Because the PAA-based polymer was highly hydrophilic, there was no way to efficiently coat hydrophobic quantum dots unless the native nonpolar

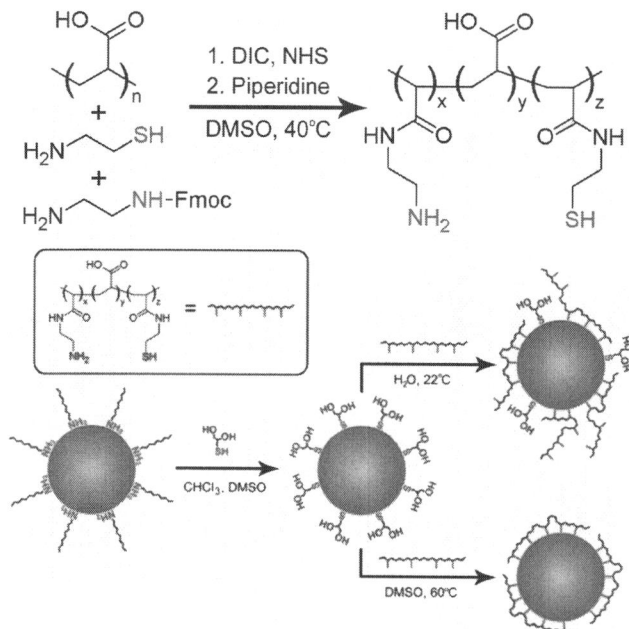

Fig. 10.1. *Upper*: Synthesis of multidentate polymer ligands with a balanced composition of thiol (−SH) and amine (−NH₂) coordinating groups grafted to a linear polymer chain. *Lower*: Procedures for achieving self-assembly of multidentate ligands on quantum dots

ligands were first displaced with a polar ligand (thioglycerol). This allowed the dissolution of the quantum dots in a polar solvent, in which thioglycerol could be replaced with the tightly binding multidentate ligand. Performing this process in water, however, resulted in large nanocrystals with extensive aggregation, likely due to crosslinking of the quantum dots through the multidentate ligand. Instead, it was found that robust, compactly coated quantum dots could only be produced after heating (60–70 °C) for 1–2 h in an aprotic solvent (DMSO) under inert conditions. This observation is in accord with the "loops, trains, and trails" model of polymer surface adsorption [22–24]. Although it is thermodynamically favorable for the linear polymer maximize its adsorption (train domains), self-assembly of this highly ordered structure does not readily occur at room temperature. The kinetics for the closure of loops and trails are slow, and thus elevated temperatures are needed to expedite this process.

Figure 10.2 compares the optical properties and hydrodynamic sizes of CdTe quantum dots (2.5 nm) coated with a traditional amphiphilic polymer (octylamine-modified polyacrylic acid) or the mixed thiol/amine multidentate ligand. Although the amphiphilic polymer and the multidenate ligand were prepared from the same molecular-weight polyacrylic acid backbone, the

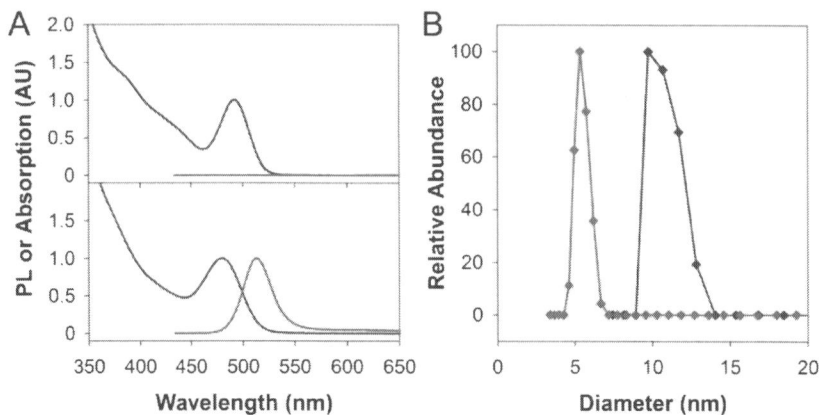

Fig. 10.2. Comparison of optical and hydrodynamic properties of CdTe quantum dots (2.5 nm) solubilized in water with an amphiphilic polymer (octylamine-modified polyacrylic acid) or a multidentate polymer ligand. (**a**) Absorption (*blue curves*) and fluorescence emission (*red curves*) spectra of CdTe quantum dots with amphiphilic polymer (*upper*) or multidentate polymer (*lower*) coatings. (**b**) Dynamic light scattering size data of quantum dots with amphiphilic polymer (*blue curve*) and multidentate polymer (*green curve*) coatings. *PL* Photoluminescence, *AU* Arbitrary units. All samples were dissolved in phosphate buffered saline

quantum dots coated with the multidentate ligand are considerably smaller in size and also much brighter in fluorescence. Dynamic light scattering measurements show that the multidentate polymer coating is only 1.5–2 nm in thickness, and a lack of aggregation is verified via TEM. This compact shell matches the geometric predictions of a polymer conformation with a high degree of adsorption on the quantum dot surface, enabled by its high affinity and low molecular weight. In comparison, the coating thicknesses are on the order of 4–7 nm for amphiphilic polymers and even some monovalent molecular ligands [5–7, 12, 14–21]. It is also worth noting that the CdTe quantum dot is not protected with an electronically insulating inorganic shell (e.g., ZnS or CdS) and its fluorescence is retained with the multidentate polymer, but nearly completely quenched by the amphiphilic polymer.

10.2.1 Molar Capping Ratio

As shown in Fig. 10.3, the fluorescence quantum yield, monodispersity and photostability of these polymer-coated quantum dots are strongly dependent on the molar capping ratio (MCR), which is calculated by dividing the sum of basic groups (amine or thiol) on the polymer by the sum of cadmium and tellurium atoms on the quantum dot surface. When the MCR values are below 1.0, the amount of polymer is insufficient to completely coat 2.5-nm

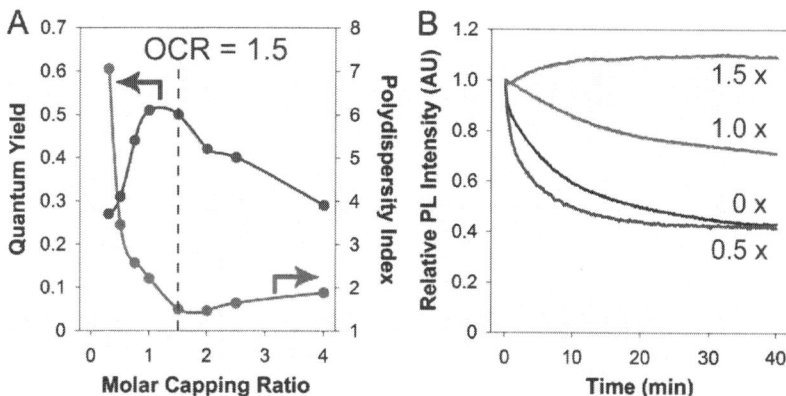

Fig. 10.3. Effects of polymer capping ratios on quantum dot properties. (a) Fluorescence quantum yield (*blue curve*) and polydispersity index (*red curve*) of 2.5 nm CdTe quantum dots as a function of molar capping ratio. Polydispersity indices were calculated from gel filtration chromatograms. (**b**) Photostability data at various capping ratios (MCR = 1.5, 1.0, or 0.5) and in the absence of polymer (MCR = 0)

CdTe quantum dots, resulting in polydisperse nanocrystals. Polydispersity was quantitatively assessed from the polydispersity index (PDI) in gel filtration chromatograms. When the MCR values are above 2.0, the excess polymer leads to better monodispersity and colloidal stability, but a reduced fluorescence quantum yield. Between these two limits is the optimal capping ratio (OCR) of approximately 1.5, yielding small, monodisperse nanocrystals (PDI < 1.5) with bright fluorescence ($\sim$ 50% quantum yield) and exceptional photostability.

The OCR is dependent on the size of the quantum dot, and its value changes to 1.0 for 3.0 nm cores and to 0.5 for 4.0 nm cores. This trend is indicative of the size-dependent differences in nanocrystal surface curvature, the intrinsic degree of flexibility of the polymer, and the increasing availability of more than one free orbital per surface atom with decreasing nanocrystal size. The OCR can be semiquantitatively determined using the aqueous ligand coating procedure. In this method, the multidentate ligand is added to thioglycerol-coated quantum dots at room temperature. Addition of a small excess of polymer above the OCR value results in complete precipitation of the nanocrystals from solution, and the quantum yield trend with the MCR was found to be similar for the aqueous procedure.

10.2.2 Minimum Size

The smallest nanoparticles that could be prepared were 5.5 nm in hydrodynamic diameter, encapsulating a 2.5-nm core. This 2.5-nm core size is also the smallest size that can be made reliably using the coordinating solvent synthesis described herein. In order to prepare smaller nanocrystals, these particles were oxidatively etched in a slow, controlled process to yield monodisperse nanocrystals with a diameter of 1–2 nm. However, after coating with the multidentate polymer, these nanocrystals were actually found to be larger in size, with a diameter of 6–6.5 nm, which is incidentally similar to the length of the fully outstretched polymer ($\sim$6.3 nm). We attribute this interesting effect to the extremely high surface curvature of these nanocrystals, which is not conducive to multivalent interactions with a linear polymer. Although cores smaller than 2.5 nm were colloidally stable after coating with the polymer, they were found to have significant deep trap emission, and were prone to photobleaching.

10.2.3 Stability

The multidentate polymer-coated quantum dots are stable at room temperature for over 6 months after purification, with no significant changes in gel filtration chromatograms. The quantum yield is retained under these conditions when stored in the dark, but gradually decays to $\sim 20\%$ with continual exposure to room light. In comparison, purified CdTe quantum dots coated with monovalent ligands generally precipitate within 2 days at room temperature, and are even unstable when stored in excess ligand, and CdTe nanocrystals coated in amphiphilic polymers completely dissolve over the course of 1–2 weeks when stored at room temperature. The quantum dots coated with multidentate ligands could also undergo dialysis for more than 1 week without deleterious effects, whereas quantum dots coated with monovalent ligands generally aggregate within 2–3 h. These nanocrystals can also withstand ultracentrifugation, and spread evenly on TEM grids when cast from aqueous solutions, unlike their aggregation-prone counterparts coated in monovalent ligands. Indeed, this multidentate polymer combines the compact size of the monovalent ligand coatings, the antioxidant properties of reduced thiols, and the colloidal stability of amphiphillic micellar coatings.

10.2.4 Size Comparison with Proteins

In order to assess the relevance of these new quantum dots for bioimaging applications, their hydrodynamic sizes were compared directly with proteins. Figure 10.4 shows a size comparison of gel filtration chromatograms of multidentate polymer-coated quantum dots (four emission colors) with globular protein standards. The results demonstrate that the green-emitting QDs (515 nm) have a hydrodynamic size slightly larger than fluorescent proteins (MW = 27–30 kDa), while the yellow-emitting quantum dots (562 nm) are slightly smaller than serum albumin (MW = 66 kDa). Even the near-infrared emitting dots (720 nm) are similar to antibodies (MW = 150 kDa) in hydrodynamic size.

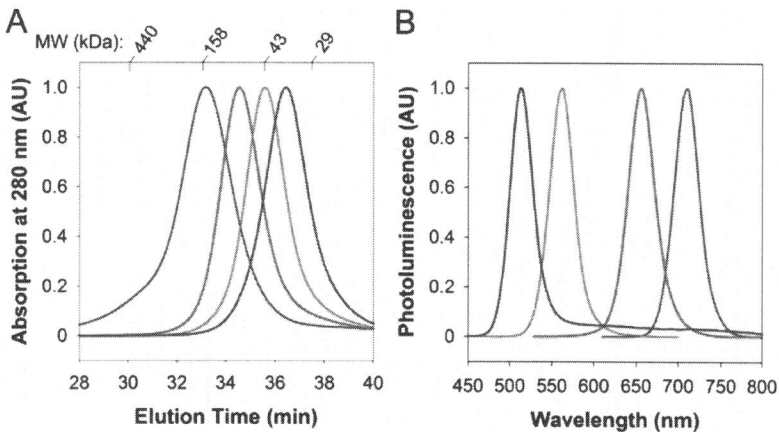

Fig. 10.4. (a) Gel filtration chromatograms of multidentate polymer coated CdTe quantum dots showing direct size comparison with protein standards ferritin (440 kDa), aldolase (158 kDa), ovalbumin (43 kDa), and carbonic anhydrase (29 kDa). (b) Fluorescence emission spectra from the corresponding quantum dots. The quantum dot hydrodynamic sizes are 5.6 nm (2.5 nm core, blue), 6.6 nm (3.1 nm core, green), 7.8 nm (4.0 nm core, red), and 9.7 nm (6.0 nm core, brown)

10.2.5 Super-Resolution Imaging

Recent advances in super-resolution optical imaging have made it possible to detect and locate single fluorescent molecules and single nanoparticle emitters at spatial resolutions far beyond the optical diffraction limit [25–28]. For single-molecule imaging in living cells, Hell and coworkers [29–31] have developed stimulated emission depletion (STED) microscopy, and have achieved focal-plane resolution as high as 10–20 nm. Single molecules and single particles have also been detected at nanometer accuracy by fitting their fluorescence intensity profiles to a two-dimensional Gaussian function or point spread function (PSF) [25–28]. Work in our own group has used color-coded nanoparticles and automated image processing algorithms for nanometer-scale structural mapping of biomolecular complexes [32]. In comparison with organic dyes and fluorescent proteins, QD probes are nearly 100 times brighter (as judged by the rate of photon emission at the same flux of excitation light) and 2–3 orders of magnitude more stable against photobleaching. In addition, QDs with different emission colors can be simultaneously excited with a single light source, avoiding the problems of par-focality and image registration often encountered in multicolor fluorescence imaging [33].

A major problem in using QDs for STED super-resolution imaging is significant overlapping between the absorption and emission spectra of conventional semiconductor nanocrystals. To address this problem, we have recently developed strain-tunable colloidal nanocrystals by using lattice-mismatched heterostructures that are grown by epitaxial deposition of a compressive shell (e.g., ZnSe or CdS) onto a soft and small nanocrystalline core (e.g., CdTe) [34]. This combination of a "squeezed" core and a "stretched" shell causes dramatic changes in both the conduction and valence band energies. As a result, we show that core-shell QDs with standard type-I behavior are converted into type-II nanostructures, leading to spatial separation of electrons and holes, extended excited state lifetimes, and giant spectral shifting. This new class of strain-tunable QDs exhibits narrow light emission with high quantum yield across a broad range of visible and near-infrared wavelengths (500–1,050 nm). As shown in Fig. 10.5, the spectral overlap between absorption and emission is dramatically reduced for type-II QDs, largely due to the physical separation of electron and holes. As noted above, this property will be important for multicolor super-resolution optical microscopy based on STED [29–31].

In summary, we have reported a new strategy to minimize the hydrodynamic size of QDs by using multifunctional, multidentate polymer ligands. A novel finding is that a balanced composition of thiol and amine groups yields

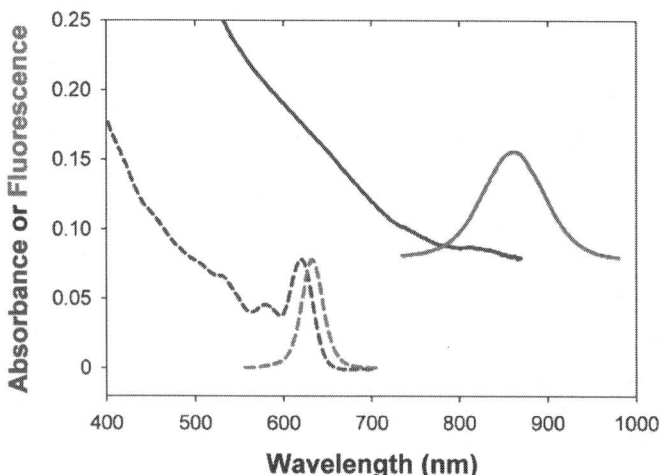

Fig. 10.5. Absorption spectra (blue) and fluorescence spectra (red) of type-I (dotted lines) and type-II (solid lines) (core)shell quantum dots. In conventional type-I quantum dots such as (CdSe)ZnS, the core material valence band is higher in energy than the shell and the conduction band energy is lower in energy, confining both of the charge carriers to the core. Thereby, shell overgrowth only marginally changes the bandgap, and the absorption and fluorescence spectra are similar to those of the core, but with an enhanced stability and fluorescence efficiency. In strain-tunable type-II quantum dots such as (CdTe)CdSe, the energy bands are staggered such that the charge carriers are spatially segregated, allowing only indirect band-edge transitions. Shell overgrowth decreases the bandgap, allowing electronic transitions at lower energy, thus red-shifting the absorption and fluorescence spectra. Discrete transitions are attenuated from the absorption spectra and the band edge oscillator strength diminishes

a highly compact coating for QDs, with a hydrodynamic thickness of only 1.5–2 nm. This has led to a new generation of highly bright and stable QDs with hydrodynamic diameters similar to proteins (5.6–9.7 nm) with tunable fluorescence emission from the visible to the near infrared. This coating technology has also been applied to strain-tunable quantum dots with enhanced optical properties for STED imaging. These size-minimized QDs open new possibilities in multicolor molecular and cellular imaging at the level of single molecules and single nanoparticles.

10.3 Experimental Section

10.3.1 Polymer Synthesis

PAA (1 g, 13.9 mmol carboxylic acids) was mixed with 25 mL DMSO in a 150-mL three-necked flask. After stirring for 24 h at 35°C, freshly prepared anhydrous solutions of cysteamine (187 mg, 2.43 mmol) and Fmoc-ethylenediamine (686 mg, 2.43 mmol), each dissolved in 6 mL DMSO, were added. The solution was protected from light and bubbled with argon for 30 min at 35°C. After the addition of an anhydrous solution of NHS (1.12 mg, 9.71 mmol) in 6 mL DMSO, DIC (736 mg, 5.83 mmol) was slowly added over the course of 40 min during vigorous stirring. Bubbling was continued for 30 min, and then the reaction was allowed to proceed for 7 days at 40°C in the dark. Piperidine (18 mL) was then added and the solution was stirred for 4 h to deprotect the primary amines. β-mercaptoethanol (501 mg, 6.41 mmol) was added to quench the reaction, and the solution was stirred for 2 h at 40°C, then cooled to room temperature and filtered. The mixture was condensed to ∼4 mL at 45°C under vacuum (∼40 Pa), and the polymer was precipitated with the addition of a 2:1 mixture of ice-cold acetone–chloroform, and isolated via centrifugation. The polymer was dissolved in ∼5 mL anhydrous dimethylformamide, filtered, and precipitated again with acetone–chloroform. This process was repeated three times, and the polymer was finally washed with acetone, dried under vacuum, and stored under argon. This modified polymer was a white powder, soluble in water, DMSO, dimethylformamide, or methanol, but insoluble in acetone, unlike PAA. If stored under air, this polymer darkened and became yellow-brown over the course of a few weeks, and also became increasingly difficult to dissolve in various solvents. This aging process coincided with a significant decrease in the number of active thiols per polymer, determined as described below, and is therefore likely due to the formation of interpolymer disulfide crosslinks.

10.3.2 Determination of Reactive Amines and Thiols

The modified polymer was assayed for reactive amines and thiols using fluorescamine and 5, 5′-dithiobis(2-nitrobenzoic acid) (Ellman's reagent), respectively. For amine determination, a 10-mg/mL solution of fluorescamine in

DMSO was freshly prepared, and glycine standards (100 nM–1 mM) were prepared in deionized water. The assay was initiated by mixing 411.3 μL water, 50 μL sample or standard, 25 μL of 1 M sodium borate buffer (pH 8.5), and 13.7 μL fluorescamine solution. After 20 min of reaction in the dark, the fluorescence intensity at 470 nm, with 380 nm excitation, was measured. The polymer was assayed immediately after dissolution at 10 μg/mL in 20 mM sodium hydroxide. For thiol determination, a 2-mM stock solution of Ellman's reagent in 50 mM sodium acetate buffer (pH 4.7), and L-cysteine standards (10 μM–100 mM) in deionized water were freshly prepared at 4°C. The assay was initiated by mixing 850 μL water, 10 μL sample or standard, 100 μL of 1 M Tris buffer (pH 8.5), and 50 μL Ellman's reagent solution. After 10 min of reaction, the optical density at 412 nm was measured. The polymer was assayed immediately after dissolution in 20 mM sodium hydroxide at 500 μg/mL. Standard curves allowed the determination of the molar amount of thiol or amine per gram of polymer. These values were converted to moles of functional group per polymer chain using the molecular weight of the modified polymer ($\sim 2,200$ Da), determined via gel filtration chromatography, which correlated strongly with theoretical calculations.

10.3.3 Ligand Exchange with Thioglycerol

Purified CdTe quantum dots (2.5 nm) in chloroform (7 mL, $\sim$150 μM) were added to a three-necked flask connected to a Schlenk line. Under intense stirring, neat 1-thioglycerol was added dropwise until the first visible sign of flocculation. Then 4 mL of DMSO was added dropwise. An excess of 1-thioglycerol (3 mL) was then added, and chloroform was removed under vacuum at 25°C. After stirring for an additional 2 h at 25°C under argon, the quantum dots were precipitated with the addition of an ice-cold mixture of acetone–chloroform (1:1, 193 mL total). Following centrifugation, the pellet was washed with acetone and dried under vacuum. It was noted that the fluorescence maximum and the first exciton peak blue-shifted if this ligand exchange procedure was performed in the presence of air and a large excess of 1-thioglycerol. The extent of blue-shifting was found to be time dependent and was substantially reduced when the reaction was performed under inert gas. This blue-shift was deemed to be the result of controlled oxidative etching of the quantum dots, rather than alternative mechanisms (e.g., formation of a $CdTe_xS_{1-x}$ alloy or core-shell structure) for several reasons. (1) This hypsochromic shift was strongly correlated with a substantial decrease in the extinction coefficient of the first exciton peak. In fact, after ~ 2 days of reaction, there was essentially no absorption at wavelengths greater than 320 nm, likely due to complete dissolution of the quantum dots. Because this etching process was uniform, the quantum dots maintained a discrete first exciton peak, allowing exact calculation of extinction coefficients of ultrasmall quantum dots with the reasonable assumption that the total number of quantum dots remained fixed. (2) This blue-shift of the optical spectra was associated with a decrease in photoluminescence efficiency, an increase in Stokes shift, and an increase in deep trap emission, common features of ultrasmall quantum dots (<2 nm). (3) Small quantum dots (2.5 nm) at high concentration could easily be detected via TEM and DLS, but after a substantial blue-shift, neither of these techniques revealed the presence of nanoparticles. This would suggest that the resulting nanoparticles were below the size limit for TEM contrast, and that their scattering intensity was reduced below their detection limit via DLS. (4) Finally, assay of the quantum dot supernatant after precipitation revealed the presence of both free cadmium and tellurium (via inductively coupled plasma–mass spectrometry) only after extended etching times.

10.3.4 Coating with Multidentate Polymer Ligands

Two techniques were employed to coat 1-thioglycerol quantum dots with the modified polymer. In the first method, CdTe quantum dots coated with 1-thioglycerol were suspended in basic water (50 mM sodium hydroxide), centrifuged at $7,000g$ for 10 min, and then filtered to remove aggregated nanocrystals. Various amounts of polymer dissolved in basic water were added to the quantum dots, which were then gently mixed. In the second method, quantum dots coated with 1-thioglycerol were suspended in DMSO and centrifuged at $7,000g$ for 10 min to remove possible nanocrystal aggregates. The nanocrystals were diluted to $\sim$5–20 µM for smaller sizes (2.5–3.5 nm) or $\sim$2–5 µM for larger nanocrystals. The nanocrystal solution was then degassed extensively at room temperature and charged with argon. An anhydrous DMSO solution of the polymer ($\sim$5 mg/mL) was added under vigorous stirring. The solution was then heated to 60°C for 90 min for smaller quantum dots (2.5–3.5 nm) or 70–75°C for 120 min for larger nanocrystals. In the absence of the polymer, the nanocrystals aggregated and precipitated from solution during heating. Indeed, the multidentate polymer greatly enhances the thermal stability of these nanocrystals, as there was no evidence of Ostwald ripening of 2.5-nm cores up to $\sim$130°C. After cooling the quantum dots to room temperature, ice-cold aqueous sodium hydroxide (50 mM, twice the volume of DMSO) was slowly added, and the solution was stirred for 2 h. The quantum dots were then extensively dialyzed against basic water for 2–3 days using 25 kDa molecular weight cutoff dialysis tubing (Spectra/Por).

10.3.5 Calculation of Molar Capping Ratio

The amount of polymer added per quantum dot was standardized with reference to the number of surface atoms on the quantum dot. This relationship was used in order to shed light on the mechanism of interaction between the quantum dots and the polymer, and to simplify the extrapolation of the polymer coating procedure to other nanocrystalline materials, without the need for extensive optimization. The MCR was reported as the number of thiol and amine groups per surface atom. Therefore,

$$\text{MCR} = \frac{n_{\text{SH}} + n_{\text{NH2}}}{n_{\text{Cd}} + n_{\text{Te}}},$$

where n_{SH} and n_{NH2} are the numbers of thiols and amines on the polymer ligand, respectively, and n_{Cd} and n_{Te} are the number of cadmium and tellurium surface atoms on the quantum dot surfaces. For example, a 2.5-nm CdTe quantum dot has ~ 95 total surface atoms (the calculation of the number of surface atoms per quantum dot is described below), and one polymer chain contains roughly 6.5 basic groups (3.5 thiols and 3.0 amines). Therefore, the optimal capping ratio (OCR) value of 1.5× denotes the addition of ~ 22 polymer chains per quantum dot, or roughly 48 mg of polymer per μmol of quantum dot. Indeed, this is a very small amount of polymer for such a large number of quantum dots. With elevated temperature this reaction is highly efficient as nearly all of the polymer binds to the quantum dots (no detectable free amines were found in the dialysate during purification).

10.3.6 Calculation of Surface Atoms Per Nanocrystal

Determination of surface atom density on nanocrystals can be difficult, and imprecise, especially for very small particles that cannot be easily characterized microscopically. Nevertheless, reasonable accuracy can be obtained by using theoretical calculations informed by empirical data. In this work, the CdTe nanocrystals that were prepared (2.5–6 nm diameter) were found to be in the zinc blende crystal structure, allowing the use of the bulk density and interplanar distances of zinc blende CdTe in these calculations. It is likely that a variety of crystalline facets are exposed on individual nanocrystals, each with a range of planar densities of atoms. It is also likely that there is a distribution of different facets exposed across an assembly of nanocrystals. Therefore, one may obtain an effective average number of surface atoms per nanocrystal by averaging the surface densities of commonly exposed facets in zinc blende nanocrystals over the calculated surface area of the nanocrystal. In this work we chose to use the commonly observed (111), (100), and (110) zinc blende planes, which are representative of the lattice structure, with both polar and nonpolar surfaces. For this calculation, we defined a surface atom as an atom (either Cd^{2+} or Te^{2-}) located on a nanocrystal facet with one or more unpassivated orbitals. Some facets, such as Cd^{2+}-terminated {111} faces, have closely underlying Te^{2-} atoms that are less than 1 Å beneath the surface plane. These atoms reside in the voids between Cd^{2+} atoms, and thus are likely to be sterically accessible from the surface, but because they are completely passivated, they were not included in this definition.

First we calculated the average distance between parallel planes of atoms for zinc blende CdTe. This average interplanar distance, d, is therefore the distance between the plane of surface atoms and the next underlying plane of atoms. In the [100] and [110] directions, all adjacent planes are equidistant, whereas this distance varies between neighboring planes in the [111] direction, and thus we calculated an "average" interplanar distance. We also calculated the planar density of atoms on each facet, although this data was not directly used in our calculation.

Next we calculated the effective volume of surface atoms within each quantum dot. We assumed a spherical geometry and used the interplanar distance d as the thickness of one monolayer of surface atoms in each nanocrystal. In this calculation, the surface volume was used, rather than the surface area, in order to yield a more realistic determination of surface atoms in very small nanocrystals ($<2\,$nm). For these high surface area nanocrystals, use of the surface area generally resulted in a surface atom number that was larger than the total number of atoms in each nanocrystal. Therefore,

$$V_{SA} = \frac{4}{3}\pi[r^3 - (r - d)^3],$$

where V_{SA} is the volume of surface atoms per quantum dot, r is the nanocrystal radius, and d is the average interplanar distance. It was assumed that this spherical shell of surface atoms was the same density as bulk zinc blende CdTe, and therefore the number of surface atoms per nanocrystal could be calculate as

$$n_{SA} = 2\frac{V_{SA} \times D \times N_A}{MW_{CdTe}},$$

where n_{SA} is the number of surface atoms per nanocrystal, D is the bulk density of zinc blende CdTe ($5.85\,$g cm^{-3}), N_A is Avogadro's number, MW is the molecular weight of CdTe, and 2 is a factor accounting for 2 atoms per molecule of CdTe.

Acknowledgements

This work was supported by grants from the National Institutes of Health (P20 GM072069, R01 CA108468, and U01HL080711, U54CA119338). A.M.S. acknowledges the Whitaker Foundation for generous fellowship support.

References

1. A.M. Smith, S.M. Nie, J. Am. Chem. Soc. **130**, 11278–11279 (2008)
2. H.S. Choi, W. Liu, P. Misra, E. Tanaka, J.P. Zimmer, B.I. Ipe, M.G. Bawendi, J.V. Frangioni, Nat. Biotechnol. **25**, 1165–1170 (2007)
3. W.H. Liu, H.S. Choi, J.P. Zimmer, E. Tanaka, J.V. Frangioni, M. Bawendi, J. Am. Chem. Soc. **129**, 14530–14531 (2007)
4. J.P. Zimmer, S.W. Kim, S. Ohnishi, E. Tanaka, J.V. Frangioni, M.G. Bawendi, J. Am. Chem. Soc. **128**, 2526–2527 (2006)
5. T. Pellegrino, L. Manna, S. Kudera, T. Liedl, D. Koktysh, A.L. Rogach, S. Keller, J. Radler, G. Natile, W.J. Parak, Nano Lett. **4**, 703–707 (2004)

6. T. Pons, H.T. Uyeda, I.L. Medintz, H. Mattoussi, J. Phys. Chem. B. **110**, 20308–20316 (2006)

7. A.M. Smith, H.W. Duan, M.N. Rhyner, G. Ruan, S.M. Nie, Phys. Chem. Chem. Phys. **8**, 3895–3903 (2006)

8. A.M. Smith, H.W. Duan, A.M. Mohs, S.M. Nie, Adv. Drug Delivery Rev. **60**, 1226–1240 (2008)

9. I.L. Medintz, H.T. Uyeda, E.R. Goldman, H. Mattoussi, Nat. Mater. **4**, 435–446 (2005)

10. W.C.W. Chan, D.J. Maxwell, X.H. Gao, R.E. Bailey, M.Y. Han, S.M. Nie, Curr. Opin. Biotechnol. **13**, 40–46 (2002)

11. A.P. Alivisatos, Nat. Biotechnol. **22**, 47–52 (2004)

12. W. Liu, M. Howarth, A.B. Greytak, Y. Zheng, D.G. Nocera, A.Y. Ting, M.G. Bawendi, J. Am. Chem. Soc. **130**, 1274–1284 (2008)

13. M. Howarth, W.H. Liu, S. Puthenveetil, Y. Zheng, L.F. Marshall, M.M. Schmidt, D.K. Wittrup, M. Bawendi, A.Y. Ting, Nat. Methods **5**, 397–399 (2008)

14. S.W. Kim, S. Kim, J.B. Tracy, A. Jasanoff, M.G. Bawendi, J. Am. Chem. Soc. **127**, 4556–4557 (2005)

15. S. Kim, M. Bawendi, J. Am. Chem. Soc. **125**, 14652–14653 (2003)

16. W. Guo, J.J. Li, Y.A. Wang, X.G. Peng, J. Am. Chem. Soc. **125**, 3901–3909 (2003)

17. F. Dubois, B. Mahler, B. Dubertret, E. Doris, C. Mioskowski, J. Am. Chem. Soc. **129**, 482–483 (2007)

18. K. Susumu, H.T. Uyeda, I.L. Medintz, T. Pons, J.B. Delehanty, H. Mattoussi, J. Am. Chem. Soc. **129**, 13987–13996 (2007)

19. I. Nabiev et al., Nano Lett. **7**, 3452–3461 (2007)

20. J. Lovric, H.S. Bazzi, Y. Cuie, G.R.A. Fortin, F.M. Winnik, D. Maysinger, J. Mol. Med. **83**, 377–385 (2005)

21. A.L. Rogach, T. Franzl, T.A. Klar, J. Feldmann, N. Gaponik, V. Lesnyak, A. Shavel, A. Eychmuller, Y.P. Rakovich, J.F. Donegan, J. Phys. Chem. C **111**, 14628–14637 (2007)

22. X.S. Wang, T.E. Dykstra, M.R. Salvador, I. Manners, G.D. Scholes, M.A. Winnik, J. Am. Chem. Soc. **126**, 7784–7785 (2004)

23. M.F. Wang, N. Felorzabihi, G. Guerin, J.C. Haley, G.D. Scholes, M.A. Winnik, Macromolecules **40**, 6377–6384 (2007)

24. A.K. Chakraborty, A.J. Golumbfskie, Annu. Rev. Phys. Chem. **52**, 537–573 (2001)

25. S.T. Hess, T.P. Girirajan, M.D. Mason, Biophys. J. **91**, 4258–4272 (2006)

26. M.J. Rust, M. Bates, X.W. Zhuang, Nat. Methods. **3**, 793–795 (2006).

27. X. Michalet, S. Weiss, Proc. Natl. Acad. Sci. USA. **103**, 4797–4798 (2006)

28. E. Betzig, G.H. Patterson, R. Sougrat, O.W. Lindwasser, S. Olenych, J.S. Bonifacino, M.W. Davidson, J. Lippincott-Schwartz, H.F. Hess, Science **313**, 1642–1645 (2006)

29. K.I. Willig, R.R. Kellner, R. Medda, B. Hein, S. Jakobs, S.W. Hell, Nat. Methods **3**, 721–723 (2006)

30. K.I. Willig, S.O. Rizzoli, V. Westphal, R. Jahn, S.W. Hell, Nature **440**, 935–939 (2006).

31. G. Donnert, J. Keller, R. Medda, M.A. Andrei, S.O. Rizzoli, R. Luhrmann, R. Jahn, C. Eggeling, S.W. Hell, Proc. Natl. Acad. Sci. USA. **103**, 11440–11445 (2006)

32. A. Agrawal, R. Deo, G.D. Wang, M.D. Wang, S.M. Nie, Proc. Natl. Acad. Sci. USA. **105**, 3298–3303 (2008)
33. T.D. Lacoste, X. Michalet, F. Pinaud, D.S. Chemla, A.P. Alivisatos, S. Weiss, Proc. Natl. Acad. Sci. USA. **97**, 9461–9466 (2000)
34. A.M. Smith, A.M. Mohs, S.M. Nie, Nat. Nanotechnol. **4**, 56–63 (2009)

Mapping Transcription Factors on Extended DNA: A Single Molecule Approach

Yuval Ebenstein, Natalie Gassman, and Shimon Weiss

Summary. The ability to determine the precise loci and distribution of nucleic acid binding proteins is instrumental to our detailed understanding of cellular processes such as transcription, replication, and chromatin reorganization. Traditional molecular biology approaches and above all Chromatin immunoprecipitation (ChIP) based methods have provided a wealth of information regarding protein–DNA interactions. Nevertheless, existing techniques can only provide average properties of these interactions, since they are based on the accumulation of data from numerous protein–DNA complexes analyzed at the ensemble level. We propose a single molecule approach for direct visualization of DNA binding proteins bound specifically to their recognition sites along a long stretch of DNA such as genomic DNA. Fluorescent Quantum dots are used to tag proteins bound to DNA, and the complex is deposited on a glass substrate by extending the DNA to a linear form. The sample is then imaged optically to determine the precise location of the protein binding site. The method is demonstrated by detecting individual, Quantum dot tagged T7-RNA polymerase enzymes on the bacteriophage T7 genomic DNA and assessing the relative occupancy of the different promoters.

11.1 Introduction

The coordinated regulation of gene expression in response to changes in cellular conditions is the basis for cell function. The construction of models that describe the differential expression of genes is an important step towards understanding of how living organisms behave. RNA polymerase (RNAP) is a key player in gene expression, transcribing DNA sequences (genes) into messenger RNA with the help of its associated factors. This transcriptional activity is under the control of transcription factors (TFs) and regulatory proteins such as inhibitors and enhancers, which may bind to DNA sequences and/or interact with the transcription complex. The specific DNA sequence to which RNAP binds in order to initiate the transcription of RNA is known as a promoter. Knowledge of promoter locations is the first step towards the construction of a full transcriptional regulatory network. It gives information on the activity of specific gene expressions and defines genomic regions for the identification of regulatory elements associated with various TFs, thus facilitating the interpretation of their binding sites. Considerable effort has been devoted to identify transcriptional networks and to map genomic regions that participate in the control of gene expression. Initial efforts have been based on in vitro DNA-binding and reporter assays such as electrophoretic mobility shift assays, DNA footprinting, and luciferase reporter systems. These tools allowed the identification of regulatory elements in the vicinity of selected genes and have elucidated binding motifs for candidate TFs. These approaches are limited by the fact that they require prior knowledge of candidate genes and permit analysis of relatively small genomic regions. In vitro approaches to the identification of TF binding sites have proven to suffer from both false positive and false negative results when compared to the live genomic system. This may partially be explained by the fact that protein–protein interactions may recruit TFs to sequences in the genome that differ from the optimal in vitro binding sites. The conformation differences between the artificial recombinant binding element and the native binding domains in their genomic context may also play an important role in dictating which binding motifs are active in vivo. Even in the primitive case of viral transcription, the binding of T7 bacteriophage RNAP to a recombinant promoter sequence has been altered when the plasmid containing the binding site has been transformed from supercoiled to linear form [1]. Chromatin immunoprecipitation (ChIP) allows the in vivo identification of TF binding sites in the context of the entire genome, and therefore avoids many of the problems associated with traditional in vitro approaches (for reviews see [2–4]) ChIP involves chemical crosslinking of DNA–protein interactions in living cells. Commonly, formaldehyde is used to "fix" TFs to their cognate binding sites in the genome. Formaldehyde is membrane permeable and forms a covalent bridge between proximal amino or imino groups such as a lysine in contact with a cytosine. The crosslinking provides a "snapshot" of the cell's transcription state under well defined conditions. The crosslinked DNA is then extracted from cells and fragmented. Antibodies against specific TFs are used to immunoprecipitate TFs together with their bound DNA fragments. In this way, only the DNA associated with the protein of interest is "fished out" of the rich protein–DNA soup, resulting in an enriched sample. Next, DNA–protein crosslinking is reversed, and enriched DNA fragments are purified for downstream analysis. Standard ChIP assays use Southern blotting, polymerase chain reaction (PCR), or quantitative real-time PCR (qPCR) as a read-out; however, these approaches still require a specific region of interest in the genome to facilitate analysis. The introduction of micro-array technology and its combination with ChIP in the form of genomic microarrays (ChIP-chip) [5,6], and more recently the direct sequencing of ChIP products using 454 or Solexa G1 sequencing platforms (ChIPseq) [7], allows large-scale, genome-wide identification of TF-binding sites. In ChIP-chip, the immunoprecipitated DNA is labeled with a fluorescent dye and used to probe a genomic array. Although ChIP-chip represents a high throughput approach, it requires microarrays of unbiased full genomic sequences. Whole-genome tiling arrays are available for all nonrepetitive regions of the human genome, but they are expensive, and for the human genome, consist of a set of roughly 38 arrays [8]. Lower coverage platforms, although useful in many cases are not unbiased, as they use arbitrarily defined promoter regions without experimental verification and exclude other regions that might be involved in transcriptional regulation, such as those in introns. ChIPseq involves the direct sequencing of ChIP-enriched DNA. Sequencing data is matched to its respective loci in the known genomic sequence, and binding sites are determined by accumulation of sequence reads above the background level. ChIPseq is not limited by array coverage, and therefore offers unbiased analysis of the entire genome. In addition, it is simple and offers higher resolution compared with array-based methods.

All of the above ChIP based techniques address one TF species at a time. Indirect assessment of cooperative binding of several different TFs is achieved by running parallel experiments on the same sample and comparing the location data for different TFs. Nevertheless, ambiguity remains regarding the exact nature of the cooperativity due to the fact that data comes from different DNA fragments. To address this issue, a second round of ChIP may be performed on an enriched sample using an antibody against a second TF of interest. This double ChIP procedure selects only DNA fragments interacting with both TFs. This approach has been termed double-ChIP, repChIP, or SeqChIP [9,10] and although powerful, it is limited to close range interactions comparable to the DNA fragment size used in the ChIP experiment (usually in the order of 500 bp). Regulation of gene expression, however, is not limited to close range interactions between TFs and especially in eukaryotes may be influenced by extremely distant cofactors brought together by random fluctuations or directed chromatin reorganization. An example of how chromosomal interactions can influence gene expression is the folding ability of a chromosomal region that can bring an enhancer and associated transcription factors within close proximity of a gene [11]. The Capturing Chromosome Conformation (3C) assay [12] and its offspring: ChIP-loop [13], 3C-on chip(4C) [14], and 3C-carbon copy(5C) [15] are approaches to detect the frequency of interaction between any two genomic loci (for review, see [16]). In these methods, long range DNA–protein interactions are captured by formaldehyde treatment. The crosslinked chromatin is subject to digestion by restriction enzymes to free the unwound chromatin from the bulk of crosslinked material. This is followed by ligation of crosslinked fragments, and then ligation frequencies are measured. In the ChIP-loop assay, immunoprecipitation enriches the sample for fragments bound by a specific protein, and restriction fragments are ligated to each other on the ChIP beads. In ChIP-loop and 3C, ligation frequencies are measured by quantitative PCR, using a unique primer set for each ligation junction analyzed. In 5C, ligation events are amplified by ligation-mediated amplification (LMA) with T7 and T3 primers, and then analyzed by large-scale sequencing or microarray. In 4C, ligation junctions are first trimmed by a frequently cutting secondary restriction enzyme, and then subjected to ligation to form circles followed by inverse PCR to amplify captured fragments. The 4C PCR product is analyzed by large-scale sequencing or microarray analysis. In general, 3C technology is particularly suited to study the conformation of genomic regions that range roughly from five to several hundred kilobases (kb) in size.

Recent years have brought remarkable advancements to the study of transcription regulation and the new tools developed in the field have already revealed a wealth of new information regarding the complex and poorly understood transcriptional networks underlying cell function. Despite these exciting methodological advancements and their capability for genome-wide studies of gene expression, there is still need for complimentary approaches. New approaches should both overcome some of the intrinsic drawbacks of the above methods, and more importantly provide independent evidence regarding the state of the transcription machinery on the genome. In the following sections, we present a single molecule approach for direct visualization of TFs bound to their genomic target and discuss its advantages and limitations in light of existing technologies. Finally, we demonstrate initial results highlighting the potential of this technique for genomic analysis of transcription.

11.2 Concept of Single Molecule ChIP: Seeing is Believing

Recent research using fluorescent quantum dot (QD) bioconjugates [17–20] suggests that they have great potential for development of novel techniques for mapping protein–protein and protein–DNA complexes. The spectral properties of QDs are ideal for this application. They have narrow, "tunable" emission spectra and thus provide a much larger array of colors than is possible with available organic fluorophores, an important consideration for the simultaneous observation of several components (multiplexing). QDs can also be excited with a single, common laser wavelength, reducing chromatic aberrations in microscopy. In addition, QDs are less subject to chemical or biological degradation and are much more photostable, resulting in an overall higher photon count from QDs compared to organic dyes. It was demonstrated that these features allow nanometer resolution localization of multicolor QDs [21]. We propose to build on these results to develop a novel technique for identifying the target sites of DNA binding proteins that employs QDs to uniquely tag both DNA and regulatory TFs. The outcome of these studies would allow high-resolution mapping of TFs to both their cognate DNA binding site and to their interacting protein partners. This technique will be useful for mapping protein complexes on full-length or partial genomic DNA, and identifying the precise positions of the transcriptional complexes within the genome. The localization precision (within ~ 20–$100\,\mathrm{bp}$) of this technique will provide significantly improved data for computational prediction of conserved binding sequences. Furthermore, the positioning of the binding sites relative to the RNAP binding site could be used to predict whether the interaction leads to repression or enhancement of transcriptional activity.

A schematic representation of the experimental procedure is depicted in Fig. 11.1. Initial sample preparation is based on standard ChIP protocols, TFs are crosslinked to DNA in vivo and the TF–DNA complex is extracted from the cell. As opposed to ChIP, the genomic DNA is kept as long as possible in order to retain maximum information on individual genomes, and therefore no sonication is performed. In the case of chromatin, the use of restriction enzymes such as in the 3C method may be necessary in order to separate useful DNA from the crosslinked bulk chromatin. In the next step, the sample is labeled with a DNA staining agent such as YOYO-1 and with QDs conjugated to antibodies against TFs of interest. Practically, 4–6 different TFs may be labeled with QDs of different colors and still be easily separated spectrally in the region between 550 and 700 nm. An example of T7-DNA labeled with four spectrally distinct QD species is shown in Fig. 11.1. For orientation along an unknown set of genomic fragments, another color of QD is used to tag a reference point in the genome in a sequence specific manner. This will be achieved using biotinylated DNA [22] or peptide nucleic acid (PNA) [23, 24] probes that will hybridize to the DNA and will be labeled with streptavidin

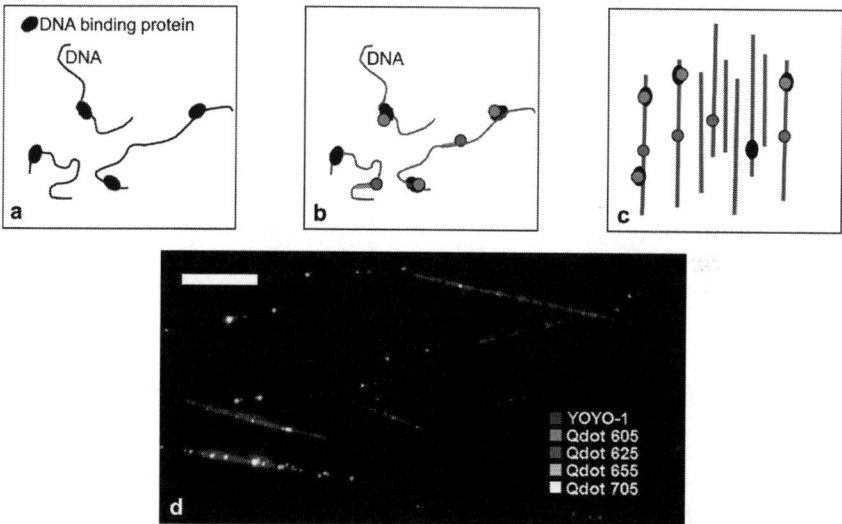

Fig. 11.1. Experimental steps for mapping DNA binding proteins. (a) Crosslinking DNA-binding proteins (black) to DNA. (b) Staining DNA (blue), QD labeling of bound proteins (green), and labeling of specific reference sequences on DNA with QDs (red). (c) Complexes are aligned on a glass coverslip and imaged by a fluorescence microscope. Image analysis provides information on protein location. (d) Pseudo color image of RNAP–biotin crosslinked to aligned DNA and bound to streptavidin-QDs of four different colors (Scale-bar 10 μm)

conjugated QDs [25]. After unbound QDs are removed from the sample, the DNA–TF–QD complexes must be aligned on a surface for optical imaging. We use a modified version of the "optical mapping" method developed by Schwartz and co workers [26]. A droplet of sample is sandwiched between two coverslips, one of which is coated with polylysine, and the induced flow causes alignment and extension of the DNA molecules on the coated surface. The resulting sample consists of linear DNA decorated with QD-labeled TFs, and is now ready to be imaged optically. A single wavelength is sufficient to excite all of the fluorescent species and color detection is done by changing filters in the emission path. For precise distance measurements, an additional image is acquired through a long pass filter eliminating image shifts due to filter changes. This enables simultaneous recording of emission from all the different QDs in the field of view with minimal chromatic aberrations, while retaining color information from the previous step. QD blinking may be utilized to distinguish between two (or more) QDs occupying the same diffraction limited region [27]. Alternatively, a multiview configuration with color registration may be used [28]. The final step consists of image analysis. Fluorescence spots are fitted to a 2D Gaussian providing nanometer resolution localization. The distance between spots related to reference sequences on the DNA and spots related to TFs on the same DNA molecule are measured and histograms are built from distance data accumulated from multiple DNA molecules. An atomic force microscope (AFM) correlated to the optical microscope may be used to "zoom-in" on regions of interest in order to validate the optical measurement. The AFM image confirms that the detected single-QD spots correspond to objects hybridized on the same DNA strand and may also provide additional data such as the presence of other, unlabeled TFs.

This simple concept may complement existing techniques and offer some unique capabilities. Data analysis lends itself easily to automation and many of the image processing issues have already been addressed, providing whole genome "shotgum analysis" [29]. Sample preparation is quick and requires almost no amplification or biological processing, and all aspects of the experiment may easily be performed "in house" using relatively basic equipment. In addition, the needed sample quantity is orders of magnitude smaller than the amount of sample commonly used for ChIP or 3C assays because the method allows monitoring individual DNA fragments. Ultimately, useful measurements can be performed on DNA from a small number of cells assuming such small quantities can be handled. The use of QDs offers several direct advantages; the samples are bright to the degree that a simple camera is sufficient to acquire data. In fact, single QDs and DNA are easily observed with the naked eye through the microscope. The use of a single excitation source simplifies data acquisition even further and allows full color multiplexing in a single shot using a color camera. With a field of view in the order of a $100\,\mu m^2$, the method is ideal for TF distances ranging from 1 to 1000 kb. This addresses coordinated transcription regulation in the window of 500–5000 bp from the promoter, a size range not readily accessible by existing techniques. Another unique feature of this approach is that it is indifferent towards DNA repeats, repetitive sequences that are indistinguishable one from the other by array or sequencing based techniques. As a first step towards realization of this method to its full extent, we demonstrate a simplified, "proof of principle" experiment. The experiment is aimed to measure one QD-labeled protein species without a reference tag and concentrates on some of the more technical issues concerning sample preparation, deposition, and imaging. As a model biological system, we use the well studied T7 bacteriophage. We will specifically study the binding sites of its RNAP along the genomic DNA, for which binding site data is already available. Biotinylated T7 RNAP is incubated in vitro with the full T7 genome (40 kb). The sample is then labeled and mounted on a glass coverslip to be subjected to optical and AFM imaging. The distance of the bound QDs from the DNA ends is measured and the data is analyzed to give information on the location and occupancy of RNAP on the genome. We use the T7 system as an illustrative example but expect the approach to be extensible to larger genomes and broader type of DNA binding proteins.

11.3 Experiment and Results

T7 bacteriophage is a lytic phage that infects *Escherichia coli*. Although it is one of the most studied biological systems and one of the simplest ones, it is not yet fully understood. It has a linear double stranded genome containing 39,936 base pairs. The full sequence as well as extensive genetic characterization has been published [30]. Following infection, phage T7 produces at least 38 proteins during its life cycle. In all, up to 57 genes encoding 60 potential proteins have been found or postulated. However, only 35 of these 60 proteins have at least one known function. And, of the 25 nonessential proteins, only 12 are conserved across the family of T7-like phage [31, 32]. The T7 genome contains 17 promoters for the T7 RNAP out of which five possess the same 23-bp consensus sequence. In addition, three strong *E. coli* RNAP promoters reside on one end of the genome, harnessing the host enzyme to transcribe the early T7 genes within 4–8 min of infection. These early genes are responsible for shutting down of the host metabolism and hijacking its translation machinery. The first T7 gene encodes for the production of the T7 RNAP. This highly efficient single unit enzyme transcribes the majority of T7 genes, including those responsible for T7 DNA replication, phage assembly, and finally lysis of the host bacteria. To facilitate easy and selective binding of QDs, the T7 RNAP was biotinylated in vivo using technology from Avidity, LLC. In vitro binding of RNAP to the T7 genome was performed similar to published protocols [33] (see Sect. 11.5). The sample was then labeled with the intercalating dye YOYO-1 and with streptavidin conjugated QDs. To evaluate the binding efficiency of the various components, a representative preparation was diluted and deposited on freshly cleaved mica for AFM imaging. Figure 11.2 shows a T7 genomic DNA from the above-mentioned sample. The

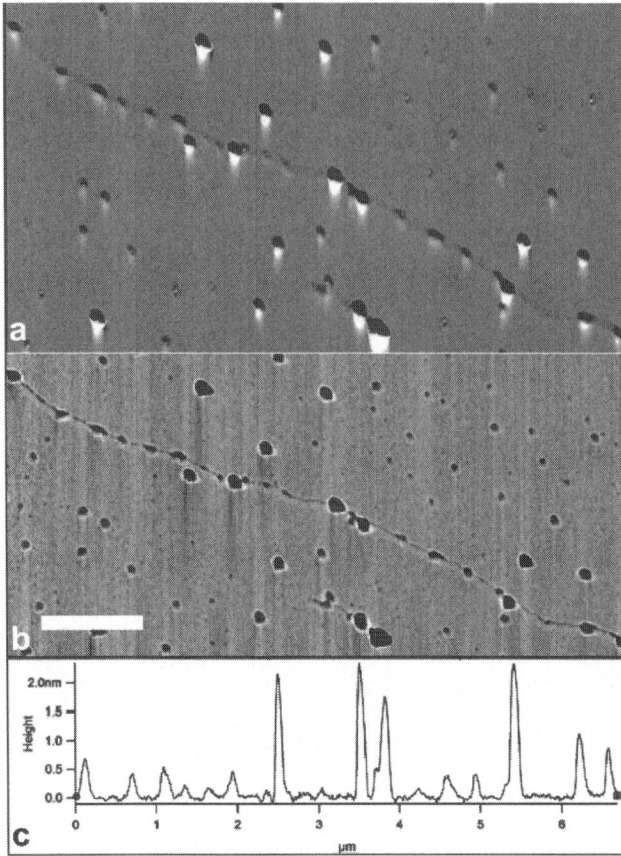

Fig. 11.2. Evaluation of RNAP and QD binding on the T7-phage genome. (a) Amplitude and (b) phase AFM scans of the T7–DNA–RNAP–QD complex. Protrusions of two typical sizes, corresponding to the RNAP and RNAP–QD are visible on the DNA backbone (Scale-bar 1 μm). (c) Cross section along the DNA backbone depicting the height of the various protrusions

DNA fiber is clearly observed with protrusions of two typical sizes along its backbone. We attribute the smaller protrusions to bare RNAP and the larger ones to QD-labeled RNAP. We next turn to optical imaging of the samples. Functionalized glass coverslips were prepared as described in Sect. 11.5 and an untreated coverslip ("out of the box") is placed on top of the polylysine surface. This top coverslip is slightly hydrophobic and serves to trap free QDs from the sample. This was verified by separately imaging the upper and lower coverslip after sample deposition. The sample, 0.5 μL, is diluted into 50 μL of 100 mM Hepes (pH 7.0) containing 0.1% of n-dodecyl-β-D-maltoside (DDM). DDM is a very mild, nonionic surfactant that does not affect the functionality of many proteins [34]. It was found to prevent nonspecific binding of

BSA and streptavidin to poly(dimethylsiloxane) (PDMS) surfaces used for microfluidic channels [35]. We found DDM to improve the extension uniformity of DNA–QD samples and to reduce nonspecific binding of free QD to the polylysine surface. The diluted sample, 3 μL, containing roughly 5 ng of DNA, was deposited on the interface between the two coverslips and immediately mobilized by capillary force, flowing between the two surfaces. The bottom surface is imaged through a 60×, oil immersion objective (PlanApo, 1.45NA, Olympus) and additional magnifying optics resulting in a field of view of ~50 μm², with 97 nm/pixel to allow high resolution localization of fluorescence spots. The sample is excited by a Xenon arc lamp at 470 nm and a fluorescence image is collected using an electron multiplying charged coupled device (EMCCD) (DU997, Andor). Images of the YOYO-1 stained DNA are acquired through a 50 nm wide band-pass filter centered at 535 nm. This is followed by acquisition of the QD signal using a similar filter centered at 655 nm. An automatic protocol was written to perform the sequential imaging and then overlay the separate frames as a color image (IQ, Andor). Cropped images of DNA–RNAP–QD complexes from several fields of view are seen in Fig. 11.3. The complex flow patterns caused by evaporation from all sides of the interface between the two coverslips causes the solution to flow over the same area several times with DNA molecules deposited in the direction of flow in each pass. We note that unidirectional deposition may be achieved by defining a narrow channel with a single inlet and outlet using double sided tape. QD fluorescent spots are clearly visible on many of the DNA molecules. We first evaluated the uniformity and degree of extension by measuring the length of QD tagged DNA molecules. A histogram of 70 such molecules from several fields of view is presented in Fig. 11.3a. The DNA is extended to about $85 \pm 6\%$ of its contour length. This variation in extension factor is the largest source of uncertainty in the measurement and can be significantly reduced by

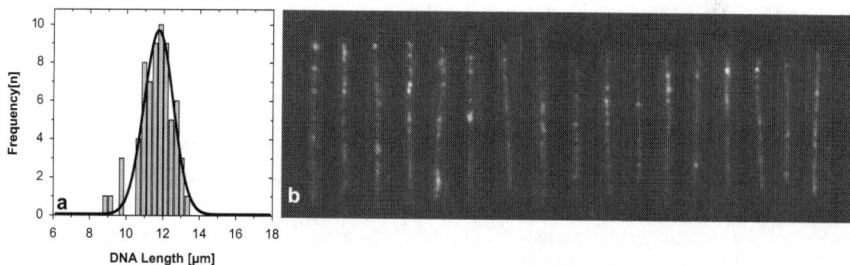

Fig. 11.3. Evaluation of stretching uniformity. (**a**) Histogram constructed from the end-to-end length of seventy T7–DNA–RNAP–QD complexes. A Gaussian fit to the data is centered at 11.75 μm corresponding to 86% of the contour length. The width of the distribution is 1.6 μm corresponding to about 12%. (**b**) Cropped fluorescence images of full length T7-phage genomes flow stretched on a polylysine coated coverslip

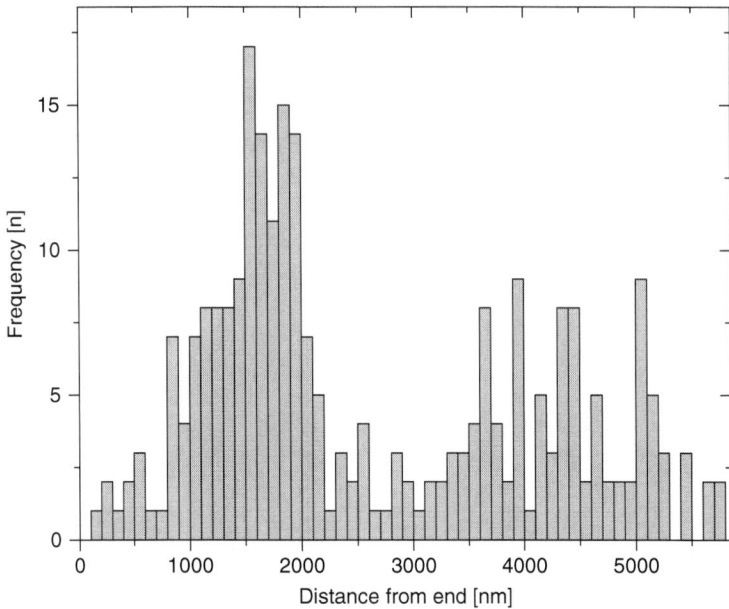

Fig. 11.4. Evaluation of binding specificity. Histogram constructed from 161 distances between a bound QD and the DNA end. To account for both possible orientations of the stretched DNA, only the distance to the closest end was considered. The nonuniform distribution of the binding profile is likely to do with specific binding to the promoter (and other "hot spots")

the incorporation of two or more sequence specific labels serving for internal calibration on every DNA molecule. We next turn to measurements of QD positions on the DNA. We first measured distances between the DNA end and a bound QD. Measurements were done manually using code written in Lab-View (National Instruments). Since this sample contains no reference marker to indicate the orientation of the extended DNA molecules, we only consider distance to the closest end. Figure 11.4 shows a histogram of these distances. Clearly, the distribution is not homogeneous, with some regions of the DNA more densely populated than others. This distribution indicates that QD positions are not random and that specific binding of RNAP is dominating the interaction under our experimental conditions. To refine the data we apply two additional steps to the distance analysis. First, we correct for variations in extension factor by dividing each distance by the total length of the individual DNA molecule; second, we assign orientation based on the best distance match to a known promoter. For this analysis both ends must be clearly defined, and therefore it is performed on 70 genomes matching this criterion. Although this analysis step is biased towards promoters, it gives a good indication on the ability of the method to correlate between measured binding events and TF binding sites. The position histogram in Fig. 11.5 shows the frequency of QD positions along the genome relative to the positions of known promoters for RNAP. Data is presented in normalized percentage units going from left to right in direction of transcription.

11.4 Discussion and Outlook

Despite the low statistics in this experiment, the histogram in Fig. 11.5 shows a considerable degree of correlation between promoters and QD clusters. Most notable is the high occupancy of the region corresponding to promoter 17 at ∼86% of the genome. We note that this region is similar in terms of distance from the genome end to the region at ∼15%, corresponding to the clustered promoters 1.1a, 1.1b, and 1.3, and thus these two regions are susceptible to mixing in our analysis. Nevertheless, they present a significant abundance of binding events relative to other regions. Interestingly, a strong terminator at 61% coincides with a relatively large cluster of recorded QDs. We could find no reported explanation for the abundance of binding events near promoter

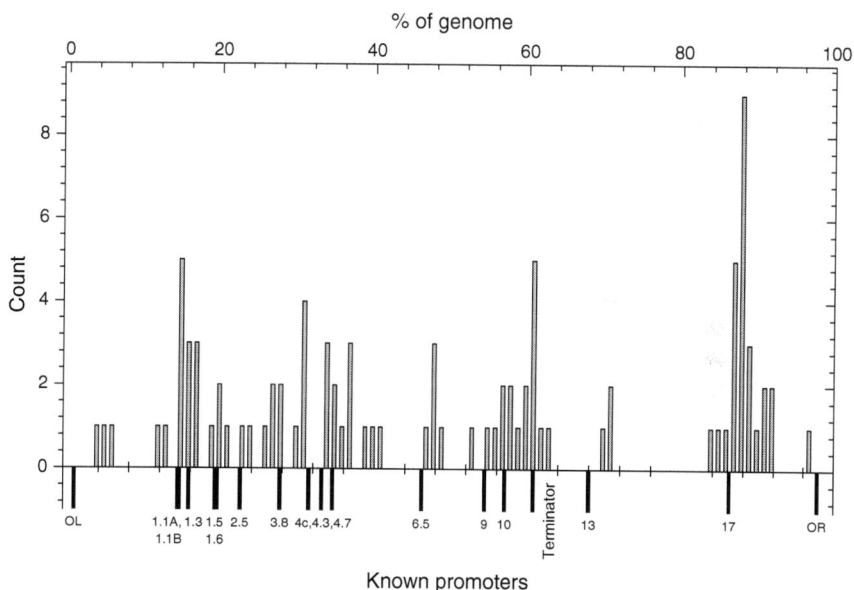

Fig. 11.5. Evaluation of promoter occupancy. Known promoter positions (black) and a histogram constructed from 70 genomes that have two visible ends (gray). Data is presented in normalized percentage units going from left to right in the direction of transcription. The distance between a bound QD and the DNA end was normalized to units of % of the total genome by taking the ratio of the distance and the total length. Orientation was determined by assigning one QD per DNA to the best matching promoter (bin=1%[400bp])

17 relative to other regions. A numerical simulation of the experiment displayed similar results when the binding probability for this promoter was set to twice that of other consensus promoters (data not shown). These observations indicate that even this simplified form of our method may yield insightful results. The interaction mechanism of T7 RNAP with the various promoters is complex and is still a subject of intensive research. Apart from the negative regulator T7 lysozyme protein, no cofactors are involved in T7 transcription, and regulation is mostly governed by selective binding [36]. Ionic strength [1], DNA conformation [1], promoter sequence variations [37], and variations in the sequence downstream [38] and upstream [39] of the promoter have all been shown to have significant effects on binding and transcription efficiency. Most of these experiments were done in vitro using PCR amplified restriction fragments and in vivo utilizing promoter regions cloned into plasmids. Many of the results did not fully correspond to observations from the native phage systems [38]. Furthermore, it has been shown that for example, the in vitro utilization of the promoter at 14.6% increased 10-fold when the fragment containing the promoter was further cleaved by an additional restriction endonuclease (*Hpa* I). This effect was present, although the promoter was located far from the cleavage site [40] pointing out a potential caveat in such assays. To this extent, even in this simplified form, our method may prove a useful tool to study transcription in the full genomic context. Further improvements such as the incorporation of sequence specific tags, microfluidic sample handling, and automated data acquisition and analysis may result in a reliable, high resolution and high throughput mapping tool on the single molecule level.

11.5 Methods

11.5.1 Protein Expression and Purification

A fragment from pDL19 [33] containing the T7–RNAP sequence was cloned into the vector pAN-4 containing an NH2-terminal BiotinAviTag peptide. The plasmid, pNG301, was transformed into AVB101 cells where the N-terminal peptide tag is recognized by an overexpressed biotin ligase which introduces a single biotin molecule at the terminal tag. The purified RNAP is biotinylated and may be readily tagged by a commercially available streptavidin-conjugated QD (qdot, Invitrogen).

11.5.2 Binding and Labeling Reaction

In vitro binding of RNAP to the T7 genome was performed similar to published protocols [34]. Samples were prepared by incubating 3 nM of T7 genome (Boca Scientific) with 30–167 nM of biotinylated T7 RNAP (total volume of about 5 μL) for 20 min at 37 °C in T7 binding buffer (30 mM Hepes pH 7.0, 25 mM K-Glutamate, and 15 mM Mg-acetate). Since T7 RNAP binds weakly to DNA [41] transcription was allowed to initiate by feeding the reaction with 1 μL of 1 mM GTP, UTP, and CTP and incubating at 37 °C for an additional 5 min. When large excess of RNAP was used (>60 nM), nonspecifically bound RNAPs were removed by treatment of 0.6 μL of 1 mg/mL heparin-sepharose (GE Health sciences) for 30 s at 37 °C. The sample was then centrifuged at 2 kg in a benchtop centrifuge for 1 min, and 5 μL was removed from the supernatant. Typical reactions contained 1 μg of DNA to which 0.3 μL of a 1 mM stock solution of YOYO-1 (Invitrogen) in DMSO was added. Streptavidin conjugated QDs (qdot 655, Invitrogen) were diluted to 100 nM and filtered by a 100-kDa pore size membrane (YM-100, Centricon) to eliminate free streptavidin. The flow-through was discarded and the QDs were resuspended in Hepes buffer to a final concentration of 20 nM. QD solution, 5 μL, was added and the sample was incubated in the dark for 30 min.

11.5.3 Surface Preparation

Functionalized glass coverslips were prepared as follows: Glass coverslips (22 mm × 22 mm × 0.13–0.17 mm, Fisher) were washed with 1% (w/v) fresh Alconox (Fisher) solution and sonicated for 15–30 min with heating. Coverslips were then washed with copious amounts of deionized water and baked at 500 °C for 2–3 h to remove any organic contamination. A 10-μL solution of 5–6 μg/mL polylysine in water was sandwiched between two cooled coverslips, and allowed to dry overnight. Before imaging, sandwiched coverslips are separated and an untreated coverslip ("out of the box") is placed on top of the poly-lysine surface.

Acknowledgements

This work was supported by the UCLA-DOE Institute for Genomics and Proteomics. Y.E thanks the Human Frontier Science Program for their support.

References

1. S.P. Smeekens, L.J. Romano, Nucleic Acids Res. **14**(6), 2811–2827 (1986)
2. A. Kirmizis, P.J. Farnham, Exp. Biol. Med. **229**(8), 705–721 (2004)
3. E.C. Massie, G.I. Mills, EMBO Rep. **9**(4), 337–343 (2008)
4. J. Wells, P.J. Farnham, Methods **26**(1), 48–56 (2002)
5. B. Ren et al., Science **290**(5500), 2306–2309 (2000)
6. M.J. Buck, J.D. Lieb, Genomics **83**(3), 349–360 (2004)
7. G. Robertson, et al., Nat. Methods **4**(8), 651–657 (2007)
8. J.S. Carroll, et al., Nat. Genet. **38**(11), 1289–1297 (2006)
9. J.V. Geisberg, K. Struhl, Nucleic Acids Res. **32**(19), e151 (2004)
10. K.M. Scully et al., Science **290**(5494), 1127–1131 (2000)

11. A. Murrell, S. Heeson, W. Reik, Nat. Genet. **36**(8), 889–893 (2004)
12. J. Dekker et al., Science **295**(5558), 1306–1311 (2002)
13. S.i. Horike et al., Nat. Genet. **37**(1), 31–40 (2005)
14. M. Simonis et al., Nat. Genet. **38**(11), 1348–1354 (2006)
15. J. Dostie, J. Dekker, Nat. Protoc. **2**(4), 988–1002 (2007)
16. M. Simonis, J. Kooren, W. de Laat, Nat. Methods **4**(11), 895–901 (2007)
17. X. Michalet et al., Science **307**(5709), 538–544 (2005)
18. A.P. Alivisatos, Nat. Biotechnol. **22**(1), 47–52 (2004)
19. E.R. Goldman et al., Anal. Chem. **76**(3), 684–688 (2004)
20. X. Michalet et al., Single Mol. **2**(4), 261–276 (2001)
21. T.D. Lacoste et al., Proc. Natl. Acad. Sci. U S A. **97**(17), 9461–9466 (2000)
22. K. Keren et al., Science **297**(5578), 72–75 (2002)
23. P.E. Nielsen et al., Science **254**(5037), 1497–1500 (1991)
24. E.Y. Chan et al., Genome Res. **14**(6), 1137–1146 (2004)
25. L.A. Bentolila, S. Weiss, Cell Biochem. Biophys. **45**(1), 59–70 (2006)
26. X. Meng et al., Nat. Genet. **9**(4), 432–438 (1995)
27. B.C. Lagerholm et al., Biophys. J. **91**(8), 3050–3060 (2006)
28. L.S. Churchman et al., Proc. Natl. Acad. Sci. U S A. **102**(5), 1419–1423 (2005)
29. A. Lim et al., Genome Res. **11**(9), 1584–93 (2001)
30. J.J. Dunn, F.W. Studier, J. Mol. Biol. **166**(4), 477–535 (1983)
31. L.Y. Chan, S. Kosuri, D. Endy, Mol. Syst. Biol. **1**, (2005)
32. R. Calendar, T.S. Abedon, *The Bacteriophages*. (Oxford University Press, USA, 2006)
33. Z. Gueroui et al., Proc. Natl. Acad. Sci. U S A. **99**(9), 6005–10 (2002)
34. G.W. Stubbs, H.G. Smith,, B.J. Litman, Biochim. Biophys. Acta (BBA) – Biomembranes. **426**(1), 46–56 (1976)
35. B. Huang et al., Lab Chip **5**(10), 1005–1007 (2005)
36. R.P. Bandwar et al., Biochemistry **41**(11), 3586–95 (2002)
37. R.A. Ikeda, C.M. Ligman,, S. Warshamana, Nucleic Acids Res. **20**(10), 2517–2524 (1992)
38. L.K. Jolliffe, A.D. Carter, W.T. McAllister, Nature **299**(5884), 653–656 (1982)
39. G.Q. Tang, R.P. Bandwar, S.S. Patel, J. Biol. Chem. **280**(49), 40707–40713 (2005)
40. G.A. Kassavetis, M.J. Chamberlin, J. Virol. **29**(1), 196–208 (1979)
41. R.A. Ikeda, C.C. Richardson, Proc. Natl. Acad. Sci. U S A. **83**(11), 3614–3618 (1986)

Part V

Molecular Motion of Contractile Elements and Polymer Formation

Single-Molecule Measurement, a Tool for Exploring the Dynamic Mechanism of Biomolecules

Toshio Yanagida

12.1 Introduction: Fluctuation and Single-Molecule Measurements

Biomolecules fluctuate in response to thermal agitation. These fluctuations are present at various biological levels ranging from single molecules to more complicated systems like perception. Despite thermal fluctuation often being considered noise, in some cases biomolecules actually utilize them to achieve function. How biomolecules do this is necessary to understand the mechanism underlying their function. Thermal noise causes fast, local motion in the time range of picosecond to nanosecond, which drives slower, collective motions [1]. These large, collective motions and conformational transitions are achieved in the time range of microsecond to millisecond, which is the time needed for a biomolecule to exceed its energy barrier in order to switch between two coordinates in its free-energy landscape. These slower conformational or state changes are likely rate limiting for biomolecule function.

Single-molecule measurements are a tool that can monitor the relationship between the thermally fluctuating behavior and function of biomolecules. In ensemble measurements, dynamic changes that occur spontaneously in individual molecules are hidden in the average values measured over a large number of molecules. Single-molecule measurements circumvent this problem by observing the biomolecule of interest directly, often by imaging and manipulation [2–5]. In manipulation measurements, single biomolecules are attached to large probes such as a glass micro-needle or a laser-trapped bead (probes). In this design, the biomolecule's mechanical properties can be measured by allowing it to interact with other molecules at locations controlled by the experimenter by manipulating the needle or bead attached. Motion of the biomolecules can be measured with nanometer accuracy by monitoring changes in the position of the fixed probes by using detection devices such as a quadrant photodiode [6–9]. When biomolecules like molecular motors move actively, an opposing force generated by the laser trap or micro-needle exerted onto the probe acts against the movement. Thus, it is possible to measure the mechanical properties of the single molecules using this opposing force. For best measurements, the thermal fluctuation of the position of the probe is minimized. The mean square displacement, $<x^2>$, of a probe is related to the stiffness of the measurement system by an equi-partition law such as $K<x^2>/2 = k_\mathrm{B}T/2$. Here, the root mean square distance decreases as the spring constant increases or the system becomes stiff.

In the single-molecule fluorescence imaging, fluorescent probes attached to biomolecules are used. Total internal reflection fluorescence (TIRF) microscopy is a popular technique for imaging the active behavior of the biomolecule of interest on a slide glass surface [10]. Single molecules are visualized by observing fluorescent spots which distribute wide due to the diffraction of the fluorescent light. By analyzing the photon-detection distribution of individual spots, the motion of moving molecules can be traced with nanometer accuracy. To monitor the structural changes of the biomolecules, fluorescence spectroscopy techniques have been combined with single-molecule imaging. For example, fluorescence resonance energy transfer (FRET) is a technique that measures the distance between two different probes in a 2–10 nm range [11]. Single-molecule FRET has been used to measures the dynamics of protein interactions and structures [12]. Fluorescence intensity or the number of photons detected fluctuates. This fluctuation reflects the stochastic nature of photon emission and detection. In fluorescence techniques such as fluorescence spectrum, fluorescence polarization, FRET and position detection, the number of photons are described as a function of wavelength, polarization direction, and pixel position. However, the fluctuation of the photon number makes results ambiguous when the number of photon detected is small. The resolution of the fluorescence profiles increase when the total number of photons increase, for example, fluorescence profiles around the probe molecules position approximate a Gaussian distribution when large number of photons are detected, allowing the position determination with nanometer order accuracy [13]. Note that the acquisition of a larger number of photons requires time, and this compromises time resolution. Thus, the stochastic nature of photon detection fluorescence measurements and the compliance of the connections between the large probes and biomolecules make accurate measurements difficult, especially when the signal changes are small. When the signals from the biomolecules also change in a stochastic manner, the experiments must be designed and analyzed to discriminate the dynamic changes originating from the experimental system and the biological one.

For events that occur in a stochastic manner, single event measurements do not necessarily characterize the system. A single event normally represents one of several possibilities. One needs to repeat the measurement many times and analyze the data statistically. Comparison with data of ensemble measurements is one of the test of the single molecule measurements. The single-molecule data could provide insightful information regarding sequence and distribution of the function, which could not be measured otherwise. The basic equation used in kinetic analysis from ensemble studies uses molecule concentrations or counts. In single-molecule measurements, these same equations are used and interpreted as equation of probabilities of the biomolecules of interest instead of concentration.

12.2 Biased Brownian Movement of Muscle Myosin

In 1957, Huxley proposed a model for muscle contraction in which the Brownian movement of myosin molecules plays a key role [14]. Muscle contraction is described as a sliding movement between interdigitated myosin and actin filaments. The sliding is generated by individual myosin molecules operating independently. Myosin molecules are allowed to thermally oscillate around a binding site along an actin filament, where they undergo a cycle of attachment to and detachment from the actin filaments coupled with the hydrolysis of ATP. Random thermal oscillation is biased to one direction as described by the direction-dependent rate of attachment and detachment; myosin molecules thermally oscillate both forward and backward direction, but attach to actin binding sites preferentially in the forward direction resulting in directional movement. This model has been a paradigm for the mechanism used by most molecular motors. Taking into account the fact that the input energy of ATP hydrolysis is not different order of thermal noises, the idea that molecular motors utilize thermal fluctuation for the movement and force generation is fascinating.

Brownian motion-dependent step movement in myosin was suggested by direct observation of the movement of single-myosin heads molecules attached to a scanning probe in 1999 [8]. The movement of muscle myosin (myosin II) has been measured by monitoring the changes in the position of either an actin filament interacting with a few myosin molecules immobilized on a glass surface or a single myosin head interacting with actin filaments immobilized on a glass surface (Fig. 12.1a, b). In the actin manipulation system, actin filaments were manipulated, for example, by trapping beads attached to both ends of the filament by a laser. Single-myosin molecules were manipulated by attaching them to a tip of a cantilever fixed to a glass microneedle (a scanning probe). The movement of a single myosin molecule was measured by monitoring the position change of the cantilever in response to the myosin moving along actin filaments. These different systems gave similar results regarding the movement coupled to the hydrolysis of ATP [15]. In general, myosin made a single step when a single ATP molecule was hydrolyzed (Fig. 12.1c).

Using myosin manipulation or scanning probe methods, these ATP hydrolysis coupled steps have been resolved into multiple stochastic substeps (Fig. 12.1d). The laser trap method, however, has not been able to resolve substeps. We attribute this to the large scanning probe slowing the substeps to a time within our system's time resolution. The size of the substeps was 5.5 nm, corresponding to the interval of actin monomers on a protofilament in a two-stranded filament. Substeps occurred both in the forward and back-

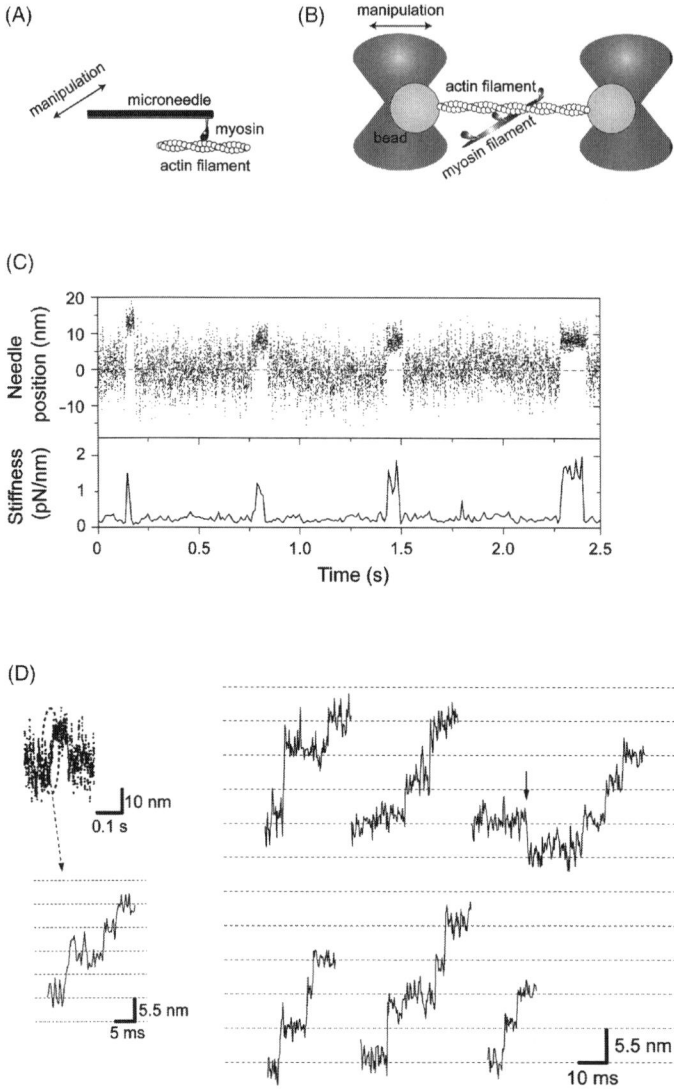

Fig. 12.1. Manipulation of molecular motors and step movement of myosin II.
(a) Manipulation of a single-myosin molecule with a microneedle. A myosin molecule
captured at a tip of the microneedle is allowed to interact with an actin filament
placed on a glass surface. (b) Manipulation of an actin filament by a laser trap. Two
beads attached at both ends of an actin filament are trapped by a laser. The actin
filament is manipulated to interact with myosin molecules on the filament placed on
a glass surface. (c) Nonprocessive movement of myosin II. (*Top*) The displacement of
the microneedle attached to single myosin II head (S1) was measured as a function
of time and the binding of myosin to actin was measured by stiffness calculated
from the displacement record (*bottom*). (d) Expansion of the rising phase of the
displacement record above

ward direction. The forward and backward substeps occurred stochastically but overall movement was performed in one direction, which was thus interpreted as evidence for biased Brownian movement. The substeps occurred thermally, while the direction was biased by some mechanism. The energy released from ATP hydrolysis was most likely used for the bias rather than movement itself. The ratio of forward and backward substeps was 10:1 at no load, but decreased with external load. Using this ratio, the difference in the activation free energy between the forward and backward substeps was 2–3 $k_B T$. Several models have been proposed to explain this asymmetric potential [16,17]. A conformational change in actin filament is one possibility. Another possibility is steric compatibility in the interaction between myosin and the actin filament. The substeps constituted a step coupled with hydrolysis of a single ATP molecule. The number of the substeps within a regular step was variable and less than 7. The number of actin monomers to which the myosin molecule attached to the scanning probe can interact with was limited to a half pitch on the protofilament. Shifting attachment to the other actin protofilament is unfavorable since this requires the myosin head to bend and rotate. Thus, the myosin is restricted to a half pitch of the filament.

12.3 Forward and Backward Step Movement of Kinesin

Thermal Brownian movement is random in direction. Assuming that myosin uses Brownian motion to drive its motility, understanding how Brownian movement is biased is important when trying to reveal the mechanism used by molecular motors. Processive motors, which move long distances without dissociating from their protein tracks, unlike muscle myosin, are popular models for the single-molecule measurements because processivity makes single-molecule measurements easier. Kinesin is a processive molecular motor that transports vesicles along microtubules. By attaching a purified kinesin single-molecule to a bead trapped by a laser and measuring its steps along microtubules immobilized on a glass surface showed that steps directed primarily in one direction when a small external load was applied and that the frequency of the backward steps increased as the external load against movement increased (Fig. 12.2a, b). In fact, at excessively high external loads, successive back steps have been seen [18].

Kinesin has two head, or motor, domains, which include the ATP and microtubule binding sites, connected at the tail domain. The two heads alternately step on the microtubule in 8-nm intervals corresponding to a unit structure. This is commonly referred to as the hand-over-hand mechanism [19,20]. Starting from the state that one head is attached, the detached head diffuses back and forth searching for the next binding site. The detached head attaches to a neighboring binding site in either direction, followed by the detachment of the other head to complete one cycle of the step movement. However, as stated above, kinesin binds preferentially to one direction. Therefore, how kinesin does this is a fundamental question for the mechanism driving directional movement.

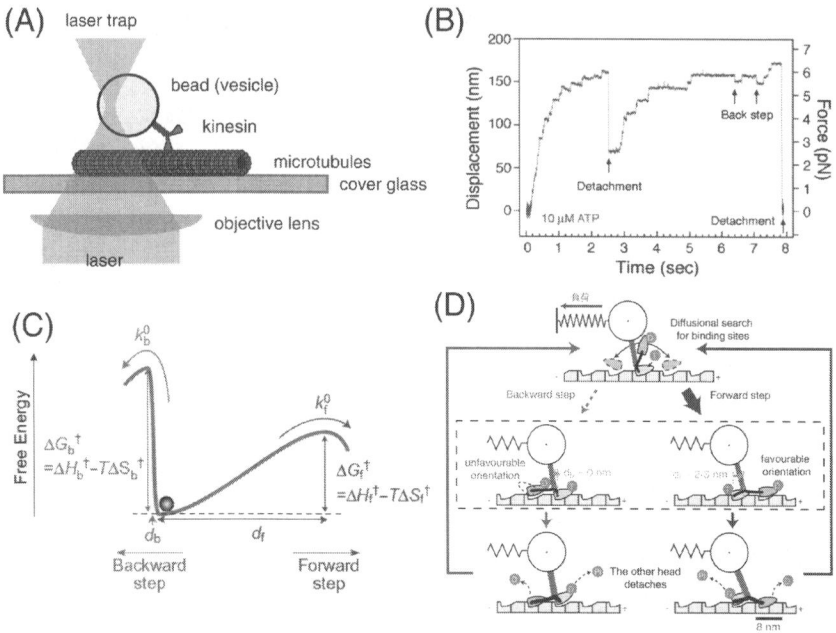

Fig. 12.2. Forward and backward step movement of kinesin. (a) A single kinesin molecule moving on microtubules. A kinesin molecule is attached to a bead trapped by a focused laser as a cargo for measurements. (b) Time trajectory of displacement and force of kinesin. In the laser trap measurement, when kinesin moves the trapping force or external load increases indicted. (c) Energy landscape for the forward and backward step of kinesin. The thermodynamic parameters obtained for the forward and backward step movement of kinesin are included in the figure. The rates for the forward and backward steps, $k_\pm$, are related to the activation energy $U_\pm$ and external load F,

$$k_\pm = \text{constant } \exp[-(U_\pm + Fd_\pm)/k_B T],$$

where $d_\pm$ is a characteristic distance and $\pm$ denotes forward and backward steps. $U_\pm$ was break down to enthalpic contribution $H_\pm$ and entropic contribution $S_\pm$, $U_\pm = H_\pm - TS_\pm$. The difference in the activation energy is mainly explained by the difference in entropy. (d) A model for preferential binding of kinesin to one direction based on the experimental results summarized in (c)

From the step movement trace of kinesin, which was obtained by laser trap measurements, the interval time between steps and directionality of the step were calculated as a function of ATP concentration, load, and temperature [19, 20]. The data were analyzed statistically and interpreted thermodynamically since the mechanochemical processes underlying the step movement of kinesin occur in accordance with thermodynamic rules. At no load limit, the probability of a backward step was estimated as 1 of 2,000, which corresponded to the difference in activation free energy ($\sim 6\,k_B T$) in the energy landscape of forward and backward step movement (Fig. 12.2c). From structural studies, it has been suggested that the conformational docking of a short region called a neck linker, which is located between the head and coiled-coil tail domain, biases the step movement of the detached head toward the forward position [21]. However, the energy reported for docking ($1\sim2k_BT$) is not sufficient to explain the difference in the free energy barrier between forward and backward steps [22].

The activation free energy U was divided into an enthalpic contribution, H, and an entropic contribution, S, such that $U = H - TS$. The temperature dependence of the step movement of kinesin showed that the difference in the activation free energy between the forward and backward movement was mainly entropic rather than enthalpic [22]. What is the entropy difference in the binding between the forward and backward direction? One possibility is the sterically restricted geometry between the kinesin head and the binding sites (Fig. 12.2d). In the forward position, the most favorable orientation of the kinesin head coincides with the orientation of the binding site. In contrast, when attempting to move backward, the favorable orientation of the kinesin head is incompatible with the orientation of the binding site. Thus, preferential binding to the forward site can be explained by a geometrical asymmetry between the orientation of the kinesin head and the kinesin binding site on the microtubules.

12.4 Directional Movement of Processive Myosin

In the case of kinesin, the orientation of the head is confined by the short neck region, resulting in steric effects on binding. In the case of myosin, however, myosin heads, which are connected to long flexible chains, orient freely. Therefore, a different mechanism may apply. Owing to the development of molecular and cell biology, various types of myosin have been discovered in addition to muscle myosin. Among them, myosin V was found to move processively like kinesin and has been studied using single-molecule measurements extensively. Like kinesin, two-headed myosin V moves in a hand-over-hand manner: one head steps while the other head stays bound to the actin filament [23]. The stepsize of myosin V is large, 36 nm, corresponding to a half pitch of the helical structure of the two-stranded actin filament. The large stepsize is thought to relate to a long and stiff neck domain called a lever arm, which is located between the head and tail domains. The Brownian search of the myosin heads for the next binding sites was observed by tracing the position and orientation of the head [24, 25]. One of the models for the directional steps is directional rotation of the lever arm, which has been supported by nucletide-dependent atomic structure of myosin head (lever arm). In this model, the step size is proportional to the lever arm length [26].

Two-headed myosin VI is also a processive motor with a large step size of 36 nm, similar to that of myosin V [27, 28]. However, the neck domain of myosin VI is too short to explain this large step, suggesting a diffusion mechanism for step movement. Furthermore, it was found that single-headed myosin VI moves processively when it is attached to a cargo, whereas without a cargo it does not [29]. Generally, single-headed motors are expected to diffuse away from their filaments when they step. Single-headed kinesin including Kif1a and peptide-fused kinesin avoids this due to an additional interaction with microtubules [30, 31]. In myosin VI, however, there are no additional interactions. Instead, the cargo, for which experimentally a 200 nm polystyrene bead is used, makes the movement of the single-headed myosin VI processive. This is attributed to differences in diffusion between the cargo and myosin VI mostly because of size difference. The diffusion constant of the 200-nm diameter bead is 60-fold lower than that of the myosin VI head in solution. Thus, the bead diffuses slowly such that when the myosin VI head detaches from one binding site, it still remains in proximity.

Given that myosin VI generally moves in one direction, the diffusion must be biased to one direction. The mechanism for this is likely strain-dependent, where a "sensor" that exists within the head detects a strain thus triggering the ATPase reaction and the interaction with actin. When the detached myosin head is pulled backward, the release of Pi and ADP from myosin and the binding of the myosin head to actin are accelerated. When the attached myosin head is pulled forward, the binding of ATP and the dissociation of the myosin head are accelerated. In support of this theory, it has been reported that the binding of ATP to myosin VI is load-dependent [32]. Similarly, we showed that myosin VI heads bind more frequently when the myosin head moves against the trap force in laser trap measurements (Fig. 12.3) [33]. Thus, a strain-sensor coordinates the dissociation and association of the directional and processive movement achieved by myosin VI. In double-headed myosin VI, the strain acts between the two heads allowing the two heads to cooperatively perform their ATPase reactions. This is an experimental evidence for the Huxley 1957 model.

12.5 Conformational Fluctuation of Actin and Activation of Myosin Motility

Biomolecules work in an environment where a large number of molecules are packed and interacting. Properties including thermal fluctuations observed in in vitro single-molecule measurements with isolated molecules are modulated by their interaction with other molecules in physiological environments. Later, we will introduce a few examples where the coordination of the fluctuating properties of biomolecules is used for higher level systems.

Actin filaments are the tracks for myosin movement. Biochemically, actin activates the ATP hydrolysis of myosin. Recently, it has been shown that myosin motility is activated through actin conformational changes. Single-molecule FRET from doubly labeled actin monomers in the filament has revealed that actin has multiple conformations and spontaneously changes between them with time (Fig. 12.4a, b) [34]. The multiple conformations are

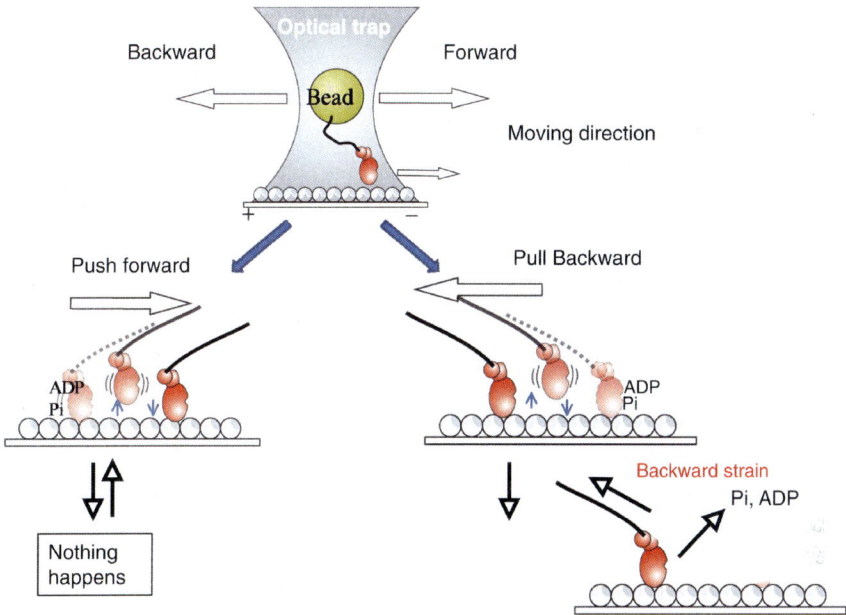

Fig. 12.3. Strain-sensor model for myosin VI

between them with time (Fig. 12.4a, b) [34]. The multiple conformations are grouped into at least two states, which can be interpreted as the state in which myosin motility is activated and the state in which myosin motility is inhibited (Fig. 12.4c). In the presence of myosin, actin favorably takes the active conformation state. The inhibition conformation state is attained when actin is crosslinked chemically. In this state, although myosin motility is inhibited, the actin-activated ATPase and the binding to actin are not affected. Thus, actin spontaneously fluctuates between the two conformational states that activate and inhibit myosin motility, and the binding of myosin activates the myosin motility by shifting the population of the actin conformational states (Fig. 12.4d).

Multiple conformational states correspond to local minima in the free energy landscape of the protein molecule of interest. When thermal motion is allowed, protein molecules dynamically change conformation accordingly on the free energy landscapes. The structures fluctuate around a stable structure when the thermal energy is small or when the proteins have only one free energy minimum. As thermal energy increases high enough to exceed the energy barriers between several local minima, protein structures vary between multiple structures. Thus, ligand binding changes the conformation by shifting pre-existing equilibrium between multiple conformations. Generally, two models have been debated for the mechanism how ligand binding affects protein structures [35]; ligand binding either induces structural changes or

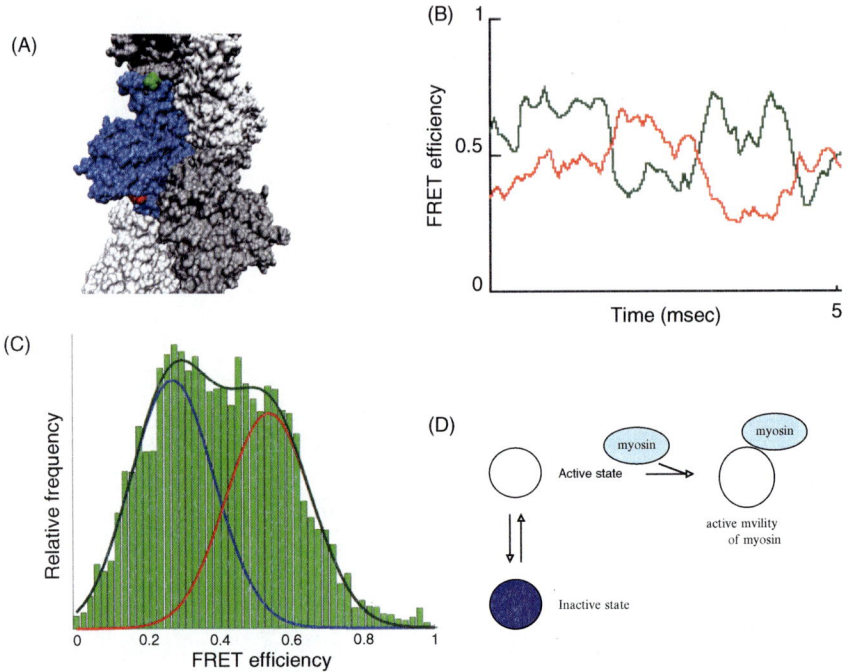

Fig. 12.4. Spontaneous conformational transition of actin and activation of myosin motility through actin conformational changes. (**a**) An actin molecule (colored blue) is labeled specifically with donor (*green*) and acceptor (*red*) and polymerized with large excess of unlabeled actin (colored *grey*). (**b**) Structural dynamics of actin was measured using fluorescence resonance energy transfer (FRET). The time trajectories of the donor (*green*) and acceptor (*red*) fluorescence changed due to the changes in FRET; that is, the donor fluorescence increased when the acceptor fluorescence decreased and vice versa. (**c**) Distribution of the FRET efficiency ($I_A/(I_D + I_A)$) when I_D and I_A are donor and acceptor fluorescence intensity, respectively, from actin. The distribution was fit to two Gaussian distributions corresponding to the high FRET efficiency state or active state in which actin activates the motility of myosin and the low FRET efficiency state or inactive state in which actin inhibits the motility of myosin. (**d**) Schematic model for actin activation of myosin motility by actin conformational transition. Actin conformation fluctuate between active and inactive states and myosin preferentially binds to actin in activated state to activate myosin motility

shifts preexisting dynamic equilibrium between the two structures [36, 37]. Our single-molecule actin experiments support latter argument.

Actin monomers assemble to form actin filaments which then associate to form higher level structures like bundles, networks, and gels dynamically in cells. Assembly and disassembly takes place spontaneously with a variety of actin-binding proteins involved in these processes. It has been reported that activated structures preexist as spontaneously fluctuating structures even in the absence of external stimuli. In *Listeria monocytogenes*, an intracellular pathogenic bacterium, actin has been seen to spontaneously assemble into polarized filaments even in the absence of external stimuli in cells [38]. The basic scheme is analogous to the scheme describing conformational fluctuation as described earlier.

12.6 Biased Brownian Mechanism of Myosin Movement and Muscle Contraction

Muscle myosin works in a specific and highly ordered array in muscle fibers. How the characteristic properties of isolated myosin molecules are modulated in muscle system remains a challenge for single-molecule measurements. What new features emerge when stochastic motors like myosin muscle assemble? How is the biased Brownian mechanism advantageous when a large number of myosin molecules work together? Experiments with multiple myosin molecules, myosin filaments, or muscle have suggested that the average distance of displacement per hydrolysis of single ATP molecules is greater than 60 nm, much more than a single molecule can achieve independently [39, 40]. Cooperative action of myosin heads arranged in a myosin filament is likely to explain the enhancement of the distance per ATP molecule. Since the experimental setup for the scanning probe system most resembles muscle as compared to other systems, the results from scanning probe measurement are most readily applicable to a model for muscle. Although single-myosin heads move a distance less than a half pitch of the actin helix, it has been demonstrated in model simulation that they move more than the half pitch when many myosin heads interact simultaneously [41]. In other words, the presence of other myosin molecules helps a given head to switch between adjacent protofilaments.

12.7 Multiple Conformations of Ras and Switching Signals

Similar conformational fluctuations between multiple states have been suggested for the signaling protein Ras. Ras, a proto-oncogene product, is a molecular switch for complex signaling networks. Cells perform a variety of functions in response to extracellular stimuli such as hormones and neurotransmitters. The signals triggered by the binding of signal molecules on the cell surface are transmitted through signal transduction processes to the nuclei inside the cell. Recently, many factors involved in these processes have been identified and several signaling pathways have been elucidated. Ras stimulated by the binding of GTP can bind to and activate varieties of target proteins, triggering respective downstream signaling processes. However, it is not known how Ras identifies the appropriate protein when switching signals.

The time trajectory and histogram of FRET efficiency for Ras showed multiple conformational states which changed spontaneously (Fig. 12.5a) [42]. Among them, a low FRET refractory state was distinguished from other high FRET activated states (Fig. 12.5b). The conformational transition between them occurred in the timescale of seconds. The effectors Ras, Raf, and RalGDS preferentially bound to the activated state to shift the population of conformational states from the inactivated state to the activated ones. Transitions between active conformational states occurred in 20 ms, but in the presence of the effectors no transitions was observed, suggesting that Ras remains in one conformational state. Thus, the binding of specific effecters may select a corresponding conformational state out of the multiple conformational states preexisting before binding. Thus, the multiple conformations of Ras may be closely related to the fact that this protein can interact with a variety of target proteins to achieve function [43].

12.8 Multiple States of Signaling Proteins in Living Cells

Single-molecule imaging has been extended to live cells, allowing signaling processes to be observed at the single-molecule level. In living cells, multiple states of Ras were observed in the kinetics of the activation process. The activation of Ras to the membrane was visualized by monitoring the recruitment of fluorescently labeled Raf1 that binds to Ras on the plasma membrane in response to activation stimulated by the binding of epidermal growth factor (EGF) [44]. It was found that there were at least two kinds of binding sites for Raf1 to the plasma membrane: faster (0.4 s) and slower (1.6 s) dissociation sites (Fig. 12.5c). The distribution of the Ras kinetic states was not homogeneous, but localized. Raf1 binds to the faster-dissociation site on the bulk membrane, while it binds to both the faster- and slower-dissociation sites at the membrane ruffle regions. That is, the slower-dissociation sites are localized specifically at the membrane ruffle regions where cells form active protrusions for migration. These results suggest that Ras has multiple states for Raf1 binding and one specific state is selected at the membrane ruffle regions. Multiple conformations of Ras may be a molecular basis for such heterogeneous activation in living cells.

12.9 Kinetic Heterogeneity of Cell Signaling Processes in Living Cells

Heterogeneity in the kinetic states of signaling molecules was also found in ligand binding reactions in chemotaxing *Dictyostelium* cells [45]. *Dictyostelium* cells exhibit chemotaxis in response to cyclic adenosine $3', 5'$-monophosphate (cAMP). *Dictyostelium* cells move randomly in the absence of a cAMP concentration gradient. In the presence of a gradient, however, movement is

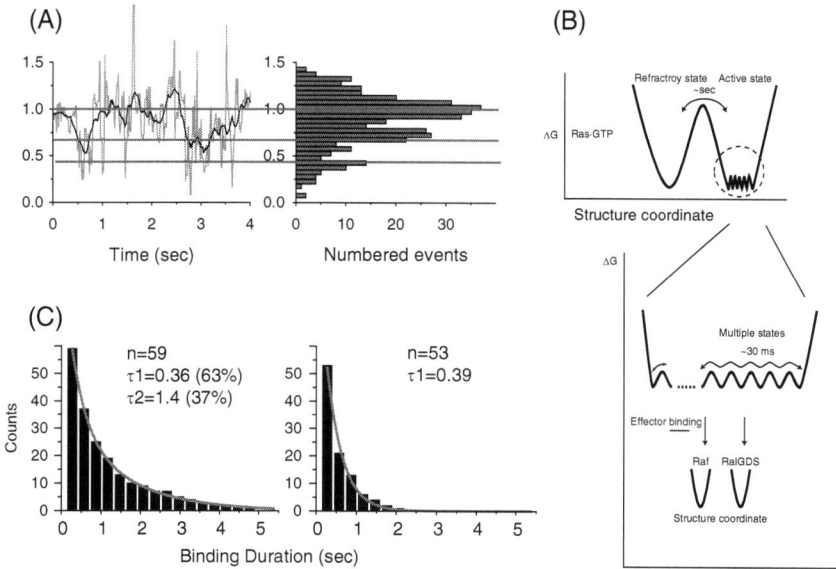

Fig. 12.5. Multiple conformational and kinetic states of Ras in vitro and in vivo. (a) Conformational dynamics of Ras in vitro. The FRET efficiency changes with time among several FRET efficiency values. Right is the replot of the FRET efficiency in a histogram. (b) Energy landscape model for Ras based on the single-molecule measurements of conformational dynamics. (*Top*) Refractory state is distinguished from other multiple active states. (*Middle*) Among the active states, there are transitions with timescale of ∼30 ms. (*Bottom*) One of the states is selected upon the binding to effectors. (c) Imaging of the binding of Raf to Ras after stimulation with EGF in living cells. Accumulation of fluorescence or GFP labeled Raf was observed (*right*). After photo bleaching, fluorescence spots (*arrows*) were observed (*right*). The binding duration of Raf onto the plasma membrane was measured at accumulation areas (*left*) and bulk membrane (*right*). The distribution of the duration time was fit to one or two exponential curves and the decay times determined are indicated

preferentially directed toward the higher cAMP concentration. This directional response has been found in various biological processes, including immunity, neuronal patterning, morphogenesis, and nutrient finding. To elucidate how cells sense the gradient, we have prepared a fluorescent labeled cAMP (Cy3-cAMP) and visualized the binding to the receptor at the single-molecule level in living cells. Statistical analysis of lifetimes of Cy3-cAMP bound to the receptor showed that the dissociation rates of Cy3-cAMP were dependent on the pseudopod and tail regions of chemotaxing cells. Cy3-cAMP dissociated faster from the pseudopod regions of the cells than from the tail by a factor of approximately 3. That is, the ligand binding properties of the receptors were not uniform but sensitive to the external chemical gradient in living cells, although the receptor itself is distributed uniformly. The differences in receptor states appear to reflect coupling with the downstream molecule heterotrimeric G protein. Overall, it is likely that the receptors can transduce signals at the pseudopod regions more efficiently than at the tail regions of chemotaxing cells.

The number of signal molecules bound to the membrane receptor fluctuates, normally by about 5% of the total cAMP concentration. But the cAMP concentration gradient between the pseudopod and tail of the cells only differs by 2%. Thus, the input signal is smaller than the noise level. An input signal hidden within noise also occurs in cells that proliferate in response to the binding of epidermal growth factor (EGF). The number of bound EGF molecules visualized at the single-molecule level upon attaching fluorescent probes to EGF was related to the intracellular calcium response downstream of the EGF signal pathway. We found that only 300 EGF molecules were sufficient to induce the calcium response, despite an abundance of EGF receptor molecules present on the cell surface [46,47]. The small input signals are analogous to molecular motors in which input ATP energy is not much greater than thermal noise where the energy released from ATP hydrolysis is thought to be $\sim 18 k_{\mathrm{B}} T$. It is likely that signaling processes depend on a mechanism that utilizes subtle input signals similar to how molecular motors utilize thermal noises.

12.10 Direct Determination of the Input–Output Relation in Protein Molecules

It is important to experimentally describe how biomolecules function in response to input signals. One of the advantages of single-molecule measurements is to directly relate two different parameters from single events in single molecules. For example, when the chemical energy from ATP is converted to mechanical work done by molecule motors, the binding of fluorescently labeled ATP to myosin and its dissociation after hydrolysis was visualized by fluorescence imaging, while the mechanical events conducted by the same myosin molecule were recorded by a laser trap measurement at the same time (Fig. 12.6) [48,49]. It was found for muscle myosin that in half the cases, the mechanical event occurred at the same time ADP was released and in the other half the mechanical event took place after ADP dissociated.

In the case of ion channels, the binding of ligands causes gate opening and ion current. Simultaneous optical and electrical single channel recordings have been done with a self-standing bilayer. The fluorescently labeled Ca^{2+}-activated K-channel was incorporated into the bilayer by the vesicle fusion technique [50]. An increase in the fluorescence from a single channel was monitored while the ion current increased. Direct determination of the relationship between input signals, structural changes of the channel proteins, and the ion current is interesting. Protein conformation and ion current were observed to fluctuate but it is not known how they are related to each other.

(A)

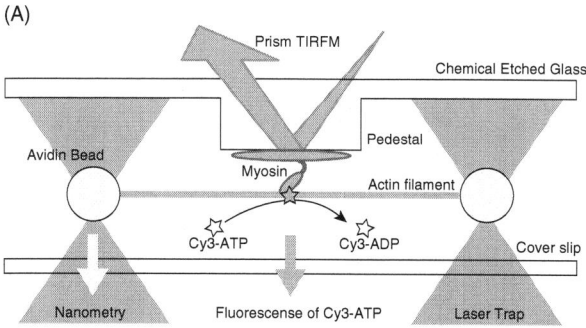

(B)

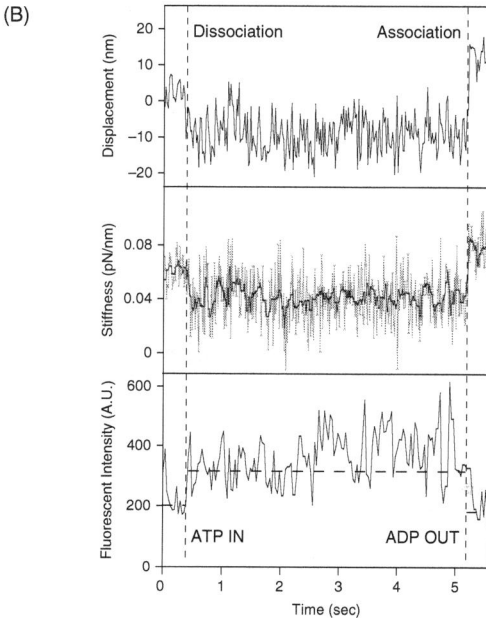

Fig. 12.6. Direct determination of mechano-chemical coupling of myosin. (a) An experimental arrangement for simultaneous measurement. Single-headed myosin in the co-filament with an excess myosin rod is immobilized on the pedestal of the slide glass. An actin filament is manipulated by a trapped laser through two beads attached at both ends to interact with a myosin head in the correct arrangement. The ATP turnover is measured by monitoring the fluorescence from Cy3-ATP (Cy3-ADP) associated to and dissociated from the myosin head using TIRF microscopy. (b) Time trajectory of the displacement of the myosin head and the ATP turnover. The upper, middle, and bottom trace show the time course of displacements, changes in stiffness, which was calculated from the variance of the thermal motion of the beads, and changes in the fluorescence intensity of Cy3-nucleotide at the position of the myosin head

12.11 Future Perspective

In summary, we have shown that stochastic fluctuations are involved in various biological activities at various hierarchical levels. Single molecule detection techniques have revealed that molecular motors utilize thermal fluctuation for their functions. This mechanism is advantageous for machines that the input energy level is not much greater than thermal noise level. Random Brownian motion is biased to one direction by preferential binding to one direction. Similarly, random conformational and kinetic state fluctuation is biased by preferential interaction with other proteins to one state for protein activation and signal processing. Thus, thermal fluctuation is biased to one direction, and the bias can be manipulated by having the molecule interact with other protein molecules. When these molecules self-assemble to higher biological systems, unique properties such as cooperativity, memory effects, and spatial asymmetry emerge to make systems function more effectively and flexibly. In such higher systems, it is advantageous for various states to preexist before stimulation [51].

Finally, it is worth mentioning that similar mechanism has also been suggested in perception processes that occur in complex networks of neurons in brain. For example, it has been shown that perception, which involves spontaneously altering between two different precepts, involve stochastic processes and can be described using the same equations used for thermal activation of biomolecules [52] (Murata, private communication). This homology between perception and molecular processes leads us to speculate that dynamics at different hierarchical levels have the same origins. Therefore, molecular dynamics are likely basis for understanding biological activities in various levels. Single-molecule measurements are a useful and powerful tool to study dynamic properties and function of biomolecules. The measurements will be performed in various levels of biological systems in vitro and in vivo in more complicated systems like live cells and whole bodies hopefully in near future. We will see dynamic events in complicated biological systems using single molecules as probes. We will know how the behavior of biomolecules observed in an isolated system is modified in more real biological systems. We will see what the role of the molecular rules in the hierarchic biological systems is.

References

1. K.A. Henzler-Wildman, M. Lei, V. Vu Thai, S.J. Kerns, M. Karplus, D. Kern, Nature 450, 913 (2007)
2. Y. Ishii, T. Yanagida, HFSP J. 1, 15 (2007)
3. A. Ishijima, T. Yanagida, Trends Biochem. Sci. 26, 438 (2001)
4. T. Yanagida, Y. Ishii, Single Molecule Dynamics in Life Science (Wiley-VCH, 2008)
5. P.R. Selvin, T. Ha Single Molecule Technique (Cold Spring Harbor Laboratory, 2008)

6. J.T. Finer, R.M. Simmons, J.A. Spudich, Nature 368, 113 (1994)
7. A. Ishijima, Y. Harada, H. Kojima, T. Funatsu, H. Higuchi, T. Yanagida, Biochem. Biophys. Res. Commun. 199, 1057 (1994)
8. K. Kitamura, M. Tokunaga, A. Iwane, T. Yanagida, Nature 397, 129 (1999)
9. K. Svoboda, C.F. Schmidt, B.J. Schnapp, S.M. Block, Nature 365, 721 (1993)
10. T. Funatsu, Y. Harada, M. Tokunaga, K. Saito, T. Yanagida, Nature 374, 555 (1995)
11. J.R. Lakowicdz, Principles of Fluorescence Spectroscopy (Kluwer Academic/Plenum, New York, 1999)
12. S. Weiss. Science 283, 1676 (1999)
13. A. Yildiz, J.N. Forkey, S.A. McKinney, S.T. Ha, Y.E. Goldman, P.R. Selvin Science 300, 2061 (2003)
14. A.F. Huxley, Prog. Biophys. Biophys. Chem. 7, 255 (1957)
15. H. Tanaka, A. Ishijima, M. Honda, K. Saito, T. Yanagida, Biophys. J. 75, 1886 (1998)
16. T.P. Terada, M. Sasai, T. Yomo, Proc. Natl. Acad. Sci. U S A 99, 9202 (2002)
17. S. Esaki, Y. Ishii, T. Yanagida, Proc. Jpn. Acad. 79, 9 (2003)
18. N.J. Carter, R.A. Cross, Nature Cell Biol. 4, 790 (2002)
19. M. Nishiyama, H. Higuchi, T. Yanagida, Nat. Cell Biol. 4, 790 (2002)
20. Y. Taniguchi, M. Nishiyama, Y. Ishii, T. Yanagida, Nat. Chem. Biol. 1, 346 (2005)
21. S. Rice, A.W. Lin, D. Safer, C.L. Hart, N. Naber, B.O. Carragher, S.M. Cain, E. Pechatnikowal, E.M. Wilson-Kubalek, M. Whitaker, E. Pate, E.R. Vook, E.W. Taylar, R.A. Milligan, R.D. Vale, Nature 402, 778 (1999)
22. S. Rice, Y. Cul, C. Sindelar, N. Naber, M. Matuska, R.D. Vale, R. Cooke, Biophys. J. 84, 1844 (2003)
23. A.D. Mehta, R.S. Rock, M. Rief, J.M. Spudich, M.S. Mooseker, R.E. Cheney Nature 400, 590 (1999)
24. A.R. Dunn, J.A. Spudich, Nat. Struct. Mol. Biol. 14, 246 (2007)
25. Y. Komori, A.H. Iwane, T. Yanagida, Nat. Struct. Mol. Biol. 14, 968 (2007)
26. I. Rayment, H.M. Holden, M. Whittaker, C.B. Yohn, M. Lorenz, R.C. Holmes, R.A. Milligan, Science 261, 58 (1993)
27. R.S. Rock, S.E. Rice, A.L. Wells, T.J. Purcell, J.A. Spudich, H.L. Sweeney, Proc. Natl. Acad. Sci. U S A 98, 13655 (2001)
28. S. Nishikawa, K. Homma, Y. Komori, M. Iwaki, T. Wazawa, A.H. Iwane, J. Saito, R. Ikebe, E. Katayama, T. Yanagida, M. Ikebe, Biochem. Biophys. Res. Commun. 290, 311 (2002)
29. M. Iwaki, H. Tanaka, A.H. Iwane, E. Katayama, M. Ikebe, T. Yanagida, Biophys. J. 90, 3643 (2006)
30. Y. Okada, N. Hirokawa Science 283, 1152 (1999)
31. Y. Inoue, A.H. Iwane, T. Miyai, E. Muto, T. Yanagida, Biophys. J. 81, 2838 (2001)
32. D. Altman, H.L. Sweeney, J.A. Spudich, Cell 116, 737 (2004)
33. M. Iwaki, A.H. Iwane, T. Yanagida Biophys. J. (suppl) 2245-Pos (2008)
34. J. Kozuka, H. Yokota, Y. Arai, Y. Ishii, T. Yanagida, Nat. Chem. Biol. 2, 83 (2006)
35. C.S. Goh, D. Milburn, M. Gerstein Curr. Opin. Struct. Biol. 14, 104–119 (2004)
36. D.E. Koshland Jr, G. Nemethy, D. Filmer Biochemistry 5, 365 (1966)
37. J. Monod, J. Wyman, J. Changeux J. Mol. Biol. 12, 88 (1965)

38. A. van Oudenaarden, J. Theriot, Nat. Cell Biol. 1, 493 (1999)
39. A. Harada, K. Sakurada, T. Aoki, D.D. Thomas, T. Yanagida, J. Mol. Biol. 216, 49 (1990)
40. H. Higuchi, Y.E. Goldman, Nature 352, 352 (1991)
41. K. Kitamura, M. Tokunaga, S. Esaki, A.H. Iwane, T. Yanagida, Biophysics 1, 1 (2005)
42. Y. Arai, A.H. Iwane, T. Wazawa, H. Yokota, Y. Ishii, T. Kataoka, T. Yanagida, Biochem. Biophys. Res. Commun.343, 809 (2006)
43. Y. Ito, K. Yamasaki, J. Iwahara, T. Terada, A. Kamiya, M. Shirouzu, Y. Muto, G. Kawai, S. Yokoyama, E.D. Laue, M. Walchli, T. Shibata, S. Nishimura, T. Miyazawa, Biochemistry 36, 9109 (1997)
44. K. Hibino, T. Watanabe, J. Kozuka, A.H. Iwane, T. Okada, T. Kataoka, T. Yanagida, Y. Sako, Chem. Phys. Chem. 4, 748 (2003)
45. M. Ueda, Y. Sako, T. Tanaka, P. Devreotes, T. Yanagida, Science 294, 864 (2001)
46. Y. Sako, S. Minoguchi, T. Yanagida, Nature Cell Biol. 2, 168–172 (2000)
47. T. Uyemura, H. Takagi, T. Yanagida, Y. Sako, Biophys. J. 88, 3720 (2005)
48. A. Ishijima, H. Kojima, T. Funatsu, M. Tokunaga, H. Higuchi, H. Tanaka, T. Yanagida, Cell 92, 161 (1998)
49. T. Komori, S. Nishikawa, T. Ariga, A.H. Iwane, T. Yanagida, Biophys. J. 96, L4 (2009)
50. T. Ide, Y. Takeuchi, T. Yanagida Jpn. J. Physiol. 52(5), 429 (2002)
51. M.J. Kirshner, J. Ferhart, T. Mitchison, Cell 100, 79 (2000)
52. T. Murata, N. Matsui, N. Miyauchi, S. Kakita, T. Yanagida, Neuroreport 14, 1347 (2003)

13

Viral DNA Packaging: One Step at a Time

Carlos Bustamante and Jeffrey R. Moffitt

Summary. During its life-cycle the bacteriophage $\varphi 29$ actively packages its dsDNA genome into a proteinacious capsid, compressing its genome to near crystalline densities against large electrostatic, elastic, and entropic forces. This remarkable process is accomplished by a nano-scale, molecular DNA pump – a complex assembly of three protein and nucleic acid rings which utilizes the free energy released in ATP hydrolysis to perform the mechanical work necessary to overcome these large energetic barriers. We have developed a single molecule optical tweezers assay which has allowed us to probe the detailed mechanism of this packaging motor. By following the rate of packaging of a single bacteriophage as the capsid is filled with genome and as a function of optically applied load, we find that the compression of the genome results in the build-up of an internal force, on the order of $\sim$55 pN, due to the compressed genome. The ability to work against such large forces makes the packaging motor one of the strongest known molecular motors. By titrating the concentration of ATP, ADP, and inorganic phosphate at different opposing load, we are able to determine features of the mechanochemistry of this motor – the coupling between the mechanical and chemical cycles. We find that force is generated not upon binding of ATP, but rather upon release of hydrolysis products. Finally, by improving the resolution of the optical tweezers assay, we are able to observe the discrete increments of DNA encapsidated each cycle of the packaging motor. We find that DNA is packaged in 10-bp increments preceded by the binding of multiple ATPs. The application of large external forces slows the packaging rate of the motor, revealing that the 10-bp steps are actually composed of four 2.5-bp steps which occur in rapid succession. These data show that the individual subunits of the pentameric ring-ATPase at the core of the packaging motor are highly coordinated, with the binding of ATP and the translocation of DNA temporally segregated into two distinct phases of the mechanochemical cycle of the entire ring. Because this ring-ATPase is a member of the ASCE superfamily of ATPases, these results may have implications for a broad and diverse family of cellular motors.

13.1 Introduction

The cell milieu is neither isotropic nor homogeneous and, as a result, many biological processes often involve directional movement and transport of chemical species. This transport takes place across membranes against electrochemical potential gradients, along linear tracks, or from large into small compartments. Unlike biological processes that simply involve second order reactions in which the reactants collide by diffusion, the directional nature of these processes cannot simply rely on random encounters between chemical species but depend instead on tiny machine-like devices that operate as molecular motors, i.e., converting chemical energy either in the form of bond hydrolysis or chemical gradients into force generation and displacement. But these molecular entities are unlike macroscopic engines in that, because of their dimensions, their many small parts must operate at energies only marginally higher than those of the thermal bath, and thus, are coupled to the bath and subjected to large fluctuations. How these devices carry out this energy conversion is of great interest in biophysics. During the last two decades, biologists have described and studied a large number of molecular motors and motor-like proteins whose functions in the cell range from ATP synthesis, cell motility, internal transport, and maintenance of the cytoskeletal architecture, to the central processes of recombination, gene replication, transcription, and translation. These studies have revealed a surprising variety in the mechanisms underlying the operation of these motors, from the two-headed *hand-over-hand* motion of kinesins and the rotary movement of ATP synthase, to the *rowing* movement of myosins and the crawling motion of polymerases. In spite of this diversity, these same studies suggest that many of these motors share a number of molecular mechanisms and a common set of underlying design principles [1–5].

The advent of methods of single molecule manipulation [6, 7] and single molecule detection [8–10] have made it possible for the first time to follow the molecular trajectories of these motors and describe in increasing detail their dynamics. The variables that are more easily detected by these methods are force, displacement, and time. These quantities are also the ones of greatest functional value to understand the mechanochemical operation of these motors.

In this paper, we describe the study of one such system, the DNA packaging motor from bacteriophage φ29, a multimeric ring-ATPase involved in the packaging and organization of the phage genome during viral assembly inside the bacterial host *B. subtilis* [11]. Multimeric ring-ATPases of the ASCE (Additional Strand, Conserved E) superfamily constitute a structurally homologous yet functionally diverse group of proteins that are involved in such varied tasks as ATP synthesis, protein unfolding, protein degradation, and DNA translocation [12–18]. As is the case for all molecular motors, these enzymes must couple a chemical reaction (ATP binding and hydrolysis) to a mechanical task. Despite their ubiquitous distribution in the cell, their step size, their mechanism of force generation, and the coordination between the hydrolysis cycles of the individual and often identical subunits that compose these ringed-proteins are poorly understood. Recent crystallographic and bulk biochemical studies [14, 16, 17, 19] suggest various models of coordination in which subunits act sequentially and in order [3, 20–28], simultaneously and in concert [29], or independently and at random [30]. In a broader sense, the importance of the packaging motor lies in that being the best studied of ring ATPases, it represents an excellent model system for understanding the operation of many other related ring-ATPases.

13.2 The Bacteriophage Packaging Motor

The question of how the dsDNA bacteriophages get their genomic DNA on the inside of the capsid, has been on the minds of virologists since Salvador Luria and Thomas Anderson made the first electron micrographs of phage T2 and saw tadpole-shaped objects that they likened to spermatozoa [31, 32]. Initially, it seemed obvious that the DNA must first condense into a compact form, after which the capsid proteins would come together to form a shell around it. The problem was framed in its current form about 25 years ago when it was shown – astonishingly at the time – that during the latter stages of a phage infection, an empty protein shell is assembled first and then the DNA is somehow transported across the shell into the interior [33–35]. Soon it was shown that a tiny motor placed at the base of the capsid carries out the job of packaging the DNA [36]. This mechanism is common to other tailed bacteriophages and animal viruses such as Herpes, Poxvirus, and Adenovirus [37].

The mechanical feat of the $\varphi29$ motor can be best appreciated when one compares the size of the virus icosahedral capsid (roughly a cylinder 40 nm in diameter and 50 nm in height) with a total volume $\sim0.065\,\mu m^3$ to the dimensions of the viral genome of ~19.2 kbp [11]. In the packaging process, the DNA molecule is compacted some 6000 times from its free volume in solution and attains an intracapsid concentration of roughly 500 mg ml^{-1}. Clearly, many forces arising from the electrostatic DNA self-repulsion, DNA bending rigidity, and entropy loss must oppose the packaging process.

The DNA packaging motor of bacteriophage $\varphi29$ is made up of three concentric rings [11] (see Fig. 13.1): (1) the head–tail connector, a dodecamer of a 36-kDa protein that fits in the pentameric opening at one of the ends of the capsid [38]; (2) a ring made up of five molecules of RNA, each 174 nt long and whose function is still unknown [38]; and (3) a pentameric ring [39] of gp16, an ATPase that belongs to the FtsK/HerA family of ASCE superfamily of AAA + ATPases [40, 41]. Initially, it was believed that the connector was the core component of the motor, packaging the DNA by ATP hydrolysis driven rotation, much like a nut on a screw [11, 42, 43]. However, the field has recently shifted to a ATPase-centric view, driven in part by our

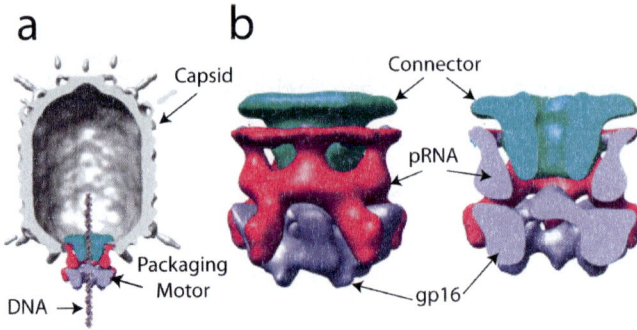

Fig. 13.1. *The packaging motor of the bacteriophage φ29.* (**a**) Cryo electron microscopy (EM) reconstruction of the capsid with the packaging motor complex. The DNA has been modeled in to set the relatively scale. (**b**) Cyro-EM reconstruction of the different components of the packaging motor: the dodecameric connector, gene product 10 (gp10), the pentameric pRNA ring, and the pentameric ATPase ring, gp16. Modified from [39]

experimental observation that the connector does not rotate during active packaging [44], but also due to the increasing evidence that phylogentically related ring-ATPases, such as the dsDNA translocase FtsK, do not need structures analogous to the connector to translocate on dsDNA [21, 45, 46]. For these reasons we focus our following discussion on the ATPase ring of the packaging motor.

13.3 A Single Molecule Packaging Assay

Biologists have speculated for a long time about the forces involved in this process, but even their order of magnitude of was not known. To address this question and to investigate the mechanical properties of the motor, we developed a single molecule optical tweezers assay to follow the packaging process of a single bacteriophage. Figure 13.2a depicts our initial experimental geometry [38]. As shown in the figure, a single capsid that has not completed the packaging of its DNA is attached via antibodies to the surface of a polystyrene bead held by suction atop a micropipette. The distal or free end of the DNA is attached to another bead held in an optical trap. With this geometry it is possible to devise two types of experiments: a constant force experiment in which, as the packaging proceeds, the tension on the DNA is kept constant at a pre-set value via a feedback operation that moves the pipette towards the bead in the optical trap, and a passive experiment in which the feedback is turned off and as the motor packages the DNA, it pulls the bead away from the center of the trap, generating increasing load.

In addition to the ability to apply opposing forces to the packaging motor, this optical tweezers assay allows the length of DNA remaining outside of the

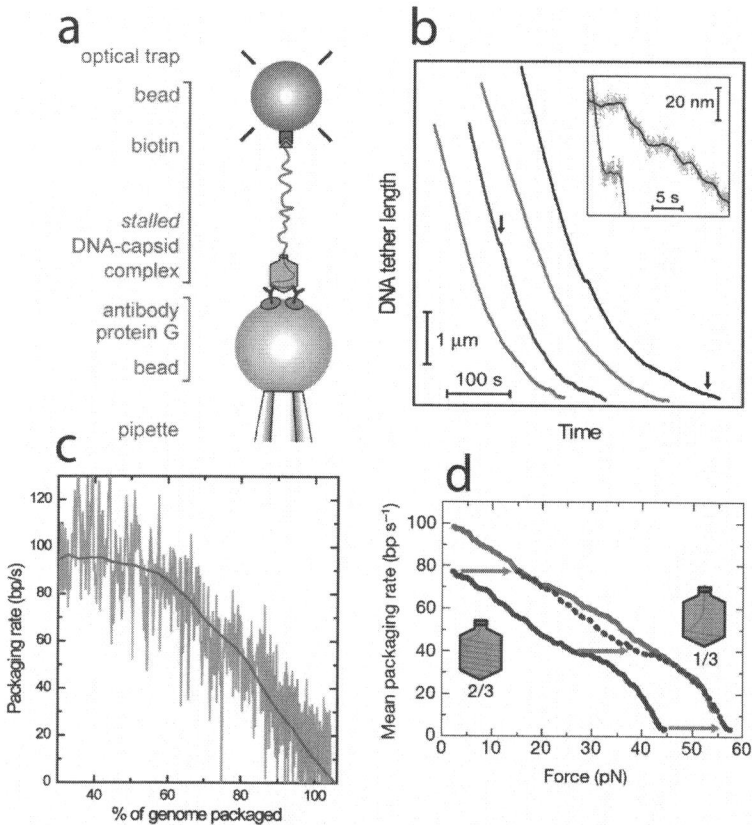

Fig. 13.2. *Following the packaging of dna by a single prohead-motor complex.* (**a**) Schematic diagram of a single molecule optical tweezers assay to follow the packaging of a single prohead-motor complex. An optical trap (hash marks) holds a micron sized polystyrene bead coated with streptavidin (square). This allows a non-convalent bond to be made to the biotinylated end (triangle) of a single piece of DNA. The other end of this DNA molecule is held by a single prohead-motor complex which is immobilized on another polystyrene bead via antibodies to the major capsid protein (gene product 8). This bead is itself immobilized on a micropipette. The entire experiment is done in a micro-fluidics chamber in aqueous buffer containing ATP. (**b**) Sample packaging traces. The length of DNA remaining between the two beads can be followed as a function of time. Arrows highlight regions shown on a larger scale in the insert. (**c**) The packaging velocity versus the percentage of the genome that has been packaged. The packaging velocity is the slope of packaging traces such as those in (**b**). Gray corresponds to the instantaneous velocity while black is a running average. (**d**) Packaging velocity versus opposing load for complexes with different amounts of DNA packaged. On the right a complex with 1/3 of the genome packaged is plotted while on the left a complex with 2/3 of the genome packaged is plotted. Remarkably, the shape of these curves differs by a constant force offset, suggesting that the affect of the extra 1/3 genome is to add a force opposing packaging. Modified from [38]

capsid, i.e., not yet packed, to be monitored in real time. Using the worm-like chain theory of polymer statistics [47, 48], it is thus possible to convert this distance to the number of base pairs at any given force, as shown in Fig. 13.2b. The slope of these plots, corresponding to the derivative of the number of base pairs with time, is the instantaneous packaging rate of the motor. As shown in Fig. 13.2c, this rate is relatively constant until roughly 50% of the genome has been packaged, at which point the packaging rate starts decreasing until it goes to zero when ∼100% of the DNA has been packaged [38, 49]. Since in these experiments packaging occurs under constant external force, this result strongly suggests that as the DNA fills the capsid, the motor experiences an opposing force due to the confined DNA that slows the motor and eventually stalls it.

To determine the magnitude of this internal force, we must establish how the motor packaging rate varies with force, i.e., we must perform experiments in which the load experienced by the motor increases as the packaging proceeds. Figure 13.2d shows the velocity–force relationship of the motor obtained under saturating [ATP]. Note that external forces as high as ∼55 pN must be applied to the motor to stall it, making this motor one of the strongest known in nature. By comparison the stall forces for myosin and kinesin are only between 3 and 6 pN. From the data in Fig. 13.2c and d, it is possible to estimate the internal force developed inside the capsid due to the confinement of the DNA. By the time the full genome has been packaged inside the capsid, the internal force reaches ∼50 pN, a value just below the average stalling force of the motor. This force can be used to estimate the average pressure inside the capsid. These analyses yield a value of the DNA internal pressure of ∼6 MPa or ∼60 atm of pressure, in reasonable agreement with values estimated from other techniques [50–52] and from theoretical studies [53]. For comparison, this pressure is about 10 times the pressure of dissolved CO_2 inside a bottle of Champagne.

Why should such pressures develop in the course of DNA packaging and at the expense of a great deal of ATP consumption? In fact, if 50% of the DNA can be packaged without a significant internal force development, it follows that 100% of the DNA could be packaged at very low energy cost if the head of the virus were simply twice as big. Why should not evolutionary pressure lead to the emergence of larger, more energy efficient viral capsids? The likely answer is that bacteriophages such as φ29 have evolved to take advantage of these high intra-capsid pressures in order to insure an efficient overall infective cycle. During phage assembly, the virus uses the host ATP that is freely available in the intracellular medium to package its DNA. This energy, however, is not lost nor dissipated. Instead, by confining the DNA into a small volume, the virus converts and storages this chemical energy as mechanical potential energy, similar to that stored in a compressed spring. Once the assembly has been completed and the newly synthesized phages are released into the extracellular medium – where ATP is scarce – the phages can use the stored energy to effectively inject the pressurized DNA into the new host cell. Of course, the fact that significant pressure only builds up after $\sim$50% of the genome is packaged implies that the generated pressure can only be used to inject $\sim$50% of the genome. Remarkably, transcription of one of the proteins expressed early in infection, gp17, appears to pull the rest of the genome into the cell [54].

13.4 Single-Molecule Mechanochemistry

While these initial experiments established the force scale for DNA compaction, they left open the question of how the motor actually generates these remarkably high forces. Where does the motor derive the necessary energy? Is it the enthalpy gained upon ATP binding or the entropy gained upon product release? To address these questions and others, we probed the mechanochemical cycle of the motor – the coupling between the mechanical force generation and the ATP hydrolysis cycle [55]. This cycle can be written formally as a series of motor states (free, bound to ATP, bound to ADP, etc.) connected to each other via reversible or irreversible transitions to which we associate rate coefficients. The step in which the actual translocation takes place should have associated, by definition, a *force dependent* rate constant. Thus, it is possible to determine the location of the translocation step by investigating the force dependence of the packaging rate as a function of [ATP] [56,57]. The chemical identity of the kinetic transition that corresponds to the stepping transition should help determine where the motor derives the energy needed to package the genome. With the single molecule assay developed above it is possible to follow the effect of both force and [ATP] on the packaging rate of the motor; thus, this assay is well suited to address these questions [55].

Figure 13.3a shows a plot of the rate of packaging as a function of the ATP concentration under a constant force of 5 pN. This data is well described by the characteristic Michaelis–Menten behavior with a $V_{max} \sim 100 \, \text{bp s}^{-1}$ and a $K_M \sim 30 \, \mu\text{M}$. Interestingly, the fit, done to a Michaelis–Menten–Hill equation reveals a Hill coefficient $n \sim 1$, indicating that the binding of the ATP to the motor is not cooperative. These same studies revealed that ADP is a competitive inhibitor of the motor and that phosphate release should be a nearly irreversible step [55], as its concentration in solution can be varied three orders of magnitude without affecting the rate of the motor.

Figure 13.3b shows the effect of opposing load on the packaging velocity under conditions of saturating [ATP] or limiting [ATP]. To interpret these plots we must remember that when the DNA translocation step is rate-limiting in the mechanochemical cycle of the motor, the rate of packaging should be highly force-dependent. Let us assume that the actual translocation coincides with the binding of ATP to the motor, as it is thought to be the case with the F1FO ATP synthase [58]. In such a motor, the binding of the nucleotide leads to a conformational change that constitutes the power stroke. In other words, the energy for the translocation is provided by the binding energy of

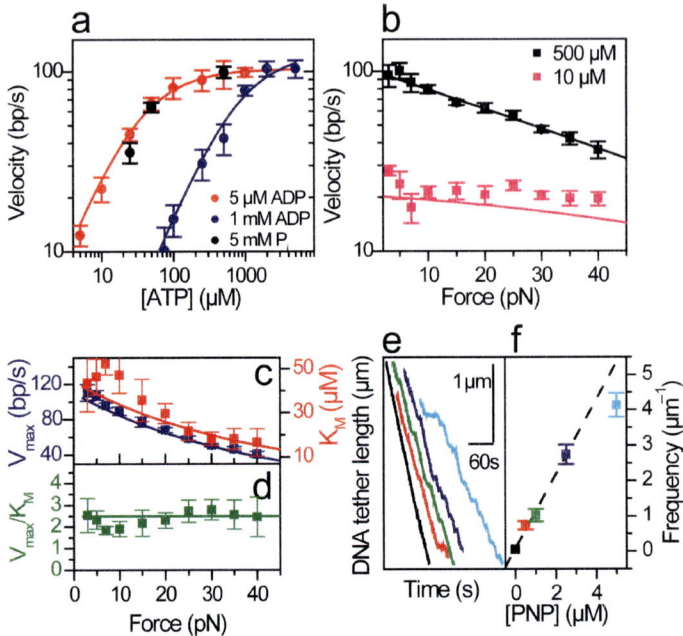

Fig. 13.3. *Packaging velocity of the bacteriophage φ29.* (**a**) Packaging velocity at different [ATP]. Red symbols correspond to low concentrations of the hydrolysis product, ADP, blue to high concentrations, and black corresponds to high concentrations of the hydrolysis product, Pi, inorganic phosphate. The solid lines are fits to the Michaelis-Menten equation. This data indicate that ADP acts as a competitive inhibitor. (**b**) Packaging velocity at different opposing forces. Black corresponds to conditions of saturating [ATP] while magenta corresponds to limiting [ATP]. The fact that the force dependence is stronger at saturating [ATP] indicates that ATP binding is not the force generating step. (**c**) Force dependence of the maximum velocity, V_{max}, and the Michaelis constant K_M. Solids lines are fits to an Arrhenius force dependence. (**d**) Ratio of V_{max} to K_M as a function of force. The fact that this ratio is force independent indicates that the stepping transition is not connected via reversible transitions to ATP binding [55]. (**e**) Sample packaging traces with increasing concentrations of the non-hydrolyzable ATP analog, AMP-PNP. No AMP-PNP, 500 nM, 1, 2.5, and 5 μM AMP-PNP are plotted in black, red, green, blue, and cyan, respectively. (**f**) Frequency of AMP-PNP induced pauses versus [AMP-PNP]. The dashed line is a linear fit. Color scheme as in (**e**). Error bars in all panels are standard deviation. Modified from [55]

ATP to the motor. In this case, hydrolysis is a step required to re-set the motor and liberate the products of the reaction. Then, in conditions in which DNA binding becomes rate-limiting, translocation should also become rate limiting and the motor should display a strong sensitivity to an external load. Conversely, in conditions in which DNA binding is no longer rate-limiting the motor should be relatively insensitive to the application of an external load. Figure 13.3b clearly shows that the packaging velocity at limiting [ATP] is largely *insensitive* to force whereas it is very force *sensitive* at saturating [ATP] – exactly the opposite behavior from that predicted for a motor in which ATP binding is the translocation step. We thus conclude that ATP binding is not the force generating chemical transition.

These arguments can be formalized by postulating a mechanochemical cycle for a single subunit within the motor:

$$E \underset{k_{\pm 1}[\text{ATP}]}{\longleftrightarrow} T \underset{k_2}{\longrightarrow} T^* \underset{k_{\pm 3}}{\longleftrightarrow} DP_i \underset{k_4}{\longrightarrow} D + P_i \underset{k_5}{\longleftrightarrow} E + D$$

where E corresponds to the empty or apo state, T to an ATP docked state, T^* to a second tightly bound ATP state, DP_i to the hydrolyzed state, and D to the ADP bound state. Here, we assume that the binding of ATP takes place in two stages (T and T^*), as has been observed for other motors [58], weak binding or "docking" of the nucleotide at the binding site followed by an irreversible step corresponding to "tight binding." (We will provide further verification for this assumption below.) Note that the step corresponding to the release of inorganic phosphate is irreversible, consistent with the observation that the packaging velocity is insensitive to $[P_i]$ over three orders of magnitude (Fig. 13.3a).

With some calculation it can be shown that this scheme predicts a packaging velocity with an [ATP] dependence described by the Michaelis–Menten expression:

$$v = \frac{V_{\max}[\text{ATP}]}{K_{\text{M}} + [\text{ATP}]}$$

In conditions where binding is rate limiting, the rate of the reaction can be written as the product of an apparent second order constant and the concentration of ATP.

$$v = \frac{V_{\max}[\text{ATP}]}{K_{\text{M}}} = k_{\text{app}}[\text{ATP}]$$

Then the force independence of the rate in these conditions (Fig. 13.3b) implies that while $V_{\max}$ and K_{M} can be themselves force dependent, their ratio should not depend on force. Figure 13.3c and d shows that this is indeed the case, while both $V_{\max}$ and K_{M} decrease with the magnitude of the external force, both do so in identical proportion throughout the range of forces studied (Fig. 13.3d). Algebraic analysis [55] of the general kinetic scheme proposed above shows that while $V_{\max}$ and K_{M} depend on rate coefficients k_2, $k_{\pm 3}$, k_4, and $k_{\pm 5}$, their ratio depends only on rate constants $k_{\pm 5}$, $k_{\pm 1}$, and k_2. Thus, the force independence of this ratio implies that the force generating step (i.e., the step that has associated with it is a force dependent rate coefficient) can only be either ATP hydrolysis or release of inorganic phosphate. Because, as shown above, the latter is a largely irreversible step and nucleotide hydrolysis does not involve a high drop in potential energy along the reaction coordinate [58], we conclude that the force generating step in the φ29 packaging motor must coincide with the release of phosphate. (These arguments can be extended to a more general kinetic scheme with the result that force generation must be separated from ATP binding by two irreversible transitions, one before binding and one after [55]. Unless additional kinetic states are postulated, this leaves inorganic phosphate release as the most likely candidate for force generation.)

13.5 Hints of Coordination Among the ATPases

Our analysis above focused on only one of the individual ATPase subunits, but the packaging motor has a pentameric ring of identical ATPases which drives packaging [39]. Are the individual hydrolysis cycles of these subunits coordinated? The above observation of a Hill coefficient of 1 suggests that there is no cooperativity in ATP binding (Fig. 13.3a). One explanation for this observation is that the individual subunits act independently of one another and in a random order, similar, perhaps to the protease ClpX [30]. This model can be tested directly by introducing trace amounts of nonhydrolyzable ATP analogs such as AMP-PNP and Gamma S-ATP, which are known to stall packaging reversibly in bulk [38, 55]. If the various ATPase subunits of the motor fire at random and their activities are independent of their neighbors in the ring, at concentrations in which on average an increasing number of subunits are bound to a non-hydrolyzable nucleotide analog, the motor will continue packaging albeit at a correspondingly slower rate. If, on the contrary, the mechanochemical cycles of the ATPases around the ring are coordinated and subunit i must package before subunit $i + 1$ can do the same, then if subunit i is bound to a non-hydrolyzable analog the motor halts until that subunit exchanges its nucleotide for ATP. Figure 13.3e shows that in the presence of AMP-PNP the motor displays pauses of variable duration whose frequency increases with the concentration of the AMP-PNP. The pause length is not affected by the concentration of AMP-PNP nor of ATP. Figure 13.3f shows that the frequency of pauses grows linearly with the concentration of the analog, indicating that a single bound analog is sufficient to halt the motor. These observations indicate that these ATP-analog-induced pauses are manifestations of discrete binding and unbinding of single AMP-PNP molecules to a given ATPase in the motor and that ATPases in the packaging motor of bacteriophage $\varphi29$ do not act independently – the individual hydrolysis cycles must be coordinated in some fashion. These conclusions are supported by previous mutation studies with the pRNA ring, which indicated that all of the pRNA molecules are used during packaging [59]. This result rules out the alternative interpretation of the above data, that only one out of five subunits is active and solely responsible for the packaging observed.

13.6 Resolving the Discrete Steps of the Packaging Motor

The above measurements indicate the minimal properties any mechanochemical model of the packaging motor must have: DNA must be packaged on the release of product and the subunits must be coordinated. Given these restrictions, there are clearly several open questions that still remain. First, what is the step size of the packaging motor – the discrete increments of DNA packaged per released phosphate? Bulk measurements suggest a value near ~2bp per ATP consumed [55, 60], but such measurements can easily be biased by background processes which consume ATP but do not result in packaged DNA. Second, the previous results [55] require that the motor subunits be coordinated in some fashion, but leave open the possibility of a variety of coordination models. A sequential model, in which each subunit completes its own mechanochemical cycle before activating an adjacent subunit, has been proposed previously [55, 59]. While consistent with the data and with the majority of suggested models for related ring-ATPases [61, 62], such a model is clearly only the simplest of many possibilities. As will be shown below, direct observation of the steps of the enzyme will allow us to address both of these outstanding questions. First, it should be possible to quantify their size directly and to determine their variability if any. Second, by resolving the steps, we can clearly delineate when a single mechanochemical cycle of the enzyme begins and ends, making it possible to carefully compile the statistics of these cycle completion times. While different coordination models may predict the same average stepping velocity, these statistics will be distinct for different models, allowing us to test the coordination model proposed in previous studies [55, 59].

Because the distance between adjacent base pairs along the contour length of dsDNA is just ~3.4 Å [63], to observe a putative step size of only 2 bp we must develop the ability to resolve Ångstrom scale length changes – a spatial resolution not possible with traditional optical tweezers [64]. Thus, to address these questions, it was necessary to first construct a more sensitive optical tweezers instrument. The construction of an optical trap with higher spatial resolution requires a detailed consideration of all the sources of noise that limits such measurements [65, 66]. This noise can be divided into two broad categories: (1) experimental and environmental noise, such as temperature fluctuations or laser power fluctuations and (2) the more fundamental Brownian fluctuations of the beads we use as molecular handles. There are several different ways to address the first of these sources of noise [64, 67–70]. For example, mechanical and acoustic vibrations and temperature fluctuations can all be minimized through careful choice of a trapping environment [65]. However, perhaps, the most successful approach is to simply decouple the measurement from these sources by introducing a second optical trap formed by the second orthogonal polarization of the same laser [68]. In this design, each optical trap attaches to one of the ends of the system of interest

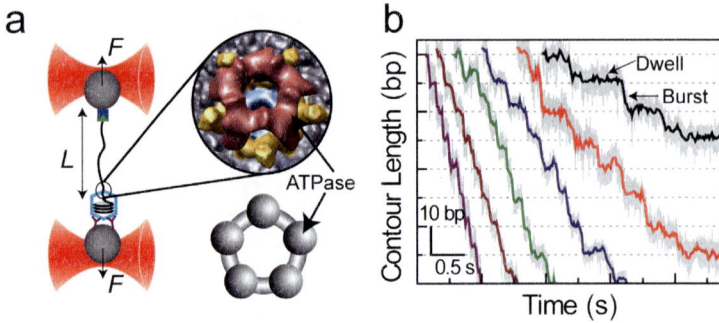

Fig. 13.4. *High resolution measurements of the dynamics of the packaging motor of the bacteriophage φ29.* (**a**) Schematic picture of the two trap geometry used for high resolution measurements. Two optical traps hold two 860-nm-diameter polystyrene beads each of which holds one end of a single DNA–prohead–motor complex. By adjusting the distance between the two traps, different forces, F, can be applied to the system. As the motor (top inset: cryo-electron microscopy reconstruction; courtesy of M. Morais; bottom inset: cartoon depiction of the pentameric ATPase) packages the DNA, the length of DNA remaining outside of the capsid, L, will decrease in time. By following this length with base-pair-scale resolution, we can infer the dynamics of the packaging motor. (**b**) Packaging dynamics observed under low opposing loads, ∼8 pN, and the full range of [ATP] (5, 10, 25, 50, 100, 250 μM in *black, red, blue, green, brown, and purple*, respectively.) Data in light gray are collected at 1.25 kHz and averaged to 100 Hz in color. Modified from [74]

[64,67,68] (see Fig. 13.4a below). This geometry not only decouples the measurement from fluctuations in the laboratory environment but it also reduces the effect of fluctuations in optical components, such as the laser, since many of these components will be common for the two traps. Indeed, because these fluctuations are shared by both trapping beams, they result in the common motion of the two traps, motions that represent an overall translation in the center of mass of the system and that do not affect the ability to measure the extension of the system of interest (vgr. the length of the DNA) between the two beads.

Even by addressing these extraneous noise sources the spatial resolution of an optical tweezers instrument is still fundamentally limited by the second source of noise, the fundamental thermal forces acting on the polystyrene beads we use as molecular handles [66, 71]. The continuous spontaneous fluctuations of the surrounding aqueous medium give rise to a fluctuating force on the trapped beads that in turn results in fluctuations in both the optically applied force and the tension of the DNA molecule. Because we must know the tension on the DNA molecule in addition to its extension in order to infer exactly the number of base pairs remaining between the two beads, these tension fluctuations give rise to an inherent uncertainty in the length of the DNA. While the introduction of a second optically trapped bead greatly improves the isolation of the system, it also introduces a second fluctuating degree of freedom subject to independent tension fluctuations. Thus, a priori it would seem that the dual trap geometry will actually degrade the signal-to-noise ratio of a single trap instrument. However, the fluctuations in the two beads are not completely independent of one another – the presence of the biological system tethered between them introduces correlations in the motion of the beads [72]. It is possible to exploit these correlations, partitioning the fluctuations into in-phase or out-of-phase modes, and discard the in-phase fluctuations which do not affect the measured end-to-end extension of the system [66]. However, this strategy is possible only if the motions of both beads are monitored simultaneously and if the extension of the system (the length of the DNA in this case) is not estimated by the position and force of a single bead but by the differential position and average force of the two beads. Remarkably, it can be formally shown that this combined measurement is less sensitive to Brownian fluctuations than a single trap system [66]. Thus, we built an instrument which combines the improved stability and isolation afforded by the dual trap geometry with the reduction in Brownian noise provided by a differential detection system [64–66]. This instrument holds the current record for spatial and temporal resolution with an optical trap – 1 bp length changes to a single piece of dsDNA measured with a temporal resolution of 250 ms [65].

13.7 DNA is Packaged in 10-bp Increments

Figure 13.4a shows a DNA packaging experimental set up using a dual trap optical tweezers. Here, one optical trap holds a micron sized polystyrene bead coated with streptavidin that is used to hold one end of a DNA molecule, labeled with biotin, that is being packaged into a single viral prohead. The other bead holds the prohead of a virus attached via antibodies raised against the major capsid protein, gene product 8 [11]. The initial connection between the prohead–motor complex and the free DNA can be established either in the test tube by stalling actively packaging complexes with non-hydrolyzable ATP analogs [38, 55] or in the tweezers by physically bumping a prohead–motor coated bead and a second bead coated with DNA [49, 73]. Either way, addition of ATP into the surrounding packaging buffer promotes the packaging of DNA by the motor. As above, packaging of the bacteriophage results in a decrease in the length of the DNA remaining outside of the capsid, and by measuring this length as a function of time we can infer the dynamics of the packaging motor.

Despite the fact that the packaging motor of $\varphi 29$ is a relatively slow dsDNA translocase, its maximum packaging rate of 100 bp s^{-1} [55] would still be too fast for us to resolve the expected 2-bp steps (2 bp every 20 ms).

To slow down the motor, we decreased the concentration of ATP to $5\,\mu$M, significantly below the Michaelis constant for the motor, $\sim 30\,\mu$M [55]. Figure 13.4b shows that under these conditions it is possible to clearly observe a stepwise decrease in the contour length of the DNA molecule [74]. Surprisingly, the average size of the DNA packaged in each "step" is not the expected 2 bp estimated from bulk measurements, but rather ~ 10 bp – an increment five times larger. We can quantify the spatial periodicity in these traces by using a pair-wise distance distribution (PWD) – a histogram of the distance between every two points in a trace [75]. The average PWD for all data collected under conditions of $5\,\mu$M [ATP] reveals a spatial periodicity of 10.6 ± 0.4 bp. Figures 13.4b and 13.5a show that as [ATP] is increased from limiting to saturating concentrations the motor displays the same discrete increments of packaged DNA in each cycle. Since there is no statistically significant change in the size as a function of [ATP], the average of these data produce a best estimate for the amount of DNA packaged during a cycle of the packaging motor of 10.0 ± 0.2 bp.

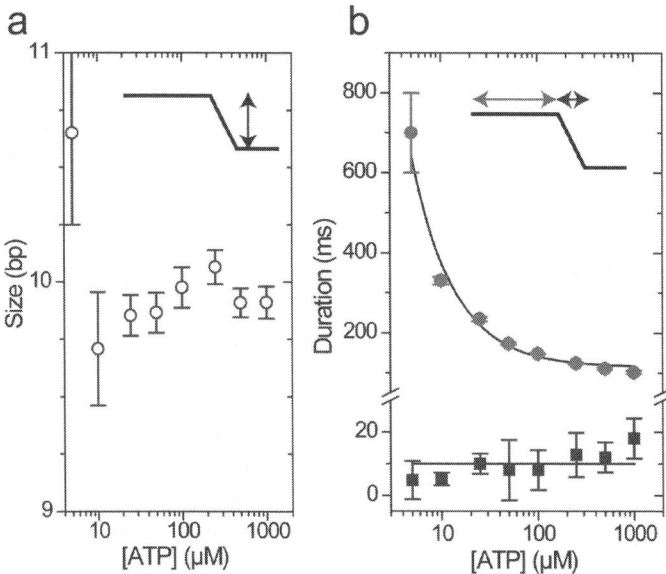

Fig. 13.5. *Mean burst size and dwell and burst duration as a function of [ATP].* (**a**) The spatial periodicity observed in the average PWD for the low force data in Fig. 13.4b as a function of [ATP]. Error bars represent the standard deviation in a linear fit to the position of each of the peaks in the average PWD. (**b**) Mean dwell and burst durations for the low force data in Fig. 13.4b as a function of [ATP]. The mean dwell time before the 10-bp bursts (*circles*) is well described by a Michaelis–Menten [ATP] dependence (*solid line*). The average burst duration (*squares*) shows no apparent [ATP] dependence, with a mean of ~ 10 ms (*solid line*). Modified from [74]

13.8 Multiple ATPs Bind Before each 10-bp Step

To further elucidate the mechanism of the packaging motor, we also quantified the time the motor paused or dwelled at a single DNA location before the 10-bp steps. In contrast to the PWD analysis above, quantifying the dwell times requires the identification of the exact time at which the motor transitions from one constant position to the next, i.e., the exact time at which a step occurs. Many different algorithms have been developed to find such stepping transitions, including variance filters, velocity thresholds, or more exotic step finders [76]. However, in each of these cases the exact choice of threshold or the final averaging bandwidth of the data is not well defined, lending a bit of art to the exact performance of the step finder [76]. To address this issue we modified a step finding algorithm based on the Student's t-test [74, 77]. The t-test is the appropriate statistical method to determine if two different sets of Gaussian-distributed data with equal variance have the same mean. This is exactly the situation encountered in our experiments – the positional noise is Gaussian distributed with a variance that changes – due to stiffness variation in the DNA – but not over the distance scale of a single step. Thus, this statistical test is well suited for our purposes. To employ this test we select two regions of data before and after a given time point, calculate the t-value – the difference in the mean divided by the standard error of the mean, and then calculate the probability of observing this t-value given the null hypothesis – i.e., that no step was taken between the two data sets. Figure 13.6 illustrates that the t-value peaks when a step is taken and that the probability of observing that t-value drops dramatically. To select our step transitions, we then set a probability threshold of 1e-4. One advantage of the t-test is that this probability threshold has a clear interpretation – if no step were present we would only score a false step one in 10,000 times. Moreover, because the t-value is the difference in the means scaled by the standard error of the mean, a quantity which is largely independent of bandwidth, the t-value is not sensitive to any pre-averaging or down sampling of the data, removing one free parameter from data analysis.

Using this technique, we can calculate the average duration of the dwell before each 10-bp packaging step as a function of [ATP]. Figure 13.5b shows that the average duration follows a simple, Michaelis–Menten-like [ATP] dependence. Namely, $\langle t_{\text{dwell}} \rangle = \left(K_{1/2} + [\text{ATP}] \right) / k_{\max}[\text{ATP}]$, where $K_{1/2}$ is the [ATP] at which the dwell time adopts twice its minimum value, $1/k_{\max}$. If we consider a more general [ATP] dependence, $\langle t_{\text{dwell}} \rangle = \left(K_{1/2}^n + [\text{ATP}]^n \right) / k_{\max}[\text{ATP}]^n$; i.e., a Hill-like expression [78], we find a Hill coefficient consistent with 1, $n = 0.7 \pm 0.3$. Clearly, the dwell times that separate the 10-bp steps show a strong [ATP] dependence, indicating that nucleotide binding occurs during this dwell.

In addition to the dwell durations, we noticed that many stepping transitions take a finite time to complete – a time larger than the sampling time or averaging time of our measurement. We quantified this duration using the

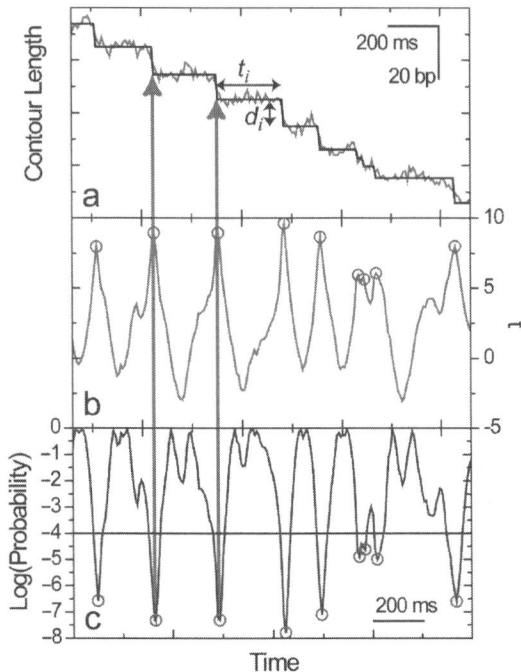

Fig. 13.6. *Student's t-test step finding algorithm.* (**a**) A sample stepping trace in gray with a staircase fit in black determined from the location of stepping transitions. The step size of a given step i is determined from the difference in the mean of two adjacent regions between stepping transitions while the corresponding dwell time, t_i, is the time between the given transition and the preceding transition. (**b**) The t-value calculated for 20-ms regions of data before and after a given point. Circles correspond to identified stepping transitions. (**c**) The logarithm (base 10) of the probability of observing a given t-test given the assumption that no step occurred between the two 20-ms windows. The line corresponds to the detection threshold of 1 in 10,000. The exact location of the stepping transition is determined from the position of a local minima within each set of contiguous points below this threshold. The burst duration is calculated from the duration in time for which the t-value is below the threshold

t-test analysis described above by measuring the time for each transition for which the calculated value had a probability lower than the threshold value (see Fig. 13.6). The average duration was found to be independent of the concentration of [ATP] and was typically $\sim$30–40 ms. However, this value is very near the averaging time of our typical measurement, $\sim$20 ms. Since this is the shortest duration we can measure, i.e., the dead-time of our measurement, the average stepping durations are likely to be biased by this dead-time. Thus, in Fig. 13.5b we report not the average, but the average decay rate of the distribution of stepping durations ($\sim$10 ms), a value less biased by the dead-time.

The fact that the stepping durations are independent of [ATP] indicates that no ATP binding occurs during the actual translocation of the motor.

The behavior of the mean stepping duration and the mean dwell duration as a function of [ATP] combine to produce an [ATP] dependence of the mean packaging velocity consistent with what has been observed previously [55] (see Fig. 13.3a). However, the ability to observe the actual steps allows us to compute not just the mean dwell time but to determine also the exact times at which the motor begins and ends each dwell. This analysis yields not only the mean but also the fluctuations around this mean. Why would we expect these pause durations to fluctuate, even for constant concentration of ATP? As discussed above, molecular motors, unlike their macroscopic counterparts, work at an energy scale similar to the thermal energy contained in the surrounding aqueous media. Thus, they operate as open systems, actively exchanging energy with the surroundings. As a result, their dynamics are dominated by the random interactions with the thermal bath that provide the energy necessary to overcome transition barriers and that give these motors their stochastic behavior [56, 79]. Interestingly, while we cannot see the exact time at which all of the internal kinetic transitions occur – since we only observe the transitions which generate a length change in the DNA molecule – it is still possible to infer properties of these internal kinetic states from these fluctuations [79]. In particular, since the full dwell duration is the sum of the individual lifetimes of the internal and hidden kinetic states, the statistics of these hidden lifetimes have measurable effects on the statistics of the full lifetime. For example, if we can adjust one of the kinetic lifetimes so that it is much larger than the rest of the cycle, by lowering the concentration of ATP perhaps, then the statistics of the lifetime of the empty state will dominate the statistics of the full cycle completion time.

In general, transitions from one kinetic state to the next are Markov processes, i.e., the system has no memory of the state it came from or how long it has been in the current state. The result is that the probability of leaving a given kinetic state is constant in time, leading to lifetimes that are exponentially distributed [79]. To be concrete, if a single ATP were to bind to the motor before each of the 10-bp "steps," we would expect the statistics of the dwells under conditions of limiting [ATP], 5 μM, for example, to be dominated by the statistics of the time needed to bind ATP. Thus, we would expect a lifetime described by a single exponential decay. Figure 13.7a shows the dwell time distributions observed for all [ATP]. Note that under no conditions, limiting or saturating [ATP], are the distributions well described by a single exponential decay. Rather, all of the distributions are peaked functions.

What is the meaning of these peaked dwell time distributions? Mathematically, the dwell time distribution is the convolution of the individual exponential distributions governing the underlying state lifetimes [79]. If one of these lifetimes is much longer than the others, we would expect to see a distribution dominated by a single exponential decay. However, if multiple kinetic states have lifetimes of comparable duration, these states can be thought of

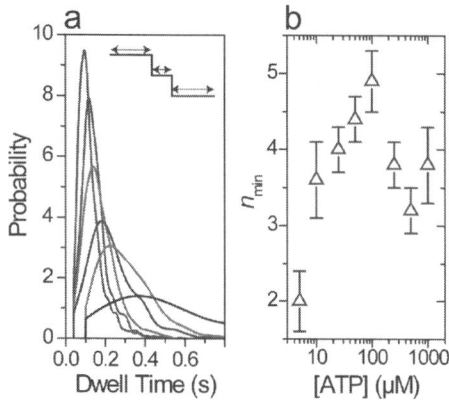

Fig. 13.7. *Fluctuations in the dwell time before the 10-bp bursts.* (**a**) Probability distributions for the dwell times before the 10-bp bursts in Fig. 13.4b as a function of [ATP]. Peaked distributions are indicative of multiple rate-limiting kinetic transitions. [ATP] decreases from 250 to 5 μM left to right. (**b**) The effective number of rate-limiting transitions, n_{min}, as a function of [ATP]. The values at low and high [ATP] indicate the minimum number of ATP binding events and of non-binding events that must occur during each dwell, ~2 and ~4 respectively. Modified from [74]

as "equally" rate-limiting, and the final distribution will be a convolution of the multiple lifetime distributions. The convolution of multiple exponential decays is a peaked function. In general, the more the exponentials, the more sharply peaked the distribution. Thus, the fact that the dwell time distributions are peaked across the full range of [ATP] implies that for all [ATP] the dwell time duration is governed by multiple rate-limiting kinetic events.

We can estimate the number of kinetic events that are effectively rate-limiting in any given process by quantifying the degree to which the distribution of its duration is peaked. The dimensionless ratio of the mean duration squared over the variance, $n_{min} = \langle t \rangle^2 / \left(\langle t^2 \rangle - \langle t \rangle^2 \right)$, a quantity related to the inverse of the randomness parameter [79], does exactly this. It is possible to formalize the arguments above and show that $n_{min} \leqslant n$ where n is the number of unique kinetic states that occur within a single dwell. The value of n_{min} is lowered from the actual number n if the lifetimes of these states are different or if the transitions between these states are reversible [79,80]. This remarkable result implies that by simply measuring the degree to which these distributions are peaked, we can place strict limits on the number of underlying kinetic events which are rate-limiting under each [ATP]. Figure 13.7b plots this statistic for the dwell time distributions observed in Fig. 13.7a. At limiting concentrations of ATP, we find that these values asymptote to ~2 indicate that under these conditions at least two kinetic events are rate-limiting. Because these are conditions of limiting [ATP], these kinetic events must be the binding of ATP. The conclusion is that no less than two ATP molecules bind to the motor before each dwell. As the concentration of ATP is increased, we expect that the time to bind ATP will decrease, becoming comparable to the other lifetimes in the kinetic cycle. Thus, n_{min} should increase, exactly as is observed. At some [ATP] concentration the time it takes to bind ATP will be comparable to the time it takes to do everything else in the cycle, and n_{min} should reach a maximum, corresponding to the maximum number of rate limiting steps in the cycle. Finally, as [ATP] is further increased, the time it takes the motor to bind nucleotide will become negligible compared to the time it takes to carry out these other events, and n_{min} should decrease and saturate to a constant value. Again, this is exactly what is observed in Fig. 13.7b with n_{min} attaining a value of ~ 4 under conditions of saturating [ATP]. This limit implies that in addition to binding events, no less than four additional kinetic events must occur within each dwell. In summary, this analysis demonstrates that each dwell is a remarkably complex process consisting of no less than six distinct kinetic states in total – no less than two ATP binding states and no less than four additional non-binding states.

13.9 Packaging Occurs in Bursts of Four 2.5-bp Steps

Figures 13.5 and 13.7 present several surprising results. First, DNA is not packaged in the 2-bp increments expected from bulk measurements, but rather increments five times larger. Second, the motor does not bind a single ATP before packaging in 10-bp as would be predicted from the simple sequential models suggested previously [55, 59], but rather it binds at least 2 ATP molecules before each 10-bp increment. Moreover, the motor takes a finite, albeit short, time to package each 10-bp of DNA in contrast to what would be expected from the quick motion of a fundamental kinetic or mechanical transition. The question then arises as to whether the 10-bp steps observed constitute the fundamental translocation units of the packaging motor or they are made up of smaller translocation events which may be too fast to observe under the experimental conditions? The fact that the stepping transitions take a finite time to complete suggests the existence of intermediate kinetic transitions. Moreover, occasionally these transitions are slow enough that they produce noticeable micro-dwells that split the 10-bp steps into smaller transitions. We can identify these micro-dwells and the smaller steps they generate using the t-test together with a novel correlation analysis [81]. This analysis confirms that these smaller steps occur in groups that add to 10-bp. These results support the hypothesis that the 10-bp packaging events are burst of multiple smaller steps that add to 10-bp.

How many steps compose the 10-bp burst? This number will give us insight into the number of ATP molecules that bind during the dwell. What is needed is a way to slow down the translocation bursts. In other molecular motors, the stepping transitions are typically preceded by the binding of substrate, so lowering the substrate concentration will slow these steps [67]. However, as shown above, this is not the case for the packaging motor in φ29; the motor simply waits as long as is needed to load ATP before packaging in a burst of smaller steps. However, since the micro-dwells that precede these smaller steps must correspond to the kinetic transitions that generate motion, and thus, force, their lifetimes must be force-sensitive [56]. Thus, while decreasing the concentration of ATP does nothing to slow these steps, increasing the opposing load should slow them.

However, as shown above, φ29 is one of the strongest molecular motors known, capable of exerting forces as high as ~60 pN [38]; thus, to slow down this motor, we must use relatively high forces, i.e., forces on the order of ~30–40 pN. Forces of this magnitude require the use of high laser powers. High laser powers, in turn, result in local heating of the sample [82], which is effectively doubled by the two trap geometry we use for high resolution. Trap-related laser heating appears to have only minor effects on the dynamics of most molecular motors. However, the packaging motor in φ29 appears to be extremely sensitive to temperature, increasing its packaging velocity by ~20 bp/s per degree C (unpublished data, J. Moffitt). This effect is so significant that in our initial attempts to apply high forces to the motor, we found that the force dependent decrease in the packaging velocity was almost perfectly compensated by a temperature dependent increase in the packaging rate. The solution of this problem was to change the absorption spectra of the surrounding aqueous media. This can be done with minor perturbation to the motor by packaging not in H_2O but in D_2O, i.e., by using heavy water. The increase in the mass of the hydrogen isotope moves the absorption spectra of heavy water out of the near-IR range used for trapping thus removing the laser heating issues [83]. Moreover, control experiments confirm that while heavy water changes the average dwell time of the motor, it does not change the size of the packaging burst, ~10-bp [74].

Under these conditions it was possible to apply ~40 pN of opposing load to the motor and slow down the translocation bursts. Figure 13.8a shows that in these experiments the DNA length is still observed to decrease in step-wise increments, but that the size of these increments is now smaller than the 10-bp increments observed under low force (Fig. 13.4b). Surprisingly, the PWD analysis of these data, shown in Fig. 13.8b, reveals that these steps are 2.4 ± 0.1 bp in size, not the expected 2-bp estimated from bulk measurements. This is a particularly surprising result given the prevailing expectation that nucleic acid translocases would move in discrete numbers of base pairs. However, there is, a priori, no reason why a nucleic acid molecular motor should translocate only in steps that are integer multiples of base pairs; this observation constitutes, nonetheless, the first direct experimental evidence of such non-integer base pair steps. Moreover, this result suggests that the 10-bp bursts are composed of four 2.5-bp steps, a conclusion that is further supported by the prominent fourth peak at 10-bp observed in the PWD function (Fig. 13.8b), a remnant of the 10-bp periodicity observed under conditions of low force.

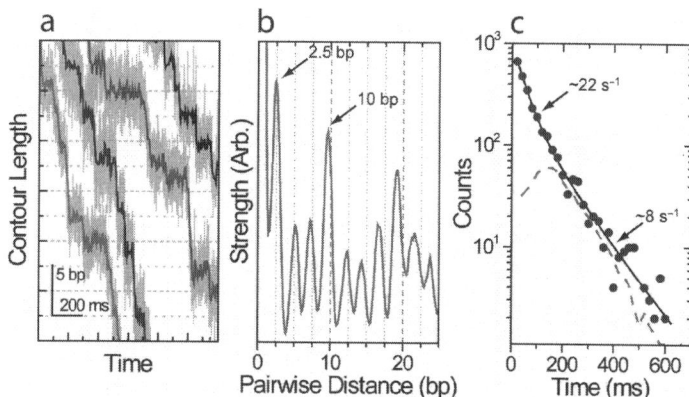

Fig. 13.8. *Packaging under high force reveals a 2.5-bp step size.* (**a**) Sample packaging traces under saturating [ATP], 250 μM, and high opposing forces, ~40 pN, in a 80% D₂O packaging buffer. Data in light gray are collected at 1.25 kHz and boxcar averaged and decimated to 100 Hz in color online. (**b**) Average PWD for packaging traces collected under high force. A clear ~2.5-bp periodicity is observed, indicating that DNA is packaged in 2.5-bp increments. The large peak at 10-bp is a remnant of the 10-bp burst structure observed at low force. (**c**) Dwell time distribution for the 2.5-bp steps under these conditions. This distribution is well described by a bi-exponential decay (*solid line*). The dashed line is the dwell time distribution observed for the dwells before the 10-bp bursts under the same buffer conditions but low force as in Fig. 13.4b scaled by 1/4. Modified from [74]

Figure 13.8c shows the distribution of dwell times that precede these 2.5-bp steps. This distribution is well described by a bi-exponential decay with a fast rate of $22 \pm 2\,\mathrm{s}^{-1}$ and a slow rate of $8 \pm 1\,\mathrm{s}^{-1}$. How do we rationalize a bi-exponential dwell time distribution? Given the burst structure suggested by the low force data, one would expect that the dwell times for the 2.5-bp steps should involve different kinetic events depending on the identity of the step within each group of four steps. In other words, the first of each set of four 2.5-bp steps should have a dwell time which is naturally longer than the dwells before the following three steps since this first dwell involves all of the kinetic events that precede a 10-bp burst at low force. In comparison, the other three steps should be preceded by shorter dwells since, presumably, they contain fewer kinetic transitions. Moreover, we would expect that – in addition to being on average longer in duration – every fourth dwell time should be governed by a peaked dwell time distribution similar to the distribution measured for the dwell times before the 10-bp bursts at low force. Remarkably, Fig. 13.8c shows that if the distribution for the 10-bp dwells under the same buffer conditions but low forces is scaled by a factor of 1/4, the resulting distribution is in excellent agreement with the slow tail observed in dwell times for the 2.5-bp steps under high force. Moreover, the expected peak in this distribution is hidden by the fast exponential decay, which is why the final distribution is well described by a bi-exponential decay. This observation not only suggests that the dwell-burst structure is maintained at high force but also that the dwell times before the other three steps are all similar with an average rate of $22\,s^{-1}$ under the conditions employed here. (Note that if each step had a different lifetime, we would not expect to observe the $1/4$ scaling for the 10-bp distribution.)

13.10 Inter-Subunit Coordination in the Packaging Motor

Taken together these results reveal a form of inter-subunit coordination for the packaging motor in $\varphi29$ not previously described for ring ATPases. The full mechanochemical cycle of the packaging motor involves the action of not a single subunit, but the coordinated action of many subunits, giving rise to a two-phase mechanochemical cycle for the entire ring. In the first phase, the "dwell" phase, the motor holds the DNA at constant length while it loads multiple ATPs. Once the ring is fully loaded, the motor enters the second phase, the "burst" phase, in which the motor packages the DNA in a concerted burst of four 2.5-bp steps. Once this cycle is completed, the motor resets and begins the process again. (See Fig. 13.9a.)

Our data also provide several constraints on the way in which ATP loads during the burst phase. First, the observation of four steps in the burst strongly suggests that the number of ATPs that bind to the ring during the dwell phase is also four – one ATP for each of the subsequent steps in the burst. The binding of four ATPs is consistent with an n_{min} at limiting ATP of 2 since reversibility in binding or different binding rates can lower this number from the actual number of binding events [79]. The binding of four ATPs per dwell would also produce a coupling constant between ATP consumption and packaging of 2.5 bp per ATP. The remaining 25% discrepancy between this value and that measured with bulk techniques, $\sim$2 bp/ATP [55,60], may be explained by the existence of processes that consume additional ATP. For example, the initiation of packaging may consume ATP or additional ATP may be required to repackaged DNA that slips from the capsid during packaging (a process observed directly in single molecule measurements [38,55]). Alternatively, it is possible that a regulatory, fifth ATP is bound to the ring each cycle and hydrolyzed futilely.

In addition, our data place strict limits on the possible mechanisms by which these ATPs load to the ring. Interestingly, these restrictions arise out of efforts to resolve an apparent contradiction in our data. In our measurements of the mean dwell time as a function of [ATP] (Fig. 13.5b), we found that the [ATP] dependence of the mean dwell time was well described by a Hill coefficient of 1, i.e., this data displayed a simple Michaelis–Menten-like [ATP] dependence. However, the requirement that multiple ATPs bind before

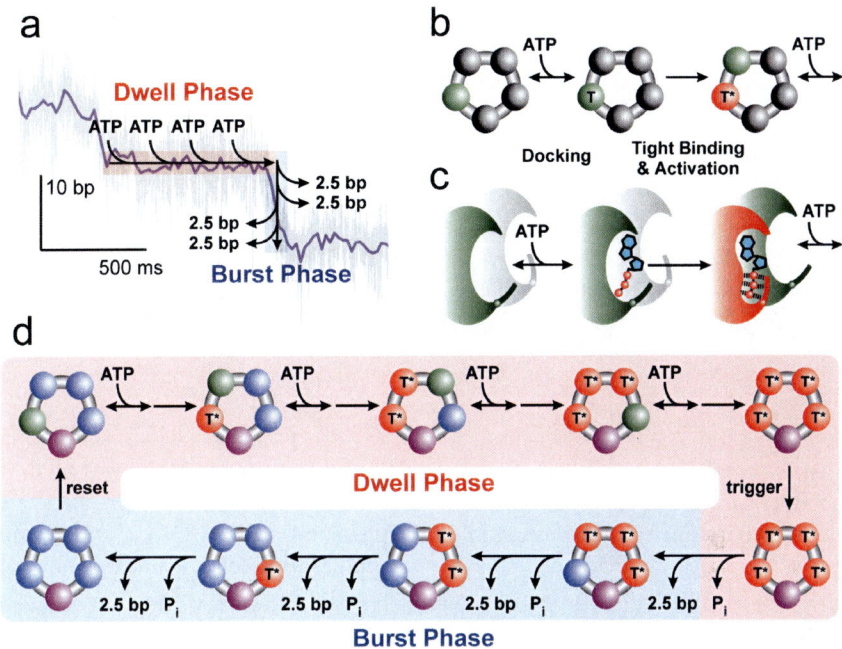

Fig. 13.9. *Inter-subunit coordination in the packaging motor.* (**a**) Schematic diagram of the two phase mechanism for packaging, overlaid on a sample packaging trace. (**b**) Detailed kinetics of ATP binding. ATP first docks reversibly to an active subunit (*green, T*) then undergoes a tight binding transition (*red, T**) which commits this ATP to the hydrolysis cycle. (**c**) The tight binding of one subunit activates the adjacent subunit, formerly inactive (*gray*), for docking. (**d**) Basic topology of the full mechanochemical cycle for φ29. Blue subunits correspond to the possible locations of ADP throughout the cycle. The purple subunit represents the one subunit which must be different than the rest to break the pentameric symmetry of the ring, generating only four steps per cycle. P_i represents inorganic phosphate. Reproduced from [74]

the formation of product (as required by our n_{min} measurements) is the classic text-book [78] example of binding cooperativity and is generally expected to produce an [ATP] dependence of the mean dwell time with a Hill coefficient greater than 1. How do we reconcile these contradictory observations? It turns out that by requiring, as suggested above, that the ATP bind to the ring not in one but in two kinetic events, that one of this transitions be essentially irreversible, and that the ATPs bind sequentially in time, the binding cooperativity inherent to a scheme with multiple ATP binding events can be effectively hidden, producing an [ATP] dependence of the mean dwell time with a Hill coefficient of 1, consistent with our data. (See Fig. 13.9b)

To illustrate how these restrictions arise consider the following kinetic scheme for a simplified two binding-pocket system:

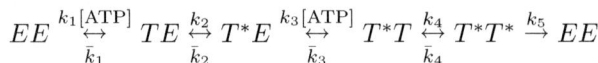

$$EE \overset{k_1[\text{ATP}]}{\underset{\bar{k}_1}{\leftrightarrow}} TE \overset{k_2}{\underset{\bar{k}_2}{\leftrightarrow}} T^*E \overset{k_3[\text{ATP}]}{\underset{\bar{k}_3}{\leftrightarrow}} T^*T \overset{k_4}{\underset{\bar{k}_4}{\leftrightarrow}} T^*T^* \overset{k_5}{\rightarrow} EE$$

where E denotes the empty or apo state, T the docking or loosely bound ATP state, and T^* a second ATP binding state, the "tightly bound" ATP state. The k_i denotes the forward rates while the $\bar{k}_i$ denote the reverse rates. The docking rates are expressed as pseudo-first order rates, i.e., $k_1[ATP]$. The mean dwell time for this simplified two binding pocket system is

$$\langle t \rangle = \left(b_0 + b_1[\text{ATP}] + b_2[\text{ATP}]^2\right) / \left(a_0[\text{ATP}]^2\right)$$

where the individual coefficients, a_i and b_i, are functions of the individual rates. Note that this expression has [ATP] terms to a higher power than 1, and thus would display a Hill coefficient greater than 1, as expected for a system which requires the binding of multiple substrates [78]. To understand how the [ATP] dependence can be simplified and the binding cooperativity of the system hidden, it is particularly instructive to consider the coefficient

$$b_0 = \bar{k}_1 \bar{k}_2 \left(k_4 k_5 + k_5 \bar{k}_3 + \bar{k}_3 \bar{k}_4\right)$$

Remarkably, if b_0 is zero, then the [ATP] dependence of the above expression will simplify to

$$\langle t \rangle = (b_1 + b_2[\text{ATP}])/(a_0[\text{ATP}])$$

i.e., to a simple Michaelis–Menten-like [ATP] dependence – an [ATP] dependence typical of systems that do not exhibit binding cooperativity.

Note that b_0 will go to zero if one of just two rates is zero, $\bar{k}_1$ or $\bar{k}_2$, i.e., if either the docking of ATP or the second binding transition, the tight binding, is irreversible. By setting one of these rates to zero, the binding of the first ATP is separated from the binding of the second ATP by at least one irreversible transition; thus, the binding events are effectively decoupled. What are the physical implications of setting either of these two rates to zero? First of all, if $\bar{k}_1$ were zero, this would imply that as soon as an ATP docks to the motor, this ATP is committed to the hydrolysis cycle. This seems unlikely from theoretical studies of similar ATPases [58] and the experimental observation that ADP readily binds and unbinds from the packaging motor [55]. Since we cannot set this rate to zero, we must introduce the second binding state, the tight-binding state as an irreversible transition. Physically, this choice implies that a single ATP can dock and undock to the motor reversibly, but eventually the catalytic pocket undergoes a second conformational change that is irreversible and which commits the bound ATP to the hydrolysis cycle. Such a tight binding transition has been proposed to explain previously experimental observations [55] and has also been suggested by theoretical studies of ATP binding. In particular, these theoretical studies suggest that during the tight binding transition a series of hydrogen bonds form, or "zipper," around the phosphate backbone of ATP [58]. Such an extensive set of hydrogen bonds would provide the necessary free energy to make this transition effectively irreversible. While it is possible that there are additional intermediate kinetic states in the ATP binding process, no additional states are required to resolve and rationalize the apparent contradiction in the data.

However, there is one additional restriction imposed by our data. In the above model, we assumed that the ATP binding events within the ring occur sequentially in time. In other words, one ATP docks, then tight binds, and only then can the next ATP dock. However, given that the individual subunits are identical, it is unclear that such an assumption is reasonable. Why cannot several subunits be involved in the simultaneous docking of ATP, each docking ATP independently of the binding state of the other subunits? We can relax the assumption that the individual subunits bind ATP in a time-ordered fashion, and, for the sake of argument, consider the following simplified two-pocket system:

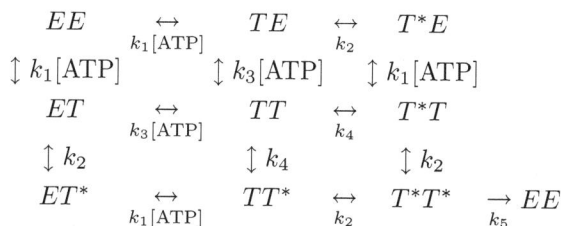

$$
\begin{array}{ccccc}
EE & \overset{\leftrightarrow}{k_1[\text{ATP}]} & TE & \overset{\leftrightarrow}{k_2} & T^*E \\
\updownarrow k_1[\text{ATP}] & & \updownarrow k_3[\text{ATP}] & & \updownarrow k_1[\text{ATP}] \\
ET & \overset{\leftrightarrow}{k_3[\text{ATP}]} & TT & \overset{\leftrightarrow}{k_4} & T^*T \\
\updownarrow k_2 & & \updownarrow k_4 & & \updownarrow k_2 \\
ET^* & \overset{\leftrightarrow}{k_1[\text{ATP}]} & TT^* & \overset{\leftrightarrow}{k_2} & T^*T^* \overset{\to}{k_5} EE
\end{array}
$$

Here all of the ATP docking events have a rate of $k_1[\text{ATP}]$ and all tight binding transitions have a rate k_2. All reverse rates (not shown) are again denoted with an overbar. By allowing a second ATP to dock before the first ATP is tightly bound (the TT state), we now allow for both subunits to dock ATP independently of one another. To facilitate analysis, we allow for different forward rates for these additional docking reactions and tight binding reactions, i.e., $k_3[\text{ATP}]$ and k_4, and for their reverse rates, $\bar{k}_3$ and $\bar{k}_4$. The mean dwell time for this system is described by the expression

$$\langle t \rangle = \left(b_0 + b_1[\text{ATP}] + b_2[\text{ATP}]^2 + b_3[\text{ATP}]^3\right)/\left(a_2[\text{ATP}]^2 + a_3[\text{ATP}]^3\right)$$

that again contains terms with [ATP] to a higher power than 1, indicative of binding cooperativity. Again, inspection of a few coefficients

$$b_0 = \bar{k}_1\bar{k}_2\left(k_2 k_5\left(k_4 + \bar{k}_3\right) + \left(k_5 + 2\bar{k}_2\right)\left(k_4\bar{k}_1 + \bar{k}_1\bar{k}_3 + \bar{k}_1\bar{k}_4\right)\right)$$

$$b_3 = k_3 k_1^2\left(k_2\left(2k_4 + k_5\right) + \left(2\bar{k}_2 + k_5\right)\left(2k_4 + \bar{k}_4\right)\right)$$

$$a_3 = 2k_3 k_1^2 k_2 k_4 k_5$$

reveals that this [ATP] dependence can be simplified by setting just two rate constants to zero, $\bar{k}_2$ and k_3. Setting the first rate to zero simply satisfies the tight binding condition discussed above. By setting the latter rate to zero, we are effectively prohibiting the simultaneous docking of ATP to multiple sites since this is the rate at which the system can dock a second ATP after the first ATP has docked but before it has been tightly bound. It can be confirmed that such a choice for the rate coefficients is the only way to convert the above expression into a simple Michaelis–Menten type equation, with an apparent Hill coefficient of 1. What are the physical implications of this result? The identical subunits within the motor cannot all be active for ATP docking simultaneously. Rather, only a single subunit can be active at a time. And only after it has committed its bound ATP, can another subunit can start the docking process. Given the large free energy change required to make the tight binding transition irreversible, this transition is a likely candidate for the kinetic transition that is responsible for activating another subunit for docking. What is the order of subunit activation? While our data cannot uniquely determine this order, the large conformational changes required for such activation in combination with the known interactions between adjacent subunits in similar rings [12, 13, 16] suggests that ATP binding occurs in an *ordinal* fashion, proceeding in the two step process depicted in Fig. 13.9b and c in a sequential manner around the ring, as depicted in Fig. 13.9d.

Figure 13.9d depicts the full mechanochemical cycle of the packaging motor of the bacteriophage φ29. During the dwell phase the motor loads four ATPs in the two-step process depicted in Fig. 13.9b and c in a temporally- and spatially-ordered fashion. Once the ring is loaded, the burst phase is triggered, the first phosphate is released, and the motor proceeds to package in a coordinated burst of four 2.5-bp steps. Just as our data do not uniquely determine the order in which the subunits bind ATP, they also do not determine the order in which the subunits translocate the DNA. However, given the cooperative nature of the burst and the fact that both the step size and the number of steps per burst appear to be highly uniform, it seems likely that the subunits fire in a well-defined and unique order. Again, the simplest model is one in which the subunits translocate the DNA in a sequential and ordinal fashion. Once the burst phase is finished, the motor is then reset, and the process begins again.

The mechanism depicted in Fig. 13.9d is only a schematic of the basic mechanochemical cycle. Fundamental chemical transitions such as ATP hydrolysis and ADP release are not depicted because our data do not uniquely locate these transitions. In a simple sequential model, in which each subunit binds ATP, hydrolyzes it, releases products, and steps, before activating the next subunit, these processes are uniquely determined by the simple requirement that ATP is hydrolyzed before products are released. However, in a more complicated kinetic model in which the individual hydrolysis cycles of the identical subunits are interwoven, such as that depicted in Fig. 13.9, there is no longer a single unique position for these important kinetic events. ATP hydrolysis, for example, might occur immediately after the tight binding of each ATP and before the next subunits is active for ATP docking.

Alternatively, all four ATPs may bind, accumulating subtle conformational changes in the ring which trigger hydrolysis only once all of the ATPs are bound. Moreover, it is also possible that the ATPs are hydrolyzed in the burst phase, immediately before each subunit generates a step. Finally, it is also possible that the hydrolysis of ATP is decoupled between the subunits and that it can occur spontaneously on its own, independent of the chemical state of the rest of the ring. The burst structure would then be maintained by another kinetic state, a bottleneck state, which would force the motor to wait until the necessary hydrolysis had occurred. A similar set of arguments can be made for the location of ADP release, with the result that ADP may be released in either the burst or dwell phase.

Our kinetic measurements require that at least four additional kinetic events occur in each dwell phase (an n_{min} of ~4 at saturating ATP; Fig. 13.7b), so it seems likely that at least one of the chemical transitions described above occur in the dwell phase if not both. However, it is also possible that these four additional kinetic events correspond to the tight binding transitions though this scenario is not favored since it is believed that these transitions occur quickly [55], and thus are not likely candidates for rate-limiting transitions. One particularly appealing scenario, which remains speculative based on our data, is that ADP release occurs within the dwell phase following the previous translocation, and that it is the release of four ADPs that provide the four rate-limiting transitions at saturating [ATP]. This scenario is appealing for several reasons. First, it would explain why only one pocket is competent to dock ATP at a time – all other pockets are filled with the ADP from the previous cycle. Second, it is believed that the release of ADP requires the un-zipping of an extensive set of hydrogen bounds around the remaining phosphates [58], a process which would require a fair amount of free energy. In this scenario, the necessary free energy would be provided by the zipping of hydrogen bonds in the tight binding reaction of the adjacent subunit; thus, the zipping of one set of hydrogen bonds would provide the free energy necessary to unzip another set. Thus, the tight binding of one ATP would pry out an ADP from an adjacent pocket, and the rate-limiting steps measured at saturating [ATP] would be the undocking of the resulting loosely bound ADP. In addition, there is some experimental evidence for related ring-ATPases which may suggest that hydrolysis also occurs with the dwell phase. In the hexameric-helicases Rho [23] and T7 gp4 [25], pre-steady state, stopped flow measurements have revealed that phosphate release lags behind ATP hydrolysis by multiple subunits, more than 2 for T7 and around 4 for Rho. This lag can be naturally explained by a two-phase burst model in which ATP hydrolysis occurs in the dwell, while phosphate release would occur in the following burst. Despite these arguments, the location of ADP release and ATP hydrolysis remain unknown for the packaging motor of φ29 and will be the subject of future experiments.

13.11 Summary and Conclusions

The first report of the *B. subtilis* bacteriophage φ29 was published in 1965 [84] and the first dramatic electron microscopy images followed quickly [85], just 1 year later. In the intervening 40 years, the genome has been sequenced, the proteins compiled and indexed, the polymerase has been commercialized, the progression of the procaspid morphogenesis has been detailed, and we are now at the level where the atomic scale motions of the remarkable set of molecular machines used throughout the bacteriophage life-cycle are now being tabulated and understood [11, 86].

In this chapter, we have summarized our contributions to the growing understanding of one important stage of the life-cycle of this bacteriophage – the packaging of its genome. By utilizing a novel optical trapping assay, we determined exactly how much force is generated when one compresses DNA to near crystalline densities into the small confines of the preassembled capsid [38]. By following the decrease in packaging rate of a single bacteriophage as a function of force and as a function of internal filling of the capsid, we could infer the apparent internal force generated by the compressed DNA. This force can then be converted into a pressure, yielding a macroscopic value of 60 atm. Remarkably, the value of the internal force generated in this process is just slightly lower than the maximum extensional force DNA can experience before undergoing an overstretching phase transition [87], suggesting that the packaging motor is as strong as the physical properties of its substrate will permit.

Once we had addressed exactly what forces were involved in the DNA compaction process, we turned our attention to exactly how the motor generates these forces, i.e., its mechanochemistry [56, 57]. By carefully measuring the average packaging velocity, using the same optical tweezers assay, we were able to determine the coupling between [ATP] and force, along with the effect of hydrolysis products and non-hydrolyzable ATP analogs [55]. We found that at limiting concentrations of ATP, the packaging velocity is insensitive to force; whereas at high concentrations of ATP, the velocity is highly force sensitive. This result indicates that it is not the binding of ATP which generates force, as was expected from studies of related molecular motors [58], but rather the release of hydrolysis products, most likely inorganic phosphate. Thus, the energy for translocation comes not from the free energy released upon ATP binding but from the entropy generated upon product release.

Measurements of the average packaging velocity as a function of force or [ATP] were not sufficient to address the next level of detailed questions, so we developed a new optical tweezers – an instrument capable of resolving the angstrom scale increments of DNA packaged by the motor in real time [64–66]. By following the packaging dynamics of a single bacteriophage with this new instrument [74], we determined that the motor packages DNA in increments of 10 bp for all concentrations of ATP, a value five times larger than the expected step size. Moreover, the observation of the discrete steps of the packaging motor allowed us to measure the full distribution of pause durations before these steps, revealing the fluctuations inherent in the function of the packaging motor. A detailed statistical analysis of these fluctuations showed that the motor loads multiple ATPs before each 10-bp step. By applying large opposing loads, we could probe the detailed structure of the 10-bp increments, demonstrating that these events are not single steps, but bursts of four 2.5-bp steps in quick succession.

Taken together these measurements suggest a novel form of coordination between the identical ATPase subunits in the packaging motor. The motor loads ATP to four of the five subunits, delaying the utilization of these molecules until the entire ring is fully loaded, and then it packages DNA in four 2.5-bp steps. A temporal segregation between the binding of ATP and its utilization suggests a degree of coordination between the individual subunits which is unprecedented among ring-ATPases, and which requires a more extensive set of interactions between the subunits than anticipated [16, 19].

While this model represents the most detailed picture we have of the inter-subunit dynamics of a homomeric ring-ATPase, it raises more questions than it answers. For example, the action of four subunits as opposed to five requires that one of the subunits be unique. What breaks the symmetry of the ring? Moreover, is this symmetry breaking reflected in the structure of the ring, through departures from a planar ring, features which have been recently revealed in crystal structures of related ring-ATPases [22, 24], or is it reflected in the chemical states of the motor, perhaps via changes in DNA affinity under different nucleotide state [55, 74]. Is this symmetry breaking permanent, or does this special subunit process around the ring, changing its identity each mechanochemical cycle? In addition, a step size that is a non-integer number of base pairs requires radically new ideas for the motor–DNA interaction [74]. While it certainly is true that integer base pair steps are not required, they were anticipated for good reason – uniform and consistent motor–DNA contacts would lock the steps of the motor to the base pair scale. Thus, this result suggests that the motor makes more complicated and perhaps variable contacts with the DNA. The answer to these questions will come from the combination of targeted mutagenesis and single molecule measurements, not to mention more detailed studies of both the nature of the motor–DNA contacts and the degree to which 10.5-bp rotational symmetry of the DNA is reconciled to the pentameric symmetry of the motor, perhaps by relative DNA-motor rotation.

Multimeric ring-ATPases of the ASCE family, of which the packaging motor is a member, perform a remarkably diverse set of cellular functions by drawing upon a relatively small set of common structural features [12, 13, 18]. Where does this functional diversity arise? Is it that small changes to the same basic structure can lead to diverse cellular functions – for example, the same structural dynamics may lead to the pumping of double stranded and single stranded nucleic acids along with peptides simply by modifications in substrate binding domains – or is it that the same structural ATPase core can be modified dramatically in its function by subtle amino acid substitutions or larger changes such as the insertion of accessory domains [12, 18] – in which case the same ATPase cycle may result in very different mechanics within rings of these core structures. High resolution measurement of the structural dynamics of these systems, such as those described here, will prove an invaluable tool in addressing these questions. Moreover, the answers will yield fundamental insights into the basic design principles that underlie the function of these remarkable molecular machines.

Acknowledgments

J.R.M acknowledges the National Science Foundation's Graduate Research Fellowship for support. This work was supported in part by NIH grants GM-071552, DE-003606, and GM-059604, and DOE grant DE-AC03–76DF00098.

References

1. C. Bustamante, J.C. Macosko, G.J. Wuite, Nat. Rev. Mol. Cell Biol. **1**, 130–136 (2000)
2. *Molecular Motors.* ed. by Schliwa M. Weinham (Wiley-VCH Verlay GmbH & Co, 2003)
3. K. Kinosita, K. Adachi, H. Itoh, Annu. Rev. Biophys. Biomol. Struct. **33**, 245–268 (2004)
4. L. Bai, T.J. Santangelo, M.D. Wang, Annu. Rev. Biophys. Biomol. Struct. **35**, 343–360 (2006)
5. K.M. Herbert, W.J. Greenleaf, S.M. Block, Annu. Rev. Biochem. **77**, 149–176 (2008)
6. C. Bustamante, L. Finzi, P. Sebring, S. Smith, SPIE Proc. Ser. **1435**, 179–187 (1991)
7. S.B. Smith, L. Finzi, C. Bustamante, Science **258**, 1122–1126 (1992)
8. W.E. Moerner, L. Kador, Phys. Rev. Lett. **62**, 2535–2538 (1989)
9. M. Orrit, J. Bernard, Phys. Rev. Lett. **65**, 2716–2719 (1990)
10. E. Betzig, R.J. Chichester, Science **262**, 1422–1425 (1993)
11. S. Grimes, P.J. Jardine, D. Anderson, Adv Virus Res. **58**, 255–294 (2002)
12. J.P. Erzberger, J.M. Berger, Annu. Rev. Biophys. Biomol. Struct. **35**, 93–114 (2006)
13. L.M. Iyer, D.D. Leipe, E.V. Koonin, L. Aravind, J. Struct. Biol. **146**, 11–31 (2004)
14. T. Ogura, A.J. Wilkinson, Genes Cells **6**, 575–597 (2001)
15. M. Latterich, S. Patel, Trends Cell Biol. **8**, 65 (1998)
16. D.E. Kainov, R. Tuma, E.J. Mancini, Cell Mol. Life Sci. **63**, 1095–1105 (2006)
17. K.P. Hopfner, J. Michaelis, Curr. Opin. Struct. Biol. **17**, 87–95 (2007)
18. N.D. Thomsen, J.M. Berger, Mol. Microbiol. **69**, 1071–1090 (2008)
19. E.J. Enemark, L. Joshua-Tor, Curr. Opin. Struct. Biol. **18**, 243 (2008)
20. D.J. Crampton, S. Mukherjee, C.C. Richardson, Mol. Cell. **21**, 165–174 (2006)
21. T.H. Massey, C.P. Mercogliano, J. Yates, D.J. Sherratt, J. Löwe, Mol. Cell. **23**, 457 (2006)

22. E.J. Enemark, L. Joshua-Tor, Nature **442**, 270 (2006)
23. J.L. Adelman, Y.J. Jeong, J.C. Liao, G. Patel, D.E. Kim, G. Oster, S.S. Patel, Mol. Cell. **22**, 611 (2006)
24. E. Skordalakes, J.M. Berger, Cell. **127**, 553–564 (2006)
25. J.C. Liao, Y.J. Jeong, D.E. Kim, S.S. Patel, G. Oster, J. Mol. Biol. **350**, 452 (2005)
26. E.J. Mancini, D.E. Kainov, J.M. Grimes, R. Tuma, D.H. Bamford, D.I. Stuart, Cell **118**, 743–755 (2004)
27. M.R. Singleton, M.R. Sawaya, T. Ellenberger, D.B. Wigley, Cell **101**, 589–600 (2000)
28. M.J. Moreau, A.T. McGeoch, A.R. Lowe, L.S. Itzhaki, S.D. Bell, Mol. Cell. **28**, 304 (2007)
29. D. Gai, R. Zhao, D. Li, C.V. Finkielstein, X.S. Chen, Cell **119**, 47–60 (2004)
30. A. Martin, T.A. Baker, R.T. Sauer, Nature **437**, 1115 (2005)
31. S.E. Luria, T.F. Anderson, Proc. Natl. Acad. Sci. U S A. **28**, 127–130 121 (1942)**
32. S.E. Luria, M. Delbruck, T.F. Anderson, J. Bacteriol. **46**, 57–77 (1943)
33. R.B. Luftig, W.B. Wood, R. Okinaka, J. Mol. Biol. **57**, 555–573 (1971)
34. D. Kaiser, M. Syvanen, T. Masuda, J. Supramol. Struct. **2**, 318–328 (1974)
35. D. Kaiser, M. Syvanen, T. Masuda, J. Mol. Biol. **91**, 175–186 (1975)
36. C. Bazinet, J. King, Annu. Rev. Microbiol. **39**, 109–129 (1985)
37. L.W. Black, Annu. Rev. Microbiol. **43**, 267–292 (1989)
38. D.E. Smith, S.J. Tans, S.B. Smith, S. Grimes, D.L. Anderson, C. Bustamante, Nature **413**, 748–752 (2001)
39. M.C. Morais, J.S. Koti, V.D. Bowman, E. Reyes-Aldrete, D.L. Anderson, M.G. Rossmann, Structure **16**, 1267 (2008)
40. L.M. Iyer, K.S. Makarova, E.V. Koonin, L. Aravind, Nucleic Acids Res. **32**, 5260–5279 (2004)
41. A.M. Burroughs, L.M. Iyer, L. Aravind, In Gene and Protein Evolution. ed. by Volff J-N: Karger; 48–65. vol **3**,] (2007).**
42. A.A. Simpson, P.G. Leiman, Y. Tao, Y. He, M.O. Badasso, P.J. Jardine, D.L. Anderson, M.G. Rossmann, Acta Crystallogr. D Biol. Crystallogr. **57**, 1260–1269 (2001)
43. R.W. Hendrix, Proc. Natl. Acad. Sci. **75**, 4779–4783 (1978)
44. T. Hugel, J. Michaelis, C.L. Hetherington, P.J. Jardine, S. Grimes, J.M. Walter, W. Falk, D.L. Anderson, C. Bustamante, PLoS Biol. **5**, e59 (2007)
45. P.J. Pease, O. Levy, G.J. Cost, J. Gore, J.L. Ptacin, D. Sherratt, C. Bustamante, N.R. Cozzarelli, Science **307**, 586–590 (2005)
46. O.A. Saleh, C. Perals, F.X. Barre, J.F. Allemand, EMBO J. **23**, 2430–2439 (2004)
47. C. Bustamante, S.B. Smith, J. Liphardt, D. Smith, Curr. Opin. Struct. Biol. **10**, 279–285 (2000)
48. C. Bustamante, J.F. Marko, E.D. Siggia, S. Smith, Science **265**, 1599–1600 (1994)
49. J.P. Rickgauer, D.N. Fuller, S. Grimes, P.J. Jardine, D.L. Anderson, D.E. Smith, Biophys. J. **94**, 159–167 (2008)
50. A. Evilevitch, L. Lavelle, C.M. Knobler, E. Raspaud, W.M. Gelbart, Proc. Natl. Acad. Sci. U S A. **100**, 9292–9295 (2003)
51. P. Grayson, L. Han, T. Winther, R. Phillips, Proc. Natl. Acad. Sci. U S A. **104**, 14652–14657 (2007)

52. P. Grayson, A. Evilevitch, M.M. Inamdar, P.K. Purohit, W.M. Gelbart, C.M. Knobler, R. Phillips, Virology **348**, 430–436 (2006)
53. P.K. Purohit, M.M. Inamdar, P.D. Grayson, T.M. Squires, J. Kondev, R. Phillips, Biophys. J. **88**, 851–866 (2005)
54. M.S.J.M.H. Víctor González-Huici, Mol. Microbiol. **52**, 529–540 (2004)
55. Y.R. Chemla, K. Aathavan, J. Michaelis, S. Grimes, P.J. Jardine, D.L. Anderson, C. Bustamante, Cell **122**, 683–692 (2005)
56. C. Bustamante, Y.R. Chemla, N.R. Forde, D. Izhaky, Annu. Rev. Biochem. **73**, 705–748 (2004)
57. D. Keller, C. Bustamante, Biophys. J. **78**, 541–556 (2000)
58. G. Oster, H. Wang, Biochim. Biophys. Acta. **1458**, 482–510 (2000)
59. C. Chen, P. Guo, J. Virol. **71**, 3864–3871 (1997)
60. P. Guo, C. Peterson, D. Anderson, J. Mol. Biol. **197**, 229–236 (1987)
61. I. Donmez, S.S. Patel, EMBO J. **27**, 1718–1726 (2008)
62. D.E. Kainov, E.J. Mancini, J. Telenius, J. Lisal, J.M. Grimes, D.H. Bamford, D.I. Stuart, R. Tuma, J. Biol. Chem. **283**, 3607–3617 (2008)
63. A. Wynveen, D.J. Lee, A.A. Kornyshev, S. Leikin, Nucleic Acids Res. **36**, 5540–5551 (2008)
64. J.R. Moffitt, Y.R. Chemla, S.B. Smith, C. Bustamante, Annu. Rev. Biochem. **77**, 205–228 (2008)
65. C. Bustamante, Y.R. Chemla, J.R. Moffitt, in *Single-Molecule Techniques: A Laboratory Manual.* ed. by Selvin PR, Ha T (Cold Spring Harbor Laboratories, 2008)pp. 297–324
66. J.R. Moffitt, Y.R. Chemla, D. Izhaky, C. Bustamante, Proc. Natl. Acad. Sci. **103**, 9006–9011 (2006)
67. E.A. Abbondanzieri, W.J. Greenleaf, J.W. Shaevitz, R. Landick, S.M. Block, Nature **438**, 460 (2005)
68. J.W. Shaevitz, E.A. Abbondanzieri, R. Landick, S.M. Block, Nature **426**, 684 (2003)
69. A.R. Carter, G.M. King, T.A. Ulrich, W. Halsey, D. Alchenberger, T.T. Perkins, Appl. Opt. **46**, 421–427 (2007)
70. L. Nugent-Glandorf, T.T. Perkins, Opt. Lett. **29**, 2611–2613 (2004)
71. F. Gittes, C.F. Schmidt, Eur. Biophys. J. Biophys. Lett. **27**, 75–81 (1998)
72. J.C. Meiners, S.R. Quake, Phys. Rev. Lett. **84**, 5014–5017 (2000)
73. D.N. Fuller, J.P. Rickgauer, P.J. Jardine, S. Grimes, D.L. Anderson, D.E. Smith, Proc. Natl. Acad. Sci. U S A. **104**, 11245–11250 (2007)
74. J.R. Moffitt, Y.R. Chemla, K. Aathavan, S. Grimes, P.J. Jardine, D. Anderson, C. Bustamante, Nature **457**, 446–450 (2009)
75. S.M. Block, K. Svoboda, Biophys. J. **68**, 2305S–2415 (1995)
76. B.C. Carter, M. Vershinin, S.P. Gross, Biophys. J. **94**, 306–319 (2008)
77. N.J. Carter, R.A. Cross, Nature **435**, 308 (2005)
78. I.H. Segel, *Enzyme Kinetics.* (John Wiley & Sons, Inc., 1975)
79. M.J. Schnitzer, S.M. Block, Cold Spring Harb. Symp. Quant. Biol. **60**, 793–802 (1995)
80. Z. Koza, Phys. Rev. E. **65**, 031905 (2002)
81. Y.R. Chemla, J.R. Moffitt, C. Bustamante, J. Phys. Chem. B. **112**, 6025–6044 (2008)
82. H. Mao, J.R. Arias-Gonzalez, S.B. Smith, I. Tinoco, Jr., C. Bustamante, Biophys. J. **89**, 1308–1316 (2005)

83. L. Kellner, Proc. R. Soc. Lond. Ser.Math. Phys. Sci. **159**, 0410–0415 (1937)
84. B.E. Reilly, J. Spizizen, J. Bacteriol. **89**, 782–790 (1965)
85. D.L. Anderson, D.D. Hickman, B.E. Reilly, J. Bacteriol. **91**, 2081–2089 (1966)
86. W.J.J. Meijer, J.A. Horcajadas, M. Salas, Microbiol. Mol. Biol. Rev. **65**, 261–287 (2001)
87. S.B. Smith, Y. Cui, C. Bustamante, Science **271**, 795–799 (1996)

14

Chemo-Mechanical Coupling in the Rotary Molecular Motor F_1-ATPase

Kengo Adachi, Shou Furuike, Mohammad Delawar Hossain, Hiroyasu Itoh, Kazuhiko Kinosita, Jr., Yasuhiro Onoue, and Rieko Shimo-Kon

Summary. F_1-ATPase is a molecular motor in which the central γ subunit rotates inside the cylinder made of $\alpha_3\beta_3$ subunits. The rotation is powered by ATP hydrolysis in three catalytic sites, and reverse rotation of the γ subunit by an external force leads to ATP synthesis in the catalytic sites. Single-molecule studies have revealed how the mechanical rotation is coupled to the chemical reactions in the three catalytic sites: binding/release of ATP, ADP, and phosphate, and hydrolysis/synthesis of ATP.

14.1 Introduction

F_1-ATPase, a water-soluble portion of ATP synthase, has been predicted [1,2] and proved [3] to be an ATP-driven rotary molecular motor in which the central γ subunit rotates inside a hexameric cylinder made of alternately arranged three α and three β subunits [4]. When the rotor subunit γ is rotated in reverse by the application of an external force, the motor turns into a generator and synthesizes ATP from ADP and inorganic phosphate (Pi) in the catalytic sites [5,6]. F_1 is thus a reversible chemo-mechanical energy converter as its physiological role implies [7–11].

We have been working on F_1 of thermophilic origin (TF_1) for about 10 years, during which our views about its rotary mechanism have changed significantly. We have been, and still are, seeking for the simplest interpretations of existing data, but Nature seems not always simple. Here, we present our current views, together with critical evaluations of our work, to help the reader make his/her own judgments.

14.2 Rotation Scheme

We observe rotation under an optical microscope by attaching β subunits of F_1 to a glass surface through the engineered histidine tags (Fig. 14.1A). Sole two cysteines in our TF_1 (only S107C in early studies) on the protruding portion

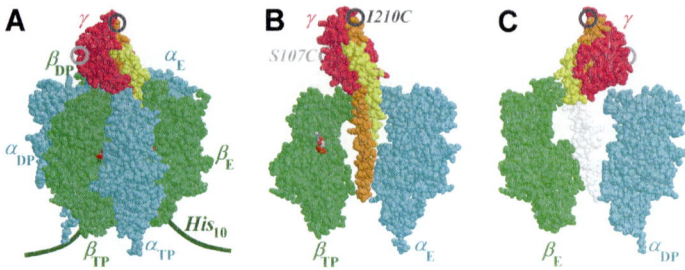

Fig. 14.1. A crystal structure [12] of mitochondrial F_1 (MF$_1$). α and β subunits are designated according to the nucleotides in catalytic sites found in the original structure [4]: TP site between β_{TP} and α_{TP}, largely in β_{TP}, bound an ATP analog, DP site ADP, and E site none, while noncatalytic sites in the other three interfaces bound the ATP analog. TF$_1$ that we use has a similar structure [13], to which we introduced two cysteines on γ (*circles*) and ten histidines at the N-terminus of β. N- and C-terminal α-helices of the γ subunit are shown in *yellow* and *orange*, respectively. Nucleotides are in *CPK* colors. (**A**) An overall view. (**B**) A side view showing an opposing $\alpha - \beta$ pair. (**C**) An "axle-less" construct in which the white portion was deleted [14]

of γ (Fig. 14.1) are biotinylated, to which a probe for imaging γ rotation, a micron-sized actin filament [3] or a spherical bead (or its duplex) with a size of 1 μm down to 40 nm is attached through biotin–avidin linkage.

14.2.1 Data Selection

We typically infuse ∼0.1 nM of F_1 into an observation chamber of height ∼50 μm. If all are attached, the surface density would be of the order of 1 molecule/μm^2. In fact, only on lucky days do we observe as many as tens of rotating probes in a field of view some $(300\,\mu m)^2$. Most of the motors are inactive, or fail to bind the probe which is huge compared to the motor size of ∼10 nm, in a configuration that allows unhindered rotation. Of those that rotate, we select those few that rotate fast, smooth, and with 120° symmetry (see below). Due to this subjective selection, ensemble statistics in our work is unreliable though we make every effort to increase the number of selected to convince ourselves that we are not reporting artifacts. Our statistics relies on repeated behaviors of a selected molecule: we analyze many consecutive revolutions without eliminating an event in the sequence.

14.2.2 120° Step Per ATP

When ATP concentration, [ATP], is low, the rotation proceeds in steps of 120°, with stepping frequency proportional to [ATP] [15]. This indicates a 120° step per ATP hydrolyzed, as expected from the presence of three basically equivalent catalytic sites. Bulk rate of ATP hydrolysis, however, has always been somewhat lower than the stepping rate under a negligible load [16]. Our explanation is that part of F_1 in bulk solution is inhibited or less active in some way, but other interpretations cannot be excluded. F_1 is known to be inhibited when it binds MgADP tightly [17, 18]. We thus prepare nucleotide-free F_1 [19, 20], but ADP is formed during catalysis. Phosphoenolpyruvate used in ATPase assay has been found to be a competitive inhibitor (R. Shimo-Kon, unpublished).

14.2.3 Substeps

With a 40-nm gold bead for which viscous friction is not too high, the 120° step is further resolved into 80–90° and 40–30° substeps [16, 21], which we refer to here as 80° and 40° (see Fig. 14.2A below). The substep behavior can be resolved only in selected molecules, and those selected molecules do not necessarily show substeps in the entire time course observed. But we have seen many examples where a molecule shows several hundreds of consecutive substeps, and thus we regard this as a genuine property of F_1. We notice momentary pauses in other angles under certain conditions, but we are much less certain if these are also genuine.

14.2.4 Direction of Rotation

In all cases, the direction of ATP-driven rotation is counterclockwise when viewed from above in Fig. 14.1. It is actually a mere presumption that in all rotating molecules the His-tags are attached to the glass surface. Indeed, rotation can be observed on a clean glass surface [16], without functionalizing the surface with Ni-NTA that is supposed to bind histidine specifically. In no experiment is there guarantee that all three His-tags are bound tightly on a surface. Sometimes, the positions of ATP-waiting angles that are separated by 120° suddenly shift during the course of observation, suggesting detachment and reattachment of a His-tag(s). When we manipulate the γ angle by magnets [21], therefore, we confirm the ATP-waiting angles both before and after a measurement by infusing nanomolar ATP, and letting the F_1 undergo spontaneous stepping rotation (a laborious and often irritating procedure). Also, a probe such as a plastic bead can attach to γ by nonspecific binding. There are cases, though rare, where we believe that γ is attached to a glass surface and a probe on the $\alpha_3\beta_3$ cylinder. The sense of rotation, though, remains counterclockwise even in this upside down configuration.

We strive to make objective selection of data and to report "facts," but we are too aware that a fact "described" by words is no longer a fact and, worse, that we often make false judgments.

14.3 Chemo-Mechanical Coupling

Here, we discuss the relation between the rotary angle of γ and chemical reactions in the three catalytic sites. ATP synthesis by manipulation of the γ-angle alone suggests a γ-dictator mechanism where the state of the motor, or rather the probability a state is realized, is determined basically by one parameter, the γ-angle (in addition to environmental variables such as [ATP]). For simplicity, we assume that this holds to a first approximation. We also simplify chemical kinetics and decompose it into a minimum of four pairs of events, binding and release of ATP, ADP, and Pi, and hydrolysis and synthesis of ATP.

14.3.1 ATP Binding

In Fig. 14.2A, we show schematically a rotation time course involving sub-steps. The duration of ATP-waiting dwells, on multiples of 120° in Fig. 14.2A, is inversely proportional to [ATP] below K_m of about $15\,\mu M$ down to nM [22] for wild-type TF_1, and become negligibly short ($\leqslant 0.1$ ms) at saturating [ATP]s. Because the 80° substep that follows is complete in 0.1 ms, short on the time scale of protein functions, we proposed that the 80° rotation is not only triggered but also driven (powered) all the way by ATP binding [16]. A later study [21], however, has indicated that ADP release also occurs during the 80° substep and that ADP release would also drive the substep (see below).

14.3.2 Hydrolysis of ATP

Histograms of the dwell times at 80° indicate that at least two reactions take place at 80° [16]. One is likely the ATP hydrolysis, because the 80° dwell becomes longer in the β-E190D mutant in which a glutamate presumed to attack the terminal phosphate is replaced with aspartate, or when the medium contained ATP-γ-S that is presumably hydrolyzed slowly [23]. Cy3-ATP, a fluorescent analog of ATP, also prolongs the 80° dwell when the Cy3-ATP is bound 200° ago [24], suggesting that binding of ATP and cleavage of that ATP is separated by 200° (see, e.g., the fate of purple ATP in Fig. 14.2). The β-E190D mutant also points to this 120° scenario [25]. The evidence, though, is not perfect in that hydrolysis has not been confirmed directly.

Oxygen exchange studies by Boyer and colleagues have shown that, at low [ATP]s, bound ATP undergoes rounds of reversible hydrolysis/re-synthesis before ADP and Pi are eventually released into the medium [7, 26]. The reversals must occur at ATP-waiting angles, possibly accompanying angular fluctuations of γ. In our scheme in Fig. 14.2, it is the blue "ATP" at 0° that would undergo reversible hydrolysis. The reversals would be biased toward synthesis at 0°, but almost completely toward hydrolysis at 80°. Whether the rate-limiting step at 80° is merely this shift in equilibrium or a process that somehow ensures completion of hydrolysis, e.g., movement of an "arginine finger" in the catalytic site [27], is not known. An unexpected finding that

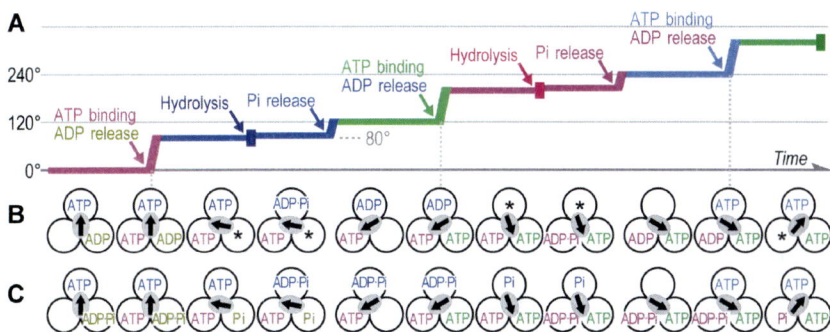

Fig. 14.2. Proposed scheme for coupling between catalysis and rotation. (**A**) Schematic time course of rotation. Colors indicate the site at which the rate-limiting reaction is to occur in B. (**B** and **C**) Alternative schemes suggested by [21], differing in the timing of Pi release by 120°. Chemical states of three catalytic sites (*circles*) and γ orientation (*central arrows*) are shown. More recent work (R. Shimo-Kon, unpublished) points to B, where the empty site with an asterisk is freely accessible to medium nucleotides

the long dwell at 80° in the β-E190D mutant disappears when the preceding ATP-waiting dwell at 0° is long (seconds) suggests that the reversals may end in hydrolysis at 0° by subsequent Pi release if sufficient number of reversals are allowed [28].

For efficient ATP synthesis in cells, coupling between rotation and hydrolysis is advantageous because, then, reverse rotation of γ will ensure ATP synthesis [10]. This is fulfilled if the shift in equilibrium from predominantly ATP to predominantly ADP + Pi occurs over the 80° rotation, but then the nature of the slow reaction at 80° observed with ATP-γ-S, Cy3-ATP, or β-E190D is unclear. Or, the shift begins after γ has rotated into the 80° position and may accompany a small amount of further rotation, e.g., from 80° to 90°. In this regard, we note that substep angle close to 80° has been observed in situations where hydrolysis reaction is presumed to be slow (ATP-γ-S, Cy3-ATP, β-E190D), whereas high-speed (and thus noisy) observation of normal rotation is compatible with any angle around 80–90° [16,21].

14.3.3 Pi Release

Another event at 80° interim is Pi release [21]: medium Pi specifically prolongs the 80° dwell, indicating that Pi release at least triggers the next 40° substep. Furthermore, with Pi at hundreds of millimolar, frequent reversals of 40° substeps are observed, showing that Pi release drives the 40° rotation. Two possibilities, however, have remained for the timing of Pi release: either immediately after hydrolysis (Fig. 14.2B) or after further 120° of rotation (Fig. 14.2C). Our recent work (R. Shimo-Kon, unpublished) suggests that the site indicated by an asterisk in Fig. 14.2B is accessible to medium nucleotides including ATP, and that medium Pi competes with the nucleotide binding.

Thus, Fig. 14.2C where this site is occupied by product Pi (Pi formed by ATP hydrolysis) is unlikely.

The analysis of 80° dwells by itself indicates only the presence of two (or more) rate-limiting reactions, which may be sequential or parallel (but both must complete before the next 40° substep). The Pi results above indicate that hydrolysis (with the reservations in 14.3.2) and Pi release occur in the same site in this order, taking ~0.2 and ~1.2 ms, respectively, at room temperature [21]. Involvement of another reaction(s), however, cannot be excluded.

14.3.4 ADP Release

Initially, we suggested, without evidence, that the 40° substep is likely driven by ADP release that occurs after 200° of rotation since that ADP was bound as ATP at 0°. We were influenced by the crystal structures that show hinge bending of ~30° in β_{TP} or β_{DP} compared to β_E (Fig. 14.1): if the 80° substep is driven by ATP binding that would bend a β, then the rest of 40° rotation would likely be driven by ADP release that should unbend the β [29]. We also thought, in those days, that the site occupancy, the number of catalytic sites occupied by a nucleotide, would alternate between one and two (see below), and hence ADP should be released before 240° of rotation.

Imaging of binding and release of Cy3-ATP [24], however, showed that the fluorescent nucleotide remains on a mutant F_1 for at least 240° of rotation. Precise timing of (Cy3-)ADP release could not be determined, but it must have been between 240° and 360°. Later, retention of Cy3-AT(D)P for $\geqslant$ 240° was confirmed with wild-type F_1: when rotation was artificially slowed by magnets, release occurred at ~240°, and, in free rotation under the conditions where the 80° dwells were long (β-E190D or in the presence of ATP-γ-S), release occurred between 240° and 320° [21]. Practical temporal resolution in these measurements, however, was ~0.1 s, and thus, in normal rotation where the 80° dwells take only 1–2 ms, release may be delayed until the end of the 320° dwell or even beyond 320° (but certainly not beyond 360°). Also, there is no guarantee that Cy3-ATP behaves in the same way as unlabeled ATP, and thus, until recently, we thought that unlabeled ADP might be released before 240°.

Now, we think that release of unlabeled ADP occurs within ~1 ms from the start of the substep from 240° to 320°: either during the substep rotation or in an early part of the dwell at 320°. Evidence comes from the site occupancy study described below.

14.3.5 Site Occupancy

Rotation in one direction requires broken symmetry. Because the crystal structures suggested a conformational change of the β subunits (bending/unbending) that depends on the presence or absence of a bound nucleotide, we thought that bi-site catalysis which would warrant maximal asymmetry (in terms of the occupancy of the three catalytic sites) would be the natural choice [8,9]. Studies on MF$_1$ also indicated bi-site catalysis [30, and references therein].

Senior and colleagues have introduced a reporter tryptophan in the catalytic sites of *Escherichia coli* F$_1$ (EF$_1$) and directly measured the site occupancy. They have shown that the ATP hydrolysis activity parallels the occupancy of the third site, implicating a tri-site mechanism where the site occupancy alternates between two and three, being three during the rate-limiting step of the catalysis [31,32]. Studies on TF$_1$ basically corroborated the results [33,34].

Our recent work using a reporter tryptophan in a mutant TF$_1$ (R. Shimo-Kon et al., unpublished) has also shown that the occupancy rises to three as [ATP] increases. The rise, however, is preceded by the rise in hydrolysis activity: at the [ATP] where the activity is half maximal (K_m), the occupancy is only slightly above two. At K_m, F$_1$ spends half of its time at the 0° (ATP-waiting) position and the other half at 80° (Fig. 14.2). For the time-averaged occupancy to be close to two, ADP has to be released during the 80° substep or immediately (well within the 80° dwell time of ∼2 ms) after the 80° rotation. At [ATP] ≫ K_m, the asterisked vacated site in Fig. 14.2 can bind medium nucleotide weakly, allowing the occupancy to rise to three. The study has also shown that, at least for the particular mutant studied, bi-site activity is virtually absent.

14.4 Energetics of Coupling

Our view here is that the free-energy drops in the chemical reactions, ATP binding, hydrolysis, Pi release and ADP release, drive rotation [10]. Some of these partial reactions may be uphill, depending on the ligand concentrations in the medium and the load if present, but the rotation goes counterclockwise as long as the overall free-energy drop is greater than the mechanical work F$_1$ has to do against a conservative load (such as a spring). For the process of Pi release, the energetics has been worked out [21].

14.4.1 Pi Release

From high-speed imaging of unloaded rotation at various [Pi], we have determined the rate constants for Pi release and Pi rebinding at 80° and 120° (vertical arrows in Fig. 14.3B) and rotation between these angles (oblique arrows). The rate constants in turn allow us to calculate free energy levels as shown in Fig. 14.3B. The free energy for a given chemical state (F$_1$ · ADP · Pi representing the state still binding the Pi to be released, or F$_1$ · ADP obtained after Pi release) must be a function of rotary angle, as in the simplified linear diagram in Fig. 14.3A. For the analysis, however, we chose the two-state approximation in Fig. 14.3B to determine a minimal number of rate constants.

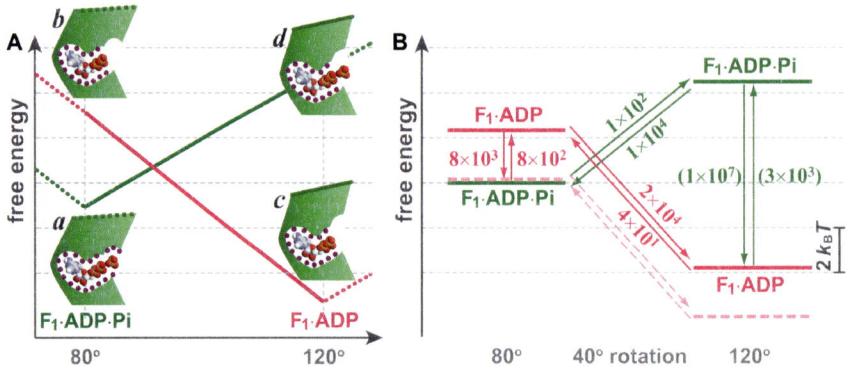

Fig. 14.3. Free-energy diagrams for the 40° rotation accompanying Pi release [21].
(**A**) Simplified potential energies for the rotation of γ in the state $F_1 \cdot ADP \cdot Pi$ before
Pi release and in $F_1 \cdot ADP$ after Pi release (*green* and *magenta lines*, respectively).
The lines show, very approximately, how the free energy of the system changes
when γ is rotated while the chemical state is fixed (see [10]). The *green blocks a–d*
are schematic representations of the two conformations (80° and 120°) of F_1 and the
catalytic site from which Pi is released. ADP and Pi are shown in *space filling* models.
Small *violet dots* represent interactions (mainly hydrogen bonds) through which the
protein part (*green*) and the ligand attract each other. (**B**) Two-state approximation
of the angle dependent free energies in (**A**). Four energy levels (*horizontal bars*) are
considered for given [Pi], and the transition rates (in s^{-1}) have been determined
from experiments, except for the parenthesized numbers for which only their ratio is
experimental [21]. *Solid lines* with numerical values are for 50 mM Pi. *Dashed lines*
are for 5 mM Pi. Diagrams for other [Pi] can be obtained by shifting the magenta
bars vertically in proportion to $k_B T \cdot \ln[Pi]$, where $k_B T = 4.1$ pN · nm is the thermal
energy at room temperature. Only the rates of Pi binding change with [Pi]

Solid horizontal bars in Fig. 14.3B show energy levels at 50 mM Pi. At this
[Pi], release of Pi at 80° is mostly unsuccessful because rebinding rate is high.
Once in about ten release events, though, the $F_1 \cdot ADP$ formed by Pi release
rotates into the 120° conformation, where the probability of Pi rebinding
is extremely low and thus $F_1 \cdot ADP$ is stable. At 5 mM Pi (dashed lines),
Pi release at 80° is easier. At 500 mM Pi, on the other hand, the energy
levels for $F_1 \cdot ADP$ are shifted upward, leading to a finite probability of Pi
rebinding at 120° followed by rotation back to 80°, as was experimentally
observed [21].

The dissociation constant for Pi, $K_d{}^{Pi}$, is calculated from the rate constants to be 4.9 mM at 80° and ~200 M at 120°, implying >10^4-fold reduction
in the affinity for Pi upon rotation from 80° to 120°. $K_d{}^{Pi}(80°)$ of 4.9 mM
is close to physiological [Pi], but this rather low affinity is due to the
two-state approximation. Actual $K_d{}^{Pi}(80°)$ in the angle-dependent diagram
(Fig. 14.3A) will be smaller, assuring efficient Pi binding during ATP synthesis
by reverse rotation.

If Pi release drives rotation from 80° to 120°, then rotation from 80° to 120°, whether spontaneous or by an external force, must accompany a reduction in the affinity for Pi. This is the law of action and reaction, proved for this case by the experiment above [21]. Cartoons in Fig. 14.3A illustrate this, where the green blocks represent the protein part and space filling models represent ADP and Pi. In a, Pi, together with ADP, pulls the protein toward themselves. When Pi leaves (b), the pulling force reduces and then the protein springs back to a less bent (relaxed) conformation (c). In this conformation, the affinity for Pi is low, because bond formation is difficult due to the widening of the catalytic site (d). If we start from d, the transition toward a is a process of "induced fit" [35] where the catalytic site, and hence the protein as a whole, adapt their conformations to tightly accommodate the bound Pi. The protein and the ligand pull each other, observing the law of action and reaction. The reverse of this process may be called "induced unfit."

14.4.2 Power Stroke vs. Diffusion and Catch

The term induced fit apparently implies that ligand binding is the cause of protein conformational change (Fig. 14.3A $c \rightarrow d \rightarrow a$). In fact, the protein conformation fluctuates thermally and thus one can expect that, when the protein adopts thermally a conformation suitable for binding, the ligand binds and stabilizes that conformation ("conformational selection"). In this latter process ($c \rightarrow b \rightarrow a$), conformational change is apparently the cause of binding. As long as one compares the start (c) and end (a), however, the two processes are equivalent: either ligand binding or conformational change can be the cause of the other. The two processes are distinguished only kinetically (through d or through b).

The equivalent of induced fit vs. conformational selection for the case of a motor protein is power stroke (a force-generating conformational change driven by a chemical reaction) vs. diffusion and catch (thermal fluctuation, even against an external force, followed by stabilization by chemical reaction). As we have discussed above, the distinction is meaningful only kinetically. For protein machines, the energy involved is not much greater than the thermal energy, and hence one cannot expect that one of the two processes overwhelms the other.

Furthermore, the two processes likely cooperate with each other. The seesaw energy diagram in Fig. 14.3A (the linear angle dependence is simplification) indicates that forward (from 80° to 120°) fluctuation in the state $F_1 \cdot$ ADP $\cdot$ Pi will decrease the affinity for Pi and thus assist Pi release, and that, once Pi is released, the downhill slope in $F_1 \cdot$ ADP will assist forward rather than backward fluctuations. Power stroke vs. diffusion and catch is more of conceptual distinction rather than practical.

In our analysis of the 40° substep by Pi release, we presumed for simplicity the power stroke ($a \rightarrow b \rightarrow c$) scenario rather than diffusion and catch

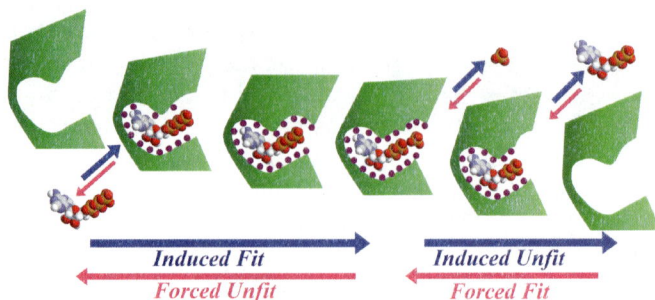

Fig. 14.4. Chemo-mechanical and mechano-chemical energy conversion through a series of fitting and unfitting processes. *Blue arrows*, reaction driven by free energy of hydrolysis; *pink arrows*, reverse reactions driven by an external force

$(a \rightarrow d \rightarrow c)$. The energy levels we arrived at (Fig. 14.3B) are consistent with this assumption, but this does not imply that the power stroke view is correct.

14.4.3 Chemo-Mechanical Coupling by Induced Fit and Unfit

We think that protein (or RNA) machines that convert free energy of chemical reactions into mechanical work operate by a series of induced fit and induced unfit (Fig. 14.4). Here, we define fit and unfit in a broad sense and do not distinguish them from conformational selection (we discuss energetics rather than kinetics).

In the case of F_1, all reactions toward right (overall hydrolysis) are likely downhill under the laboratory conditions of high [ATP] and low [ADP] and [Pi], except for the hydrolysis step for which the energy involved may be small [10]. Counterclockwise rotation under these conditions are smooth and fast, exceeding 700 revolutions s^{-1} for TF_1 at high temperatures [36].

F_1-ATPase is a reversible molecular machine [5]. ATP synthesis by clockwise rotation driven by an external force would proceed as a series of forced fit and unfit (Fig. 14.4). Whether synthesis follows the hydrolysis pathway (Fig. 14.2B) precisely in reverse is yet to be clarified.

14.5 Structural Basis of Rotation

The relation between chemical reactions and rotation has basically been worked out. The scheme in Fig. 14.2B may apply only to TF_1, or the scheme may turn out to be wrong in its details. There is, however, one coupling scheme (Fig. 14.2B) that we can propose, at least as a working hypothesis. For the structural changes underlying rotation, in contrast, we cannot propose, at present, even a possible mechanism. We thought that the push–pull mechanism suggested by Wang and Oster [29] was reasonable, but it is at best part of the actual mechanism (see below).

14.5.1 Crystal Structures

Walker and colleagues have solved many crystal structures of F_1 (mostly MF_1) that have served as the basis for the understanding of the mechanism of catalysis and also for designing experiments including single-molecule studies. Important differences among the crystals have been found in the catalytic sites, but overall structures closely resemble each other, except for one in which all three catalytic sites are filled with a nucleotide and β_E adopts a different shape [37]. All others bind two catalytic nucleotides, as in our coupling scheme (Fig. 14.2B), and one catalytic site is widely open. Even in the structure with three catalytic sites filled, subunits other than β_E adopted configurations similar to those in other crystals.

Many of the crystals were prepared in the presence of azide, which is known to enhance the MgADP inhibition [17], and azide has been resolved in a crystal [38]. Thus, the mutually similar structures, including that of EF_1 [39] which was also prepared in the presence of azide, must all represent, basically, the MgADP-inhibited state. In this state, γ is orientated at $80°$ [18], as in the $80°$ intermediate during active rotation. Fluorescence energy transfer between probes on a β and rotating γ has indicated that, of the two active rotation intermediates at $0°$ and $80°$, the crystal structures are much closer to the $80°$ intermediate [40]. At $80°$, we expect one catalytic site to be open to the medium (asterisks in Fig. 14.2B), which is consistent with the idea that the crystal structures with a fully open catalytic site mimic the $80°$ state rather than the ATP-waiting state.

The crystal structures show that the γ axle can be twisted around its axis. The bottom tip (of the longer C-terminal helix) does not differ much among the crystals, but, in the orifice region, the γ coiled coil in the three-nucleotide structure [37] is twisted clockwise by $\sim 20°$ relative to the original structure [4]. In a yeast MF_1 structure [41], on the other hand, γ is rotated $\sim 12°$ counterclockwise relative to the original structure; the latter could be due to species difference, but other two molecules in the same unit cell show clockwise, rather than counterclockwise, twists.

The twist in the three-nucleotide structure is opposite to the rotational direction, suggesting that the third nucleotide in the β_E site (ADP) may correspond to the leaving ADP in our scheme (gold in Fig. 14.2B). The tip of β_E that touches γ in the orifice region of this structure is rotated $\sim 16°$ counterclockwise [37] relative to the original structure, as though the β tip is preventing counterclockwise rotation of γ until the ADP leaves and the β tip retracts. The twist in γ, however, could be due to lattice contacts, because clockwise twist of $\sim 11°$ was also seen in the structure in Fig. 14.1 where one catalytic site is open.

14.5.2 Axle-Less Constructs

As seen in Fig. 14.1, β_{TP} and β_{DP} (with an overall structure very similar to β_{TP}) are bent toward, and apparently push, the top of the γ axle, whereas β_E

retracts and pulls γ. The combined actions could rotate the slightly bent and skewed axle while the lower tip of the axle is held relatively stationary by the $\alpha_3\beta_3$ cylinder [29]. Though this view was attractive, we have recently found that an axle-less mutant (Fig. 14.1C) can rotate in the correct direction for >100 revolutions [14]. Neither a rigid axle nor a fixed support at the bottom, essential to the push–pull mechanism, is necessary for rotation. At present, we do not understand how this axle-less construct rotates and how the γ head that apparently sits on the concave stator orifice can stay attached while undergoing many revolutions. The twist of γ seen in some crystals suggests torque generation in the orifice region alone, if it is not an artifact resulting from lattice contacts.

As we truncate the γ axle from the tip step by step, the bulk hydrolysis activity gradually diminishes, whereas the apparent torque becomes approximately half that of the wild type at the C-terminal truncation nearly level with the N-terminus and thereafter remains constant [42, 43]. It appears that a rigid axle and bottom support are needed for high-speed catalysis and generation of full torque, but orifice interactions alone can produce half the torque.

14.6 Remaining Tasks

To relate structure and function in F_1-ATPase, we need at least one more crystal structure that is grossly different from others, hopefully one that mimics the ATP-waiting state. Torque of this motor, so far inferred from the viscous friction against a probe attached to γ, needs to be better characterized. The best is to measure the stall torque against a conservative external torque, at all γ angles for each chemical state of F_1, to construct the potential energy diagram as in Fig. 14.3A for all angles and for all states. This is a formidable task, but we are still trying. ATP synthesis by reverse rotation is poorly understood. We are yet to learn, for example, at which angles ADP and Pi are picked up from the medium and when ATP is released; whether and to what extent the forced γ rotation is blocked until the expected chemical reaction has taken place. A lot remain to be clarified, even for this relatively well-understood molecular machine.

Acknowledgements

We thank the members of Kinosita and Yoshida labs for collaboration and discussion, R. Kanda-Terada for technical support, K. Sakamaki, M. Fukatsu, and H. Umezawa for encouragement and lab management. This work was supported by Grants-in-Aids for Specially Promoted Research from the Ministry of Education, Sports, Culture, Science and Technology, Japan.

References

1. P.D. Boyer, W.E. Kohlbrenner, The present status of the binding-change mechanism and its relation to ATP formation by chloroplasts, in Energy Coupling in Photosynthesis, ed. by B.R. Selman, S. Selman-Reimer (Elsevier, Amsterdam, 1981), pp. 231–240

2. F. Oosawa, S. Hayashi, The loose coupling mechanism in molecular machines of living cells. Adv. Biophys. 22, 151–183 (1986)

3. H. Noji, R. Yasuda, M. Yoshida, K. Kinosita, Jr, Direct observation of the rotation of F_1-ATPase. Nature 386, 299–302 (1997)

4. J.P. Abrahams, A.G.W. Leslie, R. Lutter, J.E. Walker, Structure at 2.8 Å resolution of F_1-ATPase from bovine heart mitochondria. Nature 370, 621–628 (1994)

5. H. Itoh, A. Takahashi, K. Adachi, H. Noji, R. Yasuda, M. Yoshida, K. Kinosita, Jr., Mechanically driven ATP synthesis by F_1-ATPase. Nature 427, 465–468 (2004)

6. Y. Rondelez, G. Tresset, T. Nakashima, Y. Kato-Yamada, H. Fujita, S. Takeuchi, H. Noji, Highly coupled ATP synthesis by F_1-ATPase single molecules. Nature 433, 773–777 (2005)

7. P.D. Boyer, The ATP synthase – a splendid molecular machine. Annu. Rev. Biochem. 66, 717–749 (1997)

8. K. Kinosita, Jr., R. Yasuda, H. Noji, F_1-ATPase: a highly efficient rotary ATP machine. Essays Biochem. 35, 3–18 (2000)

9. K. Kinosita, Jr., R. Yasuda, H. Noji, K. Adachi, A rotary molecular motor that can work at near 100% efficiency. Phil.Trans. R. Soc. Lond. B 355, 473–489 (2000)

10. K. Kinosita, Jr., K. Adachi, H. Itoh, Rotation of F_1-ATPase: how an ATP-driven molecular machine may work. Annu. Rev. Biophys. Biomol. Struct. 33, 245–268 (2004)

11. M. Yoshida, E. Muneyuki, T. Hisabori, ATP synthase – a marvelous rotary engine of the cell. Nat. Rev. Mol. Cell Biol. 2, 669–677 (2001)

12. C. Gibbons, M.G. Montgomery, A.G.W. Leslie, J.E. Walker, The structure of the central stalk in bovine F_1-ATPase at 2.4 Å resolution. Nat. Struct. Biol. 7, 1055–1061 (2000)

13. Y. Shirakihara, A.G.W. Leslie, J.P. Abrahams, J.E. Walker, T. Ueda, Y. Sekimoto, M. Kambara, K. Saika, Y. Kagawa, M. Yoshida, The crystal structure of the nucleotide-free $\alpha_3\beta_3$ subcomplex of F_1-ATPase from the thermophilic *Bacillus* PS3 is a symmetric trimer. Structure 5, 825–836 (1997)

14. S. Furuike, M.D. Hossain, Y. Maki, K. Adachi, T. Suzuki, A. Kohori, H. Itoh, M. Yoshida, K. Kinosita, Jr., Axle-less F_1-ATPase rotates in the correct direction. Science 319, 955–958 (2008)

15. R. Yasuda, H. Noji, K. Kinosita, Jr., M. Yoshida, F_1-ATPase is a highly efficient molecular motor that rotates with discrete 120° steps. Cell 93, 1117–1124 (1998)

16. R. Yasuda, H. Noji, M. Yoshida, K. Kinosita, Jr., H. Itoh, Resolution of distinct rotational substeps by submillisecond kinetic analysis of F_1-ATPase. Nature 410, 898–904 (2001)

17. J.M. Jault, C. Dou, N.B. Grodsky, T. Matsui, M. Yoshida, W.S. Allison, The $\alpha_3\beta_3\gamma$ subcomplex of the F_1-ATPase from the thermophilic *Bacillus* PS3 with the β T165S substitution does not entrap inhibitory MgADP in a catalytic site during turnover. J. Biol. Chem. 271, 28818–28824 (1996)

18. Y. Hirono-Hara, H. Noji, M. Nishiura, E. Muneyuki, K.Y. Hara, R. Yasuda, K. Kinosita, Jr., M. Yoshida, Pause and rotation of F_1-ATPase during catalysis. Proc. Natl. Acad. Sci. USA 98, 13649–13654 (2001)

19. H. Noji, D. Bald, R. Yasuda, H. Itoh, M. Yoshida, K. Kinosita, Jr., Purine but not pyrimidine nucleotides support rotation of F_1-ATPase. J. Biol. Chem. 276, 25480–25486 (2001)

20. K. Adachi, H. Noji, K. Kinosita, Jr., Single-molecule imaging of rotation of F_1-ATPase. Meth. Enzymol. 361, 211–227 (2003).

21. K. Adachi, K. Oiwa, T. Nishizaka, S. Furuike, H. Noji, H. Itoh, M. Yoshida, K. Kinosita, Jr. Coupling of rotation and catalysis in F_1-ATPase revealed by single-molecule imaging and manipulation. Cell 130, 309–321 (2007)

22. N. Sakaki, R. Shimo-Kon, K. Adachi, H. Itoh, S. Furuike, E. Muneyuki, M. Yoshida, K. Kinosita, Jr., One rotary mechanism for F_1-ATPase over ATP concentrations from millimolar down to nanomolar. Biophys. J. 88, 2047–2056 (2005)

23. K. Shimabukuro, R. Yasuda, E. Muneyuki, K.Y. Hara, K. Kinosita, Jr., M. Yoshida, Catalysis and rotation of F_1 motor: cleavage of ATP at the catalytic site occurs in 1 ms before $40°$ substep rotation. Proc. Nat. Acad. Sci. USA 100, 14731–14736 (2003)

24. T. Nishizaka, K. Oiwa, H. Noji, S. Kimura, E. Muneyuki, M. Yoshida, K. Kinosita, Jr., Chemomechanical coupling in F_1-ATPase revealed by simultaneous observation of nucleotide kinetics and rotation. Nat. Struct. Mol. Biol. 11, 142–148 (2004)

25. T. Ariga, E. Muneyuki, M. Yoshida, F_1-ATPase rotates by an asymmetric, sequential mechanism using all three catalytic subunits. Nat. Struct. Mol. Biol. 14, 841–846 (2007)

26. C.C. O'Neal, P.D. Boyer, Assessment of the rate of bound substrate interconversion and of ATP acceleration of product release during catalysis by mitochondrial adenosine triphosphatase. J. Biol. Chem. 259, 5761–5767 (1984)

27. R. Kagawa, M.G. Montgomery, K. Braig, A.G.W. Leslie, J.E. Walker, The structure of bovine F_1-ATPase inhibited by ADP and beryllium fluoride. EMBO J. 23, 2734–2744 (2004)

28. K. Shimabukuro, E. Muneyuki, M. Yoshida, An alternative reaction pathway of F_1-ATPase suggested by rotation without $80°/40°$ substeps of a sluggish mutant at low ATP. Biophys. J. 90, 1028–1032 (2006)

29. H. Wang, G. Oster, Energy transduction in the F_1 motor of ATP synthase. Nature 396, 279–282 (1998)

30. Y.M. Milgrom, R.L. Cross, Rapid hydrolysis of ATP by mitochondrial F_1-ATPase correlates with the filling of the second of three catalytic sites. Proc. Natl. Acad. Sci. USA 102, 13831–13836 (2005)

31. J. Weber, S. Wilke-Mounts, R.S. Lee, E. Grell, A.E. Senior, Specific placement of tryptophan in the catalytic sites of the *Escherichia coli* F_1-ATPase provides a direct probe of nucleotide binding: maximal ATP hydrolysis occurs with three sites occupied. J. Biol. Chem. 268, 20126–20133 (1993)

32. J. Weber, A.E. Senior, ATP synthase: what we know about ATP hydrolysis and what we do not know about ATP synthesis. Biochim. Biophys. Acta 1458, 300–309 (2000)

33. C. Dou, P.A.G. Fortes, W.S. Allison, The $\alpha_3(\beta Y341W)_3\gamma$ subcomplex of the F_1-ATPase from the Thermophilic *Bacillus* PS3 fails to dissociate ADP when

MgATP is hydrolyzed at a single catalytic site and attains maximal velocity when three catalytic sites are saturated with MgATP. Biochemistry 37, 16757–16764 (1998)

34. S. Ono, K.Y. Hara, J. Hirao, T. Matsui, H. Noji, M. Yoshida, E. Muneyuki, Origin of apparent negative cooperativity of F_1-ATPase. Biochim. Biophys. Acta 1607, 35–44 (2003)

35. D.E. Koshland, Application of a theory of enzyme specificity to protein synthesis. Proc. Natl. Acad. Sci. USA 44, 98–104 (1958)

36. S. Furuike, K. Adachi, N. Sakaki, R. Shimo-Kon, H. Itoh, E. Muneyuki, M. Yoshida, K. Kinosita, Jr., Temperature dependence of the rotation and hydrolysis activities of F_1-ATPase. Biophys. J. 95, 761–770 (2008)

37. R.I. Menz, J.E. Walker, A.G.W. Leslie, Structure of bovine mitochondrial F_1-ATPase with nucleotide bound to all three catalytic sites: implications for the mechanism of rotary catalysis. Cell 106, 331–341 (2001)

38. M.W. Bowler, M.G. Montgomery, A.G.W. Leslie, J.E. Walker, How azide inhibits ATP hydrolysis by the F-ATPases. Proc. Natl. Acad. Sci. USA 103, 8646–8649 (2006)

39. A.C. Hausrath, R.A. Capaldi, B.W. Matthews, The conformation of the ε- and γ-subunits within the *Escherichia coli* F_1 ATPase. J. Biol. Chem. 276, 47227–47232 (2001)

40. R. Yasuda, T. Masaike, K. Adachi, H. Noji, H. Itoh, K. Kinosita, Jr., The ATP-waiting conformation of rotating F_1-ATPase revealed by single-pair fluorescence resonance energy transfer. Proc. Natl. Acad. Sci. USA 100, 9314–9318 (2003)

41. V. Kabaleeswaran, N. Puri, J.E. Walker, A.G.W. Leslie, D.M. Mueller, Novel features of the rotary catalytic mechanism revealed in the structure of yeast F_1 ATPase. EMBO J. 25, 5433–5442, (2006)

42. M.D. Hossain, S. Furuike, Y. Maki, K. Adachi, M.Y. Ali, M. Huq, H. Itoh, M. Yoshida, K. Kinosita, Jr., The rotor tip inside a bearing of a thermophilic F_1-ATPase is dispensable for torque generation. Biophys. J. 90, 4195–4203 (2006)

43. M.D. Hossain, S. Furuike, Y. Maki, K. Adachi, T. Suzuki, A. Kohori, H. Itoh, M. Yoshida, K. Kinosita, Jr., Neither helix in the coiled coil region of the axle of F_1-ATPase plays a significant role in torque production. Biophys. J. 95, 4837–4844 (2008)

Part VI

Force and Multiparameter Spectroscopy on
Functional Active Proteins

15

Mechanoenzymatics and Nanoassembly of Single Molecules

Elias M. Puchner and Hermann E. Gaub

Summary. We investigated the muscle enzyme, titin kinase, by means of single-molecule force spectroscopy. Our results show that the binding of ATP, which is the first step of its signaling cascade controlling the muscle gene expression and protein turnover, is mechanically induced. The detailed determination of barrier positions in the mechanical activation pathway and the corresponding functional states allow structural insight by comparing the experiment with molecular dynamics simulations. From our results, we conclude that titin kinase acts as a natural force sensor controlling the muscle build-up. To study the interplay of functional units, we developed the single-molecule cut-and-paste technique which combines the precision of AFM with the selectivity of DNA hybridization. Functional units can be assembled one-by-one in an arbitrarily predefined pattern, with an accuracy that is better than 11 nm. The cyclic assembly process is optically monitored and mechanically recorded by force-extension traces. Using biotin as a functional unit attached to the transported DNA, patterns of binding sites may be created, to which streptavidin-modified nanoobjects like fluorescent nanoparticles can specifically self-assemble in a second step.

15.1 Introduction

Proteins, like enzymes, obtain their function through their three dimensional conformation and dynamics. Force tilts the underlying energy landscape, and thus influences the conformation of an enzyme [1], resulting in changes in its functional state. Such force-induced changes in enzymatic function play a crucial role in biological processes in which mechanical stress is translated into biochemical signals. Atomic Force Microscopy (AFM) based single-molecule force spectroscopy [2, 3] is an ideal tool to investigate the biological force sensors since it allows for the conformational control over single enzymes, for the determination of unfolding barriers with sub-nm precision, and for the detection of ligand binding through their interactions with the binding pocket.

Besides the investigation of single enzymes [4–9], it is crucial to study their interplay without losing single-molecule sensitivity [10]. For reactions within a cell to be effective, the spatial arrangement of the involved molecular components should be decisive. To mimic such arrangements, we developed the single-molecule cut-and-paste (SMCP) technique, which combines the precision of AFM with the selectivity of DNA hybridization. It allows for the mechanical one-by-one assembly and optical monitoring of single molecules to arbitrary patterns with a precision in the 10 nm range.

15.2 The Molecular Force Sensor Titin Kinase

A prime example of a biological system that needs adaptation to mechanical stress is the vertebrate muscle. The basic mechanical components of the sarcomere are the actin and myosin filaments, which generate contraction, and the giant protein titin, which keeps myosin filaments in place and provides the muscle with its passive elasticity through PEVK-regions [11–14]. In the middle of the sarcomere, titin is cross-linked with neighboring titin molecules via several proteins forming the M-band structure (see Fig. 15.1) [15,16]. This position is ideal for the detection of shear forces between neighboring myosin filaments, where the only catalytic domain of titin, the titin kinase (TK), is located.

It was shown that this kinase domain does not exhibit basal activity and that it is autoinhibited in a dual way [17]. On the one hand, the ATP binding site is blocked by the C-terminal regulatory tail (shown in red in Fig. 15.1), preventing binding of the cosubstrate ATP. On the other hand, the catalytic aspartate is inhibited by tyrosine-170 (TYR170) so that phosphorylation of

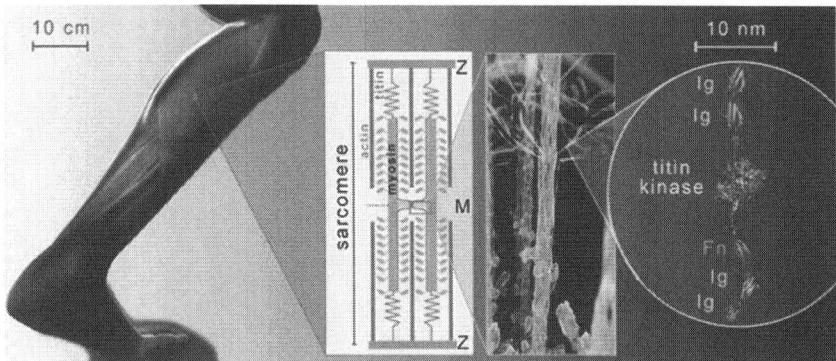

Fig. 15.1. Schematic illustration of the muscle. The force generating unit is the sarcomere with a length of about 2 μm. Contraction is caused by the myosin and actin filaments. Titin spans the whole half-sarcomere, keeps the myosin filaments in place and provides the muscle with its passive elasticity. At the M-line, where titin is cross-linked, the titin kinase is integrated with its surrounding and globular Ig and Fn domains. This position is ideal to detect the imbalances between neighboring filaments

substrates cannot take place. If, however, ATP binds to an open conformation, TYR170 gets autophosphorylated whereby the active site is activated [18] and a signaling cascade is initiated which controls muscle gene expression and protein turnover [19]. In the following, we will address the initiating step of the signaling cascade comprising the mechanical opening of the ATP binding site and binding of ATP.

To mimic the mechanical strain in the M-band, the natural protein construct consisting of the kinase domain and the surrounding immunoglobulin and fibronectin (Ig/Fn) domains (IgA168-IgA169-FnA170-TK-IgM1-IgM2) was stretched and unfolded in AFM-based force spectroscopy experiments. The force-extension traces, as shown in Fig. 15.2, exhibit a hierarchical mechanical stability of the domains: first, TK unfolds over five energy barriers below 50 pN at pulling speeds of 1 μm/s. After complete unfolding, five force

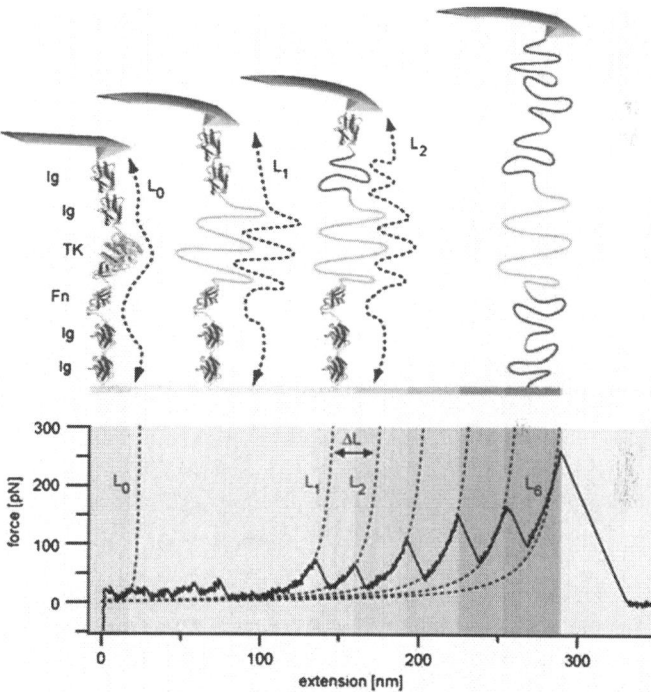

Fig. 15.2. Single-molecule force spectroscopy. The natural TK protein construct is non-specifically adhered to a gold surface and contacted with the tip of an AFM cantilever. During retraction, force-extension traces are recorded. After each rupture, the total free contour length increases so that the force drops. If further stretched, a characteristic rise in force is observed, which is due to the polymer elasticity of the unfolded polypeptide chain. In the beginning, TK unfolds below 50 pN. After complete unfolding of TK, the five structural Ig/Fn domains unfold with regular contour length increments ΔL of about 30 nm (L_1–L_6)

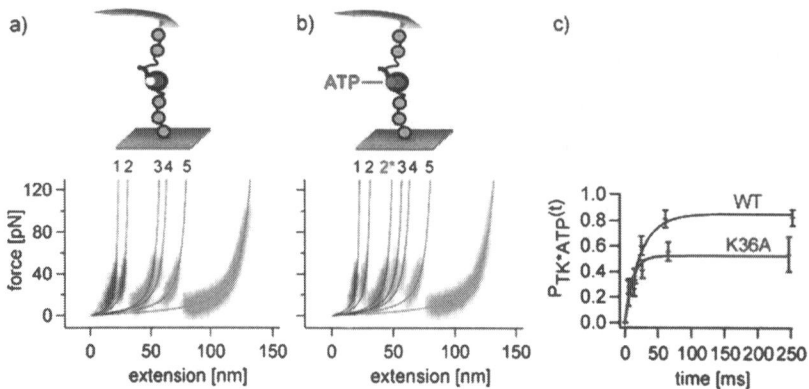

Fig. 15.3. Unfolding profile of TK and kinetics of mechanically induced ATP binding. (a) The superposition of 66 single-molecule unfolding traces of TK shows a fixed sequence of unfolding peaks (1–5). (b) Mechanically induced ATP binding leads to an additional force peak (2*) not present in the absence of ATP (44 traces). (c) The probability of ATP binding (measured by means of peak 2* at 2 mM ATP) depends strongly on the opening time of the binding pocket, which is regulated by the unfolding speed. This nonequilibrium kinetic can be fitted, yielding the kinetic constants. Mutation of lysine-36 to alanine (K36A) reduces the binding of ATP strongly

peaks are observed whose number and contour length increments correlate exactly with the expected values of the structural Ig/Fn domains. In contrast to the independently unfolding Ig/Fn domains, unfolding of TK is, despite the comparable levels of the peaks, strictly ordered and thus topologically determined.

In the presence of the cosubstrate ATP at a physiological concentration of 2 mM, a certain fraction of unfolding traces shows an additional energy barrier that is not visible without ATP (see Fig. 15.3). Since TK does not bind ATP in its autoinhibited conformation, this is already a first hint that the binding pocket is mechanically opened and that binding of ATP causes a force peak in the unfolding profile due to its interactions.

The experiment that unravels the process of ATP binding is designed according to the following idea: The different conformational states of TK are controlled by its end-to-end distance. Therefore, the time at which the binding pocket is mechanically opened would be determined by the unfolding speed. If ATP is associated with TK before the pulling cycle, i.e. not mechanically induced, then the occurrence of the ATP peak is expected not to be time-dependent. However, experiments at different opening times varying from 6 to 250 ms[1] reveal a strong dependency and nonequilibrium kinetics (see Fig. 15.3c)). At high pulling speeds, only a low fraction of traces exhibits the ATP barrier. While at slow speeds, a high saturation

---

[1] The opening times correspond to pulling speeds of 3000 and 72 nm/s respectively at an opening length of 18 nm following from MD simulations [20].

is observed since the binding pocket is opened for a long time so that the equilibrium value for ATP binding is reached. This experiment shows that ATP binds to the mechanically opened binding pocket. Fitting with nonequilibrium kinetics allows the determination of kinetic parameters which are– $k_{on} = 1.8 \pm 0.3 \times 10^4$ 1/Ms, $k_{off} = 6 \pm 3$ 1/s, and $K_d = 347\,\mu$M. The same experiment with a mutant of TK where lysine-36 is replaced by alanine (K36A) shows a more than 6-fold higher off rate and a K_d in the millimolar range. As expected from homologous kinases, lysine-36, which is located in the binding pocket, plays a crucial role for ATP binding.

Besides the mechanism and the kinetics of ATP binding, it is also crucial to know the position in the sequential activation pathway, where the ATP binding pocket is mechanically opened. Therefore, we developed a pump-and-probe type of measurement protocol that takes advantage of the mechanical control over the conformation and the nonequilibrium kinetics of ATP binding.

As shown in Fig. 15.4, TK is first unfolded to a certain conformation which is controlled by extension. This state is now prepared and "frozen" for a certain time Δt during which ATP has the possibility to bind to the pocket. After this time pulse, TK is further unfolded to determine, by means

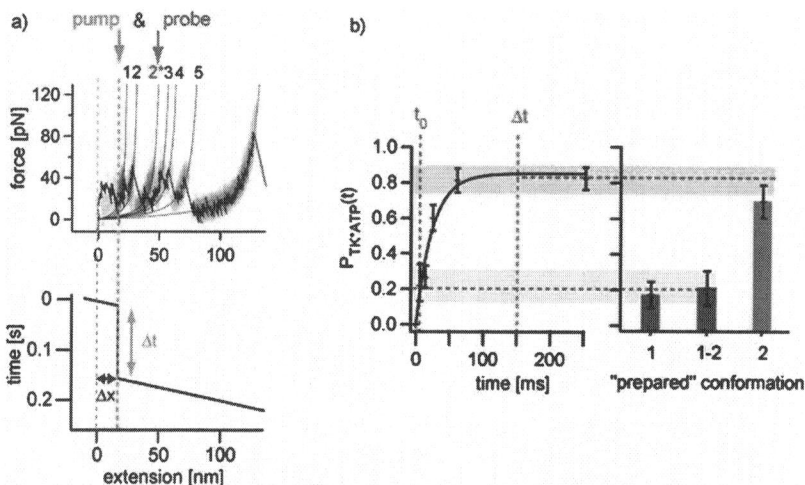

Fig. 15.4. Determination of the opened state of TK. (a) TK is unfolded to a certain state. This conformation, determined by the extension Δx, is now "frozen" for a time Δt of 200 ms. During this time, ATP has the possibility to bind. By further unfolding, it is probed with barrier 2* whether ATP is bound or not. All states before barrier 2* can be tested by varying Δx. (b) If the time pulse Δt is set before barrier 1 or between barriers 1 and 2, only the low value of ATP binding is observed which is due to the finite pulling velocity ($t_0 = 6$ ms). However, if the time pulse is set after barrier 2 and before 2*, ATP binding shows, within the experimental error, the large saturation value. Therefore, one can conclude that the binding pocket is opened after barrier 2

of the ATP barrier, whether ATP did bind or not. If a closed state of TK is prepared, then the time pulse should not influence the relative frequency of ATP binding and only a low value due to the finite pulling velocity is expected. This is experimentally observed if the time pulse is set before the first barrier or between barriers 1 and 2. However, if the state after barrier 2 and before the ATP barrier is prepared, then ATP binding is strongly enhanced and reaches the high saturation within the experimental error. This experiment shows that ATP binds to the mechanically opened conformation of TK, which is reached after barrier 2.

In order to precisely quantify the positions of the energy barriers, a description of molecular coordinates is required. The experimental variables, force and extension, at which an energy barrier is overcome, show large fluctuations due to the stochastic nature of the process, which is caused by the thermal energy $k_B T$. Furthermore, force and extension depend on experimental parameters such as pulling speed, temperature, or properties of the solvent. In contrast, the contour length is the characteristic variable for the positions of energy barriers.

The contour length can be determined by fitting force-extension traces with models for polymer elasticity [21], such as the worm-like chain model (WLC) [22]. However, in order to overcome fitting by hand and loss of information, we developed a method based on inverse models of polymer elasticity that transforms each data point of force-extension traces into force-contour length space (see Fig. 15.5) [23]. From these transformed traces, barrier position histograms can be directly created that reveal the barrier position along the contour length. In contrast to force-extension traces, barrier position histograms can be averaged and they provide the molecular fingerprint that allows comparison of different proteins and the automatic selection of traces.

Figure 15.5c shows the barrier position histograms of TK in both the absence (66 traces of Fig. 15.3a) and presence (44 traces of Fig. 15.3b) of the ATP barrier. Fitting with Gaussians allows the precise determination of energy barriers, with contour length increments of 9.1, 28.6, 7.3, 18.0, 57.9 nm and 9.1, 19.4, 10.1, 7.5, 16.4, 58.3 nm in the presence of ATP with an estimated error of only 2%. In order to verify the precision of these measurements, a contour length histogram of the Ig/Fn domains serves as an internal molecular calibration. The experimentally determined number of amino acids can be calculated by adding the diameter of the folded domain to the measured contour length increment and by dividing this value by 0.365 nm, corresponding to the separation of $C\alpha$ atoms in the polypeptide chain [24]. This can be better seen if considered backwards: If the two ends of an unfolded and completely stretched domain are approached until it is folded into its globular conformation, then the distance between the attachment points is not zero and lacking in the measured contour length increment. Therefore, this distance, which follows from the crystal structures, has to be taken into account. In this way, the mean value for the number of amino acids (aa) is determined to be $(30.45 \pm 0.6 + 4.3)$ nm/0.365 nm $= 95 \pm 2$ aa which deviates from the

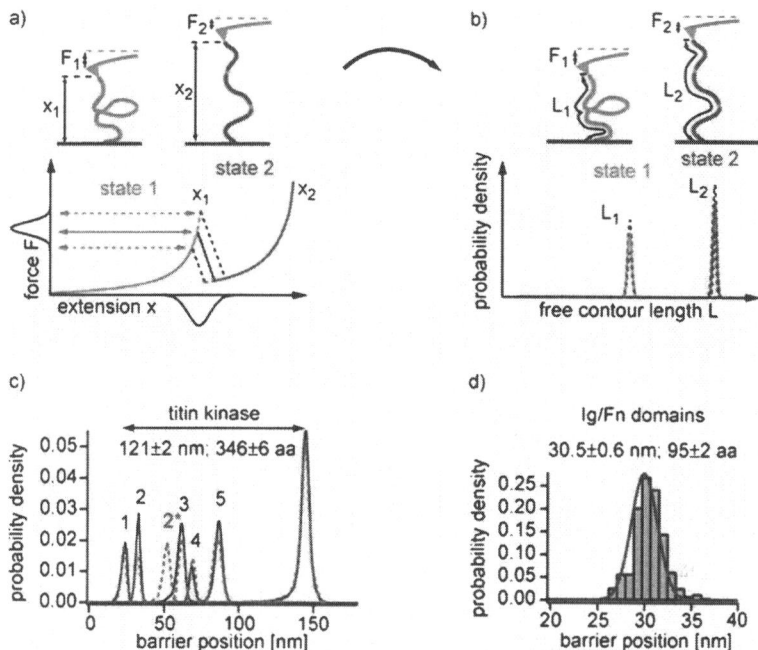

Fig. 15.5. Transformation of force-extension traces into the molecular coordinate contour length. (**a**) The rupture force and the extension x_1 and x_2 are subject to fluctuations and exhibit a broad distribution. Furthermore, they depend on experimental parameters as described in the text. The characteristic parameter of a folding state is the free contour length as illustrated in (**b**). Each data point (F_i, x_i) is transformed into force-contour length space (F_i, L_i) by means of inverse models for polymer elasticity. The transformed data points are accumulated into histograms, which directly show the barrier positions L_1 and L_2 along the contour length. (**c**) The barrier positions of TK in the absence (black) and presence (red) of ATP were determined with a relative error of 2% corresponding to only a few amino acids. The number of amino acids (346 ± 6) agrees well with the actual number (344). (**b**) Ig/Fn domains serve as an internal verification. The determined mean number of 95 ± 2 amino acids agrees again with the value of 96 aa

theoretical mean value of 96 aa by only 1%. The same calculation for the total contour length increment of TK yields $(221 \pm 2 + 5.5)$ nm/0.365 nm = 346 ± 6 residues which is again in good agreement with 344 aa of TK including its N-terminal linker.

The precise positions of the barriers along the contour length can now be correlated with molecular dynamics simulations (MD) [25] performed on TK [20]. This allows us to determine the structural states during sequential activation and unfolding of TK. A detailed discussion would go beyond the scope of this chapter and can be found in reference [18]. We only want to summarize the main results, which are in good agreement with the experiment:

Barrier 1 is caused by the unfolding of the 23 aa long N-terminal linker of TK and at barrier 2 the autoregulatory tail is removed so that the ATP binding site is opened as seen in the pump-and-probe experiment (Fig. 15.4). If ATP binds, it mainly interacts with LYS36 (as seen in the experiment) and with MET34, and thus causes the additional energy barrier.

Phosphorylation experiments with truncated TK lacking the autoregulatory tail show [18] that TYR170 gets autophosphorylated if ATP binds, whereby the second autoinhibition is abolished. At this state, TK can initiate a signaling cascade that regulates the muscle gene expression and protein turnover. The experimentally determined activation forces below 50 pN at pulling speeds in the physiological range correspond to small force imbalances between neighboring myosin filaments of four to eight motor domains and are much smaller than those unfolding the structural Ig/Fn domains. Also, the activation length between barrier 1 and 2 of 9 nm is in the physiological range. Therefore, our results show that TK can act as a force sensor, which regulates the build-up of muscle.

15.3 One-by-One Assembly of Single Molecules

We have revealed the molecular mechanism and function of a single enzyme. However, for the understanding of biological processes it is crucial to study the interplay between functional units. For many cellular processes that are diffusion limited, the spatial arrangement of the involved components is crucial for a high efficiency. To mimic and investigate such systems on the single-molecule level, new techniques are necessary that allow for the control over individual molecules and their precise assembly. For this purpose, we developed the so-called SMCP technique [26] that combines the precision of AFM [2, 27, 28] with the selectivity of DNA hybridization [29–31]. The units to be assembled are picked up with an AFM tip from a depot, where both the interaction of the unit with the depot and target surface as well as with the tip are mediated by specific DNA oligomers, allowing for a cyclic operation and thus the assembly of complex patterns of units.

The surface assembly process is schematically depicted in Fig. 15.6a. Both, depot and target areas, are functionalized through PEG spacers with DNA anchor oligomers capable of hybridizing with the so-called transfer DNA via a 30 basepair (bp) DNA sequence. In the depot area, the anchor oligomers are covalently attached with the 5' end and in the target area with the 3' end. The depot area is then loaded with the transfer DNA, which is used as a carrier for a fluorophore or other functional units such as specific binding sites. The transfer DNA is designed such that it hybridizes at its 5' end with the anchor sequence and has a 20 bp overhang at the 3' end. An AFM cantilever is covalently functionalized with a 20 bp DNA oligomer complementary to the overhang sequence. This cantilever is carefully lowered towards the depot surface allowing the tip oligomer to hybridize with the transfer DNA. While

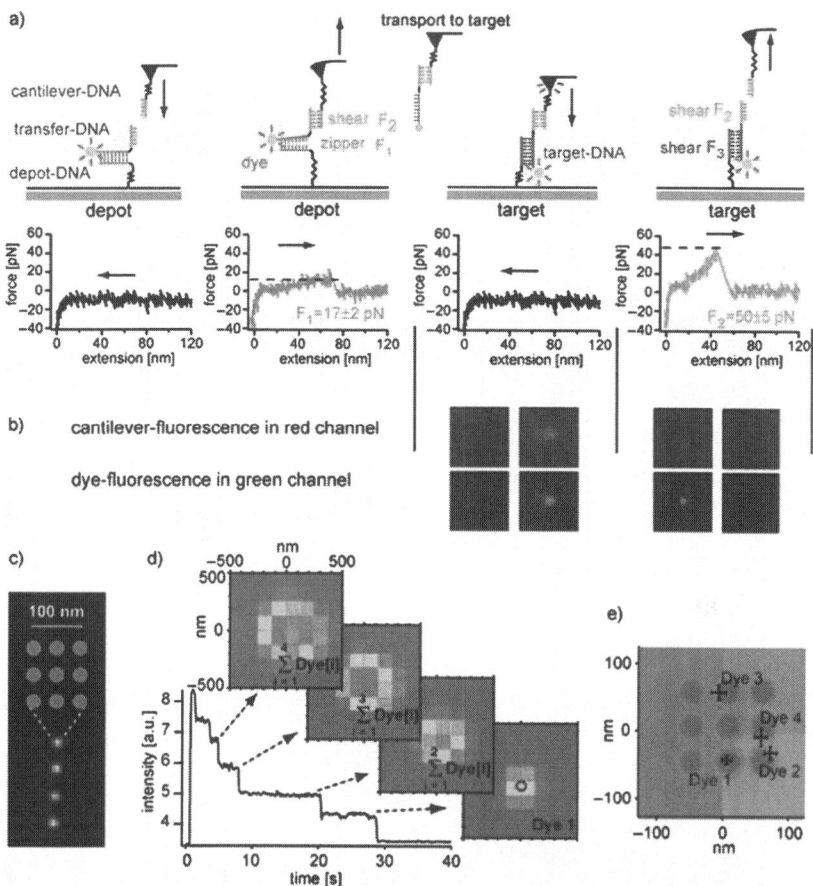

Fig. 15.6. SMCP cycle and super-resolution imaging of small structures with TIRF microscopy. (**a**) The transfer DNA, carrying a fluorescent dye, is hybridized to the depot DNA in the zipper geometry. Its overhanging sequence hybridizes to the cantilever DNA in the shear geometry, which is stronger and is picked upon retraction of the cantilever. This rupture from the depot DNA causes a characteristic force plateau at 17 pN. After transport to the target region, the transfer DNA is brought into contact with the surface where it hybridizes to the target DNA in the shear geometry. Retraction results in the bond rupture to the cantilever DNA at about 50 pN. (**b**) Simultaneous TIRF microscopy images show the cantilever and dye fluorescence in the different channels. After retraction, the cantilever is not visible any more whereas the dye stays on the surface and bleaches after a certain time. (**c**) The written patterns with a dye spacing of 50 nm appear as spots. (**d**) The stepwise bleaching allows the creation of averaged images that show the contribution of dye 1, dye 1 and dye 2, and so forth. The differences between these images contain the contributions of the individual dyes so that their positions can be determined and compared to the ones instructed to the AFM (**e**)

withdrawing the tip from the surface, the force that is built up in the molecular complex propagates through the two oligomers with different geometries. Whereas the anchor duplex is loaded in unzip geometry, the tip duplex is loaded in shear geometry. As has been shown, the unbinding forces for these two configurations under load differ significantly [32–34]. The rationale behind this effect is that the mechanical work required to overcome the binding energy is performed over paths of different length, resulting in different forces. Despite the higher thermodynamic stability of the 30 bp anchor duplex compared to the 20 bp tip duplex, the rupture probability for the anchor is higher by an order of magnitude than that of the tip duplex and results in a characteristic plateau in the corresponding force-extension trace at 17 pN. As a result, the transfer DNA with the functional unit is now bound to the tip and may be transferred to the target area.

At the target site the tip is lowered again, allowing the transfer DNA to hybridize at the chosen position with an anchor oligomer. Now, due to the different attachment, both duplexes are loaded in shear geometry when the tip is withdrawn. The longer anchor oligomer keeps the transfer DNA bound, and the tip is free again and ready to pick the next object. Again, the rupture of the DNA causes a characteristic force-extension trace so that a complete mechanical protocol is available for one cut-and-paste cycle. Simultaneously, the delivery of the fluorescent-labeled transfer DNA is optically monitored with the objective type of total internal reflection microscopy (TIRF) [35–37] as shown in Fig. 15.6b.

In this way, one molecule after the other can be assembled to arbitrarily predefined structures. To test the accuracy of the cut-and-paste process and the possibility to optically resolve small structures [38], patterns with a dye spacing of 50 nm were created [39]. These small patterns appear as spots with the size of a Rayleigh disc (Fig. 15.6c), however, subdiffraction resolution can be obtained [40, 41] e.g. by taking advantage of the time domain of the signal [42, 43]. As shown in Fig. 15.6d, one fluorophore after the other bleaches [44] and results in a stepwise intensity profile. From the number of steps it can be seen that only in two-thirds of the attempts an intact dye was delivered. The last dye can be precisely localized up to a few nanometers by fitting with a two-dimensional Gaussian. Its position can be determined by subtracting its contribution to the last dye and the dye before the last (Fig. 15.6d). In this manner, the pattern can be reconstructed as shown in Fig. 15.6e. Since the error grows with decreasing lifetime of the fluorophores, the analysis was restricted to the last four molecules. However, this might be overcome using better dyes [45].

The examination of several individual cut-and-paste cycles gives a drift corrected accuracy of about ±11 nm, which is due to the length of the involved spacers (see reference [39]). Besides these small structures, large arrays of molecules can be created. In reference [26], we assembled 10 µ sized structures with more than 5000 units and with a loss in transport efficiency of less than 10%.

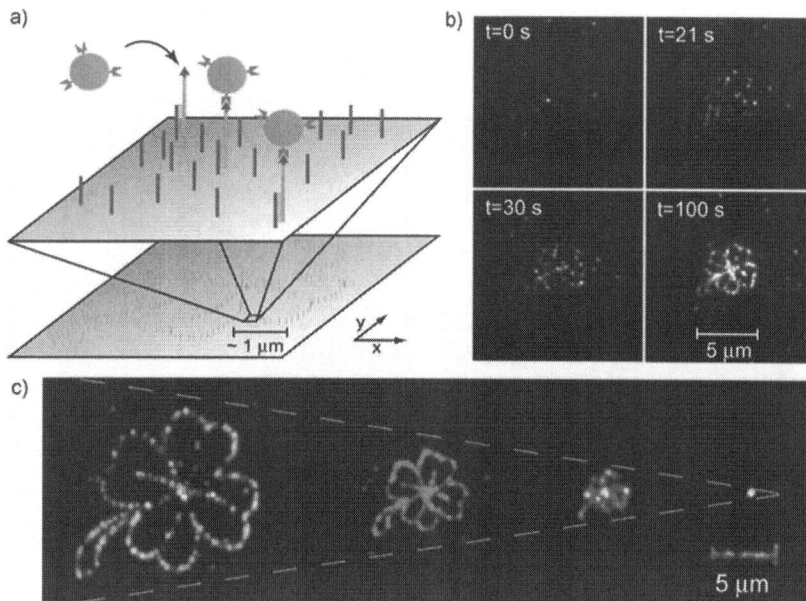

Fig. 15.7. Self-assembly of nanoparticles to patterns of binding sites. (**a**) The transfer DNA is modified with biotin, so that patterns of specific binding sites are created with SMCP. Fluorescent nanoparticles carrying streptavidin self-assemble to these patterns and form superstructures. (**b**) The formation of superstructures is observed online with TIRF microscopy. (**c**) The patterns of binding sites were created with different size and incubated with nanoparticles fluorescing at different wavelengths. Also, the scale bar is formed in this way. The pictures are standard deviations of TIRF microscopy image series recorded at 20 Hz

If larger objects to be assembled are attached to the transfer DNA, one faces the problem that they may disturb the hybridization process and thus prevent successful transport cycles. Therefore, the SMCP technique is expanded by attaching the specific binding site biotin to the transfer DNA. Now, arbitrary patterns of binding sites can be created to which any object carrying streptavidin at the surface, can self-assemble in a second step [46–49]. We placed biotins 100 nm apart from each other along the outline of a cloverleaf. This is schematically shown in Fig. 15.7a. The sample was then incubated with a 500 pM solution of fluorescent nanoparticles carrying an average of 7 streptavidins, which recognize and selectively bind to biotin [50]. Although some spots of the written pattern may consist of more than one biotin because of the chosen high cut-and-paste efficiency, binding of more than one nanoparticle is unlikely because of their large size (about 20 nm) and their low concentration. The binding process was observed online by fluorescence microscopy in TIRF excitation.

As can be seen from the picture series in Fig. 15.7b, the nanoparticles gradually assemble on the scaffold and finally decorate the outline of the cloverleaf. This self-assembly process on the predefined scaffold is completed within minutes. Because of the specific binding between biotin and streptavidin, and the low concentration of only 500 pM of the nanoparticles, non-specific adhesion was negligible as can be seen in Fig. 15.7c. It is interesting to note that not all of the positions light up, although our transfer protocols corroborate that a biotin was deposited at these optical voids. A comparison of AFM images (not shown here) with the fluorescence images demonstrated that nanoparticles had bound at these positions. Obviously, those nanoparticles had been optically inactive, a phenomenon that has been frequently described in the literature [51].

To demonstrate the versatility of this approach, we created binding patterns of different size and allowed different nanoparticles to form superstructures (Fig. 15.7c). Again a fraction of nanoparticles was inactive, and the thermal drift caused a slight distortion of the red structure. However, even the scale bar could be trustfully assembled. The expansion of this approach towards multicomponent structures is straightforward since there exist couplers with orthogonal affinities that can be linked to the transfer DNA. Whereas the assembly of planar nanoparticle structures of arbitrary design can easily be assembled this way, an expansion into the third dimension appears challenging but achievable.

15.4 Concluding Remarks

We have revealed the mechanical activation mechanism of the enzyme titin kinase, which acts as a force sensor in muscle tissue. By transforming the single-molecule force spectroscopy data into contour length space, barrier position histograms were created that allow the precise determination of the energy barriers up to a few amino acids. For the investigation and creation of functional networks on the single-molecule level, new approaches are needed. We, therefore, developed the SMCP technique that allows for the precise assembly of individual units one-by-one. Besides the mechanical protocol that is recorded, the assembly can be optically monitored and even diffraction-limited patterns can be reconstructed. Using specific binding sites as a functional unit, arbitrary patterns can be created to which nanoobjects self-assemble in a second step. The combination of single enzyme investigations in SMCP assembled networks of functional units will open a wide field in the future. This new approach towards synthetic single-molecule biology may allow us to create enzymatic cascades or to mimic other biological networks at the most fundamental level.

Acknowledgments

We thank S.K. Kufer, S. Stahl, M. Strackharn, H. Gumpp and M. Gautel, H. Grubmüller and their groups for collaborations and helpful discussions and SFB 486, Nanosystems Initiative Munich (NIM) and Center for Integrated Protein Science Munich (CIPSM) for financial support.

References

1. H. Frauenfelder, S.G. Sligar, P.G. Wolynes, Science **254**(5038), 1598–1603 (1991)
2. G. Binnig, C.F. Quate, C. Gerber, Phys. Rev. Lett. **56**(ISI: A1986A54360 0013), (1986)
3. E.L. Florin, V.T. Moy, H.E. Gaub, Science **264**(5157), 415–417 (1994)
4. Y. Komori, A.H. Iwane, T. Yanagida, Nat. Struct. Mol. Biol. **14**(10), 968–973 (2007)
5. H. Noji, R. Yasuda, M. Yoshida, K. Kinosita, Nature **386**(6622), 299–302 (1997)
6. K. Svoboda, C.F. Schmidt, B.J. Schnapp, S.M. Block, Nature **365**(6448), 721–727 (1993)
7. S. Myong, I. Rasnik, C. Joo, T.M. Lohman, T. Ha, Nature **437**(7063), 1321–1325 (2005)
8. L. Edman, Z. Foldes-Papp, S. Wennmalm, R. Rigler, Chem. Phys. **247**(1), 11–22 (1999)
9. K. Hassler, P. Rigler, H. Blom, R. Rigler, J. Widengren, T. Lasser, Opt. Express **15**(9), 5366–5375 (2007)
10. P.J. Choi, L. Cai, K. Frieda, S. Xie, Science **322**(5900), 442–446 (2008)
11. A. Sarkar, S. Caamano, J.M. Fernandez, J. Biol. Chem. **280**(8), 6261–6264 (2005)
12. W.A. Linke, M. Kulke, H. Li, S. Fujita-Becker, C. Neagoe, D.J. Manstein, M. Gautel, J.M. Fernandez, J. Struct. Biol. **137**(1–2), 194–205 (2002)
13. L. Tskhovrebova, A. Houmeida, J. Trinick, J. Muscle Res. and Cell Motil. **26**(6–8), 285–289 (2005)
14. H. Li, A.F. Oberhauser, S.D. Redick, M. Carrion-Vazquez, H.P. Erickson, J.M. Fernandez, Proc. Natl. Acad. Sci. U S A. **98**(19), 10682–10686 (2001)
15. A. Fukuzawa, S. Lange, M. Holt, A. Vihola, V. Carmignac, A. Ferreiro, B. Udd, M. Gautel, J. Cell Sci. **121**(Pt 11), 1841–1851 (2008)
16. I. Agarkova, J.C. Perriard, Trends Cell Biol. **15**(9), 477–485 (2005)
17. O. Mayans, P. van der Ven, M. Wilm, A. Mues, P. Young, D. Furst, M. Wilmanns, M. Gautel, Nature **395**, 863–869 (1998)
18. E.M. Puchner, A. Alexandrovich, A.L. Kho, U. Hensen, L.V. Schafer, B. Brandmeier, F. Grater, H. Grubmuller, H.E. Gaub, M. Gautel, Proc. Natl. Acad. Sci. U S A. **105**(36), 13385–13390 (2008)
19. S. Lange, F. Xiang, A. Yakovenko, A. Vihola, P. Hackman, E. Rostkova, J. Kristensen, B. Brandmeier, G. Franzen, B. Hedberg, L.G. Gunnarsson, S.M. Hughes, S. Marchand, T. Sejersen, I. Richard, L. Edstrom, E. Ehler, B. Udd, M. Gautel, Science **308**(5728), 1599–1603 (2005)
20. F. Grater, J. Shen, H. Jiang, M. Gautel, H. Grubmuller, Biophys. J. **88**(2), 790–804 (2005)

21. T. Strick, J.F. Allemand, V. Croquette, D. Bensimon, Prog. Biophys. Mol. Biol. **74**(1–2), 115–140 (2000)
22. C. Bustamante, J.F. Marko, E.D. Siggia, S. Smith, Science **265**(5178), 1599–1600 (1994)
23. E.M. Puchner, G. Franzen, M. Gautel, H.E. Gaub, Biophys. J. **95**(1), 426–434 (2008)
24. M. Rief, M. Gautel, A. Schemmel, H.E. Gaub, Biophys. J. **75**(6), 3008–3014 (1998)
25. M. Karplus, J. Kuriyan, Proc. Natl. Acad. Sci. U S A. **102**(19), 6679–6685 (2005)
26. S.K. Kufer, E.M. Puchner, H. Gumpp, T. Liedl, H.E. Gaub, Science **319**(5863), 594–596 (2008)
27. M. Radmacher, M. Fritz, H.G. Hansma, P.K. Hansma, Science **265**(ISI: A1994PF33600034), (1994)
28. D.J. Muller, F.A. Schabert, G. Buldt, A. Engel, Biophys. J. **68**(5), 1681–1686 (1995)
29. S.Y. Park, A.K. Lytton-Jean, B. Lee, S. Weigand, G.C. Schatz, C.A. Mirkin, Nature **451**(7178), 553–556 (2008)
30. C.A. Mirkin, R.L. Letsinger, R.C. Mucic, J.J. Storhoff, Nature **382**(6592), 607–609 (1996)
31. A.P. Alivisatos, K.P. Johnsson, X. Peng, T.E. Wilson, C.J. Loweth, M.P. Bruchez, P.G. Schultz, Nature **382**(6592), 609–611 (1996)
32. M. Rief, H. Clausen-Schaumann, H.E. Gaub, Nat. Struct. Biol. **6**(4), 346–349 (1999)
33. T. Strunz, K. Oroszlan, R. Schafer, H.J. Guntherodt, Proc. Natl. Acad. Sci. U S A. **96**(20), 11277–11282 (1999)
34. J. Morfill, F. Kuhner, K. Blank, R.A. Lugmaier, J. Sedlmair, H.E. Gaub, Biophys. J. **93**(7), 2400–2409 (2007)
35. M. Tokunaga, K. Kitamura, K. Saito, A.H. Iwane, T. Yanagida, Biochem. Biophys. Res. Commun. **235**(1), 47–53 (1997)
36. D. Axelrod, N.L. Thompson, T.P. Burghardt, J. Microsc.-Oxford **129**, 19–28 (1983)
37. W.E. Moerner, J. Phys. Chem. B. **106**(5), 910–927 (2002)
38. G. Seisenberger, M.U. Ried, T. Endress, H. Buning, M. Hallek, C. Brauchle, Science **294**(5548), 1929–1932 (2001)
39. S.K. Kufer, M. Strackharn, S.W. Stahl, H. Gumpp, E.M. Puchner, H.E. Gaub, Optically monitoring the mechanical assembly of single molecules. Nat. Nanotechnol. **4**(1), 45–9 (2009)
40. P. Dedecker, J.I. Hotta, C. Flors, M. Sliwa, H. Uji-I, M.B.J. Roeffaers, R. Ando, H. Mizuno, A. Miyawaki, J. Hofkens, J. Am. Chem. Soc. **129**(51), 16132–16141 (2007)
41. S.W. Hell, J. Wichmann, Opt. Lett. **19**(11), 780–782 (1994)
42. M.P. Gordon, T. Ha, P.R. Selvin, Proc. Natl. Acad. Sci. U S A. **101**(17), 6462–6465 (2004)
43. T.D. Lacoste, X. Michalet, F. Pinaud, D.S. Chemla, A.P. Alivisatos, S. Weiss, Proc. Natl. Acad. Sci. U S A. **97**(17), 9461–9466 (2000)
44. R. Zondervan, F. Kulzer, M.A. Kol'chenko, M. Orrit, J. Phys. Chem. A. **108**(10), 1657–1665 (2004)
45. E. Lang, R. Hildner, H. Engelke, P. Osswald, F. Wurthner, J. Kohler, Chemphyschem **8**(10), 1487–1496 (2007)

46. E.M. Puchner, S.K. Kufer, M. Strackharn, S.W. Stahl, H.E. Gaub, Nano Lett. **8**(11), 3692–3695 (2008)
47. E.V. Shevchenko, M. Ringler, A. Schwemer, D.V. Talapin, T.A. Klar, A.L. Rogach, J. Feldmann, A.P. Alivisatos, J. Am. Chem. Soc. **130**(11), 3274–3275 (2008)
48. D. Nykypanchuk, M.M. Maye, D. van der Lelie, O. Gang, Nature **451**(7178), 549–552 (2008)
49. M.M. Maye, D. Nykypanchuk, D. van der Lelie, O. Gang, Small **3**(10), 1678–1682 (2007)
50. S. Mann, W. Shenton, M. Li, S. Connolly, D. Fitzmaurice, Adv. Mater. **12**(ISI:000084981300021), 147–150 (2000)
51. J. Yao, D.R. Larson, H.D. Vishwasrao, W.R. Zipfel, W.W. Webb, Proc. Natl. Acad. Sci. U S A. **102**(40), 14284–14289 (2005)

Single Cell Physiology

Pierre Neveu, Deepak Kumar Sinha, Petronella Kettunen, Sophie Vriz, Ludovic Jullien, and David Bensimon

Summary. The possibility to control at specific times and specific places the activity of biomolecules (enzymes, transcription factors, RNA, hormones, etc.) is opening up new opportunities in the study of physiological processes at the single cell level in a live organism. Most existing gene expression systems allow for tissue specific induction upon feeding the organism with exogenous inducers (e.g., tetracycline). Local genetic control has earlier been achieved by micro-injection of the relevant inducer/repressor molecule, but this is an invasive and possibly traumatic technique. In this chapter, we present the requirements for a noninvasive optical control of the activity of biomolecules and review the recent advances in this new field of research.

16.1 Introduction

Living organisms are made of cells that are capable of responding to external signals (food, hormones, neurotransmitters, morphogens, etc.) by modifying their internal state (e.g., gene expression or protein phosphorylation patterns) and subsequently their external environment (ionic concentration, pH, release of signaling molecules or enzymes). Revealing and understanding the spatio-temporal dynamics of these complex interaction networks is the subject of a field known as systems biology [1, 2]. In multicellular organisms in particular, cellular differentiation and intracellular signaling are essential for the coordinated development and behavior of the organism. While many of the actors that play a major role in these processes are known (for example, morphogens in development and a variety of kinases in signal transduction), much less is known of the quantitative rules that govern their interaction with one another and with other cellular players (such as the type of complexes, rate constants, diffusion range, strength of feedback, or feedforward loops). To investigate these interactions (a necessary step before understanding or modeling them), one needs to develop means to control or interfere spatially and temporally with these processes.

One of the crudest and oldest means to interfere with developmental networks has been to use mutants (or to create conditional mutants) for one of the molecular actors and look for the resulting response of the organism e.g., the phenotype. While this approach has been very successful in identifying the qualitative features, such as the other actors and topology of a network (for example, the genes downstream of a given regulation factor), it cannot yield quantitative information on the network, i.e., the affinities, rate and diffusion constants, or type of nonlinearities.

Moreover, individual genes may be expressed in a temporally, spatially, and tissue specific manner and may be involved in different biological networks at different times. To study those systems, tools have been developed to control gene expression at different stages of development [3]. Systems have been engineered to induce or excise genes [4]; these constructs allow gene expression patterns to be temporally controlled. However, the quick (minutes) and fine spatial (cellular) control over protein activity or gene expression patterns is still problematic. In particular, it is impossible to analyze in vivo the specific influence of a given cell on a whole tissue or study its response to controlled alterations of the behavior of neighboring cells, phenomena that are anticipated to be extremely important in embryogenesis, organ regeneration, and cancer biology. Thus, if one could quickly and locally release or activate a given morphogen, one could perturb the associated developmental network and learn about its spatio-temporal dynamics. Similarly, if one could irreversibly label a given cell in a tissue, one might monitor its progeny and thus identify the stem cells implicated in tissue regeneration or investigate the growth of a tumor from a single cell.

To address biological processes at the single cell level in a live organism, electroporation techniques have been developed [5]. They involve electroporation by injection of various molecules (e.g., DNA, RNA or morpholinos) present in a micropipette brought in the close vicinity of the targeted cell (see Fig. 16.1). The technique is invasive, and the amount of electroporated material is unknown. Moreover, because the whole tissue is displaced by the inserted micropipette, the success rate in targeting a specific cell in a live embryo is very low. It would be desirable if a noninvasive technique could be devised to control the rapid activation of a known concentration of biomolecules in a specific cell of a live organism.

In fact, such a technique exists. It implies the photoactivation of caged molecules. Various biomolecules (neurotransmitters, hormones, RNA, proteins, etc.) can be inactivated (caged) by their covalent binding to appropriate chemical groups [6–8]. Upon illumination, the bond between the biomolecule and the caging group is broken and the molecule thus activated. To locally activate a caged molecule that has reached its target, two-photon excitation is the method of choice [9]: being a nonlinear phenomenon, it is limited to the focal point of the illuminating laser beam and molecules along the optical path are not affected. Uncaging, which implies the breaking of covalent bonds, is an energetically demanding process that usually requires UV (365 nm) light

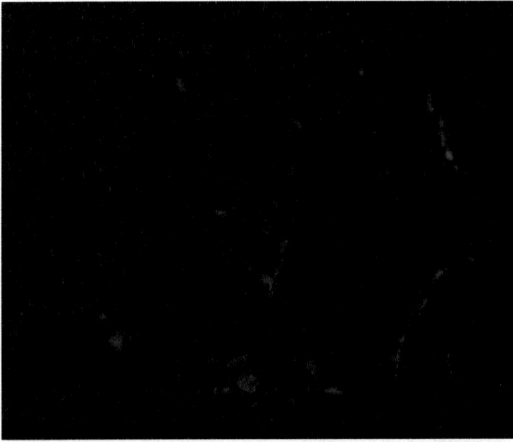

Fig. 16.1. Electroporation of red-fluorescently labeled LNAs in a single Mauthner neuron of a live zebrafish embryo. The reticulospinal neurons were labeled by retrograde transport of green-fluorescently labeled dextran injected in the tail of a 5 day old fish

to proceed. With two-photon processes, the requirement translates into near infrared (IR) pulsed laser beams (e.g., Ti-sapphire laser emission at 730 nm). These IR pulses have the added advantage over visible and UV light to be less scattered and to penetrate deeper into the tissues. To be nondetrimental to the illuminated cells, the IR beam should, however, be of the smallest possible intensity (a few mW, or less than $1\,MW\,cm^{-2}$ at the focal point [10,11]). This sets a not too severe constraint on the two-photon absorption cross section of the caging groups to be used. Finally, as far as the caging design is concerned, it will be very useful to quantify the concentration of molecules photoactivated in a single cell. With that purpose, the development of efficient caging groups that become fluorescent upon uncaging has been undertaken [12].

Among the various molecules that have been caged, one can differentiate between endogenous factors (present in the organism) and exogenous molecules (absent from it). The photo-activation of endogenous molecules such as hormones, retinoic acid, or neurotransmitters allows one to interfere with and learn about the cellular networks of the organism wherein these molecules play a role. The activation of exogenous factors allows to control the expression of genes introduced in specific transgenic animals.

The existence of caged molecules is a necessary but insufficient condition to control noninvasively processes at the single cell level in a live organism. To achieve that goal, the caged molecules have to reach their target cell(s) and therefore be sufficiently small and soluble to pass through the various physiological obstacles on their way (chorion, epithelium, cell membrane, etc.). This imposes severe restrictions on the type of molecules and caging groups that can be photo-activated in a live organism. Proteins and polynucleotides, for example, are too bulky. To reach every cell in an embryo, they have to be injected at the single cell stage. Their dilution and eventual degradation limit their usefulness to early stages in embryogenesis. As we shall see below, we have identified retinoic acid (a morphogen) and steroid-like hormones (in particular the nonendogenous cyclofen and its analogues) as good candidate molecules for caging, capable of reaching all cells in a live organism.

16.2 Caged Molecules

The caging of small molecules such as Ca^{2+}, cATP, cGTP, glutamate or serotonin has been achieved a long time ago: the first molecule to be caged and used in a cell context was cAMP[13] in 1977, see also some recent reviews [6–8]. When planning to cage biomolecules, issues of solubility, stability, cellular toxicity, uncaging cross-section, and uncaging kinetics have to be taken into consideration. Many photolabile protecting groups are presently available for caging various chemical functionalities. Despite such a large collection, caged molecules still experience restrictions to face biological situations, in particular in live organisms. In that respect, the main current development in the field of caging groups deals with improving their photophysical and photochemical properties (in particular red-shift of absorption with one-photon excitation or increase of the uncaging cross sections with one- and two-photon excitation). Another current trend concerns the reversibility of photoactivation. This feature cannot be reached with conventional caging moieties as kinetics and selectivity are not favorable to link again two fragments which originate from light-induced bond breaking. In contrast, photochromes that may exhibit major conformational changes upon illumination can be used to design molecules switching between several states of different biological activities [6–8].

Caged Ca^{2+} [14] has been the most widely used caged compound. This fact can be explained by the importance of Ca^{2+} in cellular physiology and the existence of Ca^{2+} sensitive dyes. It has led to tremendous progress in the understanding of neurotransmitter secretion.

Caged ATP is another caged compound, which led to great insights in enzymatic reactions in the past decades. Na^+-K^+ ATPase properties have been extensively studied over the years, first to determine the characteristics of Na^+ flow [15] and then the pump current properties [16]. Other significant results include the study of molecular motors among which muscle fibers relaxation kinetics [17], myosin orientation [18], and ultimately of force generation characteristics of kinesin at the single molecule level [19].

Caged cAMP has been particularly useful in neuroscience to decouple the neuron response to the natural neuronal stimulation by bypassing the activation of the pathway, especially for olfactory transduction [20, 21] or for axon guidance during the establishment of a neural map [22].

Apart from those compounds, only proof of principles have been published for most of the other caged molecules.

16.3 Optical Control in Neurophysiology

In one area, neurophysiology, much progress has been made in the study of the fast spatio-temporal dynamics of neuronal networks by using photoactivation methods to trigger the response of a single cell to an influx of calcium (using photoactivable chelators of Ca^{2+}) and neurotransmitters (using, for example, caged analogs of glutamate or serotonin) or more recently by controlling the opening/closing of light-sensitive transmembrane channels in cell cultures [23] and in live organisms [24].

Glutamate is a neurotransmitter of choice as it can stimulate almost all the different kinds of vertebrate neurons. Glutamate uncaging can mimic synaptic input, lead to the generation of action potentials, and has been used to map glutamate sensitivity or neuronal connections. Its use with two-photon excitation has led to maps of exquisite resolution [25] and has allowed the study of long term potentiation at the single spine level in a structural and mechanistic way [26–28].

More recently, light-sensitive ionic channels and pumps have been engineered [29]. Two approaches have been taken: one involves the binding of photo-sensitive chemical groups to the channels, allowing their activation/inactivation upon illumination, whereas the second directly uses the intrinsic light-sensitivity of natural channels. The first strategy [30] uses the vast conformational changes upon photoisomerization of molecules such as azobenzene: UV can trigger trans to cis isomerization, cis to trans being triggered by a longer wavelength or by thermal energy. Such a molecule can be attached to the channel ligand (and thus hide or reveal the ligand) or crosslinked to the actual protein (and thus induce conformational changes in the protein upon illumination). Light is then used at will to turn on or off the channel. The second strategy uses light-activated opsin-based channels or pumps [31]: channelrhodopsin 2 (ChR2) (or halorhodopsin (NpHR)) to depolarize (or hyperpolarize) the cell upon illumination with light at 450 nm (or 560 nm) (see Fig. 16.2). This approach is particularly promising since the channels can be expressed in transgenic animals in which they can be excited with light at the appropriate wavelength. Using two-photon excitation, it might even be possible to target specific synapses and investigate the neural response of a live animal to such very local noninvasive excitations. The coupling of these optical techniques with the existing Ca^{2+} sensitive dyes will allow the in vivo study of neural networks with unprecedented spatio-temporal resolution [32]. This technique is bound to revolutionize neurophysiology the way patch-clamp did 30 years ago. Indeed, it has already been shown in a live mouse that, upon expression of channelrhodopsin in its brain, neurons can generate spike trains the frequency

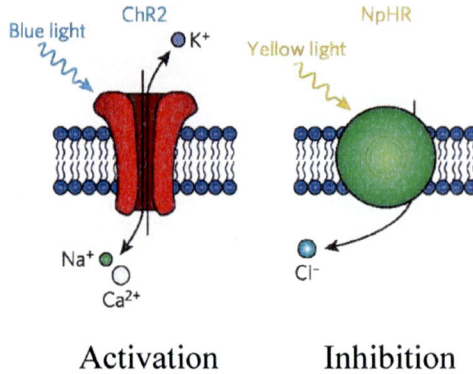

Activation   Inhibition

Fig. 16.2. Depolarization (activation) and hyperpolarization (inhibition) of a neuron by the rhodopsin-based light-activated channels ChR2 and the chloride pump NpHR. Reprinted with permission from Häusser and Smith [29]

of which is similar to the one of the excitatory light source. Moreover, the mouse can be trained to respond to a light stimulation [33]. This study shows that such a technology can be readily implemented in a freely moving mouse.

16.4 Optical Control of Gene Expression

The standard way to gain control of gene expression in a live organism is through the use of inducible tissue specific promoters. In appropriate transgenic animals, a desirable gene (for example GFP) is put under the control of a tissue specific promoter responsive to an exogenous inducer (an antibiotic such as tetracycline or doxycycline [34] or a hormone such as ecdysone [35]), the receptor of which has also been engineered in the animal. By controlling the diet of the animal (feeding it or not with the inducer molecule), the responsive gene can be turned on or off. Various inducers have been caged, e.g., ecdysone [36], doxycyclin [37], and their effect upon photo-activation in cell cultures has been demonstrated, typically by turning on the expression of GFP or β-galactosidase. So far, their photo-activation in a live organism has not been demonstrated, although it should, in principle, be possible using two-photon illumination to release the caged inducer in a single cell of the organism. Other considerations, mainly toxicity and permeability, may so far have hindered their use in a live organism.

16.5 Optical Control of RNA Expression

The optical control of RNA expression has been reported using an approach known as "statistical backbone caging" of polynucleotides [38]. In this approach, a number of sites (about 30) on a mRNA (about 1 kb long) are blocked with a photo-cleavable coumarin moiety. The mRNA was injected in a live zebrafish embryo at the single cell stage. The blocked mRNA was transcriptionally inactive but could be activated when illuminated with UV light (at 365 nm). If the RNA coded for GFP, some of the illuminated embryos would turn fluorescent, whereas if it coded for the transcription factor engrailed, some of the embryos would show developmental abnormalities (small or no eyes) after UV illumination [38] (see Fig. 16.3). However, because of the statistical nature of the caging/uncaging reactions, this approach yields results with great variability and low reproducibility. More recent work have described a method to overcome this problem by synthesizing a polynucleotide chain (e.g., a morpholino) tethered to a short complementary oligomer with a photo-cleavable linker. When injected at the one-cell stage in a zebrafish embryo, the morpholino could be released at a later stage by UV illumination. The embryo then exhibited the same phenotype as mutants lacking the gene targeted by the morpholino [39, 40].

The problem with that approach is that the caged oligo-nucleotides have to be injected at the one cell stage. As they are diluted and degraded during the normal development of the organism, their uncaging is efficient only for a few hours after fertilization.

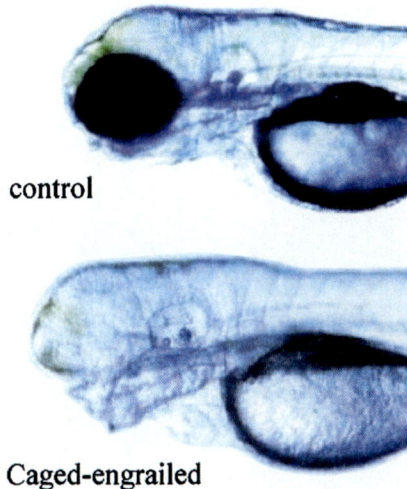

control

Caged-engrailed

Fig. 16.3. Zebrafish embryo injected at the one cell stage with a caged mRNA for the transcription factor engrailed and photo-activated 12 h post-fertilisation exhibits an absence of eye development, with permission from Ando et al. [38]

16.6 Optical Control of an Endogenous Morphogen: Retinoic Acid

The spatio-temporal control of the concentration of biomolecules is particularly attractive for the investigation of developmental networks. In these cases, cellular differentiation is determined by the interaction between gradients of morphogens and signaling molecules that directly or indirectly tune the expression of various genes. It has recently been shown that these morphogen gradients determine the fate of cells with single cell spatial resolution [41]. Therefore, controlling morphogen concentration at the single cell level may allow us to study signaling and developmental networks with unprecedented spatio-temporal control and resolution. However, the quick spatio-temporal control of these endogenous factors imposes strong constraints on their possible caging. Thus, it might be impossible to cage a secreted protein morphogen that has to undergo posttranslational modifications or get exported via a specific pathway. In this respect, retinoic acid (RA) is a target of choice for this type of investigations [42]. RA is a small lipophilic molecule whose metabolism is tightly regulated. It is synthesized from retinol (vitamin A) and can be sequestered or degraded in the cell by various proteins. As a morphogen, it plays a role in many early developmental pathways such as somitogenesis, hindbrain development, left-right symmetry, and eye and heart development [43]. Retinoic acid is easy to cage by reacting it with the alcohol of the caged moiety. The resulting ester is soluble, nontoxic, permeates the embryo, and can be uncaged to exhibit the same teratogenic effects as RA (see Fig. 16.4) [44].

To demonstrate the usefulness of cRA in the study of developmental pathways, we have used two-photon excitation to release RA in a few (4) cells of

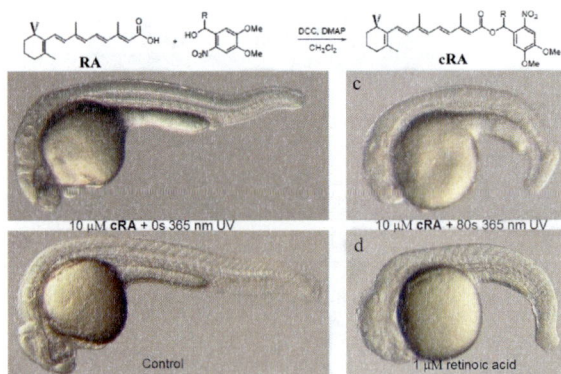

Fig. 16.4. Caged retinoic (cRA -formula on the right) and its effects on the development of a zebrafish embryo. In absence of UV illumination (**a**), it has no effect – the embryo develops normally as in the control (**b**). When illuminated 80 s with a 356 nm UV light, RA is released, and the embryo exhibits similar developmental defects (**c**) as an embryo incubated in a similar concentration of RA (**d**)

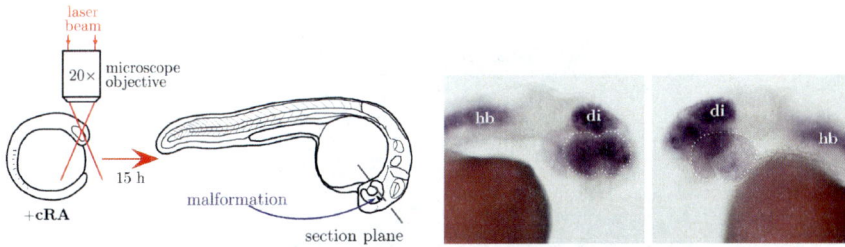

Fig. 16.5. (a) A pulsed Ti-Sa laser (750 nm) is used to uncage caged RA in a single cell of a live zebrafish embryo. The effect of this perturbation to the local concentration of RA is observed 15 h later. (b) Comparison of the right (illuminated) retina with the left (untreated) one. The dorsal part of the treated retina is deformed, and the distribution of a retina marker (pax6, *blue color*) differs markedly from that of the normal (*left*) eye (di: diencephalon, hb: hindbrain)

the dorsal part of the retina of a zebrafish embryo at the 4-14 somite stage. We chose that pathway because previous experiments with RA-soaked beads implanted in the dorsal part of the retina showed that the eye developed abnormally. In these experiments, however, the whole retina is continuously subject to a high concentration of RA, whereas in our approach, a pulse of RA is released in a few cells of the retina. Surprisingly, however, this pulse of RA is sufficient to induce a malformation of the illuminated retina (see Fig. 16.5). How this local perturbation of the RA morphogen gradient is transmitted to the other parts of the retina is an interesting issue that can now be addressed in greater detail (using RA-responsive GFP embryos and quantitative RT-PCR). Beyond the study of RA in eye development, caged RA could also be used to investigate the role of RA in somitogenesis and hindbrain development. During somitogenesis, it has been proposed that successive somites are being generated by coupling the signal of a putative "somitogenesis clock" to a bi-stable switch formed by the interaction between gradients of RA and a growth factor (Fgf8) [45]. If this mechanism is correct, one should be able to generate extra-somites by activating RA at appropriate posterior regions of an embryo undergoing somitogenesis.

16.7 Optical Control of Protein Activity

The direct caging of peptides and proteins has been achieved by many different groups [7], but rarely at the single cell level in a live organism [46]. These caged proteins have residual activity when caged, incomplete recovery of activity after uncaging, and most importantly, they must be micro-injected into the cells.

Instead of adopting this approach, we decided to control the activity of a protein through its fusion to a steroid hormone receptor ligand binding domain. In absence of its ligand, the receptor forms a cytoplasmic complex

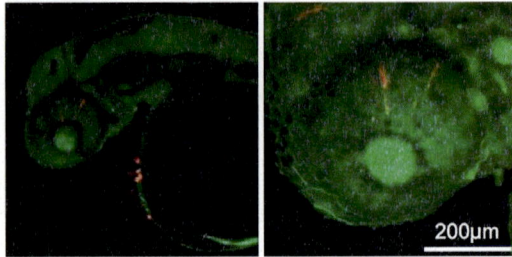

Fig. 16.6. Single cell activation of a Cre-recombinase fused with the ERT receptor upon two-photon release of its caged ligand. The transgenic embryo expresses a GFP gene flanked by two loxP sites. When these are excised, a dsRed gene is expressed in the target cell and its descendents, the *red* cells, in the eye of this embryo shown at two different magnifications

with the heat shock protein hsp90, which inactivates the fused protein [47]. Steroid hormones are small lipophilic molecules that can diffuse through the epithelial layer that surrounds the embryo. As a result, activation of the protein can be achieved by incubation of the embryos in a standard culture medium containing the hormone of choice. This system has been used to induce the activity of a large number of proteins (transcription factors such as engrailed, otx2, Gal4, p53, kinases such as raf-1 and Cre and Flp recombinases). In particular, it has gained wide acceptance as a means to irreversibly induce the expression of a gene using a Cre-recombinase fused to a mutated steroid receptor specific for tamoxifen (ERT) [48]. In the presence of the ligand, an upstream segment flanked by loxP sites is excised, allowing expression of the gene of interest. To target that approach to a single cell, we have developed a caged analog of tamoxifen (caged cyclofen) that does not isomerise upon illumination like tamoxifen and is as active when uncaged. By using two-photon illumination, we can uncage this compound in a subcellular volume of a live zebrafish embryo. As shown in Fig. 16.6, the fusion of a Cre-recombinase with this receptor performs as expected. It can be released by two-photon illumination in a few cells of a zebrafish embryo, wherein it removes a GFP gene flanked by two loxP sites, thereby turning on a RFP gene in the target cell(s) and its descendents. This method could be very useful for cell lineage by labeling cells in an embryo and following their descendents. It might be particularly interesting to investigate tumor growth and identify the stem cells in a regenerating tissue.

In conclusion, we hope to have convinced our reader that the optical control of the activity of biomolecules (neurotransmitters, ion channels, enzymes, transcription factors, morphogens, mRNA, etc.) opens up a new and exciting field of research, here called single cell physiology, in which the investigation of physiological networks (implicated in processes such as memory, development, cancer growth, etc.) can be performed at the most relevant level for an organism: the single cell.

Acknowledgements

We thank Frédéric Rosa and Shuo Lin for access to their fish facilities, and Laure Bally-Cuif and Christof Leucht for the gift of the zebrafish line used for the Cre reporter experiments. This work has been supported by Association pour la Recherche sur le Cancer. PN research is supported in part by the National Science Foundation under Grant No. PHY05-51164. DKS acknowledges support from the NABI CNRS-Weizmann Institute program and DB the partial support of a PUF ENS-UCLA grant.

References

1. J. Gerhart, M. Kirschner, *Cells, Embryos, and Evolution* (Blackwell Science, London, 1997)
2. H. Kitano, Science **295**, 1662–1664 (2002)
3. R.H. Singer, D.S. Lawrence, B. Ovryn, J. Condeelis, J. Biomed. Opt. **10**, 051406 (2005)
4. A.D. Ryding, M.G. Sharp, J.J. Mullins, J. Endocrinol. **171**, 1–14 (2001)
5. K. Haas, W.C. Sin, A. Javaherian, Z. Li, H.T. Cline, Neuron **29**, 583–591 (2001)
6. M. Goeldner, R. Givens (eds.), *Dynamic studies in biology. Phototriggers, photoswitches and caged biomolecules* (Wiley, Weinheim, 2005)
7. G. Mayer, A. Heckel, Angew. Chem. Int. Ed. Engl. **45**, 4900–4921 (2006)
8. G.C. Ellis-Davies, Nat. Meth. **4**, 619–628 (2007)
9. W. Denk, J.H. Strickler, W.W. Webb, Science **248**, 73–76 (1990)
10. P.S. Tsai, B. Friedman, A.I. Ifarraguerri, B.D. Thompson, V. Lev-Ram, C.B. Schaffer, Q. Xiong, R.Y. Tsien, J.A. Squier, D. Kleinfeld, Neuron **39**, 27–41 (2003)
11. E.J. Peterman, F. Gittes, C.F. Schmidt, Biophys. J. **84**, 1308–1316 (2003)
12. N. Gagey, P. Neveu, L. Jullien, Angew. Chem. Int. Ed. Engl. **46**, 2467–2469 (2007)
13. J. Engels, E.J. Schlaeger, J. Med. Chem. **20**, 907–911 (1977)
14. G.C. Ellis-Davies, Chem. Rev. **108**, 1603–1613 (2008)
15. J.H. Kaplan, R.J. Hollis, Nature **288**, 587–589 (1980)
16. K. Fendler, E. Grell, M. Haubs, E. Bamberg, EMBO J. **4**, 3079–3085 (1985)
17. Y.E. Goldman, M.G. Hibberd, J.A. McCray, D.R. Trentham, Nature **300**, 701–705 (1982)
18. P.G. Fajer, E.A. Fajer, D.D. Thomas, Proc. Natl. Acad. Sci. U S A **87**, 5538–5542 (1990)
19. H. Higuchi, E. Muto, Y. Inoue, T. Yanagida, Proc. Natl. Acad. Sci. U S A **94**, 4395–4400 (1997)
20. G. Lowe, G.H. Gold, Proc. Natl. Acad. Sci. U S A **92**, 7864–7868 (1995)
21. T. Kurahashi, A. Menini, Nature **385**, 725–729 (1997)
22. X. Nicol, S. Voyatzis, A. Muzerelle, N. Narboux-Nême, T.C. Südhof, R. Miles, P. Gaspar, Nat. Neurosci. **10**, 340–347 (2007)
23. M. Volgraf, P. Gorostiza, R. Numano, R.H. Kramer, E.Y. Isacoff, D. Trauner, Nat. Chem. Biol. **2**, 47–52 (2006)

24. F. Zhang, L.P. Wang, M. Brauner, J.F. Liewald, K. Kay, N. Watzke, P.G. Wood, E. Bamberg, G. Nagel, A. Gottschalk, K. Deisseroth, Nature **446**, 633–639 (2007)
25. M. Matsuzaki, G.C. Ellis-Davies, T. Nemoto, Y. Miyashita, M. Iino, H. Kasai Nat. Neurosci. **4**, 1086–1092 (2001)
26. M. Matsuzaki, N. Honkura, G.C. Ellis-Davies, H. Kasai, Nature **429**, 761–766 (2004)
27. C.D. Harvey, K. Svoboda, Nature **450**, 1195–1200 (2007)
28. J. Tanaka, Y. Horiike, M. Matsuzaki, T. Miyazaki, G.C. Ellis-Davies, H. Kasai, Science **319**, 1683–1687 (2008)
29. M. Häusser, S.L. Smith, Nature **446**, 617–619 (2007)
30. P. Gorostiza, E.Y. Isacoff, Science **322**, 395–399 (2008)
31. F. Zhang, A.M. Aravanis, A. Adamantidis, L. de Lecea, K. Deisseroth, Nat. Rev. Neurosci. **8**, 577–581 (2007)
32. S.H. Chalasani, N. Chronis, M. Tsunozaki, J.M. Gray, D. Ramot, M.B. Goodman, C.I. Bargmann, Nature **450**, 63–71 (2007)
33. D. Huber, L. Petreanu, N. Ghitani, S. Ranade, T. Hromádka, Z. Mainen, K. Svoboda, Nature **451**, 61–64 (2008)
34. M. Gossen, H. Bujard, Proc. Natl. Acad. Sci. USA **89**, 5547–5551 (1992)
35. D. No, T.P. Yao, R.M. Evans, Proc. Natl. Acad. Sci. USA **93**, 3346–3351 (1996)
36. W. Lin, C. Albanese, R.G. Pestell, D.S. Lawrence, Chem. Biol. **9**, 1347–1353 (2002)
37. S.B. Cambridge, D. Geissler, S. Keller, B. Curten, Angew. Chem. Int. Ed. Engl. **45**, 2229–2231 (2006)
38. H. Ando, T. Furuta, R.Y. Tsien, H. Okamoto, Nat. Genet. **28**, 317–325 (2001)
39. I.A. Shestopalov, S. Sinha, J.K. Chen, Nat. Chem. Biol. **3**, 650–651 (2007)
40. X.J. Tang, S. Maegawa, E.S. Weinberg, I.J. Dmochowski, J. Am. Chem. Soc. **129**, 11000–11001 (2007)
41. T. Gregor, D.W. Tank, E.F. Wieschaus, W. Bialek, Cell **130**, 153–164 (2007)
42. G.C. Vilhais-Neto, O. Pourquié, Curr. Biol. **18**, R191–R192 (2008)
43. G. Duester, Cell **134**, 921–931 (2008)
44. P. Neveu, I. Aujard, C. Benbrahim, T. Le Saux, J.F. Allemand, S. Vriz, D. Bensimon, L. Jullien, Angew. Chem. Int. Ed. Engl. **47**, 3744–3746 (2008)
45. A. Goldbeter, D. Gonze, O. Pourquié, Dev. Dyn. **236**, 1495–1508 (2007)
46. S.B. Cambridge, R.L. Davis, J.S. Minden, Science **277**, 825–828 (1997)
47. W.B. Pratt, Y. Morishima, Y. Osawa, J. Biol. Chem. **283**, 22885–22889 (2008)
48. D. Metzger, J. Clifford, H. Chiba, P. Chambon, Proc. Natl. Acad. Sci. USA **92**, 6991–6995 (1995)

Force-Clamp Spectroscopy of Single Proteins

Julio M Fernandez, Sergi Garcia-Manyes, and Lorna Dougan

Summary. Force-clamp AFM, with its remarkable ability to manipulate short recombinant proteins, has become a useful probe of protein dynamics, allowing us to sense conformational changes down to the sub-Ångström scale. The single protein data is providing a new view that will help guide the development of theories on enzyme catalysis, the statistical dynamics of folding, and *ab initio* studies of a chemical reaction while placed under a stretching force; of common occurrence in nature.

Our objective is to understand how mechanical forces, over the full biological spectrum, affect the dynamics and chemistry of proteins [1]. Using molecular biology techniques, we engineer tandem modular proteins that consist of identical repeats of a protein of interest (Fig. 17.1) [2]. When such polyproteins are extended by an AFM (Atomic Force Microscopy), their force properties are unique mechanical fingerprints that unambiguously distinguish them from the more frequent nonspecific events that plague single-molecule studies [3]. Our initial experiments were performed by extending polyproteins at constant velocity, resulting in the now familiar sawtooth pattern traces of unfolding. We made observations that could never have been obtained from the bulk protein biochemistry. For example, while under a stretching force, proteins unfold following mechanical hierarchies, thereby explaining the mechanical architecture of large elastic proteins such as titin and fibronectin [4–6]. We also discovered that the mechanical stability of a protein depends on the direction of the applied force, revealing that proteins have intrinsic "Achilles' heels" for unfolding, which may play a role in their degradation [7]. However, in our early experiments, the variables of force, length, and loading rate changed simultaneously over wide ranges, and thus yielded only qualitative results. We solved these problems by introducing the force-clamp AFM [8]. With this approach, the length of an extending polyprotein is measured while the pulling force is actively kept constant by negative feedback control. The force-clamp technique combined with polyprotein engineering has become a powerful approach to study proteins. We have investigated the force dependency of protein

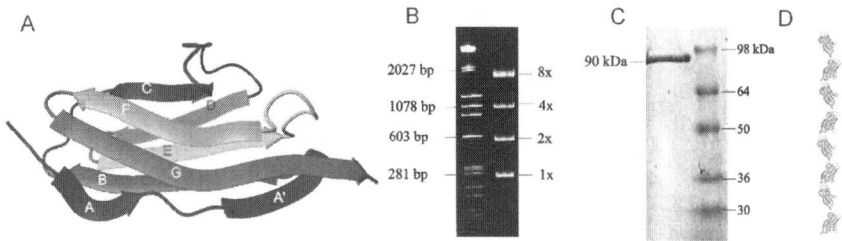

Fig. 17.1. Engineering an I27₈ polyprotein. (**a**) Diagram of the β-sandwich structure of I27 with each of the β-strands shown in different colors. (**b**) Agarose gel stained with ethidium bromide showing the size of the I27 multiples (right lane). (**c**). Coomassie blue staining of the purified I27₈ polyprotein (∼90 kDa) separated by SDS-PAGE. (**d**) Schematic representation of the resulting I27₈ polyprotein

folding [9,10], unfolding [11,12], and of chemical reactions [13–15]. From the force dependence, we extract features of the transition state of these reactions that reveal details of the underlying molecular mechanisms. What follows is a brief recapitulation of the most significant features of this technique and its promise in uncovering the physical mechanisms underlying protein dynamics.

17.1 The Importance of Polyprotein Engineering

In the most typical single-molecule AFM experiment, a modular protein adsorbed between a substrate and the cantilever tip is extended vertically at a constant rate while the resulting force is measured [4]. An early challenge was that the native modular proteins combined a wide diversity of protein modules of very different mechanical properties. Hence, it was difficult to uniquely assign each response to a specific module in the protein. We solved this problem by ligating multiple copies of the cDNA coding for a single protein domain and expressing the resultant gene in bacteria. The first such engineered "polyproteins" consisted of 8 and 12 tandem repeats of the 89 amino acid β-sandwich protein I27, cloned from human cardiac titin (Fig. 17.1) [2,16].

In addition to providing an essential mechanical fingerprint, polyproteins can be studied with the certainty that all the measured mechanical parameters result from the repeating protein module. Owing to the ease of their production using molecular biology techniques, good expression levels and high pick-up rate by AFM cantilevers, I27 polyproteins became a convenient workhorse in the laboratory. Polyprotein engineering was the enabling technology that permitted us to unambiguously quantify the properties of single proteins.

17.2 Force-Clamp Spectroscopy of Polyproteins

When a polyprotein is picked up and stretched at a constant rate by an AFM, the resulting force-extension curve has the characteristic appearance of a sawtooth pattern [2]. The sawtooth pattern results from the sequential unfolding of the protein modules, as the protein is elongated. The peak force reached before an unfolding event measures the mechanical stability of the protein module, while the spacing between peaks is a measure of the increased contour length of the protein as it unfolds. However, executing a force-extension cycle, changes the force and the loading rate on the molecule over wide ranges on short time scales. Given that the most interesting observables of a single-molecule experiment are force-dependent (see below), analysis of sawtooth pattern data could only be done through the use of highly simplified Monte Carlo models [2]. This approach was sufficient only to obtain rough estimates of the force-dependent parameters of a polyprotein. In order to solve these problems, we developed the "force-clamp" mode of stretching single proteins [8]. In the force-clamp mode, a polyprotein is stretched under a PID feedback mechanism that adjusts the length of the protein in order to keep it at a constant stretching force. In this case, unfolding events are observed as step increases in the length of the polyprotein (Fig. 17.2). The force-clamp method directly measures the force dependency of a reaction and permits complex pulse patterns of force to be applied to a protein.

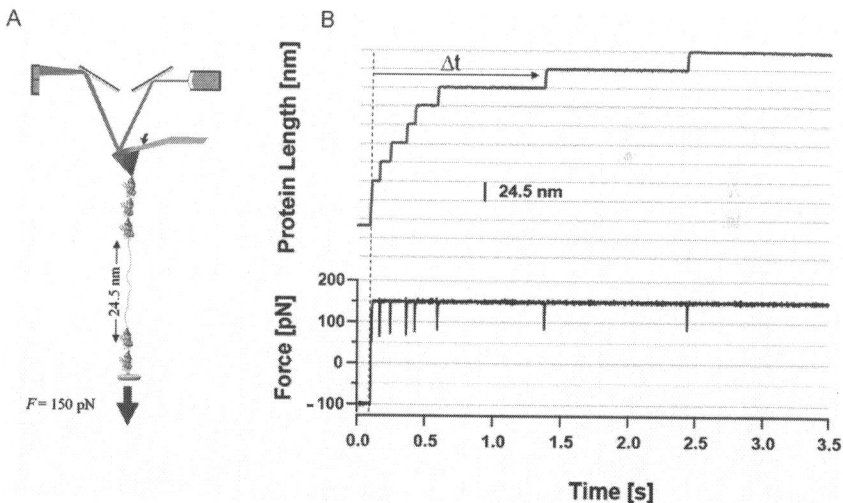

Fig. 17.2. (a) Cartoon of the unfolding of an I27 polyprotein at a constant stretching force of 150 pN. (b) A typical length versus time recordings (top trace) obtained by stretching an I27$_8$ polyprotein at a constant force of 150 pN (bottom trace). The polyprotein elongates in steps of 24.5 nm, marking the unfolding of individual proteins in the chain. The dwell time (Δt) of each unfolding event can be accurately measured

A

B

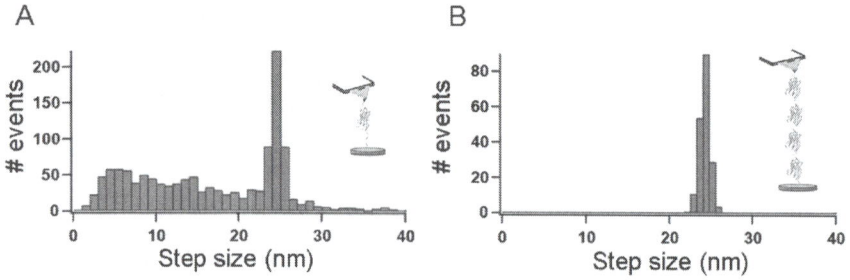

Fig. 17.3. Step size histograms obtained from force-clamp traces measured from (a) single I27 proteins where all observed steps are considered or from (b) I27$_8$ polyproteins where only staircases of three or more identical steps are considered [16]. The figure shows that the polyprotein fingerprint (b) weeds out spurious interactions (a)

The extension of a single monomer I27 protein, while possible, is less practical because the unfolding events are more difficult to distinguish from the nonspecific interactions that are always seen as the cantilever tip is pulled away from an adsorbed protein layer (Fig. 17.3) [17]. The use of polyproteins was validated by confirming that the individual repeats behaved independently [18], and that their folding and unfolding kinetics were independent of the number of repeats and indistinguishable from those of a single monomer protein [17, 19].

17.3 Force-Clamp Spectroscopy Measures the Distance to the Transition State Δx

In order to understand the effect that a mechanical force has on the reaction rate, we can make use of a simplification and assume that the mechanical work is done linearly over the reaction coordinate (length) of a simple energy barrier separating two well-defined states (Fig. 17.4). The reaction rate is then determined by the activation energy (ΔG) and the reaction length (Δx). The force-dependent rate constant is given by an Arrhenius term; $k(F) = A \exp[-(\Delta G - F\Delta x)/kT]$, where A is the attempt frequency and Δx is the distance to the transition state of the reaction, k and T are the Boltzmann's constant and the temperature [20, 21]. Determination of ΔG and the distance to the transition state, Δx, can be easily done by measuring how much the rate constant depends on the applied force. The most interesting parameter of these measurements is the distance to the transition state of the reaction Δx. This choice may seem arbitrary; however, the transition state is a very short-lived structure that limits the rate at which a reaction proceeds. From this perspective, defining the molecular/atomic structure of transition states is a most important task, essential for understanding the dynamics of a protein. Transition states are visited only briefly and is of the

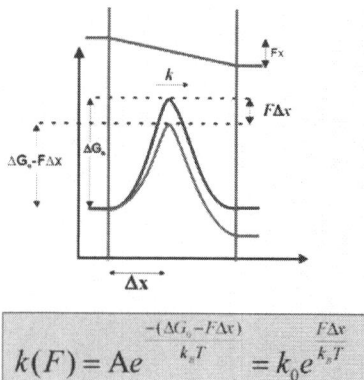

$$k(F) = Ae^{\frac{-(\Delta G_a - F\Delta x)}{k_b T}} = k_0 e^{\frac{F\Delta x}{k_b T}}$$

Fig. 17.4. Illustration of the effect of a mechanical force on a simple two-state energy landscape. The reaction is characterized by the position of the transition state, Δx, along the reaction coordinate and the height of the activation energy barrier, ΔG. A pulling force applied along the reaction coordinate accelerates the rate of the reaction, $k(F)$, exponentially by reducing the size of the activation energy barrier by an amount equal to $F\Delta x$. The attempt frequency, A, is necessary for determining ΔG

order of a single molecular vibration (e.g. $10^{-12}-10^{-15}$ s), so far accessible only through the use of femtosecond-laser spectroscopy in small molecules [22]. The transient nature of the transition state structure has made it difficult to study, particularly in large organic molecules such as proteins. An applied force does mechanical work on the transition state structure, regardless of how ephemeral. Thus, obtaining a structural/atomic perspective on the measured values of Δx provides a new tool to probe the transition state structure in protein reactions.

17.4 Force Dependency of Protein Unfolding

An example of such an experiment is shown in Fig. 17.5a, demonstrating the procedures followed to obtain the force dependency of the mechanical unfolding of the I27$_8$ polyprotein. First, we obtain a set of unfolding traces from different molecules stretched at a constant force (Fig. 17.5a). An ensemble average of a few such traces ($n = 5-20$) can be fitted with a single exponential, allowing for a direct measurement of the unfolding rate at a given force (Fig. 17.5b, [11]). An estimate of the standard error of the unfolding rate is obtained by applying a bootstrapping technique to the set of single-molecule traces [13, 14, 23]. These procedures can be repeated over a range of forces, obtaining the force dependency of the unfolding reaction (Fig. 17.5c). In the case of the I27 polyprotein, we obtain an exponential force dependency for the unfolding rate. These data can then be fitted with a simple Arrhenius term

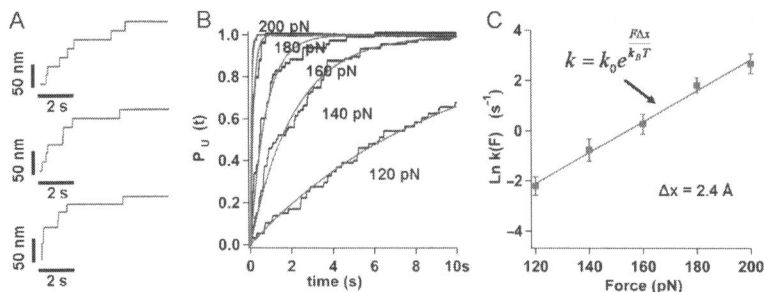

Fig. 17.5. Measurement of the force dependency of unfolding. (a) Traces of $I27_8$ unfolding at 160 pN. (b) Normalized ensemble averages of unfolding traces (>20 per force) fitted by single exponentials, measure the unfolding rate constant at each pulling force, $k(F) = 1/\tau(F)$, where $\tau(F)$ is the time constant of the exponential fits to the averaged unfolding traces. (c) Logarithmic plot of the unfolding rate constant (symbols) as a function of the pulling force. An Arrhenius fit (solid line) measures $\Delta x = 2.5$ Å

(Fig. 17.5c, solid line) to determine the values of ΔG and Δx. These force-clamp experiments show that the distance to the transition state of unfolding of the I27 protein, $\Delta x_u = 2.4$ Å, which is the size of a water molecule [24]. In addition, the value of the unfolding rate extrapolated to zero force measures the size of the activation energy barrier ΔG. However, ΔG obtained from the Arrhenius fits depends on the value used for the attempt frequency A (Fig. 17.4), which needs to be measured independently for each reaction [24]. By contrast, the value of Δx is measured from the slope of the Arrhenius plot, and thus is independent of the value of the attempt frequency A.

17.5 Molecular Interpretation of Δx in Protein Unfolding

Steered Molecular Dynamics (SMD) simulations of forced unfolding of the I27 protein [25, 26] suggested that resistance to mechanical unfolding originates from a localized patch of hydrogen bonds between the A′ and G β-strands of the protein (Fig. 17.1a). The A′ and G strands must slide past one another for unfolding to occur. Since the hydrogen bonds are perpendicular to the axis of extension, they must rupture simultaneously to allow relative movement of the two termini [25]. Thus, these bonds were singled out to be the origin of the main barrier to complete unfolding [25]. This view was experimentally validated by force spectroscopy experiments on I27 proteins, with mutations in the A′ and G β-strands of the protein [27, 28]. The SMD simulations also showed that water molecules participated in the rupture of the backbone H bonds during the forced extension of the protein [26]. Although the transition state structure could not be determined from such simulations, the

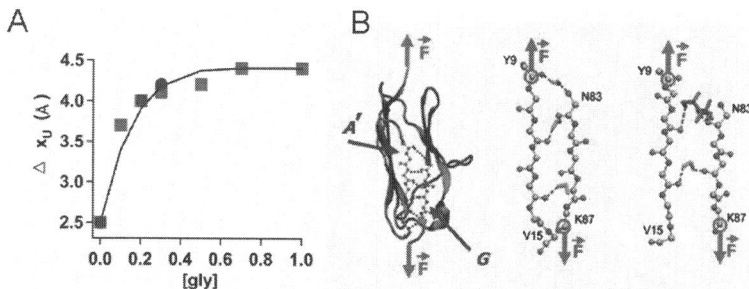

Fig. 17.6. Solvent molecules bridge the A′–G β-strands at the unfolding transition state. (**a**) Distance to the unfolding transition state Δx_U measured from I27$_8$ polyproteins under a range of aqueous glycerol solutions of varying volume fraction *[gly]*. (**b**) Cartoon of the I27 protein highlighting the location of β-strands A′ and G and the direction of the pulling forces. In water, up to six water molecules bridge the protein backbone (three are shown). If a single glycerol molecule replaces a water molecule, the backbone is forced to a larger separation at the transition state, increasing the value of Δx_U. This model correctly predicts the increase in the value of Δx_U with *[gly]* (thick line in A)

putative role played by the water molecules was highly suggestive of their part in forming the unfolding transition state structure.

We recently tested this view by using solvent substitution. In these experiments, water was gradually replaced by the larger molecule glycerol (2.5 Å vs. 5.6 Å, respectively) [29]. At each glycerol concentration, the force dependency of unfolding of I27$_8$ was measured, yielding values of Δx that grew rapidly with the glycerol concentration, reaching a maximum value of $\Delta x = 4.4$ Å, clearly showing that the value of Δx followed the size of the solvent molecule (Fig. 17.6a) [24]. We interpreted these results as an indication that at the transition state, the solvent molecules bridge the key A′ and G β-strands of the I27 protein (Fig. 17.6b) [24]. While this view needs further validation, it illustrates the utility of force-clamp spectroscopy in capturing molecular level features of the elusive transition structures of a protein. Furthermore, this technique has provided the opportunity to explore the role of solvent molecules in the mechanical unfolding transition state of a protein.

17.6 The Force Dependency of Chemical Reactions

Perhaps the most striking use of force-clamp spectroscopy so far has been in the study of the effect of force on a chemical reaction [13–15]. While mortars and pestles have been used for thousands of years to catalyze chemical reactions, it had never been possible to examine a molecule undergoing a chemical reaction when placed under a calibrated and vectorially defined force. This has now been achieved by combining protein engineering with force-clamp spectroscopy. Our basic strategy is shown in Fig. 17.7.

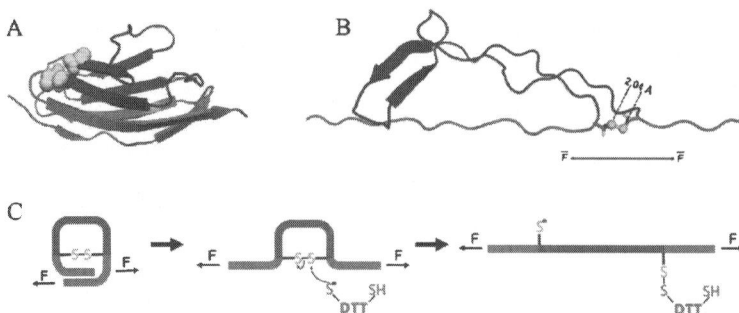

Fig. 17.7. Engineered protein designed for mechanochemistry studies. (a) A pair of cysteine residues introduced into the I27 protein (positions 32 and 75; sulfur atoms as spheres) spontaneously form a buried disulfide bond. (b). In response to an unfolding force, the protein extends right up to the disulfide bond. Unfolding exposes the disulfide bond to the solution. (c) Then, a nucleophile such as DTT can initiate a S_N2 reaction, leading to the reduction of the disulfide bond and the concomitant extension of the amino acids that were trapped behind the disulfide bond. This sequence of events unambiguously identifies individual disulfide bond reduction events, allowing for the study of a pulling force on a S_N2 chemical reaction

We engineered a polyprotein with repeats of the I27 module which were mutated to incorporate two cysteine residues (G32C, A75C; Fig. 17.7a, spheres). The two cysteine residues spontaneously form a stable disulfide bond that is buried in the β-sandwich fold of the I27 protein [30]. We call this polyprotein $(I27_{S-S})_8$. The disulfide bond mechanically separates the I27 protein into two parts: a first region of unsequestered amino acids that readily unfold and extend under a stretching force. The second region marks 43 amino acids which are trapped behind the disulfide bond (Fig. 17.7b) and can only be extended if the disulfide bond is reduced by a nucleophile such as DTT, TCEP, or the enzyme thioredoxin (Fig. 17.7c). We use the force-clamp AFM to extend single $(I27_{S-S})_8$ polyproteins. The constant force causes individual I27 proteins in the chain to unfold, resulting in stepwise increases in length of the molecule following each unfolding event. However, this unfolding is limited to the "unsequestered" residues by the presence of the intact disulfide bond, which cannot be ruptured by force alone. After unfolding, the stretching force is directly applied to the now solvent-exposed disulfide bond, and if a reducing agent is present in the bathing solution, the bond can be chemically reduced (Fig. 17.7c).

In order to study the kinetics of disulfide bond reduction as a function of the pulling force, we utilize a double pulse protocol in force clamp. With a first pulse, we unfold the unsequestered region of the $(I27_{S-S})_8$ modules in the polyprotein, exposing the disulfide bonds to the solution. With a second (test) pulse, we track the rate of reduction of the exposed disulfides at various pulling forces in the presence of various reducing agents.

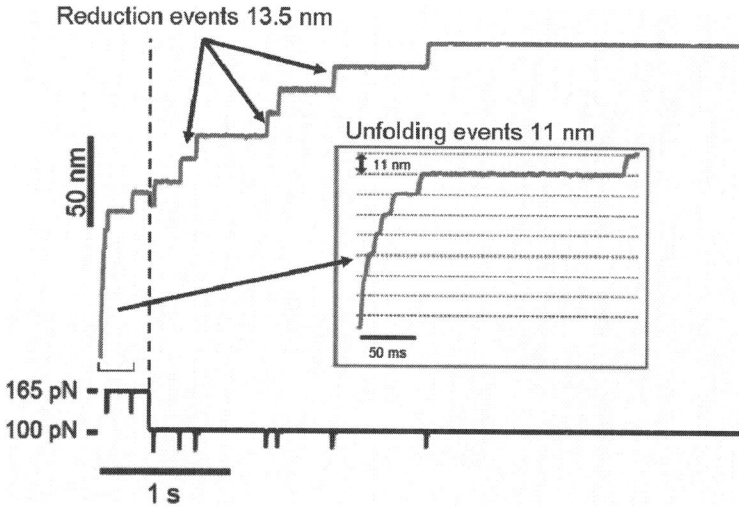

Fig. 17.8. Double pulse protocol designed to measure the force dependency of the reduction of the disulfide bonds of a $(I27_{S-S})_8$ polyprotein. The first pulse triggers unfolding, causing a series of 11 nm steps. In the presence of 8 µM *E. Coli* Thioredoxin, unfolding is followed by a series of 13.5 nm steps that mark the reduction events. A test pulse (shown to 100 pN) is used to probe the force dependency of the reaction. In the absence of thioredoxin, the 13.5 nm steps are never observed

Figure 17.8 demonstrates the use of the double pulse protocol using *E. Coli* Thioredoxin as the reducing agent. The first pulse to 165 pN elicits a rapid series of steps of ~11 nm, marking the unfolding and extension of the unsequestered residues. After exposing the disulfides by unfolding, a subsequent test pulse probes the rate of reduction at 100 pN. In the absence of Trx, no steps are observed during the test pulse. However, in the presence of Trx (8 µM), a series of ~13.5 nm steps follow the unfolding staircase. Each 13.5 nm step is due to the extension of the trapped residues, unambiguously marking the reduction of each module in the $(I27_{S-S})_8$ polyprotein.

The size of the step increases in length observed during these force-clamp experiments corresponds to the number of amino acids released, serving as a precise fingerprint to identify the reduction events. We measure the rate of disulfide bond reduction at a given force by fitting a single exponential to an ensemble average of 10–30 traces obtained during the test pulse (Fig. 17.8). A single exponential is fit to each averaged trace; we calculate the rate constant of reduction as $r = 1/\tau_r$, where τ_r is the time constant measured from the exponential fits, as was done before to measure the rate of unfolding (Fig. 17.5).

Figure 17.9 shows a plot of the rate of reduction, r, as a function of force for experiments done in the presence of human thioredoxin (diamonds), *E. Coli* thioredoxin (circles), and the reducing agent DTT (triangles). The

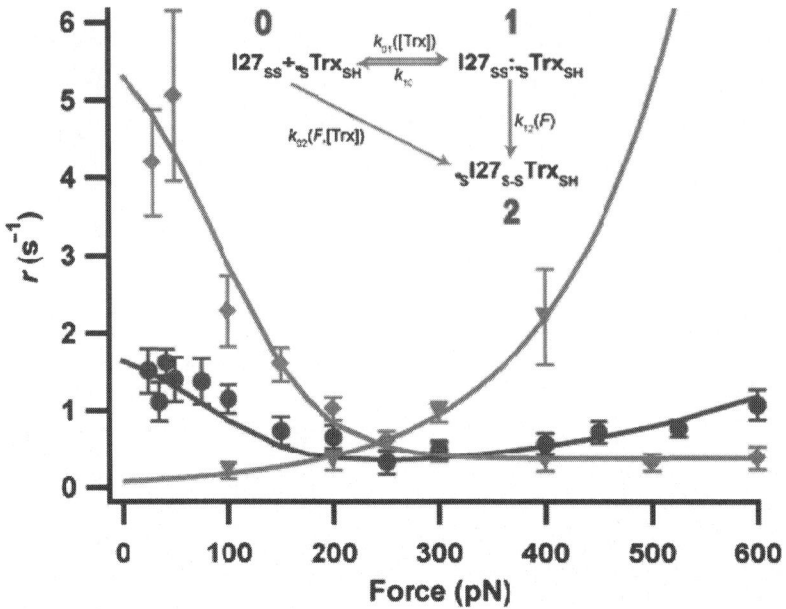

Fig. 17.9. The rate of reduction of disulfide bonds is sensitive to the force applied and the type of reducing agent. The figure shows the measured rate of reduction as a function of the pulling force for human thioredoxin (hTrx, red diamonds), *E. Coli* thioredoxin (ECTrx, circles) and the reducing agent DTT (triangles). For DTT, we observe an exponential increase in the reduction rate with force. The solid lines are fits to the data for thiredoxin with the model shown in the inset. The model shows two distinct mechanisms of reduction. The first pathway (I) proceeds through a Michaelis complex (0–1) and then through a force-dependent reduction step (1–2), with a rate that decreases with the applied force. A second pathway (II) proceeds directly to the reduced state (0–2) without necessitating a Michaelis complex. Pathway II shows a rate that increases with force. ECTrx has a biphasic rate dependency that requires both pathways whereas the more active and specific hTrx shows only pathway I. In contrast, reduction by DTT shows a simple exponential increase in rate with force

figure shows striking differences in how force affects these different types of chemical reactions. In the case of the small molecule DTT, we observe an exponential *increase* in the reduction rate with force.

17.7 Molecular Interpretation of Δx in Chemical Reactions

Through a simple Arrhenius fit to these data, we found that this force-dependent increase in the reduction rate can be explained by an elongation of the disulfide bond by $\Delta x = 0.34$ Å, at the transition state of the S_N2

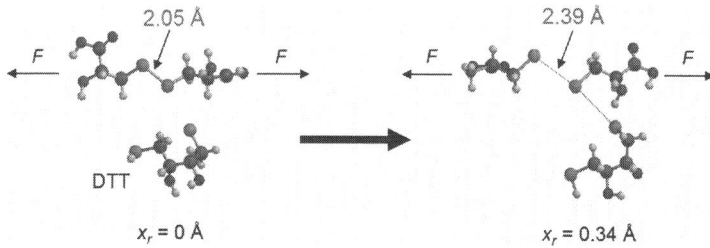

Fig. 17.10. (a) Before the substrate disulfide encounters the nucleophile the S-S bond is 2.09 Å long. (b) When a nucleophile attacks the S-S bond the disulfide bond elongates (shown elongating by 0.34 Å), as shown by quantum mechanical calculations

chemical reaction (Fig. 17.10) [13]. Other nucleophiles such as TCEP, showed a larger bond elongation of $\Delta x = 0.46$ Å at the transition state of the reaction, in agreement with quantum mechanical calculations of the transition state structures [15].

In contrast to DTT, the rate of reduction of the disulfides by the human enzyme thioredoxin (hTrx) is rapidly *inhibited* by force on the substrate disulfide bonds. We used a simple Michaelis-Menten type model with force-dependent rate constants to explain our data (Fig. 17.9, inset) [14]. Fits of the model to the data showed that we could explain the negative force dependency of the reduction rate by a value of $\Delta x_{12} = -0.76$ Å at the transition state of the Trx catalyzed reduction [14]. A simple framework for the molecular mechanism underlying the force dependency of the reaction was obtained by inspecting the structure of human Trx in complex with a substrate peptide (Fig. 17.11; PDB: 1MDI). In the structure, a peptide-binding groove is identified on the surface of Trx in the vicinity of the catalytic Cys32. It is known that the disulfide bond reduction proceeds via a $S_N 2$ mechanism where the three participating sulfur atoms form a $\sim 180°$ angle (Fig. 17.10b). Given that the disulfide bond in 1MDI forms an angle with respect to the axis of the peptide binding groove (Fig. 17.11a), it is apparent that the target disulfide bond must rotate with respect to the pulling axis in order to acquire the correct $S_N 2$ geometry (Fig. 17.11b). However, bond rotation causes a contraction of the target polypeptide, against the pulling force, by an amount of $\Delta x_{12} = -0.77$ Å for this structure [14]. In this hypothesis, bond rotation will be increasingly discouraged by the pulling force, turning off the enzyme. Although the structure of the unfolded I27$_{\text{S-S}}$ peptide bound to human Trx is unavailable, it is reasonable to assume that a similar mechanism is responsible for the force dependency of the reaction. In contrast to human Trx, the *E. Coli* form of the enzyme shows a second chemical pathway that becomes apparent only at high forces. This pathway has a force dependency that is similar to that of DTT and requires a value of $\Delta x_{02} = 0.22$ Å to be explained. We do not yet have a molecular interpretation for this catalytic mode; however, it

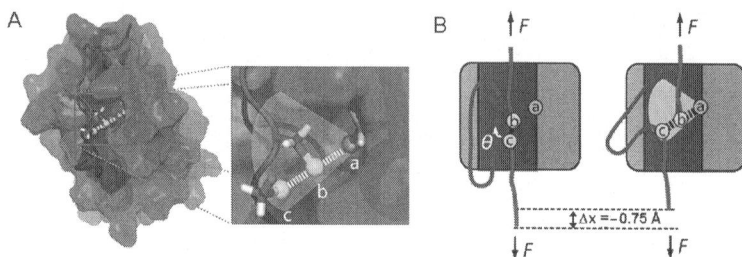

Fig. 17.11. Hypothetical model explaining the sensitivity of Trx catalysis to a pulling force applied to the substrate disulfide bond. (**a**) Structure (PDB: 1MDI) of human Trx disulfide-bonded to a peptide from NF-κB (ribbon). Inset (sulfur atoms as spheres) shows the nucleophilic attack of Trx cysteine (a) on a disulfide bond (b, c) within the peptide binding groove (darker region). (**b**) Cartoon representation of (**a**). Left; after initial unfolding of I27$_{S-S}$, the disulfide bond (b, c) is aligned within ~20° of the pulling axis. However, the disulfide bond is not in the correct orientation for S_N2 attack by the thiolate (a) of Trx. Right; the disulfide bond must rotate by an additional angle θ to acquire the correct geometry for nucleophilic attack. This rotation causes a shortening of the substrate polypeptide by an amount Δx_{12}. This rotation is opposed and eventually blocked by the pulling force. We hypothesize that this is the origin of the negative force dependency of disulfide bond reduction by this enzyme

may represent mostly prokaryotic chemistry from the early evolution of the chemical mechanisms of this enzyme.

The sub-Ångstrom resolution of the transition state dynamics of a chemical reaction obtained using force-clamp techniques makes a novel contribution to our understanding of protein-based chemical reactions. These experiments, combined with quantum chemical predictions of transition state structure, hold promise for developing a quantitative view of enzyme catalysis and other protein-based chemical reactions, at a resolution currently unattainable by any other means.

17.8 A Statistical Dynamics View of Protein Folding

Classical biochemical assays of protein folding, conducted in bulk, have provided valuable information regarding the kinetics of unfolding and folding of a great variety of proteins placed under strong chemical denaturants or temperature jumps [31]. These experiments average out over the multitude of trajectories, such that only the most thermodynamically stable conformations stand out, which often leads to an oversimplification of the protein folding scenario as a two-state process. This view was challenged by lattice and molecular dynamics simulations which described the folding of a single protein as a diffusional process over a funnel-like energy landscape, where folding proteins never followed the same trajectory and lacked well-defined

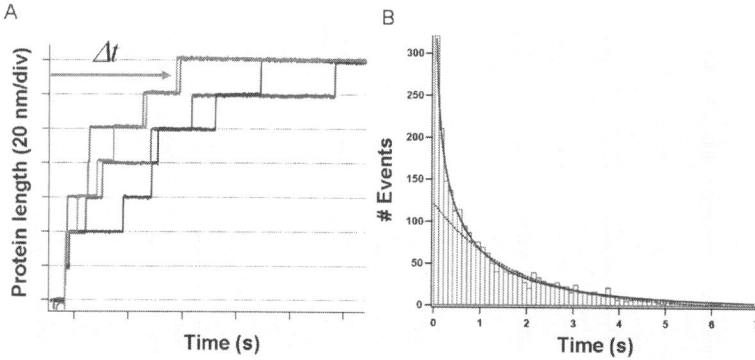

Fig. 17.12. Unfolding data obtained from ubiquitin, Ubi$_9$, polyproteins at a constant pulling force of 110 pN. **(a)** The length versus time traces show 20 nm stepwise increases in length each time a single protein domain unfolds. The time of occurrence of the unfolding events (Δt) is probabilistic. **(b)** Histogram of unfolding dwell times is measured at 110 pN. The dashed line is a single exponential fit with an unfolding rate, k, of $0.6\,\mathrm{s^{-1}}$ [10,18]. The solid line is a stretched exponential fit with $p(t) = kb(kt)^{(b-1)} \exp[-(kt)^b]$ with $b = 0.7$

thermodynamics states [32–35]. Although these theoretical studies proposed appealing statistical models of protein folding, their experimental verification remained inaccessible until now.

Ubiquitin is a fast folding protein that is ideal to study the folding and unfolding pathway of a single protein under force-clamp conditions. Ubiquitin is a 76 amino acid long $\alpha - \beta$ protein, which is used as a naturally occurring nine repeat, Ubi$_9$, polyprotein [7]. Force-clamp experiments on ubiquitin polyproteins have allowed, for the first time, the capture of individual unfolding and folding trajectories of a single protein under the effect of a constant stretching force. These observations have confirmed the scenarios predicted by the statistical theories of protein folding.

In most of our unfolding force-clamp experiments, the assumption of a simple first-order two-state kinetics works well as a first approximation to measure the mean unfolding rate at a given force. Therefore, exponential fits to ensemble averages of the single protein traces extract rate constants that are representative of the most typical reaction rate (Fig. 17.4) [11,13–15,24]. In addition to ensemble averages, we also used the well-established techniques of dwell time analysis that were developed to study the kinetics of single ion channels under voltage-clamp conditions [36]. The analysis is very similar to that shown in Fig. 17.12. We first collect a number of polyprotein unfolding traces such as those shown in Figs. 17.2 and 17.12a. Then, we measure the dwell times of each unfolding event (Δt; Fig. 17.12a). The accumulated dwell time values can then be represented as a histogram (Fig. 17.12b) [12,19,36,37]. For a simple two-state kinetic mechanism, the dwell time distribution can be fitted with a single exponential function [36,37]. Dwell time analysis of the protein ubiquitin showed that ∼80% of the observed events were well described by a single exponential. However, there were a significant number of outliers that extended the dwell time distribution and that departed from the simple exponential behavior (Fig. 17.12b). Such broad distribution of unfolding rates and dwell times was indicative that the native form of ubiquitin is rough, composed of natively folded ubiquitin structures that interconvert between states that are close in energy but different enough to have distinct unfolding transition states [12, 19].

Our results were in excellent agreement with dynamic ensemble refinement of ubiquitin structures, showing that the native state of this protein must be considered a heterogeneous ensemble of conformations [38] which are sampled over a wide range of time scales [39].

17.9 The Force-Quench Experiment

The statistical theories of protein folding proposed that an extended polypeptide evolved through progressively smaller conformational ensembles along a rough, funnel-like energy surface leading to the natively folded conformation. Such predictions are in conflict with the thermodynamic view of protein folding that emerged from bulk experiments that envisage folding as a two-state process. We designed force-clamp pulse protocols to study the individual folding trajectories of ubiquitin, and resolve this controversy. We illustrate this experimental approach, named "force-quench" [9], in Fig. 17.13. In its force-quench mode, the protein is first extended to a well-defined state and its subsequent journey towards the ensemble of native conformations is monitored as a function of length over time. This experimental approach allows us to dissect the individual folding trajectories and to understand the physical mechanisms that govern each stage involved in the folding trajectory of ubiquitin, from the fully extended state to the natively folded form.

A first force pulse of 100 pN triggers unfolding of the ubiquitin chain, as marked by step increases in length of 20 nm (Fig. 17.13). After 4 s, the force was quenched from 100 pN down to 10 pN (Fig. 17.13) in order to trigger the collapse of the extended protein. After the quench, the polyprotein was observed to spontaneously collapse in stages until it reached its folded length (Fig. 17.13). Unlike the sequential unfolding process, where the polyprotein unfolded module by module, the collapse of the extended polyprotein was highly cooperative [40]. The mean duration of the collapse trajectories was strongly dependent on the quenching force. An Arrhenius fit measured a value of $\Delta x_f = 8.2$ Å (Fig. 17.13), which is much larger than the value of Δx_u of unfolding and most likely reflects the role of distant residues and longer range force fields acting in the collapse trajectories [9]. After 12 s at the quench force, to confirm that the polyprotein had indeed folded we raised the stretching force back to 100 pN (Fig. 17.13). We observed that the ubiquitin chain extended again in steps of 20 nm, confirming that the protein had indeed

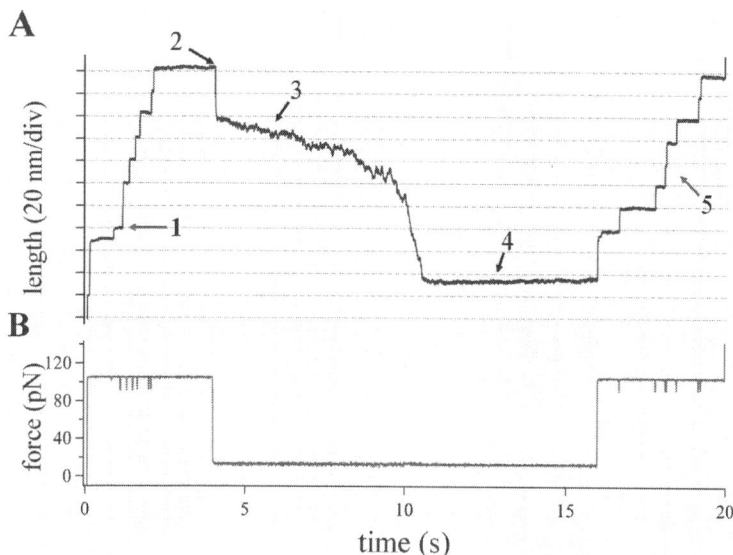

Fig. 17.13. Force quench captures the full folding trajectory of a protein, after mechanical unfolding. The end-to-end length of an Ubi$_9$ polyprotein as a function of time is shown in (a). The corresponding applied force as a function of time is shown in (b). The length of the protein (nm) evolves in time as it first extends by unfolding (1, 2) at a constant stretching force of $\sim$100 pN. Upon quenching the force to $\sim$10 pN the protein spontaneously contracted, first in a step-like manner due to elastic recoil of the unfolded polymer followed by a continuous and cooperative collapse trajectory (3). After the protein has collapsed it acquires the final native contacts that define the native fold (4). To confirm that our polyubiquitin had indeed folded, at 16 s we stretched again back to $\sim$100 pN registering a new staircase of unfolding events (5)

folded and regained its mechanical stability [9]. Hence, the spontaneous contraction of the protein observed upon reducing the force from 100 pN down to 10 pN corresponded to the folding trajectory of the mechanically unfolded ubiquitin.

The stochastic nature of the folding trajectories was also evident in these experiments. Identical force-quench pulses elicited very different responses from different molecules (Fig. 17.14). After the quench, some proteins responded by collapsing and folding over varying time scales (Fig. 17.14a), while many others contracted much less and failed to fold (Fig. 17.14b) [10]. Both the frequency and duration of the collapse were strongly dependent on the magnitude of the quench. A quench to forces above $\sim$35 pN invariably led to failures [9]. These observations were surprising in their striking departure from the widely accepted two-state folding behavior of small proteins [40, 41].

The first two stages observed in the folding trajectories corresponded to a fast entropic recoil of the extended polymer, followed by a cooperative collapse that reduces the length of the protein back to its folded length [10]. To further

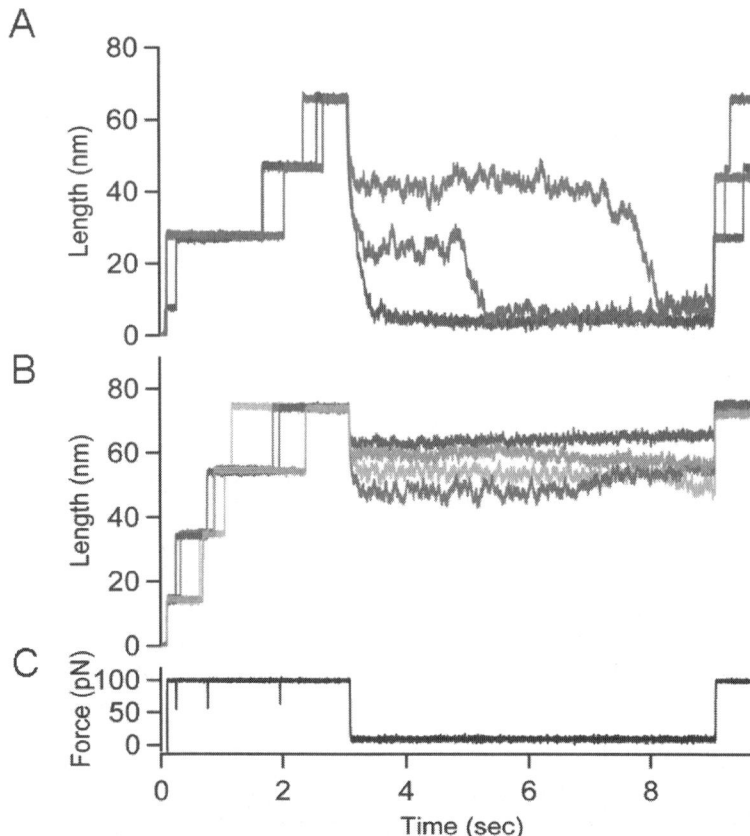

Fig. 17.14. Force-quench experiments capture the variability in the collapse behavior of an extended protein. A single polyubiquitin protein is stretched at an initial high force of 100 pN, causing unfolding of individual modules marked by step increases in length. Subsequently, the force is quenched down to 10 pN to monitor the collapse of the extended chain. To probe refolding, the protein is stretched again at 100 pN. The end- to-end length of several Ubi$_9$ polyproteins as a function of time is shown in (**a**), (**b**). The applied force, which is same in all cases, is shown in (**c**). In (**a**), the trajectories show full collapse and folding. In (**b**), the force quench fails to trigger significant collapse and no folding is observed

explore the driving forces underlying the collapse behavior of an extended protein, we used a combination of step and force-ramp protocols [10]. Following this experimental approach, a polyubiquitin was first totally unfolded and extended at 100 pN, and then the force was linearly relaxed down to 10 pN to explore the length-force relationship of the extended protein over a wide range of forces (Fig. 17.15a and b) [10].

Our experiments showed that the collapse behavior of the extended proteins varied greatly from one molecule to the next (Fig. 17.15c), and even for

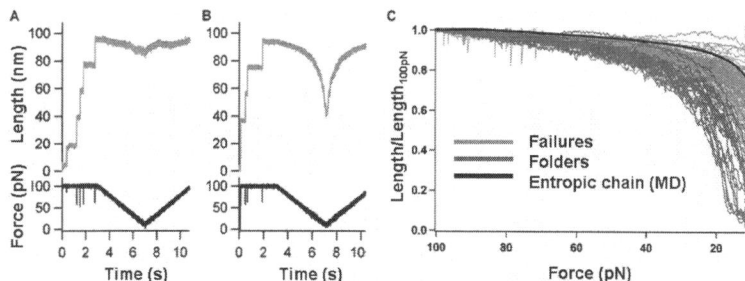

Fig. 17.15. A force-ramp protocol probes the length-force relationship of an unfolded ubiquitin polyprotein by first unfolding the protein at a high force of 100 pN then linearly decreasing the force to 10 pN. The force was subsequently ramped up to 100 pN to test for refolding. In some cases, while the force was being relaxed, the protein collapsed very little (**a**), whereas in others, the same reduction in force caused a large contraction of the extended protein (**b**). (**c**) To compare all recordings, the length during the ramp was normalized by its value for the extended conformation at 100 pN. This normalized length is shown as a function of force during the ramp down to 10 pN for proteins which successfully folded (light traces) and proteins which failed to fold (darker traces). The force-length relationship of a purely entropic chain obtained with MD simulations is shown as the black curve

the same molecule, between consecutive extensions [10]. Some molecules were observed to collapse very little upon relaxation of the stretching force down to 10 pN (Fig. 17.15a). Such molecules were well represented by the length-force relationship obtained from MD simulations of a polypeptide where enthalpic interactions had been turned off [10] (Fig. 17.15c; solid line). The more frequent and much larger collapse observed in these experiments was explained by the effect of the mostly hydrophobic interactions of the collapsing polypeptide [10]. This was further explored experimentally by placing the polyprotein in different solvent environments. The high variability in the strength of the hydrophobic collapse was explained as resulting from vastly different pathways through dihedral space, as the polypeptide collapsed down to its folded length [10]. Given that for most values of end-to-end length, a protein can visit a very large number of combinations of dihedral angles, it is clear that a collapsing protein would never follow the same collapse trajectory, which is in good agreement with our observations (Figs. 17.14 and 17.15).

In a very recent series of experiments, we studied how the collapsed polypeptide evolves into a fully folded protein [42], thus completing the full picture of the free energy landscape of a folding polypeptide. Using a variety of force pulse protocols we now demonstrate that after collapsing, the polypeptide rapidly forms an ensemble of minimum energy collapse states that exhibit lower mechanical stability [42,43] from where it proceeds to the native state through an activated barrier crossing event [44,45]. The significance of the ensemble of minimum energy collapsed states is that they greatly reduce the dimensionality of the search for the native state, allowing the protein to fold within biological time scales [42, 43].

The single-molecule force spectroscopy experiments show that the protein folding is a highly heterogeneous process where the collapsing polypeptide visits broad ensembles of conformations of increasingly reduced dimensionality. Our data showed that a folding trajectory cannot be described by well-defined thermodynamic states. These experimental results are in excellent agreement with the statistical theories of protein folding developed over a decade ago [32–35]. However, testing of these ideas remained inaccessible in bulk experiments and it is only now, at the single-molecule level, that their presence becomes apparent.

17.10 Summary

The remarkable ability of the atomic force microscopy in manipulating short engineered proteins is put to use as a new form of single-molecule spectroscopy. Force spectroscopy uncovers many novel features of protein-based reactions. For example, by measuring the force dependency of a reaction, we can estimate the actual physical distance to the transition state, Δx which can be related to well-defined molecular events; the size of a water molecule in the unfolding transition state, or the sub-Ångström S-S bond elongation during a S_N2 chemical reaction. Another useful application of force-clamp spectroscopy is in the study of a protein diffusing over a complex energy landscape during its folding/unfolding reaction. The ability to manipulate a single protein with various types of force pulses uncovered a broad diversity of folding/unfolding trajectories that can no longer be described by well-defined thermodynamic states. As these examples demonstrate, force-clamp spectroscopy provides a direct view of the physical and chemical mechanisms underlying protein-based reactions. This new knowledge combined with molecular dynamics and quantum mechanical simulations promises to establish a more realistic view of protein dynamics.

References

1. H.P. Erickson, Proc. Natl. Acad. Sci. USA. **91**(21), 10114–10118 (1994)
2. M. Carrion-Vazquez et al., Proc. Natl. Acad. Sci. USA. **96**(7), 3694–3699 (1999)
3. T.E. Fisher, P.E. Marszalek, J.M. Fernandez, Nat. Struct. Biol. **7**(9), 719–724 (2000)
4. M. Rief et al., Science **276**(5315), 1109–1112 (1997)
5. H.B. Li et al., Nature **418**(6901), 998–1002 (2002)
6. A.F. Oberhauser et al., J. Mol. Biol. **319**(2), 433–447 (2002)
7. M. Carrion-Vazquez et al., Nat. Struct. Biol. **10**(9), 738–743 (2003)
8. A.F. Oberhauser, et al., Proc. Natl. Acad. Sci. USA. **98**(2), 468–472 (2001)
9. J.M. Fernandez, H.B. Li, Science **303**(5664), 1674–1678 (2004)

10. K.A. Walther et al., Proc. Natl. Acad. Sci. USA. **104**(19), 7916–7921 (2007)
11. M. Schlierf, H.B. Li, J.M. Fernandez, Proc. Natl. Acad. Sci. USA. **101**(19), 7299–7304 (2004)
12. J. Brujic et al., Nat. Phys. **2**(4), 282–286 (2006)
13. A.P. Wiita et al., Proc. Natl. Acad. Sci. USA. **103**(19), 7222–7227 (2006)
14. A.P. Wiita et al., Nature **450**(7166), 124–+ (2007)
15. S.R.K. Ainavarapu et al., J. Am. Chem. Soc. **130**(20), 6479–6487 (2008)
16. M. Carrion-Vazquez et al., Prog. Biophy. Mol. Biol. **74**(1–2), 63–91 (2000)
17. S. Garcia-Manyes et al., Biophys. J. **93**(7), 2436–2446 (2007)
18. H.B. Li et al., Proc. Natl. Acad. Sci. USA. **97**(12), 6527–6531 (2000)
19. J. Brujic et al., Biophys. J. **92**(8), 2896–2903 (2007)
20. G.I. Bell, Science **200**(4342), 618–627 (1978)
21. E. Evans, K. Ritchie, Biophys. J. **72**(4), 1541–1555 (1997)
22. A.H. Zewail, Angew. Chem.-Int. Ed. **39**(15), 2587–2631 (2000)
23. B. Efron, *The Jackknife, the Bootstrap, and other resampling plans*, ed. by S.I.A.M. (S.I.A.M, Philadelphia, PA, 1982)
24. L. Dougan et al., Proc. Natl. Acad. Sci. USA. **105**(9), 3185–3190 (2008)
25. H. Lu et al., Biophys. J. **75**(2), 662–671 (1998)
26. H. Lu, K. Schulten, Biophys. J. **79**(1), 51–65 (2000)
27. P.E. Marszalek et al., Nature **402**(6757), 100–103 (1999)
28. H.B. Li et al., Nat. Struct. Biol. **7**(12), 1117–1120 (2000)
29. K. Kiyosawa, Biochim. Biophys. Acta. **1064**(2), 251–255 (1991)
30. S.R.K. Ainavarapu et al., J. Am. Chem. Soc. **130**(2), 436–+ (2008)
31. A. Fersht, *Structure and mechanism in protein science: a guide to enzyme catalysis and protein folding*, ed. by M.R. Julet (W.H. Freeman, New York, 1998)
32. A. Sali, E. Shakhnovich, M. Karplus, Nature **369**(6477), 248–251 (1994)
33. K.A. Dill et al., Protein Sci. **4**(4), 561–602 (1995)
34. J.N. Onuchic, P.G. Wolynes, Curr. Opin. Struct. Biol. **14**(1), 70–5 (2004)
35. P.G. Wolynes, Q Rev. Biophys. **38**(4), 405–410 (2005)
36. F.J. Sigworth, S.M. Sine, Biophys. J. **52**(6), 1047–1054 (1987)
37. R. Szoszkiewicz et al., Langmuir **24**(4), 1356–1364 (2008)
38. K. Lindorff-Larsen et al., Nature **433**(7022), 128–132 (2005)
39. M. Vendruscolo et al., J. Am. Chem. Soc. **125**(51), 15686–15687 (2003)
40. J. Brujic. J.W. Fernandez, Science **308**, 498c (2005)
41. J.M. Fernandez, H.B. Li, J. Brujic, Science. **306**(5695), 411C–+ (2004)
42. S. Garcia-Manyes et al., Proc. Natl. Acad. Sci. USA. **106**(26), 10534–10539 (2009)
43. C.J. Camacho, D. Thirumalai, Phys. Rev. Lett. **71**(15), 2505–2508 (1993)
44. D. Thirumalai, J. Phys. I. **5**(11), 1457–1467 (1995)
45. R.K. Ainavarapu et al., Biophys. J. **92**(1), 225–233 (2007)

Unraveling the Secrets of Bacterial Adhesion Organelles Using Single-Molecule Force Spectroscopy

Ove Axner, Oscar Björnham, Mickaël Castelain, Efstratios Koutris, Staffan Schedin, Erik Fällman, and Magnus Andersson

Summary. Many types of bacterium express micrometer-long attachment organelles (so-called pili) whose role is to mediate adhesion to host tissue. Until recently, little was known about their function in the adhesion process. Force-measuring optical tweezers (FMOT) have since then been used to unravel the biomechanical properties of various types of pili, primarily those from uropathogenic *E. coli*, in particular their force-vs.-elongation response, but lately also some properties of the adhesin situated at the distal end of the pilus. This knowledge provides an understanding of how piliated bacteria can sustain external shear forces caused by rinsing processes, e.g., urine flow. It has been found that many types of pilus exhibit unique and complex force-vs.-elongation responses. It has been conjectured that their dissimilar properties impose significant differences in their ability to sustain external forces and that different types of pilus therefore have dissimilar predisposition to withstand different types of rinsing conditions. An understanding of these properties is of high importance since it can serve as a basis for finding new means to combat bacterial adhesion, including that caused by antibiotic-resistance bacteria. This work presents a review of the current status of the assessment of biophysical properties of individual pili on single bacteria exposed to strain/stress, primarily by the FMOT technique. It also addresses, for the first time, how the elongation and retraction properties of the rod couple to the adhesive properties of the tip adhesin.

18.1 Introduction

Infections remain a major cause of mortality in the world. In particular, the widespread bacterial resistance to antibiotics is a ubiquitous and rapidly growing problem that needs to be addressed. There is therefore an urgent need for new anti-microbial drugs that can combat bacterial infections. It is a general consensus that the development of new drugs requires the identification of new targets in bacteria which, in turn, requires detailed knowledge of microbial pathogenic mechanisms. Since adhesion of bacterial pathogens to the host tissue is a prerequisite for infection, the adhesion mechanism is one such possible target.

A bacterial infection starts in general with the adhesion of a bacterium onto a host cell. The adhesion mechanism has turned out to be far more complex than anticipated. For example, bacteria that express their adhesins directly on the cell wall are susceptible to shear forces. A shear flow will apply a torque onto the bacteria that will induce a successive breakage of bonds and result in bacterial rolling [1, 2]. This often implies that the bacteria detach from the host cell, which, in turn, makes it less likely that they can pursue their infectious task.

However, certain bacteria, and in particular those that are exposed to various types of rinsing processes, e.g., uropathogenic *E. coli* (UPEC), have developed adhesion organelles (pili) assembled on the cell walls to deal with shear forces. These pili are composed of a number of subunits, arranged in a helix-like arrangement in the form of a rod. At the end of the helix-like rods, a short thin thread (tip fibrillum) is expressed, which anchors the adhesin that binds to the receptors expressed by host cells [3]. Moreover, UPEC bacteria can express different types of pilus. For example, P pili (pyelonephritis associated pilus, PAP) are a type that are expressed predominantly by isolates from the upper urinary tract [4,5]; in fact, they are expressed by ∼90% of the *E. coli* strains that cause pyelonephritis (severe urinary tract infection) [6]. Type 1 pili, on the other hand, are commonly found in the lower urinary tract and the bladder and can cause cystitis [7]. Both P and type 1 pili consist of a ∼ 1-μm long and 6–7-nm diameter rod that is composed of $>10^3$ subunits of similar size, PapA and FimA for P and type 1 pili, respectively, arranged in a comparable higher order structure, viz. a right-handed helix-like arrangement with 3.28 and 3.36 subunits per turn, respectively [8,9]. P pili have an adhesin known as PapG, which binds to galabiose, whereas the adhesin of type 1 piliis referred to as FimH which binds to mannose [10, 11].

It was suggested in the mid-1990s that these organelles are expandable when exposed to a force [8]. It was later shown that the pili rod can not only be significantly elongated by an unfolding process when exposed to force but it can also retract by a refolding process to its original length when it is no longer exposed to any force [12]. It has therefore been hypothesized that these adhesion systems may act as dynamic biomechanical machineries, enhancing the ability of bacteria to withstand high-shear forces originating from rinsing flows, e.g., that of urine [1]. In particular, it has been conjectured that a large flexibility of the pili could allow for a redistribution of an external shear force among a large number of pili so that each pilus is exposed to a force that can be sustained by the adhesin, making the attachment organelles of crucial importance for the ability of bacteria to withstand rinsing actions. They thereby constitute an important virulence factor and a possible target for future anti-microbial drugs. All this makes it important to assess the biophysical and biomechanical functions of different types of attachment organelle, primarily their elongation and contraction properties, including the unfolding and refolding of the rod.

Static pictures of the pili structure can be obtained from conventional microscopy studies, e.g., using AFM or electron microscopy, and from crystallographic data, using X-ray diffraction [8, 9]. However, to obtain information about the dynamic properties of adhesion organelles, the biomechanical function of pili has to be probed in real time by studies under strain or stress. Moreover, bulk studies of bacteria will not give any detailed information about the function of an individual pilus, but rather an averaged ensemble behavior of all the pili involved [13]. In order to assess in detail the function of a given adhesion system, the elongation and contraction properties need to be assessed on a single pilus level (thus on an individual bacterial cell) under controlled conditions, primarily by an appropriately sensitive force measuring technique, capable of addressing a single macromolecule at a time. Thanks to the rapid development of sensitive force-measuring techniques such studies can nowadays be performed. Several such studies have also been conducted lately, primarily by the authors, using force-measuring optical tweezers (FMOT) [12, 14–24]. This work presents the status of the field and summarizes the present knowledge about the biophysical and biomechanical properties of individual pili on single bacterial cells, in particular their elongation, unfolding, and refolding properties under strain. In addition, it addresses for the first time, how the elongation and retraction properties couple to the adhesive properties of the tip adhesin.

18.2 Instrumentation, Procedures, and Typical Force-vs.-Elongation Response of Pili

18.2.1 Force-Measuring Optical Tweezers – Instrumentation

Optical tweezers (OT) is a technique by which micrometer sized objects (inorganic as well as biological objects) can be trapped by well-focused laser light, by transfer of light momentum [25]. The basic principles of optical trapping and the first stable three-dimensional trap, based upon counter propagating beams, were established in the early 1970s by Arthur Ashkin [26, 27]. This made it possible to nonintrusively control and manipulate with great precision living biological objects in a fully controlled manner solely by light. A few years later, the first single-gradient trap was completed and the optical tweezers technique, as we know it today, was born [28]. Since then, numerous groups worldwide have constructed and developed sensitive and user-friendly optical tweezers systems for optical micro-manipulation [29].

Optical tweezers can also be used for force measurement in biological systems [30–34]. The basis is that the single-gradient trap provides an attractive force (typically on the order of picoNewton, pN) on micrometer-sized dielectrical particles (primarily nonabsorbing particles with an index of refraction larger than that of the surrounding) that confines the particles in the focal region of the light. When an external force is applied to a trapped particle, its

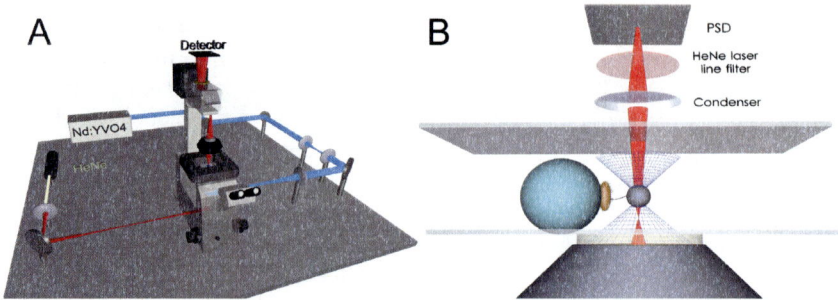

Fig. 18.1. Panel (**a**) shows a schematic illustration of the force measuring optical tweezers (FMOT) system used by the authors. An inverted microscope has been modified for introducing laser light for trapping (Nd:YVO$_4$) and probing (HeNe) through two side ports. The probing laser is fiber-coupled to reduce vibrations and to provide easy alignment. The trapping light is blocked by laser line filters so that only light from the probe laser reaches the PSD-detector. Panel (**b**) illustrates schematically the model system; a single bacterium is mounted on 9 µm large bead firmly attached to the microscope slide. A trapped bead, to which one or several pili are attached, is used as a force transducer. The elongation is created by adjustments of the position of the microscope slide. The position of the trapped bead is probed by a probe laser. The deflection of the probe light is monitored by a position sensitive detector (PSD) whose output is converted to a force by a calibration process. Copyright Wiley-VCH Verlag GmbH & Co. KGaA. Reproduced with permission from [23]

position in the trap will shift due to force balance a distance F_{ext}/κ, where F_{ext} is the externally applied force and κ the stiffness of the trap. If the position of the bead in the trap is monitored, the force to which the bead is exposed at any given time can be assessed instantaneously. For force-vs.-elongation measurements, the bead thus acts both as a handle for the tweezers and a force transducer. For studies of adhesive properties, the bead also serves the purpose of carrying the receptor sugar to which the adhesin binds. The optical tweezers instrumentation used in this work has been described previously in the literature [35–37] and Fig. 18.1a shows a schematic illustration of the set up.

18.2.2 Experimental Procedure

The biological model system has been described in some detail previously [12]. In a typical experiment, an individual free-floating bacterium is trapped by the optical tweezers (run at low power, typically a few tenth of mW at the sample) and mounted on a large (9 µm) bead that is firmly attached to the microscope slide. A small free-floating bead (3 µm) is then trapped by the optical tweezers with normal power (a few hundreds of mW) and brought to a position close to but not in direct contact with the bacterium. The system is then calibrated by using the Brownian motion technique [38]. The small bead is subsequently brought close to the bacterium in order to achieve attachment between a few pili and the bead (Fig. 18.1b). The piezo stage is then set in motion (at a constant speed) in order to provide a controlled elongation of the pilus/pili under study.

18.2.3 Typical Force-vs.-Elongation Response of Pili

The experimental procedure produces in most cases multi-pili attachment, as is illustrated in Fig. 18.2, which shows two typical force-vs.-elongation responses of *E. coli* pili. Panel (a) shows the response from a multitude of pili, whereas panel (b) illustrates the behavior from a single pilus. The black curves represent the response during elongation, whereas the gray curves correspond to contraction. Multi-pili responses will limit the amount of information that can be retrieved from the system since they will not give an unambiguous picture of the fundamental interactions in the system, primarily due to a partly unknown interplay between various pili. Single pilus experiments can be performed by first stretching the pili so that detachment sets in. By then stopping when only one pilus is attached, and finally retracting the piezo-stage to the starting position, measurement on a single (and the same) pilus can then be performed a multitude of times.

Due to its structure, a pilus has an intricate force response that differs from that of a single bond as well as those of many other types of biopolymer. As illustrated in Fig. 18.2b, a force-vs.-elongation response of a single pilus can be seen as composed of three regions; *Region I*, in which the response is basically linear, like that of a normal (elastic) spring; *Region II*, in which the

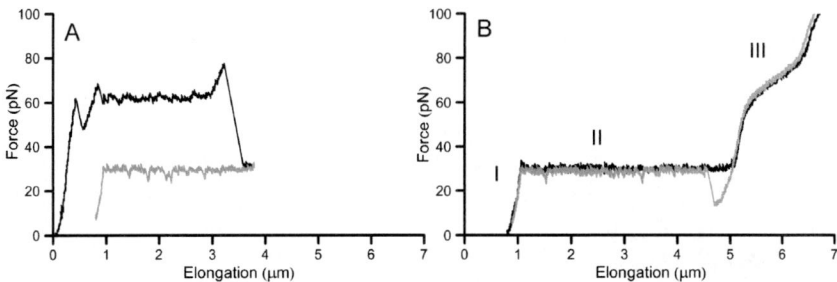

Fig. 18.2. Panel **(a)** shows by the black line the force-vs.-elongation response of a multi-pili attachment. The binding is mediated by several pili for the first 0.5 μm of the elongation (*black curve*). From around 1 μm, the attachment is sustained by two pili elongated in region II. At around 3 μm, one pilus enters its region III, after which one pilus detaches. At around 3.5 μm, only one pilus remains. The elongation is halted and reversed shortly thereafter. The retraction (*gray curve*) is mediated by a single pilus being in region I. Panel **(b)** shows a second elongation and contraction cycle of the same pilus, thus a pure single pilus force-vs.-elongation response

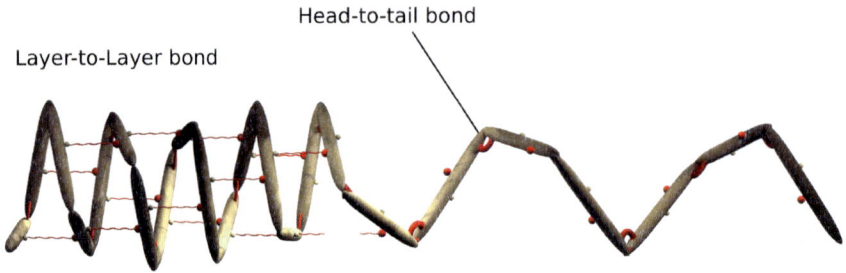

Fig. 18.3. Schematic illustration of a pilus that is partly unfolded. The head-to-tail (HT) bonds, which holds consecutive subunits together and constitutes the backbone of the pilus rod, are marked with a thick arc; whereas the layer-to-layer (LL) bonds, which are located between subunits in consecutive layers and hold the rod together, are illustrated by thin lines. Copyright Wiley-VCH Verlag GmbH & Co. KGaA. Reproduced with permission from [23]

force response is constant (elongation independent), like that of a material that undergoes plastic deformation; and *Region III*, in which the response has a monotonically increasing but nonlinear force-vs.-elongation response.

While the first region, in which the pilus is elongated a fraction of its length, can be understood as a general stretching of the pilus, but with no conformational change (no opening or closure of bonds), the other regions originate from the opening and closure of individual noncovalent bonds connecting the subunits in the pilus. As is schematically shown in Fig. 18.3 (and further discussed below), the constant force-vs.-elongation response in region II is a direct result of an unfolding of the helix-like quaternary structure of the pilus by a *sequential* opening of the layer-to-layer (LL) bonds. The reason the opening of bonds is sequential in this region is that each layer of the quaternary structure consists of several bonds (slightly more than three subunits). Since each subunit can mediate one LL bond, there are ~3 LL bonds per turn [8]. When a pilus is exposed to a force, each bond in the interior of the rod will therefore experience approximately only one third of the applied force. The bond connecting the outermost unit in the folded part of the rod, on the other hand, experiences a significantly higher force, virtually the entire force to which the pilus is exposed. This implies that the outermost LL bond of the rod will open significantly more often (or more easily) than a bond in the interior. Moreover, for an unfolding process in the interior of the rod to occur, three successive LL bonds have to open simultaneously. Since this is a very rare process, the unfolding of the rod takes place predominantly by an opening of the outermost LL bonds, which gives rise to a sequential opening procedure, sometimes referred to as a zipper-like unfolding.

In contrast, the soft wave-like force-vs.-elongation behavior in region III originates from a stochastic conformational change of the bonds between consecutive subunits of the pilus in the linearized part of the rod (referred to as head-to-tail (HT) bonds). Since these bonds, which are composed by a donor strand interaction, can change (open and close) in a *random* manner, the particular shape of this region is governed by both properties of the individual bonds and entropy of which the latter gives it its specific wave-like shape.

In order to comprehend the adhesion properties of piliated bacteria, it is necessary to acquire detailed information not only about the biophysical properties of individual pili under various conditions, in particular their behavior in regions II and III, but also the adhesin on the tip of the pilus. Moreover, to understand the unfolding and refolding properties of the pilus rod a good knowledge about the properties of individual bonds exposed to strain/stress is needed. A full understanding of adhesion properties of piliated bacteria requires finally knowledge about how several pili cooperate to deal with an external force.

18.3 Theory

18.3.1 Bonds, Energy Landscapes, and Forces

Energy Landscape Representation of a Bond

As has been described previously in the literature [39], a noncovalent bond can be described in terms of an energy landscape, which is a representation of its energy-vs.-length dependence, consisting of at least two minima, representing the bond being "closed" and "open" (referred to as states A and B), respectively, and an intermediate local maximum, called the transition state (denoted by T). A closed bond can open, and thus elongate, only if it, somehow, "passes" this transition state. The energy of the transition state, ΔV_{AT}, schematically illustrated in Fig. 18.4, thereby represents the activation energy for bond opening. For an internal bond in a macromolecule, such as the LL or the HT bonds in pili, state B is localized, whereas for a single external bond, e.g., that of an adhesin, it is nonlocalized.

Thermal Bond Dynamics

Due to the thermal energy, a bond will vibrate and each oscillation can be seen as an attempt to open. The bond opening rate can therefore, in general, be written as a product of an attempt rate, ν, and an Arrhenius factor, the latter encompassing the activation energy for opening the bond, i.e., as $\exp(-\Delta V_{\text{AT}}/kT)$, where ΔV_{AT} is the difference in energy between a closed bond and the transition state, k is the Boltzmann constant and T is the absolute temperature. A typical attempt rate for a bond in a liquid is around 10^9–10^{10} Hz, whereas a typical activation energy for a noncovalent bond can take values up to several tens of kT (which takes a value of $4\,\text{pN nm}$ under room conditions) [40].

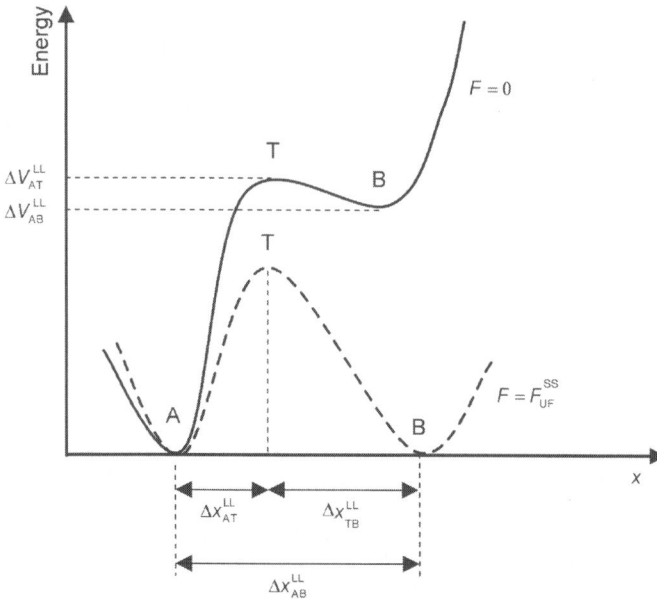

Fig. 18.4. Schematic energy landscape diagram of a bond. The state A represents a closed bond, whereas state B symbolizes an open. The position of the maximum of the energy landscape curve between states A and B is referred to as the transition state, and is denoted by T. The uppermost curve represents the energy landscape for a bond not exposed to any force, whereas the lower refers to the case when the bond is exposed to a force equaling the opening and closure rates. The bond length, Δx_{AT}^{LL}, represents the distance from the minimum of state A to the transition state, whereas Δx_{TB}^{LL} is the distance from the transition state to the minimum of state B. Δx_{AB}^{LL} represents the total elongation of the bonds along the reaction coordinate when it opens

An open bond in the localized B state can subsequently close. Also, the closure rate will take place with a rate given by the product of the attempt rate and an Arrhenius factor, this time encompassing the difference in energy between the transition state and the open state, i.e., $\exp[(-(\Delta V_{AT} - \Delta V_{AB})/kT]$, where ΔV_{AB} is the difference in energy between an open and closed bond.

A Single Bond Exposed to a Force

Applying a force to a bond (often called stress in the world of biophysics) alters the activation barriers and thereby also the bond opening and closure rates, which, in turn, affects the behavior of the bond [39–43]. This is the basis for force spectroscopy of bonds. The activation energy for bond opening is lowered by an amount equal to the product of the force and the bond length (the latter given by the distance from the closed state to the transition state), i.e., by an amount of $F\Delta x_{AT}$. The activation energy for bond closure is similarly increased to an amount equal to the product of the force and the length between the transition state and the open state, i.e., by $F\Delta x_{TB}$ [39]. This is usually envisioned as if the energy landscape is tilted with a slope given by the force, as is illustrated by the dashed curve in Fig. 18.4. The closure rate for a single bond with a nonlocalized "open" state exposed to an external force is most often negligible.

Dynamics of a Single Bond Exposed to a Force

Due to the statistical properties of thermodynamics, a single bond, exposed to an increasing force a number of consecutive times, will open at a broad distribution of forces, of which the peak position is called the bond strength [40,44]. For a soft force transducer, the force applied to the bond can be written as $\Delta L\kappa$, where ΔL is the distance the force transducer has been moved. When the distance is increased with a given speed, $\dot{L}$, a single bond will be exposed to a linearly increasing force, given by $\dot{L}\kappa$, referred to as the loading rate. As is further discussed later, a bond exposed to a constant loading rate will experience a progressively increased probability for opening. The integrated history of such a time-dependent opening probability determines the fate of the bond. This implies that the strength of a single bond will depend on the loading rate and thereby also the elongation rate. It can be shown that, under a set of fairly normal conditions, the strength of a single bond is proportional to the logarithm of the loading rate (or the elongation speed), whereas the slope provides information about the bond length [39–44].

Sequential Bond Dynamics

In contrast, if a macromolecule consists of several bonds that are arranged in a sequential configuration of parallel bonds, as is the case for a helix-like structure, the width of the bond opening distribution is decreased by a force stabilizing process and all bonds will open in succession at more or less the same force, which, in turn, gives rise to a force plateau in the force-vs.-elongation response [12, 17, 45]. As discussed later, the value of this force depends, under some conditions, on the elongation speed; it is fairly independent of the elongation speed for low speeds, but proportional to the logarithm of the elongation speed for higher speeds.

18.3.2 Rate Theory for Unfolding and Refolding of a Helix-like Polymer Exposed to Strain or Stress: Pili Elongated in Region II

Opening and Closure Rates of an Individual Bond

The net opening rate of the outermost LL bond in the folded part of a helix-like polymer exposed to a force, F, defined as the number of times an LL bond opens per unit time, $k_{AB}^{LL,net}(F)$, can be written as the difference between the bond opening and the bond closure rates under the exposed stress, i.e., as $k_{AB}^{LL}(F) - k_{BA}^{LL}(F)$, which, in turn, can be expressed as [39]

$$k_{AB}^{LL}(F) = k_{AB}^{LL,th} e^{F\Delta x_{AT}^{LL}/kT} \qquad (18.1)$$

and

$$k_{BA}^{LL}(F) = k_{AB}^{LL,th} e^{(\Delta V_{AB}^{LL} - F\Delta x_{TB}^{LL})/kT}, \qquad (18.2)$$

where $k_{AB}^{LL,th}$ is the thermal bond opening rate in the absence of force. Δx_{AT}^{LL} and Δx_{TB}^{LL} are the distances between the closed state A and the transition state T, and between the transition state T and the open state B for the LL bond, respectively, where the former is often called the bond length. These expressions are sometimes referred to as Bell's equations for unfolding and refolding [39].

The Rate-Equation for Bond Opening and Bond Closure in Region II

Due to the sequential mode of opening and closure, the bond opening rate of region II, defined as the number of bonds that open per unit time, dN_B/dt, where N_B is the number of open LL bonds at any given time, is identical to the net opening rate of the outermost bond, $k_{AB}^{LL,net}(F)$. This implies that the bond opening rate of region II can be expressed as a rate-equation according to

$$\frac{dN_B}{dt} = k_{AB}^{LL}(F) - k_{BA}^{LL}(F). \qquad (18.3)$$

The Rate-Equation for the Force in the System

Owing to the experimental procedure, the force in the system, F, can be related to the forced elongation speed of the pili, $\dot{L}$, and the bond opening rate, dN_B/dt, according to

$$\frac{dF}{dt} = \left(\dot{L} - \frac{dN_B}{dt} \Delta x_{AB}^{LL} \right) \kappa, \qquad (18.4)$$

where Δx_{AB}^{LL} is the bond opening length, defined as $\Delta x_{AT}^{LL} + \Delta x_{TB}^{LL}$ [17].

Information about the force-vs.-elongation-speed-behavior of a helix-like polymer, $F_{II,UF}(\dot{L})$, unfolding at a constant force, can then be obtained by solving (18.3) and (18.4) for $dF/dt = 0$. Equation (18.4) then shows that the (forced) bond opening rate, dN_B/dt, is equal to the ratio of the elongation speed and the bond opening length, i.e., $\dot{L}/\Delta x_{AB}^{LL}$. Solving (18.3) under such conditions gives rise to a force-vs.-elongation-speed-behavior that is schematically illustrated in Fig. 18.5a.

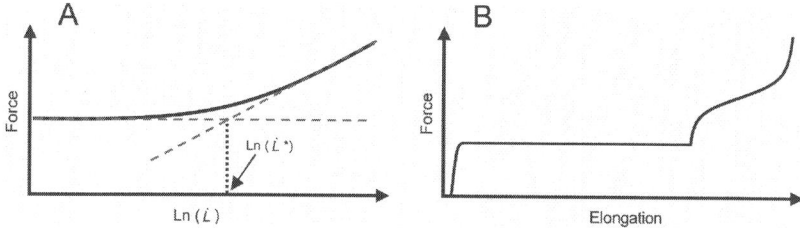

Fig. 18.5. Panel (**a**) gives a schematic illustration of the force response of a helix-like biopolymer (pili in region II) as a function of elongation speed. Panel (**b**) illustrates a numerical simulation of the rate equations for elongation of a pilus under steady-state conditions

Unfolding and Refolding Under Steady-State Conditions

As can be seen from the figure, for low speeds, up to the so-called corner velocity, $\dot{L}^*$, given by [12]

$$\dot{L}^* = \Delta x_{AB}^{LL} k_{AB}^{LL,th} e^{\left(\Delta x_{AT}^{LL}/\Delta x_{AB}^{LL}\right) V_{AB}^{LL}/kT}, \tag{18.5}$$

the force at which the rod unfolds depends only weakly on the elongation speed. For such elongation speeds, there is a balance between the bond open-ing and the bond closure rates, both being larger than the forced bond opening rate; in particular the bond closure rate is larger than the forced bond opening rate, i.e., $k_{BA}^{LL}(F) > dN_B/dt$. This implies that each bond will open and close several times with similar rates, $k_{AB}^{LL}(F) \approx k_{BA}^{LL}(F)$, which often are referred to as the balance rate, $k_{AB}^{LL,bal}$, before the neighboring bond starts to open. Measurements performed under such conditions are therefore referred to as being done at under steady-state conditions. Under such conditions, (18.3) gives rise to an expression for the steady-state unfolding force of a helix-like polymer (i.e., a pilus in its elongation region II), $F_{II,UF}^{SS}$, that reads

$$F_{II,UF}^{SS} = \Delta V_{AB}^{LL}/\Delta x_{AB}^{LL}. \tag{18.6}$$

This shows that under low-speed conditions, the force is (virtually) indepen-dent of the elongation speed and solely given by energy landscape parameters. This also implies that the refolding force, $F_{II,RF}^{SS}$, is similar to the unfolding rate.

Unfolding and Refolding Under Dynamic Conditions

For higher elongation speeds, on the other hand, i.e., those above $\dot{L}^*$, the forced bond opening rate, dN_B/dt, becomes larger than the bond closure rate, $k_{BA}^{LL}(F)$. The closure rate can therefore be neglected whereby the bond opening rate, k_{AB}^{LL}, is in direct balance with the forced elongation opening rate, $\dot{L}/\Delta x_{AB}^{LL}$. Solving (18.3) and (18.4), under these conditions, gives rise to an expression that relates the unfolding force in region II under dynamic elongation conditions, $F_{\mathrm{II,UF}}^{\mathrm{DFS}}$, to the elongation speed in a logarithmic manner, viz. as

$$F_{\mathrm{II,UF}}^{\mathrm{DFS}}(\dot{L}) = \frac{kT}{\Delta x_{\mathrm{AT}}^{\mathrm{LL}}} \ln\left(\frac{\dot{L}}{\Delta x_{\mathrm{AB}}^{\mathrm{LL}} k_{\mathrm{AB}}^{\mathrm{LL,th}}}\right). \tag{18.7}$$

Force measurements made under these conditions are commonly referred to as dynamic force spectroscopy (DFS). The advantage with DFS measurements is that it is possible to assess values to physical entities that cannot be addressed by measurements under steady-state conditions, predominantly those related to the transition state, i.e., $\Delta x_{\mathrm{AT}}^{\mathrm{LL}}$ and $k_{\mathrm{AB}}^{\mathrm{LL,th}}$ [12, 46, 47].

Under dynamic conditions, the refolding force, $F_{\mathrm{II,RF}}^{\mathrm{DFS}}$, is lower than the unfolding force under steady-state as well as dynamic conditions [45].

Unfolding Under Relaxation Conditions

As an alternative to DFS, it is possible to monitor the decay in force that follows when the elongation of the pilus is suddenly halted. This is particularly useful for the cases when the corner velocity is low, as is the case for type 1 pili (see further discussion below). An expression for the unfolding force from an elongated pilus that is suddenly halted can be derived from (18.3) and (18.4) under the condition that $\dot{L} = 0$. Again under the condition that the refolding rate can be neglected, the resulting expression becomes a separable differential equation from which an expression of the decaying unfolding force under dynamic relaxation conditions (DRC), $F_{\mathrm{II,UF}}^{\mathrm{DRC}}(t)$, can be derived [23],

$$F_{\mathrm{II,UF}}^{\mathrm{DRC}}(t) = -\frac{kT}{\Delta x_{\mathrm{AT}}^{\mathrm{LL}}} \ln\left[e^{-F_{\mathrm{II,UF}}^{0}\Delta x_{\mathrm{AT}}^{\mathrm{LL}}/kT} + \frac{\Delta x_{\mathrm{AB}}^{\mathrm{LL}}\kappa\Delta x_{\mathrm{AT}}^{\mathrm{LL}}}{kT}k_{\mathrm{AB}}^{\mathrm{LL,th}}t\right], \tag{18.8}$$

where $F_{II,UF}^{0}$ is the initial force at the time when the elongation is halted and t is the time thereafter. This expression thus describes the force-vs.-time response of a helix-like biopolymer when the applied elongation is suddenly halted. The actual form of this expression is shown below as fits to measurements.

18.3.3 Rate Theory for Elongation and Contraction of a Linear Polymer Exposed to Strain or Stress – Pili Elongated in Region III

The Rate-Equation for Bond Opening and Closure in Region III

The elongation of a linearized pilus, i.e., when the entire helix-like structure has been unfolded and the pilus resides in region III, differs from that of a helix-like structure in the respect that all bonds have the same probability to open and close, irrespective of their neighbors. Region III is therefore governed by a set of random transitions between the closed and opened states of the HT bonds, referred to as states B and C, respectively. The randomness implies that the bond opening rate is strongly affected by entropy, which gives it its soft wave-like form, sometime referred to as entropic softening. The HT bond opening rate in the linearized configuration (region III), dN_C/dt, can thereby be described by the rate equation

$$\frac{dN_C}{dt} = N_B k_{BC}^{HT}(F) - N_C k_{CB}^{HT}(F), \tag{18.9}$$

where N_B and N_C are the number of closed and opened HT bonds, respectively, and where the bond opening and closure rates are similar to those defined in (18.3).

In as much as all bonds are involved in the length regulation of region III (and not solely the outermost bond of the folded part of the rod as is the case for region II), the dynamic region is entered whenever $\dot{L}/\Delta x_{BC}^{HT} > N_C k_{CB}^{HT}(F)$. Since $N_C \gg 1$ (N_C is in most cases $\sim 10^3$), the onset of the dynamic response in region III takes place at a significantly higher speed than in region II. Region III can therefore be considered to be in its steady-state region for the entire range of speeds examined.

Steady-State Conditions

Under steady-state conditions, the force response of region III can be written as

$$F_{III}^{SS} = \frac{\Delta V_{BC}^{HT}}{\Delta x_{BC}^{HT}} + \frac{kT}{\Delta x_{BC}^{HT}} \ln\left(\frac{N_{TOT} - N_B}{N_B}\right), \tag{18.10}$$

where ΔV_{BC}^{HT} and Δx_{BC}^{HT} are the difference in energy between an open and closed HT bond (in the absence of stress) and the bond opening length, respectively, and N_{TOT} and N_B are the total number of units in the rod and the number of closed HT bonds in the linearized pili, respectively. Since the length of the pilus, L, can be related to N_{TOT} and N_B by geometrical means, (18.10) provides an expression for the force-vs.-elongation (as well as the force-vs.-contraction) behavior of pili in region III under steady-state conditions [16–18]. Although not explicitly evident from the derivation given earlier, the expression given in (18.10) includes the softening effect of entropy [12].

The fact that region III most often is in its steady-state regime, implies that some energy landscape parameters for the HT bond, e.g., Δx_{BT}^{HT} and ΔV_{BT}^{HT}, cannot readily be assessed.

18.3.4 Predicted Force-vs.-Elongation Response of a Single Pilus

As was illustrated earlier, the rate equation for the force, (18.4), can, together with the rate equations for bond opening and closure, (18.3) and (18.9), and

Bell's equations for unfolding and refolding, (18.1) and (18.2), be solved ana-
lytically under steady-state conditions, for the regions II and III, respectively,
giving rise to (18.6) and (18.10). This behavior is schematically illustrated in
Fig. 18.5b. As can be seen by a direct comparison with the force-vs.-elongation
response of a single pilus shown in Fig. 18.2b, the agreement is good.

However, it is not as simple to provide an expression for the force-vs.-
elongation behavior under general conditions, in particular not for noncon-
stant elongation speeds or elongation speeds in proximity of the corner velocity
(i.e., for $\dot{L} \approx \dot{L}^*$). The reason is that analytical solutions can only be found
under steady-state conditions or if the refolding rate is neglected, which are
good approximations when $\dot{L} < \dot{L}^*$ and $\dot{L} > \dot{L}^*$, respectively [12]. Under such
conditions, the force-vs.-elongation behavior of a helix-like polymer (i.e., a
pilus in region II) is given either by (18.6) or (18.7). The dynamic response
in the intermediate range needs therefore to be solved numerically, e.g., by
numerical integration, or can be found by Monte-Carlo simulations [22].

18.3.5 Pili Detachment Under Exposure to Stress

A pilus will detach from its receptor on the host cell whenever the adhesin
bond opens. As was shown by (18.1), also the opening rate for the adhesin
(Ad) bond, $k_{AB}^{Ad}(F)$, depends exponentially on the force to which it is exposed.
Since the adhesin is located on the tip of the pilus, the force to which it is
exposed is that exerted by the rod, which for a single pilus attachment is equal
to the external force.

It has been found that the lifetime of an adhesin is rather long (>1 s) for
forces below the steady-state unfolding force, $F_{II,UF}^{SS}$ [48]. This implies that it
is reasonable to assume that the adhesin bond will remain closed as long as
the pilus is elongated in region I.

In elongation region II, the force to which the adhesin is exposed is equal
to the unfolding force, $F_{II,UF}$. This implies that the bond opening rate for the
adhesin, $k_{AB}^{Ad}(F)$, can, for the two cases with slow or fast elongation (i.e., for
$\dot{L} < \dot{L}^*$ and $\dot{L} > \dot{L}^*$, respectively), be written as

$$k_{AB}^{Ad}(F) = k_{AB}^{Ad,th} e^{F_{II,UF}^{SS} \Delta x_{AT}^{Ad}/kT} = k_{AB}^{Ad,th} e^{\left(\Delta x_{AT}^{Ad}/\Delta x_{AB}^{LL}\right)\Delta V_{AB}^{LL}/kT} \quad (18.11)$$

and

$$k_{AB}^{Ad}(F) = k_{AB}^{Ad,th} e^{F_{II,UF}^{FDS} \Delta x_{AT}^{Ad}/kT} = k_{AB}^{Ad,th} \left(\frac{\dot{L}}{\Delta x_{AB}^{LL} k_{AB}^{LL,th}}\right)^{\rho}, \quad (18.12)$$

respectively, where $k_{AB}^{Ad,th}$ stands for the thermal bond opening rate for the
adhesin, and where we have used ρ as a short-hand notation for the ratio of
the lengths of the adhesin and the LL bonds, $\Delta x_{AT}^{Ad}/\Delta x_{AT}^{LL}$, respectively.

Since the probability of finding a bond exposed to a constant force in
the closed state is given by $\exp\left(-k_{AB}^{Ad}t\right)$, the *lifetime* of the adhesin bond in

region II, $\langle\tau\rangle_{\mathrm{II}}^{\mathrm{Ad}}$, defined as the time after which e^{-1} of all bonds remain in the closed state, can be written as the inverse of the bond opening rate. This implies that it is given by

$$\langle\tau\rangle_{\mathrm{II}}^{\mathrm{Ad}} = \frac{1}{k_{\mathrm{AB}}^{\mathrm{Ad,th}}}\mathrm{e}^{-F_{\mathrm{II,UF}}^{\mathrm{SS}}\Delta x_{\mathrm{AT}}^{\mathrm{Ad}}/kT} = \frac{1}{k_{\mathrm{AB}}^{\mathrm{Ad,th}}}\mathrm{e}^{-\left(\Delta x_{\mathrm{AT}}^{\mathrm{Ad}}/\Delta x_{\mathrm{AB}}^{\mathrm{LL}}\right)\Delta V_{\mathrm{AB}}^{\mathrm{LL}}/kT} \quad (18.13)$$

and

$$\langle\tau\rangle_{\mathrm{II}}^{\mathrm{Ad}} = \frac{1}{k_{\mathrm{AB}}^{\mathrm{Ad,th}}}\mathrm{e}^{-F_{\mathrm{II,UF}}^{\mathrm{DFS}}\Delta x_{\mathrm{AT}}^{\mathrm{Ad}}/kT} = \frac{1}{k_{\mathrm{AB}}^{\mathrm{Ad,th}}}\left(\frac{\dot{L}}{\Delta x_{\mathrm{AB}}^{\mathrm{LL}}k_{\mathrm{AB}}^{\mathrm{LL,th}}}\right)^{-\rho}, \quad (18.14)$$

for the two cases with $\dot{L} < \dot{L}^*$ and $\dot{L} > \dot{L}^*$, respectively.

For the case when a pilus detaches in region II, it is possible to define an *expected elongation length*, $\langle L\rangle_{\mathrm{II}}^{\mathrm{Ad}}$, as the length of a pilus can be elongated (in region II) before the adhesin bond opens. Under the condition that the pilus is elongated with a constant elongation speed, this implies that $\langle L\rangle_{\mathrm{II}}^{\mathrm{Ad}}$ can be expressed as the product of $\langle\tau\rangle_{\mathrm{II}}^{\mathrm{Ad}}$ and the elongation speed, i.e., as

$$\langle L\rangle_{\mathrm{II}}^{\mathrm{Ad}} = \langle\tau\rangle_{\mathrm{II}}^{\mathrm{Ad}}\dot{L} = \frac{\dot{L}}{k_{\mathrm{AB}}^{\mathrm{Ad,th}}}\mathrm{e}^{-F_{\mathrm{II,UF}}^{\mathrm{SS}}\Delta x_{\mathrm{AT}}^{\mathrm{Ad}}/kT}$$

$$= \frac{\dot{L}}{k_{\mathrm{AB}}^{\mathrm{Ad,th}}}\mathrm{e}^{-\left(\Delta x_{\mathrm{AT}}^{\mathrm{Ad}}/\Delta x_{\mathrm{AB}}^{\mathrm{LL}}\right)\Delta V_{\mathrm{AB}}^{\mathrm{LL}}/kT} \quad (18.15)$$

and

$$\langle L\rangle_{\mathrm{II}}^{\mathrm{Ad}} = \frac{\dot{L}}{k_{\mathrm{AB}}^{\mathrm{Ad,th}}}\mathrm{e}^{-F_{\mathrm{II,UF}}^{\mathrm{DFS}}\Delta x_{\mathrm{AT}}^{\mathrm{Ad}}/kT} = \frac{k_{\mathrm{AB}}^{\mathrm{LL,th}}}{k_{\mathrm{AB}}^{\mathrm{Ad,th}}}\Delta x_{\mathrm{AB}}^{\mathrm{LL}}\mathrm{e}^{(1-\rho)\frac{F_{\mathrm{II,UF}}^{\mathrm{DFS}}\Delta x_{\mathrm{AT}}^{\mathrm{LL}}}{kT}}$$

$$= \frac{1}{k_{\mathrm{AB}}^{\mathrm{Ad,th}}}\frac{\dot{L}^{1-\rho}}{\left(\Delta x_{\mathrm{AB}}^{\mathrm{LL}}k_{\mathrm{AB}}^{\mathrm{LL,th}}\right)^{-\rho}}, \quad (18.16)$$

again for the two cases with $\dot{L} < \dot{L}^*$ and $\dot{L} > \dot{L}^*$, respectively. The expected elongation length under dynamic conditions has in (18.16) been expressed alternatively in terms of the unfolding force of region II under dynamic conditions, $F_{\mathrm{II,UF}}^{\mathrm{DFS}}$, and the elongation speed, $\dot{L}$. Obviously, these expressions are valid only for the cases when $\langle L\rangle_{\mathrm{II}}^{\mathrm{Ad}}$ is smaller than the maximal elongated length of the pilus rod, $N_{\mathrm{TOT}}\Delta x_{\mathrm{AB}}^{\mathrm{LL}}$.

If the pilus remains attached to region III, in which the force increases sharply with elongation (and thus time), the probability for detachment will rapidly increase and the pilus will experience a significantly reduced lifetime.

18.4 Results and Discussion

18.4.1 Typical Force-vs.-Elongation Response of P and Type 1 Pili

Experiments have been performed on several types of pilus. The biophysical properties of both P and type 1 pili have been assessed in some detail recently by the authors using FMOT [12, 14–24, 36, 37], whereas studies of other types of pilus (e.g., S and F1C) are presently ongoing. Most of the analysis presented here will therefore concern P and type 1 pili. Figure 18.6 shows a typical force-vs.-elongation response of these two pili for an elongation speed of 0.1 μm/s, with the black and gray curves representing elongation and contraction, respectively.

The plateau in region II for P pili, which has almost identical force values for unfolding and refolding (28 ± 2 pN), agrees well with the predicted shape of the force-vs.-elongation behavior for unfolding of a helix-like structure under *steady-state* conditions given by (18.6) and shown in Fig. 18.5b. The behavior of type 1 pili, on the other hand, with dissimilar plateau values for unfolding and refolding, shows instead the predicted unfolding response of a helix-like polymer under *dynamic* conditions, given by (18.7). Also, the P pili can unfold its quaternary structure under *dynamic* conditions, although this requires a higher elongation speed than 0.1 μm/s [12, 23].

For both types of pilus, region III agrees well with the predicted shape for the elongation of a linear polymer under steady-state conditions, given by (18.10). By fitting the model equations to curves like those in Fig. 18.6, most of model parameters introduced in the models, e.g., ΔV_{AB}^{LL}, ΔV_{BC}^{HT}, Δx_{AB}^{LL}, Δx_{BC}^{HT}, can be assigned values for the two types of pilus.

However, there are two features that cannot be explained by the simple theory given earlier. The first is the fully reproducible dip in force that appears during contraction following the transition from regions III and II. This is considered to be caused by a lack of a nucleation kernel for the formation of a first layer in the quaternary structure during contraction [12,49]. Occasionally,

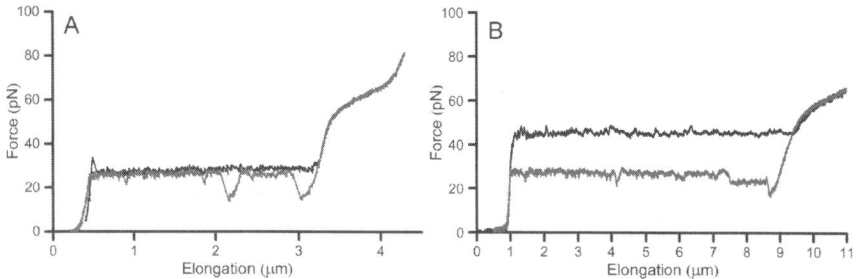

Fig. 18.6. Panels (**a**) and (**b**) show typical single pili responses from a P and a type 1 pili, respectively, for an elongation speed of 0.1 μm/s

also other dips in the contraction data can occur, as for example the case at around 1.5 μm for the P pilus (Fig. 18.6a). As is further discussed later, such dips are not reproducible and are considered to originate from sporadic misfoldings not included in the model.

The second feature is the ability of type 1 pili to refold their quaternary structure at two different force levels, ∼30 and ∼25 pN, respectively, which is illustrated by the gray curve in Fig. 18.6b. Although not yet proven, this suggests that type 1 pili can refold into two dissimilar helix-like configurations [21].

Repetitive unfolding and refolding cycles of the PapA rod have furthermore shown that a single P pilus can be elongated and contracted through its entire elongation-and-contraction cycle a large number of times ($>10^2$) without any alteration of its properties [19]. Such experiments have thus showed that it is possible to elongate a pilus numerous times without any sign of fatigue.

18.4.2 Dynamic Force-vs.-Elongation Response

Figure 18.7a shows a compilation of some typical force-vs.-elongation-speed data for P pili (circles) and type 1 pili (triangles) from measurements such as those shown in Fig. 18.6, presented as the unfolding force of region II vs. the elongation speed in a lin–log plot. For each type of pili, the data gather in two regions, following two asymptotes; one that is virtually independent of the elongation speed (for low elongation speeds) and one that is linear in a lin–log plot, thus showing an logarithmic behavior (for large elongation speeds). All this is in agreement with the theoretical predictions given in Fig. 18.5b above. The two asymptotes meet at the corner velocity. Studies such as these support the model given earlier and indicate that the corner velocity for the P pili and type 1 pili are significantly different, viz. 400 and 6 nm/s, respectively [12].

An example of a study of the dynamic response that follows a sudden halt in elongation is shown in Fig. 18.7b together with fits to (18.8), and it reveals that the force for P and type 1 pili relaxes to the steady-state force according to (18.8) for both types of pilus, although with significantly dissimilar speeds for the two types of pilus.

These types of curve can then serve the purpose of assessing values of the bond length and the thermal bond opening rate for the LL bond of the two types of pilus, i.e., Δx_{AT}^{LL} and $k_{AB}^{LL,th}$. It was found, for example, that these take the values of 0.76 nm and 0.8 Hz for P pili. The same entities take values of 0.59 nm and 0.016 Hz for type 1 pili.

The examples given in Figs. 18.6 and 18.7, show that all major properties of the force-vs.-elongation behavior of an individual pilus, with the exception of some refolding, have been captured by the simple theory and expressions given earlier.

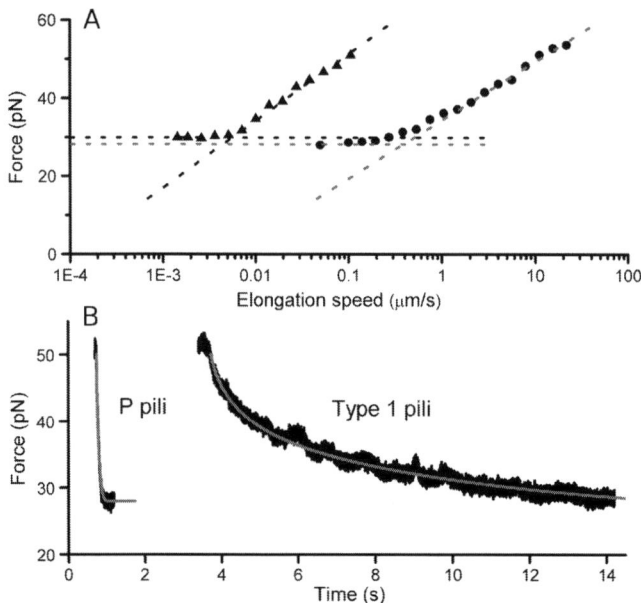

Fig. 18.7. Panel (**a**) gives the unfolding force in region II of P pili (*circles*) and type 1 pili (*triangles*) as a function of (the logarithm of) the elongation speed. The *dashed lines* represent the low- and high-elongation speed asymptotes, given by (18.6) and (18.7), respectively. The intercept of the two asymptotes represents the corner velocity, $\dot{L}^*$, whereas the intercept of the high-elongation speed asymptote with the x-axis gives an entity, $\dot{L}^0$, that provides a value for the thermal bond opening rate, $k_{AB}^{LL,th}$. The data for P pili is replotted from Fig. 18.5 in [12]. Panel (**b**) shows the response from unfolding of pili under relaxation conditions for P and type 1 pili (with an initial force of 50 pN). The data show that P pili reach their steady-state force after a fraction of a second, whereas type 1 pili reach it after ~9 s. Copyright Wiley-VCH Verlag GmbH & Co. KGaA. Reproduced with permission from [23]

18.4.3 Compilation of Model Parameters and Pili Properties

Table 18.1 presents a compilation of some bond parameters that have been assessed for P and type 1 pili. Parameters for other types of pilus, e.g., S and F1C, which are presently under scrutiny, will be presented elsewhere. In addition to measurements on P and type 1 pili performed by the authors [12,14–18,21,36,37], the table includes some parameters for type 1 pili assessed by Miller et al. [50] and Forero et al. [51] using force measuring AFM. Note though, to make a comparison possible with the results obtained by the OT-technique, some of their values have been recalculated to agree with the nomenclature used with OT.

It is known from scanning transmission electron microscopy that many types of pilus have structural similarities [9]. Although measurements of their

Table 18.1. Some bond parameters for P and Type 1 pili of *E. coli*

Parameters	P pili	Type 1		
Technique	FMOT[a]	FMOT[a]	AFM [50]	AFM [51]
$F_{\mathrm{II,UF}}^{\mathrm{SS}}$ (pN)	28 ± 2	30 ± 2	$(60)^{\mathrm{b}}$	22
$k_{\mathrm{AB}}^{\mathrm{LL,th}}$ (Hz)	0.8 ± 0.5	0.016 ± 0.009	0.05	0.5
$k_{\mathrm{AB}}^{\mathrm{LL,bal}}$ (Hz)	120 ± 75	1.2 ± 0.9	0.8	2.2
$\dot{L}^{*}$ (nm/s)	400	6		
$\Delta V_{\mathrm{AT}}^{\mathrm{LL}}$ (kT)	23 ± 1	27 ± 2	26^{c}	29
$\Delta V_{\mathrm{AB}}^{\mathrm{LL}}$ (kT)	24 ± 1	37 ± 2	16^{c}	25
$\Delta x_{\mathrm{AT}}^{\mathrm{LL}}$ (nm)	0.76 ± 0.11	0.59 ± 0.06	0.2	0.26 ± 0.01
$\Delta x_{\mathrm{AB}}^{\mathrm{LL}}$ (nm)	3.5 ± 0.1			5 ± 0.3

[a][12–18, 21, 36, 37]
[b]Measured at 1–3 µm/s, which, according to FMOT measurements, is not at steady-state
[c]Recalculated using at an attempt rate of 10^{10} Hz

physical properties to a certain extent have shown similar behaviors, they have also revealed striking differences. It has, for example, been found that the steady-state unfolding forces of the quaternary structure of P and type 1 pili, $F_{\mathrm{II,UF}}^{\mathrm{SS}}$, are comparable, 28 ± 2 pN and 30 ± 2 pN, respectively.[1] On the other hand, presently ongoing work (unpublished results) has shown that other types of pilus, e.g., F1C, have significantly lower steady-state unfolding forces, <10 pN.

Moreover, the thermal bond opening rate, $k_{\mathrm{AB}}^{\mathrm{LL,th}}$, for P pili has been assessed to 0.8 ± 0.5 Hz [12] which is significantly larger (~ 50 times) than that for type 1 pili, which has been assessed to 0.016 ± 0.009 Hz. This implies that the LL bonds for P pili open, and thereby close, thermally much more often than those for type 1 pili. This is also in agreement with the fact that the balance rate and the corner velocity are significantly larger for P pili than for type 1 pili ($k_{\mathrm{AB}}^{\mathrm{LL,bal}}$ is 120 vs. 1.2 Hz, whereas $\dot{L}^{*}$ is 400 vs. 6 nm/s for the two types of pili, respectively). This indicates that that P pili reach their stationary force balance (i.e., $F_{\mathrm{II,UF}}(\dot{L} = 0)$) much faster than type 1 pili. Moreover, since the thermal bond opening rate, $k_{\mathrm{AB}}^{\mathrm{LL,th}}$, can be written as $\nu \exp\left(-\Delta V_{\mathrm{AT}}^{\mathrm{LL}}/kT\right)$, the two-orders-of-magnitude difference in $k_{\mathrm{AB}}^{\mathrm{LL,th}}$ indicates (under the condition that the attempt rates are the same) that the LL bonds in type 1 pili have an activation energy that is $4\,kT$ larger than that of P pili (27 and 23 kT, respectively).

How a pilus reacts to stress is not only given by the height of the transition state, also the bond length, $\Delta x_{\mathrm{AT}}^{\mathrm{LL}}$, affects the functioning of a bond. It has

---

[1] The reported unfolding force of type 1 pilus by Miller et al. [50] of 60 pN was performed for an elongation speed of 1–3 µm/s. Since this is significantly higher than the corner frequency, $\dot{L}^{*}$, their value is assumed to be an assessment of the unfolding force under dynamic conditions.

been found that the LL bond length is longer for P pili than for type 1 pili
$(0.76 \pm 0.11$ vs. 0.59 ± 0.06 nm).[2] This implies that the transition barrier for
type 1 pili is both higher and steeper than for P pili. This is in line with
the fact that type 1 pili have a slower thermal unfolding rate than P pili and
suggests that they are less sensitive to external forces.

It is interesting to note that the bond opening energy of the LL bonds of
both P and type 1 pili, ΔV_{AB}^{LL}, which were assessed to 24 and 37 kT, respectively, are *larger* than those of the transition barriers, ΔV_{AT}^{LL}, 23 and 27 kT,
respectively.[3] This implies that neither type 1 pili, nor P pili, possesses a
bound unfolded state in the absence of force, as was schematically indicated
in Fig. 18.4. There is therefore no activation energy for bond closure in the
absence of force. This implies that whenever a bond has opened spontaneously,
it will close directly (with the attempt rate). None of the structures will therefore be found in an unfolded state unless an external force is applied. Hence,
static images of unfolded pili are most likely a result of the treatment during
preparation.

18.4.4 Bacterial Detachment Under Stress

As mentioned earlier, experiments performed on multi-pili bacterial adhesion
under stress will not give an unambiguous picture of the fundamental interactions in the system, primarily due to a partly unknown interplay between
various pili. It is therefore necessary to first acquire a full understanding of
the combined elongation and adhesion properties of a single pilus. When this
has been done, it is possible to model how individual pili act cooperatively
in order to sustain a shear force and thereby contribute to the ability for a
multi-pili-binding bacterium to sustain the rinsing action of shear flow. Such
models can then be compared to findings from multi-pili experiments for verification. However, although the characterization of the elongation properties
of various pili has been done during the past years, and is still under way,
investigations of the properties of the adhesin have been done only recently.
There is so far only a limited amount of data about this phenomenon available. No full description of the combined action of rod and adhesin for the
total adhesive properties of a single pilus has therefore yet been developed.
On the other hand, based upon the models of elongation-and-contraction of
pili under stress that has recently been derived, e.g., those given in Sects.
18.3.1–18.3.3 earlier, it is still possible to hypothesize how the rod and the

---

[2] The value of the bond length of type 1 pili assessed by FMOT $(0.59 \pm 0.06$ nm),
was found to be significantly larger than that estimated from AFM studies $(0.26 \pm 0.01$ nm) [51]. We have presently no explanation to this discrepancy.

[3] The bond opening energy of the LL bonds for type 1 pili is in reasonable, although
not perfect, agreement with those assessed for type 1 pili by AFM, 26 kT (recalculated from 17 kT with an attempt rate of 10^6 to an attempt rate of 10^{10} Hz which
has been used throughout this work) by Miller et al. [50] and 29 kT reported by
Forero et al. [51].

adhesin of a single pilus jointly act in response to an external force as well as the expected response from a multi-pili attachment. Such hypotheses can then be verified by experiments.

Single Pilus Detachment

Section 18.3.5 provided expressions for the expected adhesion lifetime and elongation length of a pilus elongated in region II. For the case with low-elongation speed, i.e., when $\dot{L} < \dot{L}^*$, the expected adhesion lifetime is given by (18.13), whereas the expected elongation length is given by (18.15). These expressions indicate that for low speeds, i.e., for $\dot{L} < \dot{L}^*$, the adhesion lifetime is independent of the elongation speed and the expected elongation length is directly proportional to the elongation speed. This implies that the faster a pilus is elongated in region II (although still for $\dot{L} < \dot{L}^*$) the longer it can be elongated before being detached.

In the high-speed elongation case, $\dot{L} > \dot{L}^*$, on the other hand, (18.14) indicates that the expected lifetime of a single pilus decreases with increasing elongation speed, which is the normal behavior of a single bond. However, (18.16) shows that the expected elongation length is independent of the force for the case when the lengths of the LL and the adhesin bonds are equal. Interestingly, it also shows that if the bond length for the adhesin is shorter than that for the LL bond (i.e., $\rho < 1$), the expected elongation length *increases* with force. For other cases, the expected elongation length will scale exponentially with the applied force.

It has recently been found in an ongoing work [53] performed at three different speeds, and thereby for three different unfolding forces, that the expected elongation length for the PapG-galabiose binding of P pili indeed increases with the applied force, as suggested by this hypothesis. This implies, according to (18.16), that the length of the adhesin bond is shorter than that of the LL bond for P pili, i.e., $\rho < 1$. It also implies that the higher the loading rate, the longer the pilus will be elongated before it detaches. This increases, in turn, the possibility for the bacterial system to adopt a multi-pili binding (since other pili might become elongated into their region II before the adhesin of the first pilus rupture).

It has been shown recently that the typical lifetime of a PapG-galabiose bond, when exposed to the steady-state unfolding force (28 pN), is in the order of 10 s [48]. Since the corner velocity for a P pilus is 0.4 μm/s, and the length of region II is typically a few μm, it can be estimated that a single P pilus exposed to the steady-state unfolding force and elongated at an elongation speed close to the corner velocity will have a reasonable chance of detaching in region II. If the elongation is performed at a slower pace, the probability for detachment during the elongation in region II will increase. On the other hand, if the pilus does not detach in region II, it will enter region III where the force increases rapidly. This implies that the adhesin bond will open reasonably fast (within a time ≈ 10 s).

Multi-Pili Detachment

Multi-pili attachment is presumably of more importance than single-pilus attachment since this allows for an external force, F_{ext}, to be distributed among a multitude of pili. The exact distribution of forces will depend on a number of factors, e.g., the number of pili as well as their lengths and surface distribution, therefore only some general conclusions will be drawn here.

As long as the number of pili is sufficiently large, primarily above $F_{\text{ext}}/F_{\text{II,UF}}$, no pili will be elongated into region III. If the force is above $F_{\text{II,UF}}$, it will redistribute among several pili elongated either into region I or II. All pili in region II (N_{II}) will then experience the same expected lifetime, namely $\langle\tau\rangle_{\text{AB}}^{\text{Ad}}$, given by either (18.13) or (18.14). The expected lifetime before the first pilus detaches is then given roughly by $\langle\tau\rangle_{\text{AB}}^{\text{Ad}}/N_{\text{II}}$. When one pilus has detached, the force will be rapidly redistributed among the remaining pili. As long as there are a sufficient number of pili remaining, this procedure will be repeated.

However, when the remaining pili are few, approximately below $F_{\text{ext}}/F_{\text{II,UF}}$, at least one pilus will be elongated into region III, whereby it will experience a significantly shorter lifetime. When this takes place, there will be a rapid sequence of pili detachments, until the entire bacterium is detached.

Moreover, multi-pili attachment is far from intuitive. For example, if the first pilus to support the force is being elongated rapidly (i.e., into its dynamic regime), it is possible that a second pilus can become elongated and take up some of the force before the first pilus detaches. The force will then be distributed among the two pili, which, in turn, leads to detachment rates for the two pili that are lower than that of the first pilus elongated slowly. A fast elongation can therefore give rise to a longer lifetime time for bacterial adhesion than a slow elongation.

The details of multi-pili attachment have recently been understood [52], and work is therefore presently in progress to characterize these important phenomena. It is clear though that multi-pili mechanism can induce a strong binding to the host cells and may be an important key for the rinsing-resistance of a bacterium.

18.5 Summary and Conclusions

Two types of uropathogenic *E. coli* bacterial pili, P and type 1, which are predominantly expressed by *E. coli* in the upper and lower urinary tracts, have been scrutinized in some detail using FMOT. A model for the bond opening of the LL bond in the helix-like structure as well as the opening of the HT bond between consecutive subunits in the rod, based upon an energy landscape model and a kinetic model (of sticky-chain type) for the bond opening, has been developed. It has been possible to characterize virtually all parameters in this model by the use of single pilus force-vs.-elongation measurements on individual pili. This has given rise to a model for the elongation properties of a single pilus under strain/stress. Lately, also information about the adhesin at the tip of the attachment organelle has been assessed. The first step toward a combined description of the elongation and adhesion properties of pili-expressed bacterial adhesion has thereby been taken. This analysis indicates that the unique properties of the pili provide the bacteria with an extraordinary ability to sustain significant shear forces, forces that can widely supersede the binding force of a single (nonpiliated) adhesin. Moreover, the results for the different pili provide information that presumably can be correlated to the particular environment in which they are found. For example, the steeper potential of type 1 pili may be considered a consequence of the fact that the structure has evolved to support higher forces during shorter time events, which would correlate with the irregular urine flow in the bladder and the urethra as compared to the more constant urine flow in the upper urinary tracts. It is therefore possible that the much stiffer bond potential for type 1 pili is optimized for a fast shock damping effect. All this indicates that different types of pilus can have significantly dissimilar biophysical properties and that they therefore presumably are optimized for dissimilar environments.

Further work with other pili as well as analysis of multi-pili responses of the studied systems is on its way. With this information on hand, a significantly improved understanding of adhesion by piliated bacteria can be gained. Based on this knowledge, the search for new targets for anti-adhesion drugs in bacteria which possibly could lead to new means to combat bacterial infections, including those from antibiotic resistant bacterial strains [20,54,55], can be initiated.

Acknowledgments

This work was supported by the Swedish Research Council under the project 621–2005–4662 and performed within the Umeå Centre for Microbial Research (UCMR). The authors acknowledge economical support for the construction of a force-measuring optical tweezers system from the Kempe foundation and from Magnus Bergvall's foundation. The "Fondation Pour La Recherche Médicale" is also acknowledged for the French post-doctoral fellowship awarded to Mickaël Castelain (grant no. SPE20071211235).

References

1. W.E. Thomas, E. Trintchina, M. Forero, V. Vogel, E.V. Sokurenko, Bacterial adhesion to target cells enhanced by shear force. Cell 109, 913–923 (2002)
2. B.N. Anderson, A.M. Ding, L.M. Nilsson, K. Kusuma, V. Tchesnokova, V. Vogel, E.V. Sokurenko, W.E. Thomas, Weak rolling adhesion enhances bacterial surface colonization, 10.1128/JB.00899–06. J. Bacteriol. 189, 1794–1802 (2007)

3. F. Jacobdubuisson, J. Heuser, K. Dodson, S. Normark, S. Hultgren, Initiation of assembly and association of the structural elements of a bacterial pilus depend on 2 specialized tip proteins. EMBO J. 12, 837–847 (1993)
4. J.R. Johnson, T.A. Russo, Uropathogenic escherichia coli as agents of diverse non-urinary tract extraintestinal infections. J. Infect. Dis. 186, 859–864 (2002)
5. T.A. Russo, J.R. Johnson, Medical and economic impact of extraintestinal infections due to Escherichia coli: focus on an increasingly important endemic problem. Microbes Infect. 5, 449–456 (2003)
6. G. Källenius, S.B. Svenson, H. Hultberg, R. Möllby, I. Helin, B. Cedergren, J. Winberg, Occurrence of P-fimbriated Escherichia coli in urinary tract infections. Lancet. 2, 1369–1372 (1981)
7. J. Ruiz, K. Simon, J.P. Horcajada, M. Velasco, M. Barranco, G. Roig, A. Moreno-Martinez, J.A. Martinez, T.J. de Anta, J. Mensa, J. Vila, Differences in virulence factors among clinical isolates of Escherichia coli causing cystitis and pyelonephritis in women and prostatitis in men. J. Clin. Microbiol. 40, 4445–4449 (2002)
8. E. Bullitt, L. Makowski, Structural polymorphism of bacterial adhesion pili. Nature 373, 164–167 (1995)
9. E. Hahn, P. Wild, U. Hermanns, P. Sebbel, R. Glockshuber, M. Haner, N. Taschner, P. Burkhard, U. Aebi, S.A. Muller, Exploring the 3D molecular architecture of Escherichia coli type 1 pili. J. Mol. Biol. 323, 845–857 (2002)
10. M. Forero, W.E. Thomas, C. Bland, L.M. Nilsson, E.V. Sokurenko, V. Vogel, A catch-bond based nanoadhesive sensitive to shear stress. Nano Lett. 4, 1593–1597 (2004)
11. B. Lund, F. Lindberg, B.I. Marklund, S. Normark, The papg protein is the alpha-D-galactopyranosyl-(1- 4)-beta-D-galactopyranose-binding adhesin of uropathogenic escherichia-coli. Proc. Natl. Acad. Sci. USA. 84, 5898–5902 (1987)
12. M. Andersson, E. Fällman, B.E. Uhlin, O. Axner, Dynamic force spectroscopy of the unfolding of P pili. Biophys. J. 91, 2717–2725 (2006)
13. W.E. Thomas, L.M. Nilsson, M. Forero, E.V. Sokurenko, V. Vogel, Shear-dependent 'stick-and-roll' adhesion of type 1 fimbriated Escherichia coli. Mol. Microbiol. 53, 1545–1557 (2004)
14. J. Jass, S. Schedin, E. Fällman, J. Ohlsson, U. Nilsson, B.E. Uhlin, O. Axner, Physical properties of Escherichia coli P pili measured by optical tweezers. Biophys. J. 87, 4271–4283 (2004)
15. E. Fällman, M. Andersson, S. Schedin, J. Jass, B.E. Uhlin, O. Axner, Dynamic properties of bacterial pili measured by optical tweezers. SPIE 5514, 763–733 (2004)
16. E. Fällman, S. Schedin, J. Jass, B.E. Uhlin,O. Axner, The unfolding of the P pili quaternary structure by stretching is reversible, not plastic. EMBO Rep. 6, 52–56 (2005)
17. M. Andersson, E. Fällman, B.E. Uhlin, O. Axner, A sticky chain model of the elongation of Escherichia coli P pili under strain. Biophys. J. 90, 1521–1534 (2006)
18. M. Andersson, E. Fällman, B.E. Uhlin, O. Axner, Technique for determination of the number of PapA units in an E coli P pilus. SPIE. 6088, 326–337 (2006)
19. M. Andersson, O. Axner, B.E. Uhlin, E. Fällman, Optical tweezers for single molecule force spectroscopy on bacterial adhesion organelles. SPIE. 6326, 1–12 (2006)

20. V. Åberg, E. Fällman, O. Axner, B.E. Uhlin, S.J. Hultgren, F. Almqvist, Pili-cides regulate pili expression in E-coli without affecting the functional properties of the pilus rod. Mol. Biosyst. 3, 214–218 (2007)
21. M. Andersson, B.E. Uhlin, E. Fällman, The biomechanical properties of E. coli pili for urinary tract attachment reflect the host environment. Biophys. J. 93, 3008–3014 (2007)
22. O. Björnham, O. Axner, M. Andersson, Modeling of the elongation and retraction of Escherichia coli P pili under strain by Monte Carlo simulations. Eur. Biophys. J. Biophys. Lett. 37, 381–391 (2008)
23. M. Andersson, O. Axner, F. Almqvist, B.E. Uhlin, E. Fällman, Physical properties of biopolyrners assessed by optical tweezers: Analysis of folding and refolding of bacterial pili. Chemphyschem. 9, 221–235 (2008)
24. M. Castelain, A.E. Sjöström, E. Fällman, B.E. Uhlin, M. Andersson. Unfolding and refolding properties of S pili on extraintestinal pathogenic Escherichia coli. Eur. Biophys. J. DOI: 10.1007/s00249-009-0552-8 (2009)
25. K.C. Neuman, S.M. Block, Optical trapping. Rev. Sci. Instr. 75, 2787–2809 (2004)
26. A. Ashkin, Acceleration and trapping of particles by radiation pressure. Phys. Rev. Lett. 24, 156–159 (1970)
27. A. Ashkin, Optical levitation by radiation pressure. Appl. Phys. Lett. 19, 283–285 (1971)
28. A. Ashkin, J.M. Dziedzic, J.E. Bjorkholm, S. Chu, Observation of a single beam gradient force optical trap for dielectric particles. Opt. Lett. 11, 288–290 (1986)
29. M.J. Lang, S.M. Block, Resource letter: LBOT-1: Laser-based optical tweezers. Am. J. Phys. 71, 201–215 (2003)
30. S.M. Block, Making light work with optical tweezers. Nature 360, 493–495 (1992)
31. K. Svoboda, S.M. Block, Biological applications of optical forces. Annu. Rev. Biophys. Biomem. 23, 247–285 (1994)
32. W.J. Greenleaf, M.T. Woodside, E.A. Abbondanzieri, S.M. Block, Passive all-optical force clamp for high-resolution laser trapping. Phys. Rev. Lett. 95, 208102 (2005)
33. C. Bustamante, J.C. Macosko, G.J.L. Wuite, Grabbing the cat by the tail: Manipulating molecules one by one. Nat. Rev. Mol. Cell Biol. 1, 130–136 (2000)
34. C. Bustamante, Z. Bryant, S.B. Smith, Ten years of tension: single-molecule DNA mechanics. Nature 421, 423–427 (2003)
35. E. Fällman, O. Axner, Design for fully steerable dual-trap optical tweezers. Appl. Opt. 36, 2107–2113 (1997)
36. E. Fällman, S. Schedin, J. Jass, M. Andersson, B.E. Uhlin, O. Axner, Optical tweezers based force measurement system for quantitating binding interactions: system design and application for the study of bacterial adhesion. Biosens. Bioelectron. 19, 1429–1437 (2004)
37. M. Andersson, E. Fällman, B.E. Uhlin, O. Axner, Force measuring optical tweezers system for long time measurements of Pili stability. SPIE. 6088, 286–295 (2006)
38. F. Gittes, G.F. Schmidt, Signals and noise in micromechanical measurements. Meth. Cell Biol. 55, 129–156 (1998)
39. M.G. Bell, Models for the specific adhesion of cells to cells. Science 200, 618–627 (1978)
40. E. Evans, Probing the relation between force - lifetime - and chemistry in single molecular bonds. Annu. Rev. Biophys. Biomem. 30, 105–128 (2001)

41. E. Evans, K. Ritchie, Dynamic strength of molecular adhesion bonds. Biophys. J. 72, 1541–1555 (1997)

42. E. Evans, Energy landscapes of biomolecular adhesion and receptor anchoring at interfaces explored with dynamic force spectroscopy. Faraday Discussions 1–16 (1998)

43. E. Evans, Looking inside molecular bonds at biological interfaces with dynamic force spectroscopy. Biophys. Chem. 82, 83–97 (1999)

44. E. Evans, K. Ritchie, Strength of a weak bond connecting flexible polymer chains. Biophys. J. 76, 2439–2447 (1999)

45. I. Jäger, The "sticky chain": A kinetic model for the deformation of biological macromolecules. Biophys. J. 81, 1897–1906 (2001)

46. R. Merkel, P. Nassoy, A. Leung, K. Ritchie, E. Evans, Energy landscapes of receptor-ligand bonds explored with dynamic force spectroscopy. Nature 397, 50–53 (1999)

47. T. Strunz, K. Oroszlan, I. Schumakovitch, H.J. Guntherodt, M. Hegner, Model energy landscapes and the force-induced dissociation of ligand-receptor bonds, Biophys. J. 79, 1206–1212 (2000)

48. H. Nilsson, P pili elongation generates a dynamic force response suggesting a slip-bond mechanism for the PapG-Galabiose adhesin-receptor pair, Master thesis, Umeå University, 2008

49. R.A. Lugmaier, S. Schedin, F. Kuhner, M. Benoit, Dynamic restacking of Escherichia coli P-pili. Eur. Biophys. J. Biophys. Lett. 37, 111–120 (2008)

50. E. Miller, T.I. Garcia, S. Hultgren, A. Oberhauser, The mechanical properties of E. coli type 1 pili measured by atomic force microscopy techniques. Biophys. J. 91, 3848–3856 (2006)

51. M. Forero, O. Yakovenko, E.V. Sokurenko, W.E. Thomas, V. Vogel, Uncoiling mechanics of escherichia coli type I fimbriae are optimized for catch bonds. PLoS Biol. 4, 1509–1516 (2006)

52. O. Björnham, H. Nilsson, M. Andersson, S. Schedin. Physical properties of the specific PapG-galabiose binding in E. coli P pili-mediated adhesion. Eur. Biophys. J. 38, 245–254 (2009)

53. O. Björnham, O. Axner. Multipili attachment of bacteria exposed to stress. J. Chem. Phys. 130, 235102 (2009)

54. A. Svensson, A. Larsson, H. Emtenas, M. Hedenstrom, T. Fex, S.J. Hultgren, J.S. Pinkner, F. Almqvist, J. Kihlberg, Design and evaluation of pilicides: Potential novel antibacterial agents directed against uropathogenic Escherichia coli. Chembiochem. 2, 915–918 (2001)

55. V. Åberg, F. Almqvist, Pilicides—small molecules targeting bacterial virulence. OrganicBiomol. Chem. 5, 1827–1834 (2007)

Part VII

Nanoscale Microscopy and High Resolution
Imaging

Far-Field Optical Nanoscopy

Stefan W. Hell

Summary. Since the discovery of the diffraction barrier in the nineteenth century, it has been commonly accepted that a lens-based (far-field) optical microscope cannot discern structural details much finer than about half the wavelength of light ($\lambda/2$). However, in the early 1990s, a quest toward higher resolution began, which led to the discovery that the diffraction barrier of far-field fluorescence microscopy can be radically overcome using basic molecular transitions. This chapter discusses the initial and more recent concepts that can provide far-field optical resolution down to the molecular scale. It is shown that all concepts reported and implemented to date exploit a transition between a bright and a dark state to switch the fluorescence capability of molecules such that adjacent objects or molecules emit sequentially in time. Some of these concepts can be extended to signal-giving mechanisms other than fluorescence. Likewise, purely transition-based concepts, such as stimulated emission depletion (STED) microscopy, can in principle be extended to explore the molecule itself. Emergent far-field fluorescence nanoscopy will impact not only the life sciences but also other areas that benefit from nanoscale three-dimensional (3D) mapping with conventional lenses and propagating light.

19.1 Introduction and Overview

By providing a spatial resolution down to the atomic scale, electron and scanning probe microscopy have revolutionized our understanding of life and matter. Nonetheless, optical microscopy has maintained its key role in many fields, in particular in the life sciences. This stems from a number of rather exclusive advantages, such as the noninvasive access to the interior of (living) cells and the specific and highly sensitive detection of cellular constituents through fluorescence tagging. As a matter of fact, lens-based fluorescence microscopy would be almost ideal for investigating the three-dimensional (3D) cellular interior if it could resolve details far below the wavelength of light. However, until not very long ago, obtaining a spatial resolution on the nanometer scale with an optical microscope that uses lenses and focused visible light was considered unfeasible [1, 2].

Focusing a propagating light wave means causing it to interfere constructively at a certain point in space, called the geometrical focal point (0,0,0). Due to diffraction a focal intensity pattern $I(x, y, z)$ emerges around (0,0,0), which is also referred to as the intensity point-spread-function (PSF) of the lens. $I(x, y, z)$ features a central maximum called the focal spot (Fig. 19.1a) whose full-width-half-maximum (FWHM) is $\Delta r \approx \lambda/(2n \sin \alpha)$ in the focal plane and $\Delta z \approx \lambda/(n \sin^2 \alpha)$ along the optic axis [3]. λ is the wavelength of light, α denotes the semi-aperture angle of the lens, and n is the refractive index of the object medium (Fig. 19.1a). Discerning similar objects lying within this spot is usually precluded because they are illuminated in parallel and hence give off (fluorescence) photons in parallel. Likewise, the propagation of the emitted (fluorescence) light that is collected by a lens and focused to an image plane is governed by a similar function $I_{em}(x, y, z)$, describing the blur of the coordinate from where the photons originated.

A logical consequence of the fact that light cannot be focused more sharply than the diffraction limit was to give up propagating waves and lenses and confine the light by means of a subdiffraction-sized aperture or tip. Placing this aperture in sub-$\lambda/2$ proximity to the object and scanning it across the object renders optical images with subdiffraction resolution. Proposed by Synge [4] in 1928 and again by Ash and Nicholls in 1972 [5], this concept was invigorated in the wake of the invention of the scanning tunneling microscope as near-field optical microscopy [6,7]. A fundamental difference of near-field optical microscopy to its lens-based (far-field) counterpart is that it relies on non-propagating, evanescent light fields fading out exponentially within distance $\sim\lambda/2$ from the object. Giving up focusing therefore comes at a high price: one is bound to imaging surfaces.

Because it also relies on collecting and amplifying evanescent waves, the same practical limitation applies to the recently introduced lens of negative refractive index [8]. In its currently most sophisticated version, called hyperlens [9,10], the evanescent waves are converted into propagating waves forming a magnified image of the sample on a distant screen, which is why one may think that it is far-field imaging. However, the projection to a distant screen does not change the fact that the hyperlens relies on the sample's near-field. Hence, in its current state of development, a hyperlens is not a far-field imaging device [11], but a fascinating non-scanning concept of employing the near-field.

In the twentieth century, several ideas have been put forth to address the resolution problem in the far-field as well. For example, in 1956, Toraldo di Francia suggested shrinking the central focal spot by applying an elaborate phase pattern in the entrance pupil of the objective lens [12]. Unfortunately, the creation of smaller central spots is accompanied with giant sidelobes rendering this concept impractical. In 1966, Lukosz [13] suggested that the use of gratings for the illumination and/or the detection pathway should improve the resolution in reflection imaging, but this concept yields a factor of two at most [5]. While he suggested that the resolution can be improved by reducing the microscope's field of view (which is also inherent in Synge's near-field idea),

a Confocal   **b** 4Pi   **c** STED

Single Point Versions

$\dfrac{\lambda}{2n\sin\alpha}$ → ← 200nm r

~500nm z α

Exc. STED Eff.Spot

y x r_i → ← ~20nm

z x $\dfrac{\lambda}{2n\sin\alpha\sqrt{1+I/I_s}}$

Lens

~90nm α=64-74°

d   **e**

Parallelized Versions

RESOLFT: STED, GSD   SPEM/SSIM   PALM/STORM

$>\lambda/2n$ Switch Ensembles

$p_A(r)$
$1-p_A(r)$
$I(r)$
Object

r_i r_i r

A ⟷ B   Lens System   B → A → B

Camera   Left-out Signal

Offline Deconvolution

Σ   Result

Switch Single Molecules

$>\lambda/2n$ B A B'

r

Centroid

Targeted Read-Out   r   Stochastic Read-Out r

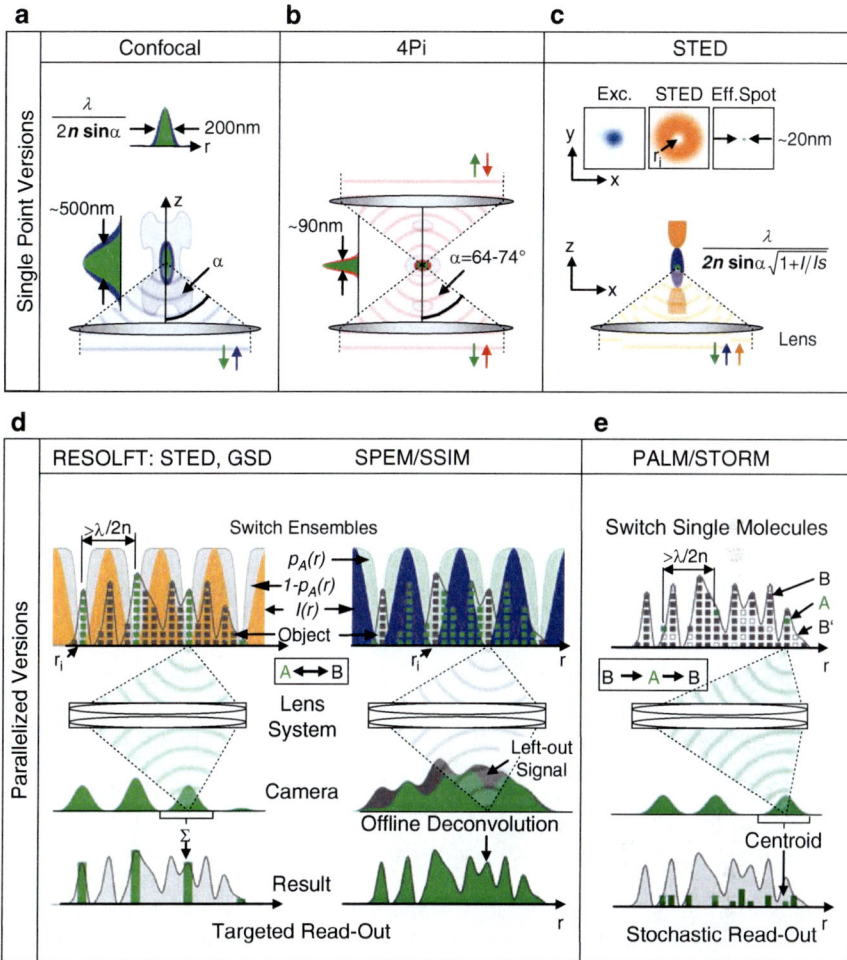

Fig. 19.1. Optical layouts and concepts for far-field fluorescence nanoscopy. (a) Confocal microscopy is shown here as a diffraction-limited reference. The excitation light wave (*blue*) is formed by the lens to a spherical wavefront cap that results in a 3D diffraction spot exciting the fluorophores in the focal region. A point-like detector (not shown) collects the fluorescence primarily from the main diffraction maximum shown in *dark green*, thus providing a slightly improved resolution over conventional fluorescence microscopy. Yet, the resolution of a confocal microscope is limited by diffraction to a full-width-half-maximum (FWHM) >200 nm in the focal plane (x, y) and to >450 nm along the optical (z) axis. (b) 4Pi microscopy improves the z-resolution by coherently combining the wavefront caps of two opposing lenses; the concept renders a main spot featuring an FWHM of 70–150 nm along the z-axis. (c) A typical single-point implementation of a STED microscope uses a focused excitation beam (*blue*) that is superimposed by a doughnut-shaped STED beam (*orange*) to keep molecules dark by quenching excited molecules through stimulated emission. In regions where the STED beam intensity is beyond a threshold I_s, the STED beam essentially switches the fluorophores off by nailing them down to the ground state.

no concrete indication was given as to how a reduction by more than 2-fold could be realized with focused light. In 1978, it was speculated that a hypothetical elliptic "4π point hologram" would be able to focus light waves to a subdiffraction-sized spot or simply to a point, thus overcoming the diffraction

---

Fig. 19.1 (Continued). By ensuring that the doughnut intensity I exceeds I_s in a large area, the spot (*green*) in which the fluorophore can still be bright and active is confined to subdiffraction dimensions. Here, a measured 20-nm diameter spot is shown, which is approximately ten times below the diffraction barrier. Scanning such a subdiffraction sized spot across the sample yields subdiffraction images. (**d**) The concepts STED, GSD, (left-hand side) and SPEM/SSIM (right-hand side) can be viewed as special cases of a more general concept called RESOLFT. A hallmark of this generalized concept is that it utilizes focal light distributions $I(r)$ with zero-intensity points at positions r_i, to confine either a bright (A) or a dark fluorophore state (B) in space. The zeros are preferably $> \lambda/(2n)$ apart in the focal plane. Two examples of this generalized concept are shown. *Left*: the intensity drives a transition $A \rightarrow B$ to confine the *bright* state A in space. This is the case for a parallelized STED, GSD, or a RESOLFT approach using reversibly photoactivatable proteins or photochromic dyes. *Right*: in the SPEM/SSIM concept the intensity $I(r)$ drives a transition $B \rightarrow A$ that confines the *dark* state B in space. In both cases, the positions of state A or B are predefined in space by $I(r)$ and r_i. When imaged onto a camera the steep regions of state A (*left*) or state B (*right*) become blurred. However, the diffraction blur can be dealt with (as shown in the left-hand panel STED, GSD) by allocating the signal (from the diffraction blob) to the known coordinate r_i of the zero in the sample space. The image is gained by scanning the array of zeros (r_i) across the sample and recording the fluorescence for each step. The diffraction blur can also be dealt with for SPEM/SSIM (right-hand panel) because the superresolved data is encoded in the steeply confined dark regions around r_i of state B. Since SPEM/SSIM initially produces a "negative data set," the SPEM/SSIM image is finally gained by mathematically converting the negative data set into a positive one. The small boxes in the sketches symbolize the fluorophore molecules that make up the object. $p_A(r) \leqslant 1$ defines the normalized probability of occurrence of state A. Although all these RESOLFT concepts are suitable to detect single molecules, they generally operate with ensembles. Since the position at which the fluorophores are in A or B and hence emitting is predefined by the zero intensity points r_i, the RESOLFT strategy has also been named the "targeted" read-out mode. (**e**) The single molecule switching concepts (PALM/STORM) do not define the region where signal is emitted, but read out the fluorophores of the object stochastically, molecule by molecule. Individual fluorophores are sparsely switched to a specific bright state A that is able to emit $m \gg 1$ photons before the molecule returns to B. The detection of $m \gg 1$ photons enables the calculation of the centroid of the diffraction blob of individual molecules when imaged onto a camera. Thus it is possible to assemble an image consisting of centroid tickmarks with a statistically variable resolution depending on m. The concepts (**c–e**), i.e., STED, GSD, RESOLFT, SPEM/SSIM, and PALM/STORM are not limited by diffraction, meaning that they can resolve similar molecules at nanometer distances

barrier [14]. However, as no near-field component is relayed in this case, a convergence to a subdiffraction spot or even a point is impossible.

Confocal fluorescence microscopy has also been connected with resolution improvement (Fig. 19.1a) [15–17]. Illuminating with a diffraction limited focused spot and detecting with a symmetrically arranged point detector, the effective focal spot, i.e., the effective PSF of this microscope is described by the product of the diffraction pattern for illumination and for detection: $I(x, y, z) I_{em}(x, y, z) \approx I^2(x, y, z)$. The multiplication of intensities and the nearly quadratic dependence on the intensity in this formula reduces the FWHM of its central spot by $\sim \sqrt{2}$. In the Fourier domain, the multiplication expands the optical bandwidth of spatial frequencies even by a factor of two. In practice, however, $I(x, y, z)$ and $I_{em}(x, y, z)$ are not really identical and the detector is not point-like [16, 17], meaning that the process is not really quadratic and the limited bandwidth expansion by a factor of two not practically realized. However, even if it were, the newly gained higher frequencies are heavily damped, which is why confocal microscopy did not really provide a higher resolution. Its actual benefit was the improved 3D-imaging and superb background rejection [16, 17].

A genuine quadratic dependence of the measured fluorescence signal on the focal illumination intensity $I(x, y, z)^2$ is provided by (nonlinear) two-photon excitation [18–20]. Exciting a fluorophore from its ground state S_0 to its fluorescent state S_1 requires two photons of half of the difference in energy between the two states, i.e. light of 2λ. The doubling in wavelength means that the size of the diffraction spot of excitation light is also doubled [20], which, unfortunately, is not compensated for by the $\sim \sqrt{2}$ FWHM reduction stemming from the quadratic nonlinearity. As a result, the resolution of a two-photon excitation microscope is usually slightly poorer than its one-photon counterpart, and the spatial frequency bandwidth is not expanded in absolute terms [21]. The same arguments are valid for m-photon absorption processes, because they usually require an even longer wavelength $m\lambda$, let alone the requirement for huge intensities and the low cross-section. However, even if m-fold longer wavelengths were not needed (as is the case for higher order scattering events), as long as m is finite, the resolution barrier is only shifted, not "broken." "Breaking" implies that the limiting role of diffraction is lifted and that the resolution can be increased, at least conceptually, to the molecular scale or even beyond. Clearly, multiphoton and confocal microscopes do not fulfill this criterion.

Purely mathematical approaches using the PSF of the system and/or a priori object information rarely exceeded a factor of two [17, 22, 23] and were prone to producing artifacts. They can be augmented through additional a priori constraints, such as the objects featuring different absorption or emission spectra [24]. In this case, the resolution problem can become almost trivial, because objects with different spectra can be separated with suitable spectral filters. However, because of the difficulty to mark all features in a sample with different labels, reducing the resolution problem to a color separation followed by data computation was not an effective pathway either. In my view, the inability of all these superresolution ideas to provide noteworthy improvements cemented the belief of the twentieth century that apart from a factor of two, perhaps, at the end of the day, the resolution of any lens-based light microscope is limited to $\Delta r > 200\,\text{nm} \approx \lambda/2$ and $\Delta z > 450\,\text{nm} \approx \lambda$.

Therefore, until the early 1990s the general belief was that in order to seriously improve the resolution in the visible range one has to discard lenses and resort to near-field optics [25,26]. Consequently, major efforts were undertaken to develop this technique, including for biological imaging [27,28]. While near-field optics became very useful in many areas [29], its restriction to surfaces remained a drawback. At the same time, the importance of lens-based fluorescence microscopy grew, due to the advent of 3D imaging [17] and a myriad of fluorescent markers, including that of the green-fluorescent-protein (GFP) and its derivatives [30,31].

For these reasons, attaining nanoscale resolution with focused light became worthwhile to pursue [32–35]. Moreover, in light of the notion that it was considered impossible, exploring ways to realize nanoscale resolution with regular lenses became a fascinating scientific quest.

The breaking of the diffraction barrier of far-field optical microscopy as we know it today was born out of the insight that the key to far-field nanoscale resolution is held by the fluorophore itself and its interstate transitions [34–36]. This approach was in stark contrast to that of near-field optics which was built on the modification of the propagation of light. The philosophy conveyed in these papers was that there must be transitions in a dye that, when properly implemented in the image formation, should neutralize the limiting role of diffraction [34]. Viewing fluorophores as facilitators of nanoscale far-field optical resolution was a major change in the perception of the fluorophore's role and capability in microscopy, because until then fluorophores were primarily regarded as indicators of molecular species or of physiological parameters such as ion concentrations.

Based on this philosophy, stimulated emission depletion (STED) microscopy [35] and ground state depletion (GSD) microscopy [36] emerged as the first concrete and viable physical concepts to fundamentally overcome the limiting role of diffraction in a lens-based optical microscope. In a nutshell, STED and GSD use a selected pair of bright (fluorescent) and dark fluorophore states to restrict the bright state to subdiffraction dimensions. This is accomplished by resorting to optical transitions that transiently switch off the ability of the dye to fluoresce by confining the dye to a dark state. The transition is effected with a light intensity distribution featuring a zero, switching the fluorescence off everywhere except at zero where the fluorophore is still allowed to be bright. Translating the zero across the specimen switches the signal of adjacent features sequentially on and off, allowing their separate registration. The spatial confinement of molecular states rather than of the light (as in near-field optics) neutralized the limiting role of diffraction in a natural way. Thus, STED and GSD microscopy radically departed from the superresolution approaches mentioned earlier. They disentangled the wavelength from its resolution-limiting role and suggested that "infinite resolution" [35] was possible without eliminating diffraction per se.

Other proposals to improve *far*-field fluorescence microscopy resolution [21, 37] have followed, which also placed fluorophore transitions at the center stage. Concretely, the concept of STED and GSD microscopy was expanded to photoswitching molecules, specifically of synthetic organic molecules or photoactivatable fluorescent proteins [38–40], in an approach dubbed RESOLFT [41, 42], standing for reversible saturable/switchable optical linear (fluorescence) transitions. A hallmark of all these concepts is that they yield images without mathematical processing and without any assumptions about the object or the performance of the lens. They are purely "physical" or "physicochemical" concepts, since the superresolution image is a direct consequence of the molecular transition employed. Another hallmark is that they define, by the position of the intensity zero of the light used, where the molecules are "on" and where they are "off," in other words, where the bright and where the dark states are established. They operate with any number of molecules, from single to many.

The concept of switching is also essential in more recent far-field fluorescence nanoscopy approaches [43–45], which differ from the previous ones by the fact that they switch molecules stochastically in space and utilize mathematics to assemble the image. Also very powerful, these concepts complement the earlier approaches. Therefore, after a brief excursion to an important all-optical improvement of the microscope's axial resolution, I will review the field of far-field optical nanoscopy with emphasis on the breaking of the diffraction barrier. In particular, I will show that all fluorescence nanoscopy concepts realized so far utilize a transition between two distinguishable states of a marker, a bright and a dark state, to record fluorescent objects at sub-$\lambda/2$ distances sequentially in time. The transition between these states is operated as a fluorescence switch. Additionally, I will classify these concepts according to the states used and show that they differ on whether the switching and sequential recording occurs at predefined sample coordinates (and thus inherently on molecular ensembles) or stochastically in space, molecule by molecule.

19.2 Improving the Axial Resolution by Combining the Aperture of Two Lenses

The z-resolution of any standard far-field light microscope is at least three times poorer than that in the focal plane which is particularly limited in 3D-imaging transparent objects such as cells. Therefore, in the quest for nanoscale resolution in far-field optical microscopy, it was most natural to start out with the axial resolution problem.

The reason for the poorer axial resolution is that the focal diffraction spot is elongated along the optic axis ($\Delta z > \Delta r$). This elongation stems from the fact that the focal spot is formed by the self-interference of a spheri-cal wave front cap [3]. If the wavefront were a sphere instead of a cap, it would yield a spot that would be nearly spherical and hence a z-resolution that would be similar to its lateral counterpart. In other words, the lack of symmetry in focusing with respect to the focal point leads to a poorer z-confinement of the diffraction maximum. The same consideration holds for collecting the nearly spherical wavefront of fluorescence emitted from the dye [32, 46]. Expanding the wavefront for illumination or fluorescence collection is equivalent to expanding the microscope's aperture. It is the key physical element in spot-scanning 4Pi microscopy [32, 33, 46] and also in the widefield I^5M [47].

4Pi microscopy improves the total aperture of a far-field fluorescence microscope through coherently adding the light fields of the spherical wave-front caps of two large-angle lenses for excitation or detection, or for both (Fig. 19.1b) [32, 46]. The two wavefront caps of excitation light interfere con-structively at the common focal point, whereas the two emerging fluorescence wavefront caps interefere constructively at a common point of detection. Had each of the lenses a semi-aperture angle, $\alpha = 90°$, the central focal diffraction spot would approach a spherical shape with a diameter of $\sim \lambda/3n$. Unfor-tunately, the available lenses feature only $\alpha \approx 68° < 90°$ which means that part of the nearly spherical wavefront is missing. The axially sharpened main diffraction spot of $\Delta z \approx \lambda/3n$ is accompanied by smaller yet pronounced side-lobes above and below the focal plane [46, 48], making the axial separation of objects ambiguous. Removing the (effect of the) lobes and making the 3D imaging unambiguous was the actual scientific challenge in this quest [49].

The first solution to this problem was to excite the fluorophore by two-photon absorption [33], because, among other effects, the quadratic dependence of the fluorescence on the excitation intensity rendered the con-tributions from the lobes so small that they could be unambiguously removed in the data by mathematical deconvolution [50]. As a result, two-photon exci-tation 4Pi microscopy provided images with substantially improved axial resolution in far-field fluorescence imaging [48, 51]. Microtubules or actin fibers, that were mingled in a confocal xz-image of a cell ($\Delta z \approx 500$–600 nm) could be clearly separated by 4Pi recording ($\Delta z \approx 140$ nm) [51–53]. Further developments, including that of multiple spot scanning and of a compact com-mercial instrument, provided a range of $\Delta z = 70$–150 nm in the 3D-imaging both in fixed and living cells [54–56]. 4Pi instruments have provided axially superresolved 3D images of H2AX chromatin clusters in the nucleus [57] in fixed cells, as well as of the mitochondrial 3D network and the Golgi apparatus in living cells [54, 58].

Novel field-corrected lenses of $\alpha = 74°$ have recently enabled dual-color 4Pi recordings also with regular one-photon excitation [59, 60]. Still applying two-photon excitation with such lenses yielded a solitary central spot of $\Delta z \sim \lambda/3n$ with negligible lobes [61] (Fig. 19.1b). The creation of such a nearly spherical solitary focal spot in 3D just by physics and its application to imaging has been a longstanding goal in the quest for far-field axial superresolution.

To provide widefield imaging, I^5M [47,62,63] illuminates with plane (unfocused) interfering waves, rather than focused spherical wavefronts. Plane waves yield a standing wave pattern with many equidistant $(\lambda/2n)$ layers of equal brightness along the z-axis. These multiple peaks render z-separation ambiguous, unless for the trivial case that the object is thinner than $< \lambda/2n$. This is why the so-called standing wave microscope (SWM) [64] strictly fails to provide axial superresolution for an arbitrary object in 3D [49,58,65,66]. In the spatial frequency domain, the ambiguity is explained by the missing frequency bands inside the optical transfer function (OTF). I^5M alleviates this problem by collecting the fluorescence like a 4Pi microscope, i.e., by coherently adding the spherical wavefronts caps of fluorescence light at a common detection point [49]. Compared to 4Pi microscopy, I^5M trades off focused 4Pi-like illumination for widefield imaging capability. While it thus is more prone to ambiguities [67], I^5M, unlike SWM, clearly is a self-consistent concept for axial resolution improvement.

The practical z-resolution increase from 400–800 nm down to 70–150 nm, i.e., by 3–7-fold, constituted the first substantial resolution improvement in far-field optical microscopy in many decades [48, 51–53, 63]. As a result, 4Pi and I^5M imaging most visibly challenged the notion of the time that far-field resolution was a closed matter and near-field optics the only way to go.

On the same note, 4Pi microscopy features the largest aperture and hence the smallest focal spot. Yet it does not break the diffraction barrier; it just pushes diffraction to its limits.

19.3 Breaking the Diffraction Barrier

Let us assume an unknown number of tiny fluorescent objects or molecules that are $D < 200$ nm apart. If they all have different colors, one can separate them with the right excitation or emission filters. Discerning objects or molecules having distinct spectral characteristics is not really challenged by diffraction. The problem was to discern an *arbitrary* number of "identical" features at *arbitrary* distances $< \lambda/(2n \sin\alpha)$.

So, which phenomenon provides a solution for separating "identical" (fluorescent) objects closer than $\lambda/(2n \sin\alpha)$? The answer is switching the fluorescence (or more generally the signaling) capability of these adjacent objects on and off so that they can be registered separately. This is exactly how scanning STED microscopy resolves objects that are closer than the diffraction limit, whereby the switching of the fluorescence capability of the molecules is accomplished with a (STED) beam [35]. As a matter of fact, fluorescence switching or modulation is used in all far-field fluorescence microscopy concepts with "diffraction-unlimited" resolution currently employed (Fig. 19.2).

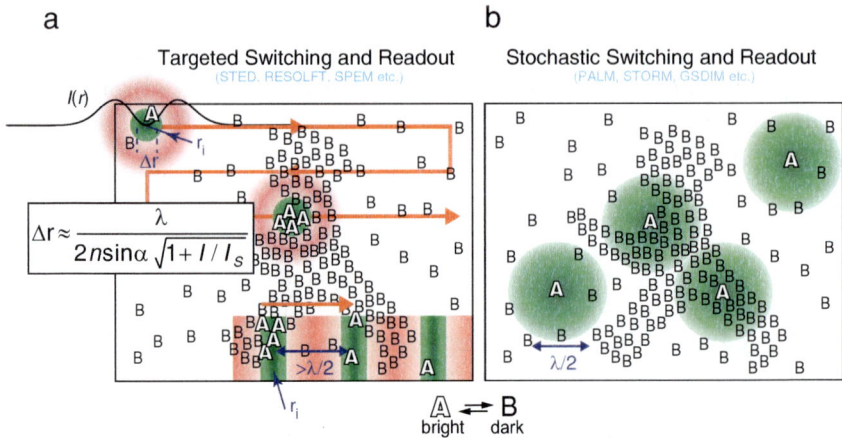

a

Targeted Switching and Readout
(STED, RESOLFT, SPEM etc.)

b

Stochastic Switching and Readout
(PALM, STORM, GSDIM etc.)

$$\Delta r \approx \frac{\lambda}{2n\sin\alpha \sqrt{1+I/I_s}}$$

$$A \rightleftharpoons B$$
bright   dark

Fig. 19.2. All far-field fluorescence nanoscopy concepts realized so far switch the fluorophore between two distinguishable states, a bright state A and a dark state B, to construct subdiffraction images. The sketched object consists of molecules initially residing in B. The task of the superresolving concept is to bring features or molecules that are closer than the diffraction limit sequentially in the bright state A. This task can be accomplished either by establishing A at specific (targeted) spatial coordinates, or by letting the state A emerge stochastically. In both cases, the neighboring molecules have to be kept dark (in B). (**a**) Targeted readout mode: state A is established in a subdiffraction-sized spot of diameter Δr that is centered around the coordinate r_i. Featuring a zero at r_i, the role of the intensity distribution $I(r)$ (sketched in red) is to ensure that the molecules remain in B (when eliciting the transition $B \rightarrow A$). All the fluorophores from the spot with diameter Δr are read out simultaneously. The number of read-out (state A) fluorophores is determined by Δr and the local concentration of the molecules at the coordinate which is targeted by the spot. The image is assembled by translating the zero in space. As the zero is translated across the object, the molecules undergo several times the transition $B \rightarrow A \rightarrow B$, which is why this concept requires a reversible transition $A \leftrightarrow B$. The zero can also be line-shaped or an array of zero-lines, preferably of distance $> \lambda/(2n)$. (**b**) The stochastic readout mode detects single fluorophores from a random position within the diffraction zone. To this end, a molecule is transferred to a state A which is able to emit $m \gg 1$ photons in a row. The neighboring molecules remain in the dark, owing to an inherent inhibition preventing $B \rightarrow A$. The closest molecule in state A should be further away than the diffraction limit $\lambda/(2n)$. The detection of $m \gg 1$ photons allows the calculation of the coordinate of emission from the centroid of the diffraction fluorescence spot formed on a camera. After the recording, the molecule is switched off to B in order to allow the recording of an adjacent molecule. If it is sufficient to record a single picture, the stochastic read-out requires each molecule to cycle only once $B \rightarrow A \rightarrow B$. Both strategies assemble an image by registering fluorophores or fluorophore ensembles sequentially in time and both concepts utilize a mechanism that keeps the neighboring molecules dark (in state B). States A and B can also be reverted

To switch fluorescence, it takes two states: a bright (fluorescent) state A and a dark state B that are connected by a transition which is the actual switch. Several states in a fluorophore are suitable for such transitions (Fig. 19.3). The fluorescent singlet state S_1 and the ground state S_0 used in STED microscopy is the most basic and obvious pair of bright and dark states [39], but other examples will be given later, when discussing various nanoscopy implementations.

Fluorescence switching can be employed in two ways [68]. In the first option, the switching is carried out in a spatio-temporally controlled way, that is, one defines the coordinates where the states (A or B) are created in the sample using a dedicated spatial distribution of light. As one knows where the signal comes from, assembling an image becomes straightforward. This strategy is conceptually the most general one because it yields subdiffraction resolution images just by inducing molecular transitions (between a signaling and a non-signaling state). It is general, because no specific conditions are required for the way the state A has to signal. It is called the *targeted switching and read-out mode* [68]. The second option here called *stochastic switching and read out* [68] is to switch on and off individual molecules stochastically in space. This strategy is viable under the condition that the state A is able to emit $m \gg 1$ photons so that these photons can be associated with the same molecule, e.g. in a burst. In this case, the coordinate of molecular emission has to be found out (mathematically) after registering the photons with a camera. The two switching and read-out options are sketched in Fig. 19.2 and will be discussed in Sect. 19.3.1.

19.3.1 Targeted Switching and Read-out Mode: STED, GSD, SPEM/SSIM, RESOLFT

In the targeted read-out mode, the bright state A is established at coordinate r_i by driving an optical transition $A \rightarrow B$ with a light intensity distribution $I = I(r)$ featuring an intensity minimum, ideally a zero, at coordinate r_i (Figs. 19.1d and 19.2a). Applying $I(r)$ transfers the markers virtually everywhere to B, except at the zero-intensity point r_i where the molecules can still remain in A. The rate of the transition $A \rightarrow B$ is given by $k_{AB} = \sigma I$, with σ denoting the optical cross-section for $A \rightarrow B$. In order to effectively switch the molecule to B, the optically induced rate k_{AB} must outperform any competing spontaneous transitions between A and B. Since these spontaneous rates are given by the inverse lifetimes $\tau_{A,B}$ of the states A and B, we obtain: $k_{AB} = \sigma I \gg (\tau_{A,B})^{-1}$. Therefore, applying an intensity I that is much larger than the "saturation intensity" $I_s = (\sigma \tau_{A,B})^{-1}$ shifts the molecule everywhere to B except in the proximity of the zero-intensity point r_i of $I(r)$. Thus, we obtain a narrowly confined region $r_i \pm \Delta r/2$ in which the molecule can still be in A. The width Δr of this region or spot is readily calculated as

$$\Delta r \approx \frac{\lambda}{2n \sin\alpha \sqrt{1 + aI_{\max}/I_s}}. \tag{19.1}$$

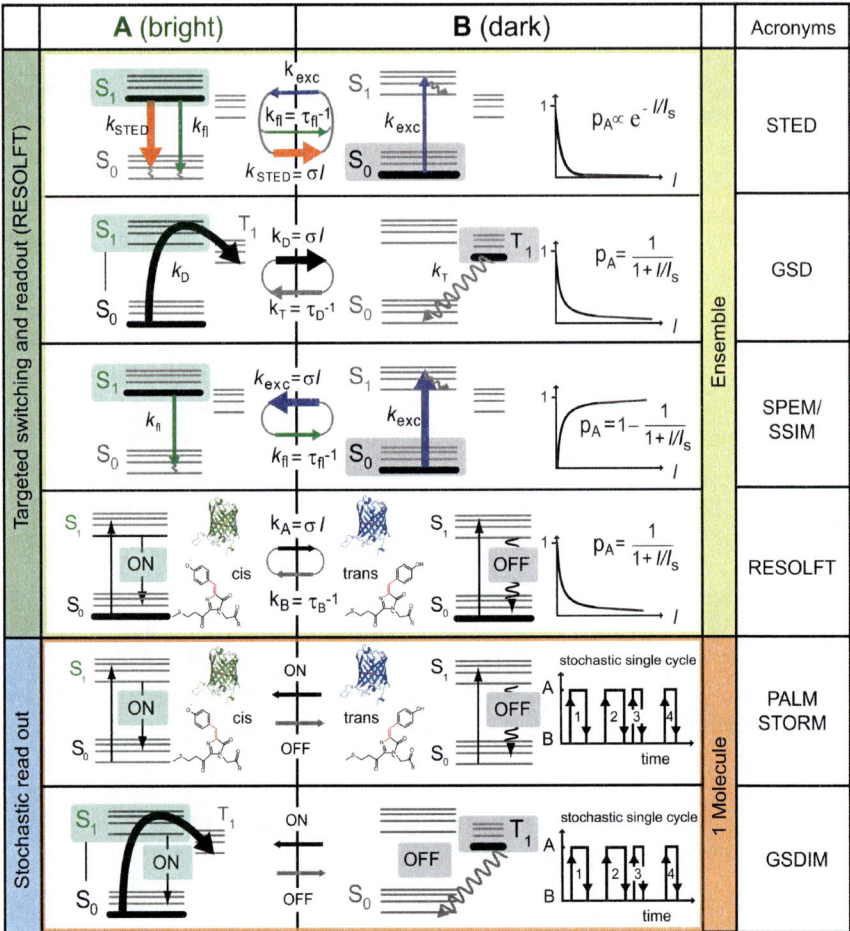

Fig. 19.3. Molecular transitions and states utilized to break the diffraction barrier. Each nanoscopy modality resorts to a specific pair of bright and dark states. Several concepts share the same states, but differ by the direction in which the molecule is driven optically (say $A \to B$ or $B \to A$) or by whether the transition is performed in a targeted way or stochastically. The targeted read-out modality drives the transition with an optical intensity I and hence operates with probabilities of the molecule of being in A or B. This probability depends on the rates k of the transitions between the two states and hence also on the applied intensity I. The probability p_A of the molecule to remain in A typically decreases as indicated in the panel. $p_A \ll 1$ means that the molecule is bound or "switched" to the state B. This switching from A to B or vice versa allows the confinement of A to subdiffraction-sized coordinates of extent Δr at a position r_i where $I(r)$ is zero. In the stochastic read-out mode, the probability that state A emerges in space is evenly distributed across the sample and kept so low that the molecules in state A are further apart from each other than the diffraction limit. An optically nonlinear aspect of the stochastic concept is the fact that the molecules undergo a switch to A from where they suddenly emit $m \gg 1$ detectable photons in a row

I_{max} denotes the intensity of the peak enclosing the zero [39,41,69,70], whereas the optional prefactor a considers its shape [70]. For $I_{max}/I_s \gg 1$, the spot Δr becomes much smaller than the diffraction limit.

A subdiffraction image with resolution Δr can now be readily assembled by scanning the zero-intensity point r_i across the object (Figs. 19.1d and 19.2a). In the process, the beam switches all molecules to the dark state B except those lying in the coordinate range $r_i \pm \Delta r/2$. As a result, the signal must originate from this range. Fluorophores closer together than Δr can be in the bright state A at the same time and hence can emit simultaneously. They cannot be resolved. However, fluorophores that are further away than Δr cannot be simultaneously in A; they are registered sequentially in time and can be separated in the image. More generally formulated, narrowly spaced features or molecules are resolved in this scheme because features further away than the subdiffraction distance Δr are forced to reside in different states A and B. The prototype of this scheme is STED microscopy, where one feature can be in $A = S_1$ and its neighbor is forced by the STED beam to reside in $B = S_0$.

The subdiffraction resolution is given by Δr as defined in (19.1) [39, 41]. Since it is a far-field approach, the resolution still scales with λ, however, it is no longer *limited* by λ, because $I_{max}/I_s \to \infty$ leads to $\Delta r \to 0$ [39,41,69]. The reason for the square-root law is that, in first approximation, the intensity $I(r)$ increases quadratically when departing from the zero-intensity point r_i.

Since diffraction precludes only the separation of objects lying closer than $\lambda/(2n \sin \alpha)$, the imaging process can be parallelized by implementing many zeros with distance $> \lambda/(2n \sin \alpha)$ from each other. In this case, the fluorescence can also be conveniently imaged on a camera (Fig. 19.1d). The whole scheme can be inverted [68] by exchanging A with B. Thus, it is also possible to confine the dark state B rather than the bright state A by applying a transition $B \to A$ (Fig. 19.1d).

Sequential readout at targeted coordinates in space is not only a hallmark of STED [35] but also of GSD microscopy [36], and other concepts exploiting reversible saturable or photoswitchable transitions $A \leftrightarrow B$ [42], including those that utilize many zeros in parallel such as saturated pattern excitation microscopy (SPEM) [71], which is also called saturated structured illumination microscopy (SSIM) [72]. They have been generalized under the acronym RESOLFT [39, 40] which is in fact synonymous with the targeted read-out mode. The resolution of all these concepts is governed by equation (19.1). Their OTF features an "infinitely expanded" frequency passband with an FWHM that scales with $1/\Delta r$ and hence with $\sqrt{I_{max}/I_s}$ [73]. Again, the reason for the square-root law in the expansion of the passband is the quadratic rise of the intensity in the proximity of local intensity zeros.

As already indicated, in the targeted read-out mode, the fluorophores located within distance Δr remain indiscernible because they are in the same state (A or B) at the same time. If it is the bright state A (as in STED microscopy) this is clearly an advantage, because all the fluorophores located within Δr add their signals. So, while the targeted read-out mode can certainly operate with and resolve single molecules, as such it is an ensemble concept dealing with any number of molecules. The number of molecules does not matter since the resolution is physically predefined by Δr which can be tuned through I_{max}/I_s following (19.1). By the same token, the tuning of Δr allows one to adjust the average number of simultaneously recorded fluorophores which is very useful in many applications [74].

The power of this concept is underscored by the fact that several molecular states are able to take the role of A and B: basic electronic states, such as the S_0, S_1, the first triplet state T_1, or "chemical" states, such as conformational or binding states of the fluorophore [39,41] (Fig. 19.3). Since the resolution and performance of a practical microscope is strongly determined by the actual choice of states A and B, we will discuss the various approaches on the basis of the states employed.

19.3.2 STED Microscopy

STED microscopy uses the most elementary states possible: the S_1 as A and the S_0 as B (Figs. 19.1 and 19.3). Most STED microscopy implementations have so far utilized a focused excitation beam and a red-shifted, doughnut-shaped "STED beam" for de-exciting potentially excited fluorophores back to the ground state $S_1 \rightarrow S_0$ by stimulated emission (note that the many photons of the stimulating beam and the handful stimulated photons are discarded in the process) [35, 75–78]. The only role of the STED beam is to switch off the ability of the dye to fluoresce by confining it to the S_0. Even if these molecules encounter an excitation photon, they will not fluoresce, because the STED beam keeps them "off." This fluorescence off-switching is, of course, absent within the range Δr around the doughnut zero.

To switch off the dye by STED, the stimulated emission rate $A \rightarrow B$ has to outperform the spontaneous decay rate of the S_1 which is given by its inverse lifetime $\tau_{fl} \approx 10^{-9}$ s. With $\sigma \approx 10^{-16}$ cm^2, $I_s = (\sigma\tau_{fl})^{-1}$ typically amounts to 3×10^{25} photons cm^{-2} s^{-1}, i.e. $\sim$ 3–10 MW cm^{-2}. If exposed to the STED beam, the probability of a molecule to reside in S_1 decreases in first approximation as $\exp(-I/I_s)$. To confine S_1 in space, the STED beam is prepared to feature a zero; usually it forms a doughnut with crest intensity I_{max}. Following (19.1), $I_{max} \gg I_s$ confines the area in which the fluorophore can reside in S_1 to subdiffraction dimensions Δr around the doughnut zero. Scanning the zero across the sample registers features with subdiffraction resolution Δr sequentially in time.

Because $I_s = 3$–10 MW cm^{-2}, STED microscopy operates with focal intensities $I = 0.1$–1 GW cm^{-2}. While these intensities are 10–100 times larger than the typical intensities used for excitation, regarding photodamage or excitation by the STED beam one has to keep in mind that the wavelength of the STED beam is adjusted to the red edge of the emission spectrum of the dye, that is to a range where the molecule has a $> 10^4$ times reduced excitation cross-section. Besides, the intensities required for STED are by 20–1,000 times lower than the $200\,\mathrm{GW\,cm^{-2}}$ required for live-cell compatible multiphoton microscopy in (sub)picosecond pulses [20,79]. The reason is that stimulated emission requires just a single photon. The need for elevated intensities just stems from the requirement that $I \gg I_s$, that is the state S_1 has to be deactivated within its nanosecond lifetime τ_{fl}. Therefore, STED microscopy can be effectively implemented both with pulsed and with continuous wave (CW) lasers [80].

The local minimum, i.e. the 'zero', need not be formed as a doughnut; it could also be a groove [69, 78, 81] in which case the resolution would be improved just in the direction perpendicular to the groove. In fact, early demonstrations utilized just a laterally offset STED beam [76, 82], because of technical simplicity. Making special "doughnuts" having a strong peak above and beneath the focal plane (Fig. 19.1c), rendered $\Delta z = 100\,\mathrm{nm}$ with a single lens [77], but the narrower STED zeros obtained from using 4Pi systems made $\Delta z = 33$–$60\,\mathrm{nm}$ possible [83, 84]. Displaying an ~ 15-fold improved axial resolution over confocal microscopy in the imaging of microtubules in fixed mammalian cells [84], these early STED-4Pi combinations demonstrated the potential of a far-field fluorescence microscope to operate in the tens of nanometers range [39].

Experiments using single molecules as test objects displayed a resolving power of 28–40 nm [81] and later $\Delta x = 16\,\mathrm{nm}$ in the focal plane [69]. These experiments confirmed equation (19.1) and showed that a focal plane resolution of $\sim \lambda/45$ was achievable [69]. Applying this resolution in immunofluorescence imaging has initially been hampered by photobleaching, but allowing long-lived fluorophore dark states to relax, enabled $\Delta r = 20$–$30\,\mathrm{nm}$ also in cells [85, 86]. A recent STED-4Pi combination called isoSTED demonstrated a nearly spherical, isotropic 3D-resolution of 40–45 nm [87]. Setting the current benchmark, such novel combinations of STED and 4Pi microscopy are likely to push all-molecular-physics-based isotropic resolution to values $< 15\,\mathrm{nm}$.

Although STED and confocal microscopy are easily combined to each others advantage in the same setup, STED is not an extension of confocal microscopy, because it does not require the imaging of the fluorescence onto a pinhole. In principle, one could detect the fluorescence signal right at the sample. Therefore, parallelized camera-based STED microscopy will also be possible with arrays of doughnuts or lines (Fig. 19.1d) [88].

STED microscopy has been applied to study the fate of synaptic vesicle proteins during exocytosis. The study revealed that, when the vesicles fuse with the presynaptic membrane, the protein synaptotagmin I forms integral nanosized clusters at the synapse [89]. Since the clusters are similar in size and molecular density as integral vesicles, the study indicated, that entire protein clusters are taken up from the neuronal membrane when forming new vesicles. By the same token, this initial application of far-field fluorescence nanoscopy to a biological problem demonstrated the potential of this emerging field for solving longstanding problems in the life sciences.

In another application, STED microscopy also revealed the ring-like structure of the protein *bruchpilot* at synaptic active zones in the drosophila neuromuscular junction [90]. Further studies included the visualization of the spatial distribution of the SNARE protein syntaxin [91], the nuclear protein SC35 [85], and the nicotinic acetylcholine receptor [92]. IsoSTED microscopy resolved the tube-like 3D-distribution of the TOM20 protein complex in the mitochondria in a mammalian cell [87].

The viability of STED microscopy with living cells was demonstrated in its early stages [77]. However, focusing on optical aspects of the concept, those experiments were carried out with slow piezo-scanning stages. Recent fast galvanometer-scanning implementations of STED microscopy [93] visualized the rapid motion of dense synaptic vesicles at the synapse of living hippocampal neurons at video rate [94]. These results showed that nanoscale resolution and the visualization of rapid physiological processes can be reconciled, on the basis of existing physical principles and with available technology.

The nanosized detection area Δr or volume created by STED also extends the power of fluorescence correlation spectroscopy (FCS) and the detection of molecular diffusion [74, 95]. For example, STED microscopy has probed the diffusion and interaction of single lipid molecules on the nanoscale in the membrane of a living cell (Fig. 19.6). The up to ~ 70 times smaller detection areas created by STED (as compared to confocal microscopy) revealed marked differences between the diffusion of sphingo- and phospholipids [74]. While phospholipids exhibited a comparatively free diffusion, sphingolipids showed a transient (~ 10 ms) cholesterol-mediated "trapping" taking place in a < 20-nm diameter area, which disappeared after cholesterol depletion. Hence, in an unperturbed cell putative cholesterol-mediated lipid membrane rafts should be similarly short-lived and smaller.

Stimulated emission occurs in all dyes investigated. Yet, just as in experiments using single molecules, the utility of a number of dyes will be precluded by bleaching. Nonetheless, several suitable organic dyes were found in each part of the spectrum. In any case, the general demand for increased photostability is leading to the design of new labels with increased potential for STED microscopy, including fluorescent proteins. STED on yellow fluorescent proteins [96] has recently been used in an application that quantified morphological changes in dendritic spines in living organotypical hippocampal brain slices upon external stimulation [97].

STED microscopy has important applications outside biology as well. For example, it currently is the only method to locally and noninvasively resolve the 3D assembly of packed nanosized colloidal particles [98, 99]. In the realm of solid-state physics, STED microscopy has recently imaged densely packed fluorescent color centers in crystals, specifically charged nitrogen vacancy (NV) centers in diamonds [100]. NV centers in diamond have attracted attention, because of their potential application in quantum cryptography and computation, but also for nanoscale magnetic imaging [101, 102]. Since NV centers do not bleach or blink upon excitation, nanoparticles of diamond containing NV centers are also being developed as nonbleaching labels for bioimaging [103, 104].

Figure 19.4 shows that these centers enable a virtually ideal implementation of the STED concept. The population of the bright state decreases with the intensity I, as one would expect from theory: $\exp(-I/I_s)$. If $I \gg I_s$, the linear representation of the exponential fluorescence "depletion curve" appears to be "rectangular" [35], meaning that one can have a very narrow intensity range $I < I_s$ in which the NV centers are "on" and a broad intensity range $I \gg I_s$ where the center is "off." As a result, STED microscopy was capable of imaging NV centers with a resolution of $\Delta r = 16\text{--}18\,\text{nm}$ in raw data. Once separated by STED, the position of the NV centers can then be calculated with Ångström precision. Recording could be continually repeated without degradation of resolution or signal, proving far-field nanoscale imaging without photobleaching. At the same time, these results underscore the potential of NV containing diamond nanoparticles for biolabeling. Last but not the least, increasing I yielded $\Delta r = 5.8\,\text{nm}$, demonstrating an "all-physics-based" resolving power exceeding the wavelength of light by 2 orders of magnitude [100]. These experiments corroborate the prediction of the original paper [35] that "rectangular depletion curves" would allow increasing the far-field resolution to "infinity."

19.3.3 Ground State Depletion (GSD) Microscopy

GSD microscopy [36, 75] was the second concretely laid out concept to overcome the diffraction barrier. Although related to STED microscopy, it uses an entirely different mechanism for switching off fluorescence. To allow for much lower intensities, the fluorophore is switched off by transiently shelving it in the metastable triplet state T_1 serving as the dark state B. The bright state A is now the S_0, or more precisely, the dye's singlet system (Fig. 19.3). The transfer to the T_1 is accomplished by repeated $S_0 \rightarrow S_1$ excitation, so that the dye is eventually "caught" in the T_1. Reading out the fluorescence of A is performed at the same wavelength. Due to the fact that the T_1 lifetime of $\tau_T = 1\,\mu\text{s} - 1\,\text{ms}$ is now by 3–6 orders of magnitude longer than τ_{fl}, I_s is reduced by the same factor. Using this mechanism, the dye can be switched off at $10^3\text{--}10^6$ times lower intensities than with STED, still giving similar resolution [36, 75].

GSD microscopy has been shown to image immunolabeled protein clusters on the plasma membrane of fixed cells with $\Delta r = 50\text{--}90\,\text{nm}$ resolution using a CW laser power of a few kilowatt per square centimeter [105]. However, plain GSD microscopy is currently challenged by the fact that the repeated population of the triplet state or similar dark states augments photobleaching [105]. Nonetheless, this early concept was important for the development of far-field optical nanoscopy for a number of reasons. First, it highlighted

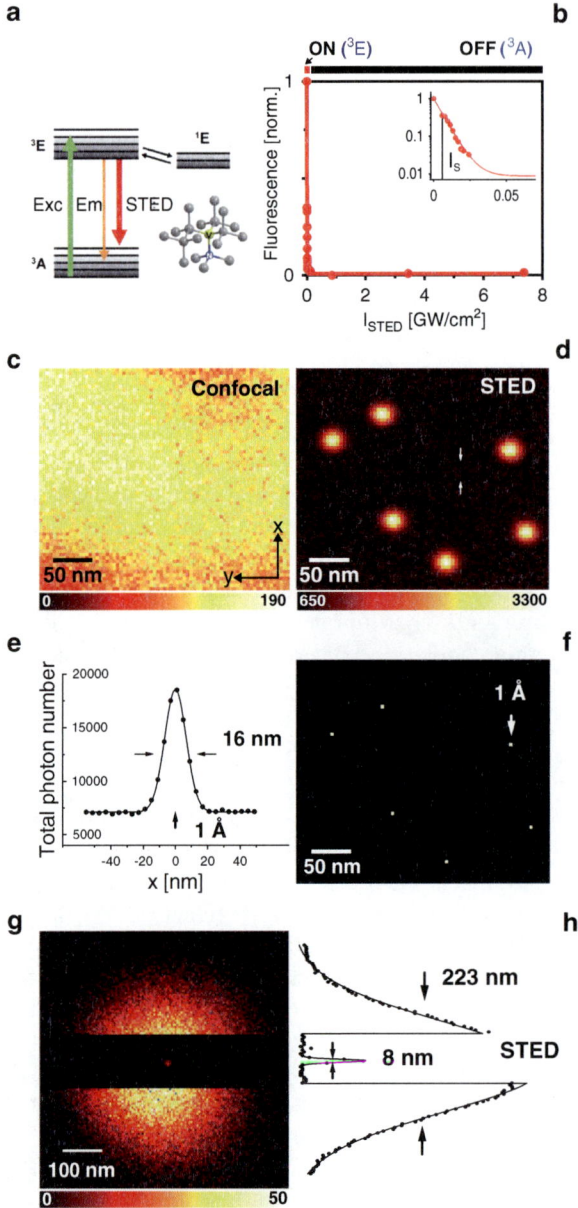

Fig. 19.4. Stimulated emission depletion (STED) microscopy reveals densely packed charged nitrogen vacancy (NV) color centers in a diamond crystal. (**a**) State diagram of NV centers in diamond (see inserted sketch) showing the triplet ground (^{3}A) and fluorescent state (^{3}E) along with a dark singlet state (^{1}E) and the transitions of excitation (Exc), emission (Em), and stimulated emission (STED). (**b**) The steep decline in fluorescence with increasing intensity I_{STED} shows that the STED-beam is able to "switch off" the centers almost in a digital-like fashion. This nearly "rectangular"

that other processes but stimulated emission – in fact, "any transition in a fluorophore that nondestructively inhibits fluorescence" [75] (i.e. keeps the molecule dark) – could be used to overcome the limiting role of diffraction. Second, whereas switching off the fluorescence capability of the dye by STED requires the presence of light, the GSD concept is the first to use a "genuine" molecular switch: flipping an electron spin switches the dye to a relatively long-lived dark state in which it is deactivated. Back-flipping, i.e., the return to its singlet states, means that the dye is switched on again. Thus, within a period τ_T, GSD provides optical bistability.

19.3.4 Saturated Pattern Excitation or Saturated Structured Illumination Microscopy

SPEM/SSIM [71, 72] differs from GSD or STED in that it confines the dark state B rather than the bright state A (Fig. 19.1d), thus producing "dark" regions in which the dye remains in B (S_0) that are steeply surrounded by areas in which it is switched to A (i.e., S_1) [39, 68]. Applying $I_{max} > I_s$ confines these dark regions to subdiffraction dimensions Δr following (19.1). Since it produces "negative data," the images have to be reconstructed computationally. Recording is also performed by scanning an array of line-shaped zeroes in the direction orthogonal to the lines and reading out the data with a camera for each scanning step. In order to cover all directions, the array is rotated an adequate number of times [72]. I_s is similar in magnitude as in the STED concept, because it relies on the same states. SPEM/SSIM

Fig. 19.4 (Continued). excited state depletion curve testifies a close to ideal implementation of the STED effect. The half-logarithmic inset representation of the depletion curve confirms the exponential optical suppression of the excited state. For $I_{STED} > 20\,\mathrm{MW\,cm^{-2}}$, the center is in essence deprived of its ability to fluoresce, i.e., switched off. The "on–off" optical switching facilitates far-field optical separation of NV centers on the nanoscale. While the confocal image (**c**) from the very same crystal region is blurred and featureless, the STED image (**d**) reveals individual NV centers. The notion that these are single color centers is supported by the fact that they are similar in brightness and appearance. The spot produced by the individual centers represents the effective point-spread-function (PSF) of the STED recording. An exemplary y-profile of the PSF is shown in (**e**), revealing a lateral resolution $\Delta y = 16.1\,\mathrm{nm}$. Once the NV centers are resolved, and provided that scanning errors can be neglected, the location of each center can be calculated with Ångström precision, as exemplified in panel (**f**) which should then be contrasted with panel (**c**). Panel (**g**) shows data from a similar experiment, demonstrating a 777-fold sharpening of the effective focal spot area through STED. As depicted by the profile in (**h**) the spot diameter is decreased from 223 nm down to 8 nm. Note that the increase in resolving power is a purely physical effect, i.e., just based on state transitions. The steep optical off-switching of the NV centers depicted in panel b indicates that optimizing the process is bound to improve the far-field optical resolution further

displayed a lateral resolution of 50 nm with beads (obtained after the mandatory computation) [72]. SPEM/SSIM can also be explained in the spatial frequency domain with the OTF [71, 72]. Like STED and GSD, the OTF of SPEM/SSIM is "unlimited," featuring an FWHM scaling with $1/\Delta r$ and hence with $\sqrt{(I_{\max}/I_{\mathrm{s}})}$ [73].

19.3.5 Photoswitching in the Generalized RESOLFT Concept

The conception of GSD microscopy to reversibly switch the fluorophore to a metastable state has led to the consideration of molecular switches between states of even longer lifetimes. As a matter of fact, the ultimate saturable or switching transition occurs between two stable states [38–41]. The advantage of switching between two stable states is obvious: since there are no spontaneous interstate transitions, it follows that $I_{\mathrm{s}} \rightarrow 0$. As a result, it should be possible to implement huge values $I_{\max}/I_{\mathrm{s}}$ which, following (19.1), should yield very small Δr even at low $I_{\max}$.

Switches between long-lasting molecular states can be realized through conformational changes such as photoinduced cis-trans isomerization between a fluorescent isomeric state A and a dark counterpart B. In other words, the reversible spin flip of the GSD concept is replaced by the relocation of an atom or a group of atoms in the molecule. Other options are bistable binding events, ring-opening, or closing reactions, etc. Concrete examples of reversible photoswitching are found in reversibly photoactivatable GFP-like proteins, such as *asFP595* [106] and *dronpa* [107], and in photochromic organic compounds. For this reason, it has been proposed to utilize photoswitchable proteins and fluorescent photochromic compounds to break the diffraction barrier in the targeted read-out mode [38, 39, 41, 108]; the concept was called RESOLFT [40].

The fluorescence is gained by exciting the dye from this (conformational) state A to an electronically excited, fluorescent state A^* and recording the emission $A^* \rightarrow A$, at the same time as the surrounding molecules are kept in the dark state B. Experiments with the reversibly photoactivatable *asFP595* indeed demonstrated for the first time the overcoming of the diffraction barrier by switching photoactivatable proteins [42]. As anticipated, this could be accomplished using intensities $I_{\max} \approx 10 \, \mathrm{W \, cm^{-2}}$, which are by 6 orders of magnitude lower than those required for STED. These results also verified the prediction [38–42] about the huge potential of switching photoactivatable proteins and photochromic (photoswitchable) organic dyes for overcoming far-field microscopy diffraction barrier.

Conversely, these experiments [42] also revealed the challenge currently encountered with switching fluorophores reversibly between A and B using a light intensity featuring a local zero: it is the finite number of switching cycles. Unfortunately, a fair number of cycles are required when defining the coordinates of the states A and B in the sample (Fig. 19.2). This stems from the fact that ensuring the exclusive presence of state A at a subdiffraction-sized coordinate $r_i \pm \Delta r/2$ necessitates fluorophores outside this subdiffraction-sized region to stay in B. So, if these fluorophores happen to cross to A, they must be optically pushed back to B, in which case they have undergone a switching cycle without contributing signal. For this reason, in a targeted read-out modality (RESOLFT concept) the number of possible $A \leftrightarrow B$ cycles that a molecule can undergo before degradation has to be considered. In the simplest case, improving the resolution by a factor of m in the focal plane forces the fluorophore to undergo about m^2 switching cycles due to the m-fold finer targeting (scanning) steps. Likewise, m^3 cycles are required if the same improvement is implemented in 3D. Thus an improvement by a factor of 10 easily entails 100–1000 cycles.

19.3.6 Stochastic Switching and Read-out Mode: PALM, STORM, GSDIM

The challenge posed by repeated $A \leftrightarrow B$ cycling is alleviated when switching individual molecules stochastically in space. In this approach, molecules that are initially in the dark state B pop up individually in A at unknown coordinates r_i. A restriction on the bright state A is that it must be able to give off many photons in a row (e.g., through repeated excitation $A \leftrightarrow A^*$) so that it renders a diffraction blob of $m \gg 1$ detected photons when imaged onto a camera. (At least one must be able to associate m detected photons with the state A.) During this process, the surrounding molecules do not cross to A due to an inherent inhibition; they remain dark. The detection of $m \gg 1$ photons from the same bright molecule is a distinct requirement of the stochastic switching mode, because the molecular position r_i has to be obtained by calculating the centroid of the blob, which can be done with subdiffraction precision $\sim \lambda/(2n \sin \alpha \sqrt{m})$ [109,110]. Called localization, this calculation rests on the assumption that the blobs on the camera represent single (or few identifiable) molecules, which is the case if the distance between two A state molecules is about $> \lambda/(2n \sin \alpha)$. After the coordinate of the A state molecule has been read out it is switched off again $A \rightarrow B$, to allow reading out a neighboring molecule using the same cycle $B \rightarrow A \rightarrow B$. The image is assembled molecule by molecule.

Whereas in the targeted read-out mode, a molecule has to undergo many cycles $A \leftrightarrow B$, in the stochastic switching mode, a single cycle $B \rightarrow A \rightarrow B$ per molecule is enough to produce an image [68]. Thus, this mode largely avoids switching fatigue. If the object needs to be imaged repeatedly, the same molecule has to be engaged repeatedly. In this case, several cycles per molecule are needed. Yet the number of cycles is still smaller than in the targeted read-out mode, because the molecules are switched only when they are supposed to contribute with a signal [68].

This strategy of combining stochastic molecular switching with localization has been used in the methods called SHRIMP [111] and NALMS [112], in which the position of a small number of bright regular fluorophores (in state A) was mapped by bleaching them down stochastically to a final dark state B. However, as these methods started out from many A state (bright) molecules and hence from a bright total signal, they could accommodate only a small number of fluorophores. The strategy of combining stochastic switching and localization for an arbitrary number of molecules [43–45,113,114] has been realized in the methods called photoactivatable localization microscopy (PALM) [43] and stochastic optical reconstruction microscopy (STORM) [44] or fluorescence photoactivatable localization microscopy (FPALM) [45]. In these studies, photoswitchable proteins or photochromic compounds were optically driven from the dark state B to the bright state A by means of a photon absorption (i.e., optically activated), localized, and finally sent back to B. So, while SHRIMP and NALMS used only a half cycle out of a bright state $A \rightarrow B$, PALM and STORM now used a full cycle $B \rightarrow A \rightarrow B$. Starting out from the dark state B they could get rid of the bright initial signal of state A molecules and hence could accommodate an arbitrary number of them.

Importantly, some molecules may be localized more precisely than others, because the number of photon emissions m follows a statistical distribution. Therefore, to ensure a certain resolution, the stochastic read-out mode defines a brightness threshold (e.g., $m > M \approx 50$). Molecules with $m < M$ are discarded (bleached) without contributing to the image. In a sense the discarding of the molecules with $m < M$ is to the stochastic read-out what "switching fatigue" is to its targeted counterpart; the higher the required resolution, the more molecular events are discarded. M depends on the average number of photon emissions in the "on" state and the desired resolution, which, provided there is no background, exceeds $\lambda / \left(2n \sin \alpha \sqrt{M} \right)$. Clearly, just as the targeted read-out mode requires a large number of switching cycles for optimal performance, the stochastic mode requires a large M. Thus, at the expense of disregarding more molecules, PALM, STORM, and other variants of the stochastic read out have achieved a resolution of < 20 nm in the focal plane [43, 115–119].

PALM has first been accomplished by switching on activatable fluorescent proteins $B \rightarrow A$ using a dedicated beam of light, whereas the off-switching $A \rightarrow B'$ was accomplished through bleaching at the expense of being able to record only a single (or very few) images. Since B can be a different state than B' the requirements on the photoswitchable compounds are more relaxed [43]. In contrast, STORM was put forth with repeated cycling between the same dark states: $B = B'$ [44]. STORM imaging [115, 120] took advantage of the reversible cis-trans isomerization of organic cyanine molecules, in fact of (cyanine) molecule pairs, whereby one of them served as an activator molecule facilitating the switching of the other.

PALM and STORM inherently operate with ultra-low light levels for activating the fluorophore, because like the earlier RESOLFT experiments [42], they switch between the relatively long-lived (conformational) states of photoactivatable proteins or organic fluorophores, which entails very low I_s of < 10 W cm^{-2}. Since activating more than about one molecule within a range $< \lambda/(2n \sin \alpha)$ has to be avoided, the intensity applied for switching is further reduced by the number of molecules expected to reside within the diffraction range. In any case, the low intensity operation greatly simplifies the implementation of this concept in a widefield illumination arrangement [115, 121, 122].

PALM images of thin cryosections of lysosomal transmembrane protein in a mammalian cell displayed a resolution of $< 30\,nm$ [43]. The demonstration of PALM and STORM involved the time-sequential use of two chopped laser beams, one for switching on $(B \rightarrow A)$ and the other one $(A \rightarrow A^*)$ for producing the m photons and switching off $(A \rightarrow B)$. The pulsed action of the lasers and the time-window of the camera read-out need to be synchronized in this case. This is different in the modalities "PALM with independently running acquisition" (PALMIRA) [118, 119] and ground state depletion followed by single molecule return (GSDIM) [123] in which no activation beam is used and the fluorophores are allowed to pop up $(B \rightarrow A)$ stochastically in time (not only in space) after most of them have been pushed back to a dark state B [114, 123]. A single CW laser beam is used for generating the m photons through $A \rightarrow A^*$ and for switching the fluorophore off $A \rightarrow B$. The bursts of photons are detected by a fast freely operating camera, whereby the intensity of the laser is adjusted such that the average duration of the m-photon burst coincides within the duration of a camera frame ($\sim 1/500\,Hz = 2\,ms$). These purely stochastic concepts probably are the simplest far-field nanoscopy systems at present, because they require just uniform laser illumination, a freely running camera, and appropriate software. Even multiple color imaging [117, 123, 124] is possible using the same laser as exemplified in Fig. 19.5c.

GSDIM [123] stands out by the fact that, unlike PALM, STORM, or PALMIRA, this concept operates with fluorophores, such as rhodamines, that are considered ordinary and not "switchable." Still, to provide nanoscale resolution, GSDIM has to employ a switching mechanism. As the name suggests, the switching mechanism is that of the old GSD concept [36]: Initially residing in the ground state S_0 (state A), the fluorophores are strongly excited with a spatially uniform CW beam so that, after a number of $S_0 \leftrightarrow S_1$ transitions, they are caught in the dark triplet state T_1 or similar metastable dark state B for a time τ_T of a few milliseconds. After τ_T, individual molecules stochastically return to the singlet system $(B \rightarrow A)$ where they instantly give rise to a bunch of m photons, followed by a return to B. The advent of a concept [114] that switches the molecules to B and relies on their spontaneous return to A showed that the stochastic single molecule switching approach for nanoscopy is actually broader in scope than suggested by the first implementations. Moreover, the use of the very same switching mechanism both in a targeted (GSD) and in a stochastic way (GSDIM) underscores once more that, on a fundamental level, all the nanoscopy concepts discussed herein rest on common molecular ground.

An early description of the localization procedure is due to Heisenberg [109] who remarked that the emission of m photons from a resting emitter enables the calculation of its position with precision $\sim \lambda/(2n \sin\alpha \sqrt{m})$.

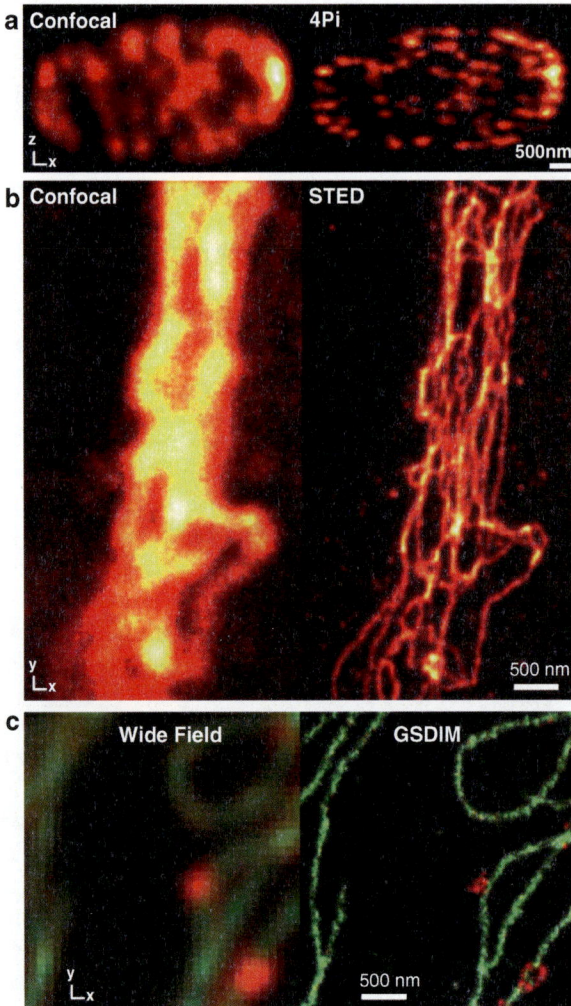

Fig. 19.5. Resolution increase exemplified in various techniques. (**a**) Confocal vs. 4Pi microscopy (xz-image) recorded from the microtubular network in a neuron, displaying an improved z-resolution of 140 nm. The 4Pi image data is due to an all-optically created narrow solitary peak, i.e., showing raw data. (**b**) Confocal vs. an STED image of immunolabeled vimentin in a mammalian neuroblastoma, after linear deconvolution, recorded using a compact supercontinuum STED-microscope with a spatial resolution < 25 nm. (**c**) Conventional widefield image compared with a GSDIM image of microtubules (*green*) and peroxysomes (*red*) in mammalian cells evidencing a substantial gain in image detail by far-field optical superresolution

However, finding out the position of an object with arbitrary precision is not the same as resolution, which is about separating similar objects at small distances. Localization per se cannot provide superresolution. This is also why, although it had been known and used for decades [109, 110] and even routinely applied to single molecules [125, 126], localization alone has not provided nanoscale images. (Note that in spite the use of localization in the 1980's and earlier, near-field optical microscopy still seemed to be the only way to attain nanoscale resolution up to the early 1990's.) Resolution clearly requires a criterion to discern objects or molecules, the simplest of which is "bright" vs. "dark."

In one particular study, it was shown that at low temperatures (1.2 K) the absorption spectra of individual pentacene molecules in a p-terphenyl crystal are so distinct (due to inhomogenous broadening) that one can address and localize (seven) fluorescent pentacene molecules individually at sub-diffraction distances by their absorption spectra. However, this interesting study left it unclear how superresolution by spectral separation could be extended to ambient conditions, to an arbitrary number of fluorophores (which is required to be a general imaging method), to other types of fluorophores, and thus become applicable [127].

In fact, assembling an image with the coordinates of fluorophores that are spectrally separated by cryogenic inhomogenous broadening had been suggested in an earlier theoretical study [128]. However, this study still proposed near-field optical microscopy as the best solution to the diffraction resolution problem. More importantly, this proposal fully relied on the *spectral* separation of molecules and did not suggest switching fluorescence on or off to record adjacent features sequentially, as is inherent in the earlier concept of STED microscopy [35] and in the GSD concept [36].

While extended molecular separation by spectral shifts is certainly interesting and may be eventually practical, the transition [111, 112] from single emitter localization to nanoscale imaging [43–45, 129] has thus not been facilitated by spectral separation, but by fluorescence switching. The reason why switching is so powerful for breaking the diffraction barrier is readily explained. Switching enables the separation of an arbitrary number of molecules or objects by stretching their signal out to an arbitrary number of time points, using just two distinguishable states A and B. Separation by color or wavelength requires a much larger – in principle arbitrarily large set of separable states (A, B, C, D, ..), which is hard to realize in an ordinary fluorophore. (In some special cases, an option could be to implement a state energy shift using an external field gradient as in magnetic resonance imaging.) Viable at room temperature and realizable with pairs of different kinds of bright and dark states, (fluorescence) switching is the essence of all nanoscopy concepts reported to date, irrespective of whether they operate with single molecules or ensembles. It is the element without which none of the nanoscopy concepts discussed herein could have produced an image.

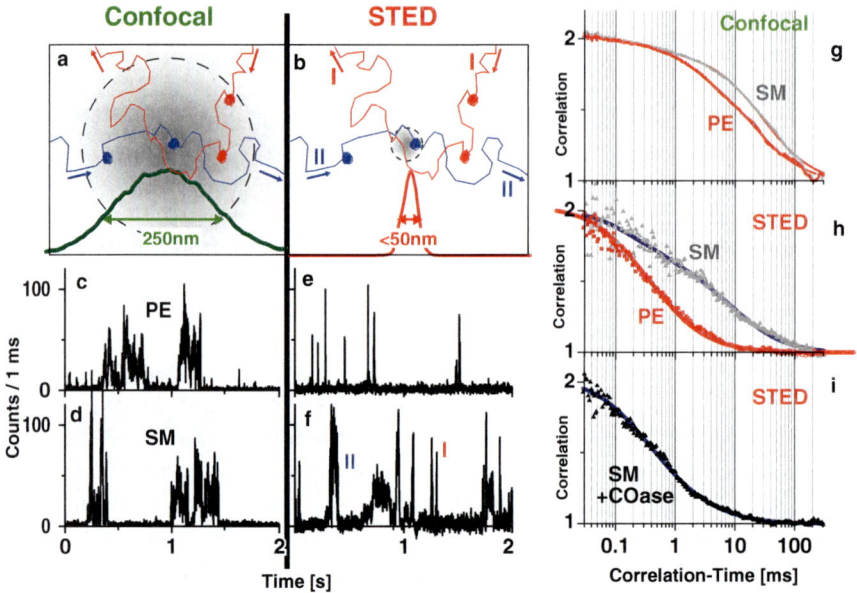

Fig. 19.6. STED microscopy discriminates the dynamics of single sphingo- and phospholipid molecules in a living cell membrane: dye-labeled sphingomyelin (SM) vs. phosphoethanolamine (PE). Panels (**a**) and (**b**) illustrate that freely moving molecules may be transiently trapped in the membrane due to interaction with another molecule. The passage time of freely diffusing molecules through the small spot created by STED is substantially reduced compared to that through a confocal microscope. c-f) Fluorescence bursts from labeled single PE and SM lipids detected with the 250-nm diameter confocal (**c, d**) and a 50-nm diameter STED spot (**e, f**) reveal that, by reducing the time that a freely diffusing molecule spends in the detection area, the smaller spot created by STED distinguishes free lipid diffusion (I) from interaction events (II) during passage through the spot. Panels (**g**)–(**i**) quantify the diffusion of PE and SM lipids on the plasma membrane using fluorescence correlation spectroscopy. Normalized correlation data of PE (*red dots*) and SM (*grey dots*) are compared for a standard confocal (**g**) and an STED recording (**h**) depicting the heterogeneous diffusion of SM. While in the confocal case the difference between the SM and PE diffusion could be explained just by a generally slower diffusion of SM, the STED recording can only be explained by a transient trapping just of SM. (**i**) Following depletion of cholesterol, the diffusion of SM (*black dots*) is similar to that of the rather freely diffusing PE. Thus, STED microscopy reveals cholesterol-assisted heterogeneous diffusion of SM on the nanoscale

19.4 Further General Aspects

The new pathway followed in STED and GSD microscopy was to spatially confine or isolate the bright molecular state by switching neighboring molecules off and keeping them dark. This change in strategy over earlier techniques is also reflected in the PSF of these microscopes. While earlier PSFs essentially described focal intensity distributions in the microscope, the PSF of a STED or a GSD microscope mirrored the spatial distribution of fluorophore states – specifically of the fluorescent state as resulting from the molecular kinetics and the imposed optical dark state transitions.

When inducing an (incoherent) optical transition $A \rightarrow B$, the probability p_A of the molecule to be in state A is proportional to $\exp(-I/I_s)$ or, if there is an equilibrating rate $B \rightarrow A$, proportional to $(1 + I/I_s)^{-1}$ as sketched in Fig. 19.3 [39]. Due to the inherently nonlinear dependence of p_A on the intensity I, one can interpret STED, GSD, SPEM/SSIM, RESOLFT as nonlinear optical concepts. However, as it stems from the transition between two states, this nonlinearity is, of course, fundamentally different from m-photon absorption or scattering processes where the nonlinearity stems from the concomitant action of m photons. Moreover, these functions p_A bring about nonlinearities of infinite order $(\gamma I)^m$ with $m \rightarrow \infty$, which allow one to "break" the diffraction barrier by obtaining "infinite" optical bandwidth. By contrast, m-photon processes are firmly limited to order m, which in practice means $m < 3$. Since I_s can be decreased by selecting long-lived states, p_A enables huge nonlinearities even at low I [38–42].

By applying $I \ll I_s$, stochastic read-out modes such as PALM, STORM, and GSDIM clearly avoid a nonlinear dependence on the switching intensity. Still, nonlinear optical aspects are inherent to these concepts as well. They stem from the molecular switching and the burst of $> m$ photons originating from a single emitter, which has been shown to be connected with diffraction-unlimited far-field optical resolution [130]. Although a nonlinearity interpretation is applicable to all current nanoscopy concepts, it clearly is not specific enough to single out the facilitator which is the switching between two states $A \rightarrow B$.

All diffraction-unlimited nanoscopy methods can provide improved axial resolution even when implemented with a single lens [77, 83, 120, 131, 132]. However, because it starts out from less favorable values, the z-resolution usually remains worse than its focal plane counterpart. The coherent use of opposing lenses pioneered in 4Pi microscopy and I^{5}M, however, facilitates an independent resolution improvement factor by 3–7-fold along the optic axis as has already been demonstrated with STED [83, 84, 87]. In the stochastic single molecule switching modalities, a similar gain in resolution will take place by the coherent use of opposing lenses [133]. Thus, while 4Pi microscopy and I^{5}M did not break the diffraction barrier, they remain cornerstones of far-field fluorescence nanoscopy in the future.

Since 1994, it has been advocated that the key to nanoscale resolution is the imaged dye itself [34–36, 39, 41], making it worthwhile synthesizing fluorophores and other markers just for the purpose of overcoming the diffraction resolution limit [39, 41]. For the stochastic single molecule switching modalities, molecules are desired that burst out large m in the shortest possible period of time, while for the targeted switching the aim should be molecules that can be switched many times between a dark and a bright state.

Importantly, in the targeted read-out mode, the state A need not be fluorescent; in fact, not even an optical emitter, but just detectable through a specific signal [41]. This is because the coordinate is defined by the zero, and that is sufficient. In the stochastic switching mode, it is just the other way around. As it has been proposed [114] and shown [123], the state A need not be activated by light; the only requirement with regard to optics is that it emits m photons that can be detected on a camera. Since there are no fundamental reasons why markers will not improve as resolution facilitators, it is likely that the performance of all these concepts will improve substantially within the coming years. Therefore, emerging optical nanoscopy is bound to transform the life sciences and also impact other fields benefiting from noninvasive nanoscale (3D) optical resolution. In any case, the concepts and results reviewed herein already broke longstanding barriers of perception of what can be accomplished with a microscope that uses just focused visible light.

Finally, it is worthwhile imagining the possible ultimate resolution limits. So, if we had the perfect switchable marker emitting a bunch of $m \gg 10^6$ photons, what would we obtain for the stochastic single molecule switching mode? We would obtain the (average) position of a molecule with a precision of a fraction of a nanometer. While this information would be invaluable for mapping the sample, it would not tell us much about the molecule itself.

However, the situation is different in a purely transition based concept such as STED, in which the molecule is optically driven between two states at defined spatial coordinates. As a matter of fact, equation (19.1) is provocative because $I_{max}/I_s \to \infty$ yields $\Delta r \to 0$. While this extreme extrapolation cannot be taken literally, it brings up the question as to what may happen if I_{max} is indeed continually increased. Let us imagine a molecule of a nanometer in size, placed at the zero-intensity point of the electric light field driving the transition. If we now steadily increase I_{max}, the light field becomes non-negligible at the periphery of the electron orbitals, while their central part would still be weakly exposed. Now, as different orbital parts experience fields of different strengths, these parts will have differently pronounced contributions to the optical transition. Hence, the extent of the molecule and its orbital can no longer be neglected with regard to the spatial structure of the light field. If we now translate the zero across the molecule, the probability of an optical transition $A \to B$, say excitation, stimulated emission, photoconversion, photo-isomerization, or whatever, will change as a function of the role of the excluded parts of the orbital in the transition. Translating the zero across the orbital should therefore yield information about the (dimensions of the) orbital and thus, about the molecule itself. No matter how quickly this "*infra*molecular" photophysics, photochemistry or exploration becomes reality, this fascinating prospect once more highlights the unexpected potential of investigating small scales in nature with freely propagating light.

References

1. E. Abbe, *Beiträge zur Theorie des Mikroskops und der mikroskopischen Wahrnehmung.* Arch. Mikr. Anat. 9, 413–468 (1873)
2. B. Alberts et al., Molecular Biology of the Cell. 4 edn. (Garland Science, New York, 2002)
3. M. Born, E. Wolf, Principles of Optics. 7th edn. (Cambridge University Press, Cambridge, New York, Melbourne, Madrid, Cape Town, 2002), p. 952
4. E.H. Synge, A suggested method for extending microscopic resolution into the ultra-microscopic region. Philos. Mag. 6, 356 (1928)
5. E.A. Ash, G. Nicholls, Super-resolution aperture scanning microscope. Nature 237 510–512 (1972)
6. D.W. Pohl, W. Denk, M. Lanz, Optical stethoscopy: Image recording with resolution $\lambda/20$. Appl. Phys. Lett. 44, 651–653 (1984)
7. A. Lewis et al., Development of a 500 A Resolution Light Microscope. Ultramicroscopy 13, 227–231 (1984)
8. J.B. Pendry, Negative refraction makes a perfect lens. Phys. Rev. Lett. 85(18), 3966–3969 (2000)
9. Z. Liu et al., Far-field optical hyperlens magnifying sub-diffraction-limited objects. Science 315(5819), 1686 (2007)
10. I.I. Smolyaninov, Y.-J. Hung, C.C. Davis, Magnifying superlens in the visible frequency range. Science 315(5819), 1699–1701 (2007)
11. V.A. Podolskiy, E.E. Narimanov, Near-sighted superlens. Opt. Lett. 30, 75–78 (2005)
12. G. Toraldo di Francia, Supergain antennas and optical resolving power. Nuovo Cimento Suppl. 9, 426–435 (1952)
13. W. Lukosz, Optical systems with resolving powers exceeding the classical limit. J. Opt. Soc. Am. 56, 1463–1472 (1966)
14. C. Cremer, T. Cremer, Considerations on a laser-scanning-microscope with high resolution and depth of field. Microscopica Acta 81(1), 31–44 (1978)
15. M. Minsky, Microscopy Apparatus. US Patent, 3,013,467 (1961)
16. T. Wilson, C.J.R. Sheppard, Theory and Practice of Scanning Optical Microscopy (Academic, New York, 1984)
17. J.B. Pawley, Handbook of Biological Confocal Microscopy, 2nd edn. (Springer, New York, 2006), p. 700
18. N. Bloembergen, Nonlinear Optics (Benjamin, New York, 1965)
19. C.J.R. Sheppard, R. Kompfner, Resonant scanning optical microscope. Appl. Optics 17, 2879–2882 (1978)
20. W. Denk, J.H. Strickler, W.W. Webb, Two-photon laser scanning fluorescence microscopy. Science 248, 73–76 (1990)
21. A. Schönle, S.W. Hell, Far-field fluorescence microscopy with repetetive excitation. Eur. Phys. J. D 6, 283–290 (1999)
22. M. Bertero, et al., Three-dimensional image restoration and super-resolution in fluorescence confocal microscopy. J. Microsc. 157, 3–20 (1990)
23. J.-A. Conchello, J.G. McNally, Fast regularization technique for expectation maximization alogorithm for optical sectioning microscopy. SPIE Proc. 2655, 199–208 (1996)
24. D.H. Burns et al., Strategies for attaining superresolution using spectroscopic data as constraints. Appl. Optics 24(2), 154–160 (1985)

25. D.W. Pohl, D. Courjon, Near Field Optics (Kluwer, Dordrecht, 1993)

26. D.W. Pohl, Near-field optics: comeback of light in microscopy. Solid State Phenom. 63–64, 252–256 (1998)

27. E. Betzig et al., Near-field fluorescence imaging of cytoskeletal actin. Bioimaging 1, 129–136 (1993)

28. A. Kirsch, C. Meyer, T.M. Jovin, Integrating of optical techniques in scanning probe microscopes; The scanning near-field optical microscope (SNOM), in Analytical Use of Fluorescenct Probes in Oncology, ed. by E. Kohen, J.G. Hirschberg (Plenum, New York, 1996), pp. 317–323

29. L. Novotny, B. Hecht, Principles of Nano-Optics (Cambridge University Press, Cambridge, MA, 2006)

30. R.Y. Tsien, The green fluorescent protein. Annu. Rev. Biochem. 67(1), pp. 509–544 (1998)

31. R.Y. Tsien, Imagining imaging's future. Nat. Cell Biol. 5, SS16–SS21 (2003)

32. S.W. Hell, Double-Scanning Confocal Microscope. European Patent, 0491289 (1990/1992)

33. S. Hell, E.H.K. Stelzer, Fundamental improvement of resolution with a 4Pi-confocal fluorescence microscope using two-photon excitation. Opt. Commun. 93, 277–282 (1992)

34. S.W. Hell, Improvement of lateral resolution in far-field light microscopy using two-photon excitation with offset beams. Opt. Commun. 106, 19–24 (1994)

35. S.W. Hell, J. Wichmann, Breaking the diffraction resolution limit by stimulated emission: stimulated emission depletion fluorescence microscopy. Opt. Lett. 19(11), 780–782 (1994)

36. S.W. Hell, M. Kroug, Ground-state depletion fluorescence microscopy: a concept for breaking the diffraction resolution limit. Appl. Phys. B 60, 495–497 (1995)

37. A. Schönle, P.E. Hänninen, S.W. Hell, Nonlinear fluorescence through intermolecular energy transfer and resolution increase in fluorescence microscopy. Ann. Phys. (Leipzig), 8(2), 115–133 (1999)

38. S.W. Hell, S. Jakobs, L. Kastrup, Imaging and writing at the nanoscale with focused visible light through saturable optical transitions. Appl. Phys. A 77, 859–860 (2003)

39. S.W. Hell, Toward fluorescence nanoscopy. Nature Biotechnol. 21(11), 1347–1355 (2003)

40. S.W. Hell, M. Dyba, S. Jakobs, Concepts for nanoscale resolution in fluorescence microscopy. Curr. Opin. Neurobio. 14(5), 599–609 (2004)

41. S.W. Hell, Strategy for far-field optical imaging and writing without diffraction limit. Phys. Lett. A 326(1–2), 140–145 (2004)

42. M. Hofmann et al., Breaking the diffraction barrier in fluorescence microscopy at low light intensities by using reversibly photoswitchable proteins. Proc. Natl. Acad. Sci. USA 102(49), 17565–17569 (2005)

43. E. Betzig et al., Imaging intracellular fluorescent proteins at nanometer resolution. Science 313(5793), 1642–1645 (2006)

44. M.J. Rust, M. Bates, X. Zhuang, Sub-diffraction-limit imaging by stochastic optical reconstruction microscopy (STORM). Nat. Meth. 3, 793–796 (2006)

45. S.T. Hess, T.P.K. Girirajan, M.D. Mason, Ultra-high resolution imaging by fluorescence photoactivation localization microscopy. Biophys. J. 91(11), 4258–4272 (2006)

46. S. Hell, E.H.K. Stelzer, Properties of a 4Pi-confocal fluorescence microscope. J. Opt. Soc. Am. A 9, 2159–2166 (1992)

47. M.G.L. Gustafsson, D.A. Agard, J.W. Sedat, Sevenfold improvement of axial resolution in 3D widefield microscopy using two objective lenses. Proc. SPIE 2412, 147–156 (1995)

48. M. Schrader, S.W. Hell, 4Pi-confocal images with axial superresolution. J. Microsc. 183, 189–193 (1996)

49. M. Nagorni, S.W. Hell, Coherent use of opposing lenses for axial resolution increase in fluorescence microscopy. I. Comparative study of concepts. J. Opt. Soc. Am. A 18(1), 36–48 (2001)

50. P.E. Hänninen et al., Two-photon excitation 4Pi confocal microscope: Enhanced axial resolution microscope for biological research. Appl. Phys. Lett. 66, 1698–1700 (1995)

51. S.W. Hell, M. Schrader, H.T.M. van der Voort, Far-field fluorescence microscopy with three-dimensional resolution in the 100 nm range. J. Microsc. 185(1), 1–5 (1997)

52. M. Schrader et al., 4Pi-confocal imaging in fixed biological specimens. Biophys. J. 75, 1659–1668 (1998)

53. M. Nagorni, S.W. Hell, 4Pi-confocal microscopy provides three-dimensional images of the microtubule network with 100- to 150-nm resolution. J. Struct. Biol. 123, 236–247 (1998)

54. A. Egner, S. Jakobs, S.W. Hell, Fast 100-nm resolution 3D-microscope reveals structural plasticity of mitochondria in live yeast. Proc. Nat. Acad. Sci. U.S.A 99, 3370–3375 (2002)

55. A. Egner et al., 4Pi-microscopy of the Golgi apparatus in live mammalian cells. J. Struct. Biol. 147(1), 70–76 (2004)

56. H. Gugel et al., Cooperative 4Pi excitation and detection yields 7-fold sharper optical sections in live cell microscopy. Biophys. J. 87, 4146–4152 (2004)

57. J. Bewersdorf, B.T. Bennett, K.L. Knight, H2AX chromatin structures and their response to DNA damage revealed by 4Pi microscopy. Proc. Natl. Acad Sci. USA 103, 18137–18142 (2006)

58. A. Egner, S.W. Hell, Fluorescence microscopy with super-resolved optical sections. Trends Cell Biol. 15(4), 207–215 (2005)

59. M. Lang et al., 4Pi microscopy of type A with 1-photon excitation in biological fluorescence imaging. Opt. Expr. 15(5), 2459–2467 (2007)

60. M. Lang, J. Engelhardt, S.W. Hell, 4Pi microscopy with linear fluorescence excitation. Opt. Lett. 32(3), 259–261 (2007)

61. M.C. Lang et al., 4Pi microscopy with negligible sidelobes. New J. Phys. 10, 1–13 (2008)

62. M.G. Gustafsson, D.A. Agard, J.W. Sedat. 3D widefield microscopy with two objective lenses: experimental verification of improved axial resolution, in Three-Dimensional Microscopy: Image Acquisition and Processing III, Proceedings of SPIE, 1996

63. M.G.L. Gustafsson, D.A. Agard, J.W. Sedat, I⁵M: 3D widefield light microscopy with better than 100 nm axial resolution. J. Microsc. 195, 10–16 (1999)

64. B. Bailey et al., Enhancement of axial resolution in fluorescence microscopy by standing-wave excitation. Nature 366, 44–48 (1993)

65. R. Freimann, S. Pentz, H. Hörler, Development of a standing-wave fluorescence microscope with high nodal plane flatness. J. Microsc. 187(3), 193–200 (1997)

66. M. Nagorni, S.W. Hell, Coherent use of opposing lenses for axial resolution increase in fluorescence microscopy. II. Power and limitation of nonlinear image restoration. J. Opt. Soc. Am. A 18(1), 49–54 (2001)

67. J. Bewersdorf, R. Schmidt, S.W. Hell, Comparison of I^5M and 4Pi-microscopy. J. Microsc. 222, 105–117 (2006)

68. S.W. Hell, Far-field optical nanoscopy. Science 316(5828), 1153–1158 (2007)

69. V. Westphal, S.W. Hell, Nanoscale resolution in the focal plane of an optical microscope. Phys. Rev. Lett. 94, 143903 (2005)

70. B. Harke et al., Resolution scaling in STED microscopy. Opt. Expr. 16(6), 4154–4162 (2008)

71. R. Heintzmann, T.M. Jovin, C. Cremer, Saturated patterned excitation microscopy – A concept for optical resolution improvement. J. Opt. Soc. Am. A 19(8), 1599–1609 (2002)

72. M.G.L. Gustafsson, Nonlinear structured-illumination microscopy: Wide-field fluorescence imaging with theoretically unlimited resolution. Proc. Nat. Acad. Sci. U.S.A 102(37), 13081–13086 (2005)

73. S.W. Hell, A. Schönle, Nanoscale resolution in far-field fluorescence microscopy, in Science of Microscopy, S.J.C.H., ed. by P.W. Hawkes (Springer, New York, 2007), p. 790–834

74. C. Eggeling et al., Direct observation of the nanoscale dynamics of membrane lipids in a living cell. Nature, 457, 1159–1163 (2009)

75. S.W. Hell, Increasing the resolution of far-field fluorescence light microscopy by point-spread-function engineering, in Topics in Fluorescence Spectroscopy, ed. by J.R. Lakowicz (Plenum, New York, 1997), p. 361–422

76. T.A. Klar, S.W. Hell, Subdiffraction resolution in far-field fluorescence microscopy. Opt. Lett. 24(14), 954–956 (1999)

77. T.A. Klar et al., Fluorescence microscopy with diffraction resolution limit broken by stimulated emission. Proc. Natl. Acad. Sci. U.S.A 97, 8206–8210 (2000)

78. T.A. Klar, E. Engel, S.W. Hell, Breaking Abbe's diffraction resolution limit in fluorescence microscopy with stimulated emission depletion beams of various shapes. Phys. Rev. E 64, 066613, 1–9 (2001)

79. W. Denk, Two-photon excitation in functional biological imaging. J. Biomed. Opt. 1, 296–304 (1996)

80. K.I. Willig et al., STED microscopy with continuous wave beams. Nat. Meth. 4(11), 915–918 (2007)

81. V. Westphal, L. Kastrup, S.W. Hell, Lateral resolution of 28 nm ($\lambda/25$) in far-field fluorescence microscopy. Appl. Phys. B 77(4), 377–380 (2003)

82. F. Meinecke, Stimulierte Emission im Fluoreszenzmikroskop: Das STED Konzept zur Überwindung der Abbeschen Beugungsgrenze (Diplomarbeit, Ruprecht Karls Universität, Heidelberg, 1996)

83. M. Dyba, S.W. Hell, Focal spots of size $\lambda/23$ open up far-field fluorescence microscopy at 33 nm axial resolution. Phys. Rev. Lett. 88, 163901 (2002)

84. M. Dyba, S. Jakobs, S.W. Hell, Immunofluorescence stimulated emission depletion microscopy. Nat. Biotechnol. 21(11), 1303–1304 (2003)

85. G. Donnert et al., Macromolecular-scale resolution in biological fluorescence microscopy. Proc. Natl. Acad. Sci. USA 103(31), 11440–11445 (2006)

86. D. Wildanger et al., STED microscopy with a supercontinuum laser source. Opt. Expr. 16(13), 9614–9621 (2008)

87. R. Schmidt et al., Spherical nanosized spot unravels the interior of cells. Nat. Meth. 4(1), 81–86 (2008)

88. J. Keller, A. Schönle, S.W. Hell, Efficient fluorescence inhibition patterns for RESOLFT microscopy. Opt. Express 15(6), 3361–3371 (2007)
89. K.I. Willig et al., STED-microscopy reveals that synaptotagmin remains clustered after synaptic vesicle exocytosis. Nature 440(7086), 935–939 (2006)
90. R.J. Kittel et al., Bruchpilot promotes active zone assembly, Ca2+-channel clustering, and vesicle release. Science 312, 1051–1054 (2006)
91. J.J. Sieber et al., The SNARE motif is essential for the formation of syntaxin clusters in the plasma membrane. Biophys. J. 90, 2843–2851 (2006)
92. R. Kellner et al., Nanoscale organization of nicotinic acetylcholine receptors revealed by STED microscopy. Neuroscience 144(1), 135–143 (2007)
93. V. Westphal et al., Dynamic far-field fluorescence nanoscopy. New J. Phys. 9, 435 (2007)
94. V. Westphal et al., Video-rate far-field optical nanoscopy dissects synaptic vesicle movement. Science 320(5873), 246–249 (2008)
95. L. Kastrup et al., Fluorescence fluctuation spectroscopy in subdiffraction focal volumes. Phys. Rev. Lett. 94, 178104 (2005)
96. B. Hein, K. Willig, S.W. Hell, Stimulated emission depletion (STED) nanoscopy of a fluorescent protein – labeled organelle inside a living cell. Proc. Natl. Acad. Sci. U S A 105(38), 14271–14276 (2008)
97. V.U. Nägerl et al., Live-cell imaging of dendritic spines by STED microscopy. Proc. Natl. Acad Sci. USA 105(48), 18982–18987 (2008)
98. K. Willig et al., STED microscopy resolves nanoparticle assemblies. New J. Phys. 8, 106 (2006)
99. B. Harke et al., Three-dimensional nanoscopy of colloidal crystals. Nano Lett. 8(5), 1309–1313 (2008)
100. E. Rittweger et al., STED microscopy reveals color centers with nanometric resolution. Nat. Photonics, 3, 144–147 (2009).
101. G. Balasubramanian et al., Nanoscale imaging magnetometry with diamond spins under ambient conditions. Nature 455, 648–651 (2008)
102. J.R. Maze et al., Nanoscale magnetic sensing with an individual electronic spin in diamond. Nature 455, 644–647 (2008)
103. C.C. Fu et al., Characterization and application of single fluorescent nanodiamonds as cellular biomarkers. Proc. Nat. Acad. Sci. U.S.A. 104(3), 727–732 (2007)
104. J.I. Chao et al., Nanometer-sized diamond particle as a probe for biolabeling. Biophys. J. 93(6), 2199–2208 (2007)
105. S. Bretschneider, C. Eggeling, S.W. Hell, Breaking the diffraction barrier in fluorescence microscopy by optical shelving. Phys. Rev. Lett. 98, 218103 (2007)
106. K.A. Lukyanov et al., Natural animal coloration can be determined by a nonfluorescent green fluorescent protein homolog. J. Biol. Chem. 275(34), 25879–25882 (2000)
107. R. Ando, H. Mizuno, A. Miyawaki, Regulated fast nucleocytoplasmic shuttling observed by reversible protein highlighting. Science. 306(5700), 1370–1373 (2004)
108. O. Haeberle, Kindling molecules: a new way to 'break' the Abbe limit. C.R. Physique 5, 143–148 (2004)
109. W. Heisenberg, The Physical Principles of the Quantum Theory (Chicago University Press, Chicago, 1930)
110. N. Bobroff, Position measurement with a resolution and noise-limited instrument. Rev. Sci. Instrum. 57(6), 1152–1157 (1986)

111. M.P. Gordon, T. Ha, P.R. Selvin, Single-molecule high-resolution imaging with photobleaching. Proc. Natl. Acad Sci. U S A 101, 6462–6465 (2004)
112. X. Qu et al., Nanometer-localized multiple single-molecule fluorescence microscopy. Proc. Natl. Acad. Sci. USA 101(31), 11298–11303 (2004)
113. H.F. Hess, E. Betzig, Optical microscopy with phototransformable labels. Patent Appl, WO 2006/127692 A2 (2005/2006)
114. S.W. Hell, Verfahren und Fluoreszenzlichtmikroskop zum räumlich hochauflösenden Abbilden einer Struktur einer Probe. German Patent, **DE 10 2006 021 317** (2006/2007)
115. M. Bates et al., Multicolor super-resolution imaging with photo-switchable fluorescent probes. Science 317, 1749–1753 (2007)
116. J. Fölling et al., Photochromic rhodamines provide nanoscopy with optical sectioning. Angew. Chem. Int. Ed. 46, 6266–6270 (2007)
117. H. Bock et al., Two-color far-field fluorescence nanoscopy based on photo-switching emitters. Appl. Phys. B 88, 161–165 (2007)
118. C. Geisler et al., Resolution of $\lambda/10$ in fluorescence microscopy using fast single molecule photo-switching. Appl. Phys. A 88(2), 223–226 (2007)
119. A. Egner et al., Fluorescence nanoscopy in whole cells by asynchronous localization of photoswitching emitters. Biophys. J. 93, 3285–3290 (2007)
120. B. Huang et al., Three-dimensional super-resolution imaging by stochastic optical reconstruction microscopy. Science 319, 810–813 (2008)
121. H. Shroff et al., Live-cell photoactivated localization microscopy of nanoscale adhesion dynamics. Nat. Meth. 5(5), 417–423 (2008)
122. H. Shroff et al., Dual-color superresolution imaging of genetically expressed probes within individual adhesion complexes. Proc. Natl. Acad. Sci. U.S.A. 104(51), 20308–20313 (2007)
123. J. Fölling et al., Fluorescence nanoscopy by ground-state depletion and single-molecule return. Nat. Meth. 5, 943–945 (2008)
124. M. Bossi et al., Multi-color far-field fluorescence nanoscopy through isolated detection of distinct molecular species. Nano Lett. 8(8), 2463–2468 (2008)
125. S. Weiss, Fluorescence spectroscopy of single biomolecules. Science 283, 1676–1683 (1999)
126. A. Yildiz et al., Myosin V walks hand-over-hand: single fluorophore imaging with 1.5-nm localization. Science 300(5628), 2061–2065 (2003)
127. A.M. van Oijen et al., Far-field fluorescence microscopy beyond the diffraction limit. J. Opt. Soc. Am. A 16(4), 909–915 (1999)
128. E. Betzig, Proposed method for molecular optical imaging. Opt. Lett. 20(3), 237–239 (1995)
129. K.A. Lidke et al., Superresolution by localization of quantum dots using blinking statistics. Opt. Expr. 13(18), 7052–7062 (2005)
130. S.W. Hell, J. Soukka, P.E. Hänninen, Two- and multiphoton detection as an imaging mode and means of increasing the resolution in far-field light microscopy. Bioimaging 3, 65–69 (1995)
131. J. Fölling et al., Fluorescence nanoscopy with optical sectioning by two-photon induced molecular switching using continuous-wave lasers. Chem. Phys. Chem. 9, 321–326 (2008)
132. M.F. Juette et al., Three-dimensional sub-100 nm resolution fluorescence microscopy of thick samples. Nat. Meth. 5(6), 527–529 (2008)
133. C.v. Middendorff et al., Isotropic 3D Nanoscopy based on single emitter switching. Opt. Expr. 16(25), 20774–20788 (2008)

Sub-Diffraction-Limit Imaging with Stochastic Optical Reconstruction Microscopy

Mark Bates, Bo Huang, Michael J. Rust, Graham T. Dempsey,
Wenqin Wang, and Xiaowei Zhuang

Summary. Light microscopy is a widely used imaging method in biomedical research. However, the resolution of conventional optical microcopy is limited by the diffraction of light, making structures smaller than 200 nm difficult to resolve. To overcome this limit, we have developed a new form of fluorescence microscopy - Stochastic Optical Reconstruction Microscopy (STORM). STORM makes use of single-molecule imaging methods and photo-switchable fluorescent probes to temporally separate the otherwise spatially overlapping images of individual molecules. An STORM image is acquired over a number of imaging cycles, and in each cycle only a subset of the fluorescent labels is switched on such that each of the active fluorophores is optically resolvable from the rest. This allows the position of these fluorophores to be determined with nanometer accuracy. Over the course of many such cycles, the positions of numerous fluorophores are determined and used to construct a super-resolution image. Using this method, we have demonstrated multi-color, three-dimensional (3D) imaging of biomolecules and cells with ∼20 nm lateral and ∼50 nm axial resolutions. In principle, the resolution of this technique can reach the molecular scale.

20.1 Introduction

Fluorescence microscopy is one of the most powerful imaging methods in modern biomedical research. Noninvasive, fluorescence microscopy allows dynamic imaging of live cells, tissues, and whole organisms. This capability for live sample imaging, combined with a large repertoire of spectrally distinct fluorescent probes and a variety of biochemically specific labeling techniques, enables the direct visualization of complex molecular and cellular processes in real time under physiological conditions.

The limited resolution of fluorescence microscopy, however, leaves many biological structures too small to be observed in detail. The resolution of conventional light microscopy is limited by diffraction to 200–300 nm in the lateral directions and 500–800 nm in the axial direction, whereas sub-cellular structures span a range of length scales from nanometers to microns. Electron

microscopy (EM) and scanning probe microscopy (SPM) offer exquisite resolution in the nanometer to even sub-nanometer range. However, these techniques
often require the fixation or freezing of samples (EM) or the use of a scanning
tip positioned within nanometers of the sample (SPM), making it difficult to image live biological samples in a noninvasive manner. To approach
molecular scale resolutions with the biochemical specificity and live-cell compatibility provided by fluorescence microscopy would open a new window for
the study of molecular interactions and biochemical processes in cells and
tissues.

Several far-field light microscopy methods have recently been developed
to break the diffraction limit. These methods can be largely divided into
two categories: (1) techniques that employ spatially patterned illumination to
sharpen the point-spread function of the microscope, such as stimulated emission depletion (STED) microscopy and related methods using other reversibly
saturable optically linear fluorescent transitions (RESOLFT) [1, 2], and saturated structured-illumination microscopy (SSIM) [3], and (2) a technique
that is based on the localization of individual fluorescent molecules, termed
Stochastic Optical Reconstruction Microscopy (STORM [4], Photo-Activated
Localization Microscopy (PALM) [5], or Fluorescence Photo-Activation Localization Microscopy (FPALM) [6]. In this paper, we describe the concept
of STORM microscopy and recent advances in the imaging capabilities
of STORM.

20.2 Results

20.2.1 Imaging Concept of STORM

STORM is based on the detection [7] and subsequent high-accuracy localization of single fluorescent molecules. The diffraction of light causes the image
of a single isolated fluorescent emitter to appear as a spot with a finite size
described by the point-spread-function (PSF). The PSF width for visible light
is approximately 200–300 nm in the lateral directions and 500–800 nm along
the axial direction. This effect is the origin of the diffraction limit of optical
image resolution. The position of the emitter can, however, be determined
to a much higher accuracy than the PSF size depending on the number of
photons detected from the emitter [8–11]. This localization accuracy is given
approximately by $s/\sqrt{N}$, where s is the standard deviation of the PSF and N
is the number of photons detected [10]. This concept has been used to track
small particles with nanometer accuracy [8, 9]. Recently, it has been shown
that even the position of a single fluorescent dye can be determined with an
accuracy as high as ∼1 nm [11].

Localization of individual fluorescent emitters with nanometer accuracy
does not, however, translate directly into nano-scale image resolution, as multiple fluorophores within a diffraction-limited area would yield overlapping

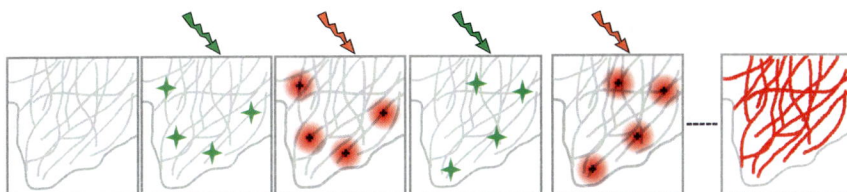

Fig. 20.1. STORM imaging. The grey filaments represent the target of interest labeled with photo-switchable probes. All fluorophores are initially placed in the dark state. In each activation cycle, the sample is exposed to a specific wavelength of light (indicated by the green arrow), causing a sparse set of fluorophores to be activated to the fluorescent state. The activated fluorophores (*green stars*) are imaged using light at a second wavelength (*red arrow*). The density of activated fluorophores is sufficiently low such that their images (*red circles*) do not overlap. The position of each activated fluorophore is then determined by fitting its image to find the centroid position (*black crosses*). In a subsequent cycle, a different set of fluorophores are activated and localized. After a sufficient number of fluorophores have been localized, a high-resolution image is constructed by plotting the measured positions of the fluorophores (*small red dots*)

images, making it difficult to unambiguously determine their individual positions. Fluorescence signals from nearby emitters could be separated based on differences in emission wavelength [12–14], the sequential photo-bleaching of each fluorophore [15, 16], or quantum dot blinking [17]. These methods have obtained high-accuracy localization for several closely spaced emitters, but are difficult to extend to densities higher than a few fluorophores per diffraction-limited area.

We have recently proposed a new super-resolution imaging method based on the sequential localization of photo-switchable fluorescent molecules, which can be switched between a nonfluorescent (dark) state and a fluorescent state by exposure to light of different wavelengths. Figure 20.1 gives an illustration of such an imaging process composed of many imaging cycles. During each cycle, a small subset of fluorophores is stochastically activated to the fluorescent state by exposing the sample to a light source at the activation wavelength. The density of activated molecules is kept sufficiently low, by adjusting activation intensity, such that the images of individual molecules do not typically overlap. This condition allows the position of each activated fluorophores to be determined with high accuracy. The fluorophores are then deactivated and this imaging cycle is repeated until a sufficient number of localizations have been recorded. A high-resolution image is then constructed from the measured positions of the fluorophores. The resolution of the final image is not limited by diffraction, but by the precision of each localization. This concept, independently developed by three research groups, has been given the names STORM [4], PALM [5], or FPALM [6].

20.2.2 Photo-Switchable Fluorescent Probes

STORM/PALM/FPALM can be realized using a variety of photo-switchable fluorophores, including dye molecules and fluorescent proteins. We have recently discovered a family of photo-switchable cyanine dyes that can be reversibly cycled between a fluorescent and a dark state in a controlled and reversible manner by exposure to light of different wavelengths [18,19]. These include many red cyanine dyes, such as Cy5, Cy5.5, and Cy7 [18,19]. Red light (e.g., 657 nm) can excite fluorescence from these dyes and also cause them to switch off into a meta-stable dark state (Fig. 20.2a). Exposure to UV light converts the dyes back to the fluorescent state (unpublished result) [20]. The red cyanine dyes can be efficiently reactivated by visible light when a second dye, e.g., Cy3, is placed in close proximity to the red cyanine dye, either by attaching both dyes to a common third molecule [4, 18, 19] or by direct covalent conjugation of the two dyes [49]. In this configuration, the dye pairs can be efficiently reactivated by green light (e.g., 532 nm) (Fig. 20.2a). In the following, we refer to the red photo-switchable cyanine dyes, such as Cy5, Cy5.5, and Cy7, as "reporters" and the Cy3 dye as an "activator." The activator–reporter pair can be rapidly switched on and off hundreds of times before permanent photo-bleaching occurs. The availability of several reporter dyes with different emission wavelengths suggests one natural approach for constructing photo-switchable probes with multiple colors. The distinct emission spectra of the Cy5, Cy5.5, and Cy7 reporters allow them to be clearly distinguished at the single-molecule level (unpublished results) [21].

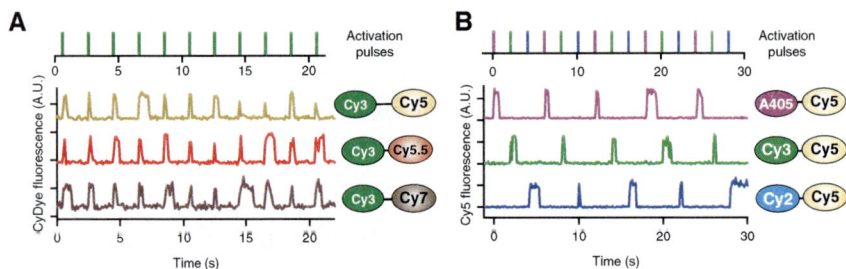

Fig. 20.2. Photo-switchable probes constructed from activator-reporter pairs. (a) Spectrally distinct reporters exhibit photo-switching behavior. The lower panel shows the fluorescence time traces of Cy5, Cy5.5, and Cy7 when paired with a Cy3 dye as the activator. The upper panel shows the green laser pulses used to activate the reporters. The red laser was continuously on, serving to excite fluorescence from the reporters and to switch them off to the dark state. (b) The same reporter can be activated by spectrally distinct activators. The lower panel shows the fluorescence time traces of Cy5 paired with different activators, Alexa Fluor 405 (A405), Cy2, and Cy3. The upper panel shows the violet (405 nm, magenta line), blue (457 nm, blue line), and green (532 nm, green line) activation pulses

Next, we tested whether spectrally distinct dyes can be used as activators for the same reporter [19]. To this end, we paired Cy3, Cy2, and Alexa Fluor 405 (A405) with the Cy5 reporter. Switching traces for individual pairs indicate efficient photo-switching in all cases, but the activation of Cy5 required different colored lasers corresponding to the absorption wavelength of the activator (Fig. 20.2b). For example, the A405-Cy5 pair was efficiently activated by a violet laser, but activation of the Cy2-Cy5 and Cy3-Cy5 pairs was much lower by the same laser. Similar results were found for blue and green activations. For each wavelength, the pair with the appropriate activator was activated with a rate at least ten times higher than the other two, indicating low false activation probabilities [19].

This wavelength-specific activation suggests a second approach for constructing multi-color photo-switchable probes. Not only can different activator–reporter pairs be distinguished by their emission color, as determined by the reporter dye, but they can also be differentiated by the color of light which activates them, as determined by the activator dye. A combinatorial pairing scheme of reporters and activators may thus allow the construction of a large number spectrally distinguishable fluorescent probes and enable STORM imaging with many colors [19].

In addition to the photo-switchable cyanine dyes, other photo-activatable dyes such as caged fluorescent compounds [5] and photo-chromic rhodamine [22], as well as photo-switchable fluorescent proteins, such as PA-GFP [23], Kaede [24] and EosFP [25], KikGR [26], Dronpa [27,28] and rsFastLime [29], and photo-switchable CFP2 [30] can also be used for STORM/PALM/FPALM imaging.

20.2.3 STORM Imaging of Biomolecular Complexes

A major factor influencing the resolution of STORM is the accuracy with which individual fluorophores are localized. The fact that cyanine dye molecules can be switched on and off many times before photo-bleaching enables a direct experimental measurement of the localization accuracy [4]. As a model system, we attached the Cy3–Cy5 dye pairs to microscope slides through a DNA linker at a low density such that individual dye pairs were resolvable. The dye pairs were repeatedly switched on and off by a green and a red laser, respectively, and the positions of individual Cy5 molecules were measured for each activation cycle from their corresponding fluorescence images by fitting each image a two-dimensional (2D) Gaussian function to determine the centroid position (Fig. 20.3a). The centroid positions determined from multiple activation cycles of the same molecule, after drift correction, follow a normal distribution with a standard deviation of 8 nm, corresponding to a full-width at half maximum (FWHM) of 18 nm (Fig. 20.3b), predicting an image resolution of ∼20 nm in the lateral directions [4]. We note that the experimentally determined localization accuracy is significantly larger than the value (∼4 nm standard deviation) calculated from the detected photon number and

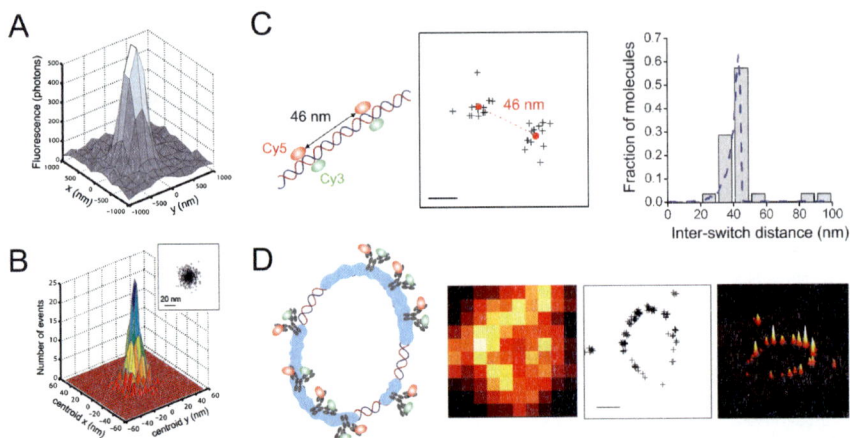

Fig. 20.3. STORM Image of biomolecular complexes with sub-diffraction-limit resolution. (**a, b**) The localization accuracy of individual photo-switchable probe. (**a**) The image profile of a Cy3–Cy5 pair during a single switching cycle. Fitting the image to a 2D Gaussian gives the centroid position. (**b**) The distribution of the centroid positions from many activation cycles after correction for stage drift. The inset shows the centroid positions. (**c**) STORM images of two Cy3–Cy5 pairs separated by a contour length of 46 nm on double-stranded DNA. A typical STORM image (*middle panel*) show two well-separated clusters of Cy5 positions (*crosses*), the center-of-mass of which (*red dots*) are separated by 46 nm. Scale bars: 20 nm. (*Right panel*) Comparison between the inter-Cy5 distances measured using STORM (*grey column*) and the predicted distance distribution (*dashed blue line*). (**d**) Imaging RecA-coated plasmid DNA. The image on the left is the conventional immunofluorescence image against RecA with Cy3–Cy5-labeled antibody. The panel on the middle shows the STORM image of the same filaments constructed from many Cy5 localizations. Scale bars: 300 nm. The *right panel* is the 3D surface plot of the STORM image constructed by convolving each Cy5 localization with a Gaussian of 18-nm FWHM

background noise [4, 10]. The discrepancy is likely due to uncertainties in drift correction and imperfect fitting of the single-molecule images. This indicates that one should be cautious about using the theoretically predicted localization accuracy as a measure of the actual image resolution.

To demonstrate that STORM can indeed resolve nearby fluorescent molecules with sub-diffraction-limit resolution, we first engineered samples with known relative positions of the fluorescent labels – double-stranded DNA labeled with two Cy3-Cy5 pairs separated by a well-defined number (135) of base pairs, corresponding to an inter-Cy5 distance of 46 nm along the contour of DNA [4]. The DNA strands were immobilized in a flat configuration to a quartz slide through multiple biotin–streptavidin linkages. The two Cy5 dyes were turned on and off, repetitively, and the image sequence was analyzed to determine the positions of individual activated Cy5 dye. We then constructed the STORM image by plotting the Cy5 localizations determined over all of the imaging cycles. The image typically shows two clusters of localizations, indicating that the two Cy5 molecules on the DNA are well-resolved (Fig. 20.3c). The measured center-of-mass distances between the clusters agree quantitatively with the engineered inter-dye distance, assuming a 50 nm persistence length for double stranded DNA (Fig. 20.3c) [31]. DNA samples with more than two Cy3–Cy5 pairs were also well resolved [4].

In principle, STORM can resolve many fluorescent molecules within a diffraction-limited spot by activating the fluorescent probes in a controlled manner. To test this capability, we prepared circular DNA plasmids coated with RecA protein and imaged them using indirect immunofluorescence with a secondary antibody doubly labeled with Cy3 and Cy5 [4]. To acquire an STORM image, a weak activation laser intensity was used such that only a small fraction of the Cy5 dyes were activated in each activation cycle. Individual isolated Cy5 images were analyzed to determine the position of the Cy5 dyes. The STORM image constructed from many Cy5 localizations accumulated over multiple activation cycles revealed the circular structure of the RecA filament with greatly increased resolution when compared with conventional wide-field images (Fig. 20.3d).

20.2.4 STORM Imaging of Cells

Applying STORM to cell imaging, we first performed immunofluorescence imaging of microtubules in mammalian cells [19]. Microtubules are filamentous structures important for cell division, intracellular transport, and many other cellular functions. To label microtubules with photo-switchable probes, we fixed and permeablized BS-C-1 cells and immunostained microtubules with primary and secondary antibodies. The secondary antibodies were doubly labeled with Cy3 and the Alexa Fluor 647 (A647), a structural analog of Cy5 with similar photo-switching properties. STORM images were generated using a green activation laser (532 nm) and a red imaging laser (657 nm) as described in Fig. 20.1.

The STORM images of the microtubule network shows a drastic improvement in resolution compared to the corresponding conventional fluorescence images of the same field of view (Fig. 20.4). In the regions where microtubules were densely packed and unresolved in the conventional image, individual microtubule filaments were clearly resolved by STORM (Figs. 20.4c and d). To determine the localization accuracy inside the cell, we identified point-like objects in the cell, depicted by small clusters of localizations away from any discernable microtubule filaments. These clusters represent individual antibodies nonspecifically attached to the cell. The FWHM of these clusters was determined to be 24 nm, providing a measure of the image resolution [19].

In addition to microtubules, we have imaged other cellular structures, such as clathrin-coated pits, actin, nuclear pore complexes, endoplasmic reticulum, and mitochondria with STORM. The STORM imaging concept applies not

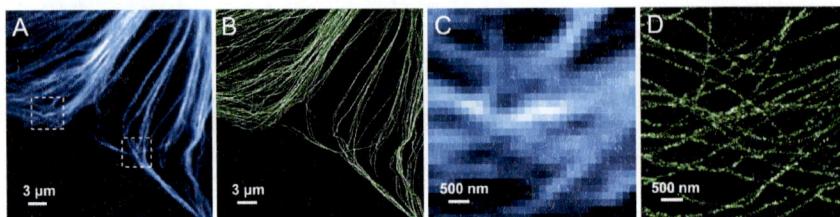

Fig. 20.4. STORM imaging of microtubules in a mammalian cell. (**a**) Conventional immunofluorescence image of microtubules in a large area of a BS-C-1 cell. (**b**) STORM image of the same area. (**c**) Conventional and (**d**) STORM images corresponding to the boxed regions in (**a**)

only to photo-switchable dye molecules but also to photo-switchable fluorescent proteins. A variety of photo-switchable fluorescent proteins have been used for STORM/PALM/FPALM imaging [5, 6, 32].

20.2.5 Multi-Color STORM

To achieve multi-color STORM imaging, we took advantage of the large number of spectrally distinct activator–reporter dye pairs that are photo-switchable. As mentioned earlier, multi-color images may be obtained either by distinguishing the emission color of the reporter dye or by differentiating the color of the activation light which depends on the activator dye. We used the selective activation scheme as an initial demonstration of multi-color STORM imaging [19].

To construct a model multi-color sample, we mixed three different DNA constructs, each labeled with a Cy3–Cy5 pair, a Cy2–Cy5 pair, or an A405–Cy5 pair and immobilized these constructs on a microscope slide. We purposely chose a high surface density, such that individual DNA molecules could not be resolved in a conventional fluorescence image [19]. To generate an STORM image, the sample was initially deactivated with a red 657 nm, and then periodically activated by a sequence of 532, 457, and 405 nm laser pulses. In between activation pulses, the sample was imaged with the red laser. The image of each activated fluorescent spot was analyzed to determine its centroid position, and a false color was assigned to each localization depending on the wavelength of the light pulse which activated it. An STORM image was then constructed by plotting the false colored localizations (Fig. 20.5a). The STORM image shows clearly separated clusters of localizations, each of which primarily contains localizations of a single color and corresponds to an individual DNA molecule (Fig. 20.5a–c). The small fractions of mis-colored localizations within each cluster provide a measure of color crosstalk (Fig. 20.5d). The FWHM of the clusters are 27, 27, and 26 nm for the three color channels, predicting an imaging resolution of 20–30 nm for three-color STORM imaging.

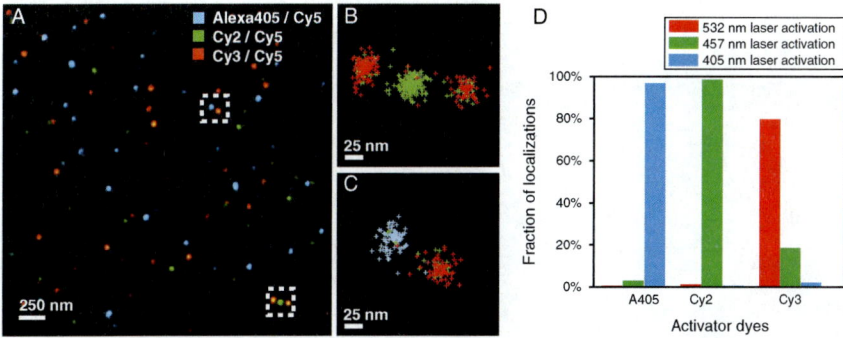

Fig. 20.5. Three-color STORM imaging of a model DNA sample. (a) STORM image of three different DNA constructs labeled with A405-Cy5, Cy2-Cy5, or Cy3-Cy5. Each colored spot in this image represents a cluster of localizations from a single DNA molecule. (b, c) Higher magnification views of the boxed regions in (a). Here, each localization (represented by a cross) was colored blue, green or red if the molecule was activated by a 405, 457, or 532 nm laser pulse, respectively. (d) Crosstalk analysis. The majority of the localizations within each cluster displayed the same color, identifying the type of activator dye (Alexa 405, Cy2, or Cy3) present on the DNA. The fractions of localizations assigned to each color channel are plotted here for the A405, Cy2, and Cy3 clusters. The crosstalk can be calculated from the ratios of incorrectly to correctly colored localizations

To demonstrate multi-color STORM imaging in cells, we simultaneously imaged microtubules and clathrin-coated pits [19], cellular structures important for receptor-mediated endocytosis. Prior to imaging, we stained the tubulin and clathrin with specific primary antibodies and secondary antibodies. The secondary antibodies used for tubulin and clathrin were labeled with the Cy2-A647 and Cy3-A647 pairs, respectively. The 457 and 532 nm laser pulses were used to selectively activate the two pairs, and each localization was false colored according to the activation light source. Figure 20.6 shows a two-color STORM image in comparison with a conventional fluorescence image of the same area. The green channel (457 nm activation) revealed filamentous structures representing microtubules. Overlapping microtubules in the conventional image are clearly resolved in the STORM image. The red channel (532 nm activation) revealed spherical structures, representing coated pits and vesicles. The diameter of these structures, 172 ± 35 nm, agrees quantitatively with the size distribution of clathrin-coated pits and vesicles determined using EM [33]. Many of the spherical CCPs appeared to have a donut shape with a higher density of localizations toward the periphery, which is consistent with the 2D projection of a 3D cage structure. This is in stark contrast to the conventional fluorescence images of CCPs, which appeared as diffraction limited spots without any discernable shape.

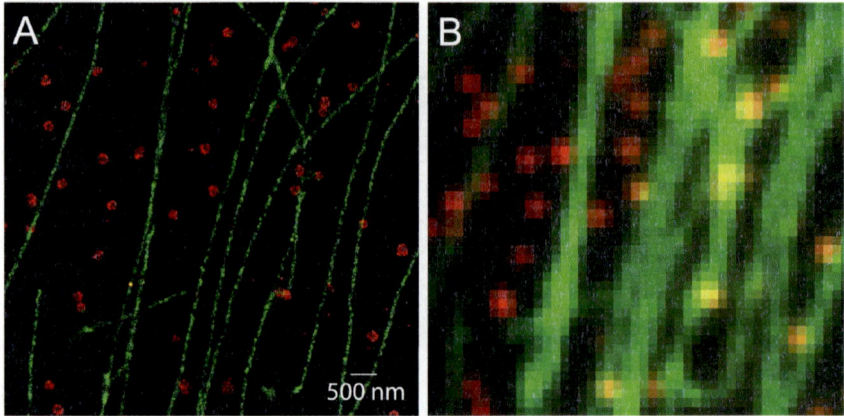

Fig. 20.6. Two-color STORM imaging of microtubules (*green*) and clathrin-coated pits (*red*) in a cell. (**a**) STORM and (**b**) conventional fluorescence images of an area of a BS-C-1 cell

In addition to using the photo-switchable cyanine dyes, multi-color super-resolution imaging can also be accomplished by using photo-switchable fluorescent proteins [34] or a combination of dyes and fluorescent proteins [35].

20.2.6 Three-Dimensional (3D) STORM

3D STORM is based on high-accuracy localization of individual fluorphores in all three dimensions. While the lateral position of a fluorescent molecule can be determined from the centroid of its image, the shape of the image contains information about the particle's axial (z) position. Nanoscale localization accuracy can be achieved in the z dimension by introducing defocusing [12,36] or astigmatism [37,38] into the image, without significantly compromising the lateral positioning capability.

We used the astigmatism approach to achieve 3D STORM [39]. In this approach, a weak cylindrical lens was introduced into the imaging path to create two slightly different focal planes for the x and y directions, such that the image has different widths in x and y. As a result, the ellipticity of a fluorophore's image varied as its position changed in z. By fitting the image with a 2D elliptical Gaussian function, we can obtain the x and y coordinates of peak position as well as the peak widths w_x and w_y, from which the z coordinate of the fluorophore can be unambiguously determined [39]. The 3D resolution of STORM is thus limited by the accuracy with which individual photo-activated fluorophores can be localized in all three dimensions during a single switching cycle. As a model system, we determined the localization accuracy of a Cy3-A647 pair attached to streptavidin molecules immobilized to a microscope coverglass [39]. A red laser (657 nm) and a green laser (532 nm) were alternated to deactivate and reactivate the A647 molecules repetitively, and their

x, y, and z coordinates were determined for each switching cycle. This procedure resulted in a 3D cluster of localizations for each molecule. The FWHM value of the cluster were $\sim$24 nm in x and y, and $\sim$47 nm in z, providing a quantitative measure of the localization accuracy in 3D. Because the image width increases as the fluorophore moves away from the focal planes, the localization accuracy decreases accordingly, particularly in the lateral dimensions. Therefore, we typically chose a z-imaging depth of about 600–700 nm near the average focal plane, within which the localization accuracy does not change substantially. The imaging depth can be increased by scanning the sample in the z direction.

Applying 3D STORM to cell imaging, we performed indirect immunofluorescence imaging of the microtubule network in cells that were immunostained with primary antibodies and then with Cy3-A647-labeled secondary antibodies [39]. The 3D STORM image not only showed a substantial improvement in resolution as compared to the conventional wide-field fluorescence image but also provided the position information in the z dimension that was not available in the conventional image (Fig. 20.7). Using the same cluster analysis as discussed earlier for the 2D STORM image of cells, the localization accuracy was determined to be 22 nm in x, 28 nm in y, and 55 nm in z inside the cell, similar to those determined earlier for individual molecules immobilized on a glass surface.

To demonstrate that 3D STORM can resolve the 3D morphology of nanoscopic structures in cells, we imaged clathrin-coated pits using a direct immunofluorescence scheme. The clathrin in the cell was stained with

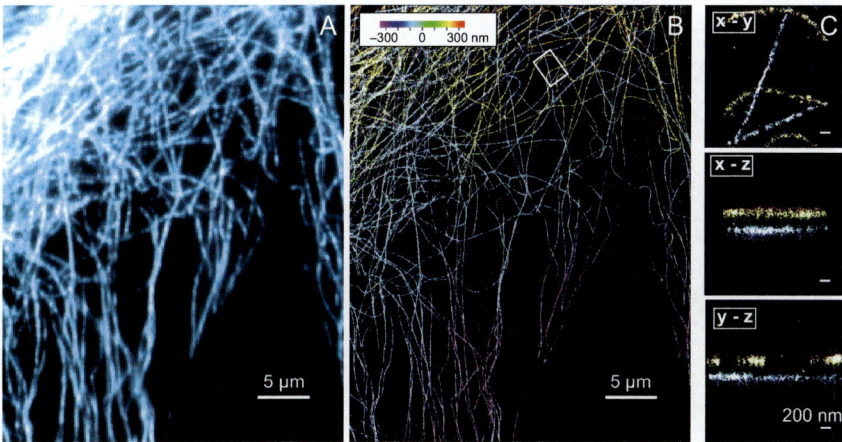

Fig. 20.7. 3D STORM imaging of microtubules in a cell. (**a**) Conventional immunofluorescence image of microtubules in an area of a cell. (**b**) The 3D STORM image of the same area with the z-position information color-coded according to the colored scale bar (**c**) The $x-y$, $x-z$, and $y-z$ cross-sections of a small region of the cell outlined by the white box in (**b**)

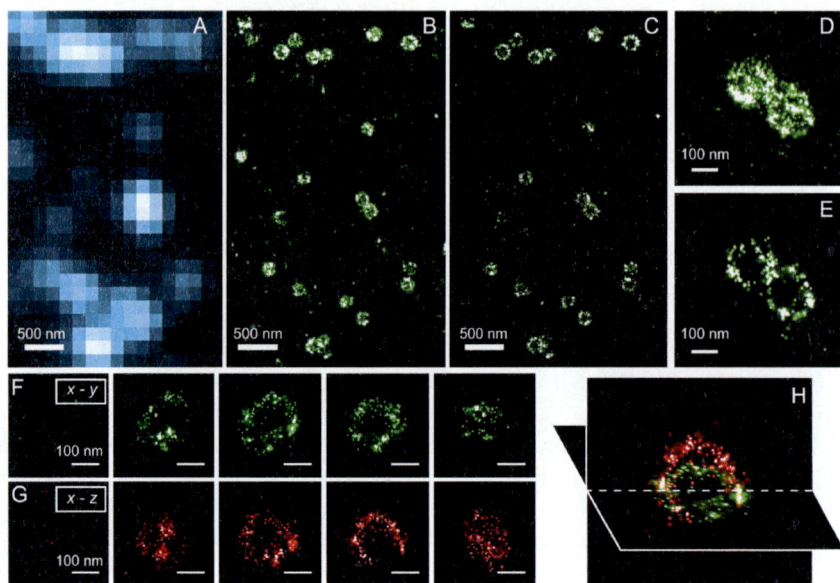

Fig. 20.8. 3D STORM imaging of clathrin-coated pits in a cell. (**a**) Conventional immunofluorescence image of clathrin in a region of a BS-C-1 cell. (**b**) The STORM image of the same area with all localizations at different z positions stacked. (**c**) An $x-y$ cross-section (50-nm thick in z) of the same area. (**d, e**) Magnified view of two nearby coated pits in 2D STORM (**d**) and their $x-y$ cross-section in the 3D image (**e**). (**f–h**) Serial $x-y$ cross-sections (each 50-nm thick in z) (**f**) and $x-z$ cross-sections (each 50-nm thick in y) (**g**) of a CCP, and an $x-y$ and $x-z$ cross-section presented in 3D perspective (**h**)

Cy3-A647-labeled primary antibodies. When imaged by conventional fluorescence microscopy, all CCPs appeared as nearly diffraction-limited spots with no discernable structure (Fig. 20.8a). In 2D STORM images, in which the z-dimension information was discarded, the round shape of CCPs was clearly seen (Figs. 20.8b and d). Including the z-dimension information allows us to clearly visualize the 3D structure of the pits. The circular ring-like structure of the pit periphery was unambiguously resolved in the $x-y$ cross-section of the image taken near the cell surface (Figs. 20.8c and e). Consecutive $x-y$ and $x-z$ cross-sections of the pits (Figs. 20.8f–h) revealed the 3D half-spherical cage like morphology of these nanoscopic structures.

3D STORM allowed us to image a 600–700-nm thick region with high-resolution in all three dimensions without requiring scanning. As the image of each fluorophore simultaneously encodes its x, y, and z coordinates, no additional time was required to localize each molecule in 3D STORM as compared with 2D STORM imaging. Thicker samples can be imaged when combined with sample scanning with ∼600 nm steps. We have acquired whole cell 3D STORM images with as few as five scanning steps in z [49]. Similar to the astigmatism approach, 3D super-resolution imaging can also be accomplished by defocusing [40], double helical PSF [50], or interference [51]. 3D STED and SIM have also demonstrated sub-diffraction-limit image resolution [41, 42].

20.3 Discussion

In summary, we have introduced a new approach to high-resolution fluorescence microscopy, which is based on photo-switchable fluorescent molecules and the high-accuracy localization of individual fluorophores. This technique, which we refer to as STORM, makes use of photo-switchable fluorescent molecules which are activated and deactivated in a controlled manner, such that at any time only a small optically resolvable subset are switched on allowing their locations to be unambiguously determined with high accuracy in all three dimensions. The overall super-resolution image is then constructed by plotting the localizations accumulated over numerous cycles of activation and deactivation. Using STORM, we have demonstrated imaging of biomolecules and cells with ~20-nm resolution in the lateral directions and ~50-nm resolution in the axial direction. This imaging capability allows nanoscale features of cellular structures to be resolved optically under ambient conditions with biochemical specificity, at a resolution previously seen only with electron microscopy.

In addition, we have discovered a family of photo-switchable probes based on photo-switchable cyanine reporters and proximal activators that facilitate the reactivation of the cyanine dyes. Variation in the activator and reporter allows multi-color imaging to be performed using two independent measures: by distinguishing the emission colors of the reporter dyes, or according to the color of the light which activates the reporter, as dictated by the activator dye. Combinatorial pairing of the activators and reporters allows the construction of many spectrally distinct photo-switchable probes. Using these photo-switchable activator–reporter dye pairs, we have demonstrated multi-color STORM imaging, which allows molecular interactions in cells and cell–cell interactions in tissues to be imaged at the nanometer scale.

As a high-resolution STORM image is constructed from localizations accumulated over many imaging frames, the imaging speed of STORM is relatively low. Currently, an STORM image at the highest resolution typically requires minutes of image acquisition time. At a compromised resolution (e.g., ~60 nm), time resolved STORM images can be acquired with a time resolution of ~10 s using cyanine dyes (unpublished results). A time resolution of 25–60 s has also be accomplished at ~60-nm resolution with photo-switchable fluorescent proteins [43]. We expect this imaging speed to be further improved with faster cameras and higher excitation power.

The spatial resolution of STORM is determined by the accuracy of each localization, the density of localizations obtained in the image, and the physical size of the fluorescent labels. Thus, in practice, several important factors affect the image resolution: (1) The brightness of the photo-switchable probes, i.e., the number of photons detected during each activation cycle of the probe directly determines the localization accuracy. (2) The contrast ratio between the fluorescent and dark state of the probe is a factor that affects the localization accuracy and the maximum density of localizations that can be achieved. Some fluorescent probes have finite fluorescence emission even in their dark state, and nearly all photo-activatable probes have a finite rate of spontaneous activation from the dark state to the bright state. Both of these properties result in an undesired background signal that impairs the localization accuracy. Eventually, at a sufficiently high label density, the background due to dark state emission or spontaneous activation may become large enough to prohibit the localization of individual probes. (3) The labeling efficiency, defined as the percentage of the target molecules that are labeled with a photo-switchable probe, also affects the density of localizations in the final image. (4) Finally, an important factor is the label size itself. Provided sufficient probe brightness and labeling density, resolution can be almost arbitrarily high. For example, the 6,000 photons detected from a Cy5 dye corresponds to a theoretical localization precision of a few nanometers, suggesting the possibility of achieving true molecular scale image resolution. At this level, the physical size of the label becomes limiting. Recently developed small-molecule labeling approaches provide a promising solution to cell labeling for super-resolution imaging. In one approach, specific peptide sequences were fused to target proteins. These peptide sequences were selected either for high affinities to specific chemical groups [44, 45], or to be substrates for enzymes that can covalently conjugate specific chemical groups to the peptide [46–48], thus allowing small organic dyes to be linked directly to target protein of interest. Combining this genetically encoded small-molecule labeling strategy with bright probes of large contrast ratios in STORM may ultimately provide molecular scale resolution in fluorescence imaging.

Acknowledgements

This work is supported by in part by the NIH (to X.Z.). X.Z. is a Howard Hughes Medical Institute Investigator

References

1. S.W. Hell, J. Wichmann, Breaking the diffraction resolution limit by stimulated emission: stimulated-emission-depletion fluorescence microscopy. Opt. Lett. 19, 780–782 (1994)
2. S.W. Hell, Far-field optical nanoscopy. Science 316, 1153–1158 (2007)
3. M.G.L. Gustafsson, Nonlinear structured-illumination microscopy: wide-field fluorescence imaging with theoretically unlimited resolution. Proc. Natl. Acad. Sci. U S A 102, 13081–13086 (2005)

4. M.J. Rust, M. Bates, X. Zhuang, Sub-diffraction-limit imaging by stochastic optical reconstruction microscopy (STORM). Nat. Meth. 3, 793–795 (2006)
5. E. Betzig, G.H. Patterson, R. Sougrat, O.W. Lindwasser, S. Olenych, J.S. Bonifacino, M.W. Davidson, J. Lippincott-Schwartz, H.F. Hess, Imaging intracellular fluorescent proteins at nanometer resolution. Science 313, 1642–1645 (2006)
6. S.T. Hess, T.P. Girirajan, M.D. Mason, Ultra-high resolution imaging by fluorescence photoactivation localization microscopy. Biophys. J. 91, 4258–4272 (2006)
7. W.E. Moerner, M. Orrit, Illuminating single molecules in condensed matter. Science 283, 1670–1676 (1999)
8. J. Gelles, B.J. Schnapp, M.P. Sheetz, Tracking kinesin-driven movements with nanometre-scale precision. Nature 331, 450–453 (1988)
9. R.N. Ghosh, W.W. Webb, Automated detection and tracking of individual and clustered cell surface low density lipoprotein receptor molecules. Biophys. J. 66, 1301–1318 (1994)
10. R.E. Thompson, D.R. Larson, W.W. Webb, Precise nanometer localization analysis for individual fluorescent probes. Biophys. J. 82, 2775–2783 (2002)
11. A. Yildiz, J.N. Forkey, S.A. McKinney, T. Ha, Y.E. Goldman, P.R. Selvin, Myosin V walks hand-over-hand: single fluorophore imaging with 1.5-nm localization. Science 300, 2061–2065 (2003)
12. A.M. van Oijen, J. Kohler, J. Schmidt, M. Muller, G.J. Brakenhoff, 3-Dimensional super-resolution by spectrally selective imaging. Chem. Phys. Lett. 292, 183–187 (1998)
13. T.D. Lacoste, X. Michalet, F. Pinaud, D.S. Chemla, A.P. Alivisatos, S. Weiss, Ultrahigh-resolution multicolor colocalization of single fluorescent probes. Proc. Natl. Acad. Sci. U S A 97, 9461–9466 (2000)
14. L.S. Churchman, Z. Okten, R.S. Rock, J.F. Dawson, J.A. Spudich, Single molecule high-resolution colocalization of Cy3 and Cy5 attached to macromolecules measures intramolecular distances through time. Proc. Natl. Acad. Sci. U S A 102, 1419–1423 (2005)
15. M.P. Gordon, T. Ha, P.R. Selvin, Single-molecule high-resolution imaging with photobleaching. Proc. Natl. Acad. Sci. U S A 101, 6462–6465 (2004)
16. X. Qu, D. Wu, L. Mets, N.F. Scherer, Nanometer-localized multiple single-molecule fluorescence microscopy. Proc. Natl. Acad. Sci. U S A 101, 11298–11303 (2004)
17. K. Lidke, B. Rieger, T. Jovin, R. Heintzmann, Superresolution by localization of quantum dots using blinking statistics. Opt. Exp. 13, 7052–7062 (2005)
18. M. Bates, T.R. Blosser, X. Zhuang, Short-range spectroscopic ruler based on a single-molecule optical switch. Phys. Rev. Lett. 94, 108101 (2005)
19. M. Bates, B. Huang, G.T. Dempsey, X. Zhuang, Multicolor super-resolution imaging with photo-switchable fluorescent probes. Science 317, 1749–1753 (2007)
20. M. Heilemann, E. Margeat, R. Kasper, M. Sauer, P. Tinnefeld, Carbocyanine dyes as efficient reversible single-molecule optical switch. J. Am. Chem. Soc. 127, 3801–3806 (2005)
21. S. Hohng, C. Joo, T. Ha, Single-molecule three-color FRET. Biophys. J. 87, 1328–1337 (2004)
22. J. Folling, V. Belov, R. Kunetsky, R. Medda, A. Schonle, A. Egner, C. Eggeling, M. Bossi, S.W. Hell, Photochromic rhodamines provide nanoscopy with optical sectioning. Angew. Chem. Int. Ed. Engl. 46, 6266–6270 (2007)

23. G.H. Patterson, J. Lippincott-Schwartz, A photoactivatable GFP for selective photolabeling of proteins and cells. Science 297, 1873–1877 (2002)

24. R. Ando, H. Hama, M. Yamamoto-Hino, H. Mizuno, A. Miyawaki, An optical marker based on the UV-induced green-to-red photoconversion of a fluorescent protein. Proc. Natl. Acad. Sci. U S A 99, 12651–12656 (2002)

25. J. Wiedenmann, S. Ivanchenko, F. Oswald, F. Schmitt, C. Rocker, A. Salih, K.D. Spindler, G.U. Nienhaus, EosFP, a fluorescent marker protein with UV-inducible green-to-red fluorescence conversion. Proc. Natl. Acad. Sci. U S A 101, 15905–15910 (2004)

26. H. Tsutsui, S. Karasawa, H. Shimizu, N. Nukina, A. Miyawaki, Semi-rational engineering of a coral fluorescent protein into an efficient highlighter. EMBO Rep. 6, 233–238 (2005)

27. S. Habuchi, R. Ando, P. Dedecker, W. Verheijen, H. Mizuno, A. Miyawaki, J. Hofkens, Reversible single-molecule photoswitching in the GFP-like fluorescent protein Dronpa. Proc. Natl. Acad. Sci. U S A 102, 9511–9516 (2005)

28. R. Ando, C. Flors, H. Mizuno, J. Hofkens, A. Miyawaki, Highlighted generation of fluorescence signals using simultaneous two-color irradiation on Dronpa mutants. Biophys. J. 92, L97–99 (2007)

29. A.C. Stiel, S. Trowitzsch, G. Weber, M. Andresen, C. Eggeling, S.W. Hell, S. Jakobs, M.C. Wahl, 1.8 A bright-state structure of the reversibly switchable fluorescent protein Dronpa guides the generation of fast switching variants. Biochem. J. 402, 35–42 (2007)

30. D.M. Chudakov, V.V. Verkhusha, D.B. Staroverov, E.A. Souslova, S. Lukyanov, K.A. Lukyanov, Photoswitchable cyan fluorescent protein for protein tracking. Nat. Biotechnol. 22, 1435–1439 (2004)

31. C. Bustamante, J.F. Marko, E.D. Siggia, S. Smith, Entropic elasticity of lambda-phage DNA. Science 265, 1599–1600 (1994)

32. A. Egner, C. Geisler, C. von Middendorff, H. Bock, D. Wenzel, R. Medda, M. Andresen, A.C. Stiel, S. Jakobs, C. Eggeling, et al., Fluorescence nanoscopy in whole cells by asynchronous localization of photoswitching emitters. Biophys. J. 93, 3285–3290 (2007)

33. J.E. Heuser, R.G.W. Anderson, Hypertonic media inhibit receptor-mediated endocytosis by blocking clathrin-coated pit formation. J. Cell Biol. 108, 389–400 (1989)

34. H. Shroff, C.G. Galbraith, J.A. Galbraith, H. White, J. Gillette, S. Olenych, M.W. Davidson, E. Betzig, Dual-color superresolution imaging of genetically expressed probes within individual adhesion complexes. Proc. Natl. Acad. Sci. U S A 104, 20308–20313 (2007)

35. H. Bock, C. Geisler, C.A. Wurm, C. Von Middendorff, S. Jakobs, A. Schonle, A. Egner, S.W. Hell, C. Eggeling, Two-color far-field fluorescence nanoscopy based on photoswitchable emitters. Appl. Phys. B 88, 161–165 (2007)

36. E. Toprak, H. Balci, B.H. Blehm, P.R. Selvin, Three-dimensional particle tracking via bifocal imaging. Nano Lett. 7, 2043–2045 (2007)

37. H.P. Kao, A.S. Verkman, Tracking of single fluorescent particles in three dimensions: use of cylindrical optics to encode particle position. Biophys. J. 67, 1291–1300 (1994)

38. L. Holtzer, T. Meckel, T. Schmidt, Nanometric three-dimensional tracking of individual quantum dots in cells. Appl. Phys. Lett. 90, 053902 (2007)

39. B. Huang, W. Wang, M. Bates, X. Zhuang, Three-dimensional super-resolution imaging by stochastic optical reconstruction microscopy. Science 319, 810–813 (2008)

40. M.F. Juette, T.J. Gould, M.D. Lessard, M.J. Moldzianoski, B.S. Nagpure, B.T. Bennett, S.T. Hess, J. Bewersdorf, Three-dimernsional sub-100 nm resolution fluorescence microscopy of thick samples. Nat. Meth. 5, 527–529 (2008)

41. R. Schmidt, C.A. Wurm, S. Jakobs, J. Engelhardt, A. Egner, S.W. Hell, Spherical nanosized focal spot unravels the interior of cells. Nat. Meth. 5, 539–544 (2008)

42. L. Schermelleh, P.M. Carlton, S. Haase, L. Shao, L. Winoto, P. Kner, B. Burke, M.C. Cardoso, D.A. Agard, M.G.L. Gustafsson, et al., Subdiffraction multicolor imaging of the nuclear periphery with 3D structured illumination microscopy. Science 320, 1332–1336 (2008)

43. H. Shroff, C.G. Galbraith, J.A. Galbraith, E. Betzig, Live-cell photoactivated localization microscopy of nanoscale adhesion dynamics. Nat. Meth. 5, 417–423 (2008)

44. B.A. Griffin, S.R. Adams, R.Y. Tsien, Specific covalent labeling of recombinant protein molecules inside live cells. Science 281, 269–272 (1998)

45. E.G. Guignet, R. Hovius, H. Vogel, Reversible site-selective labeling of membrane proteins in live cells. Nat. Biotechnol. 22, 440–444 (2004)

46. I. Chen, M. Howarth, W. Lin, A. Ting, Site-specific labeling of cell surface proteins with biophysical probes using biotin ligase. Nat. Meth. 2, 99–104 (2005)

47. M. Fernandez-Suarez, H. Baruah, L. Martinez-Hernandez, K.T. Xie, J.M. Baskin, C.R. Bertozzi, A.Y. Ting, Redirecting lipoic acid ligase for cell surface protein labeling with small-molecule probes. Nat. Biotechnol. 25, 1483–1487 (2007)

48. M.W. Popp, J.M. Antos, G.M. Grotenbreg, E. Spooner, H.L. Ploegh, Sortagging: a versatile method for protein labeling. Nat. Chem. Biol. 3, 707–708 (2007)

49. B. Huang, S.A. Jones, B. Brandenberg, X. Zhuang, Whole-cell 3D STORM reveals interactions between cellular structures with nanometer-scale resolution. Nat. Meth. 5, 1047–1082 (2008)

50. Pavani et al, Three-dimensional single-molecule fluorescence imaging beyond the diffraction limit by using a double helical point spread function. Proc. Natl. Acad. Sci. USA 106, 2995–2999 (2009)

51. Shetengel et al, Interferometric fluorescent super-resolution microscopy resolves 3D cellular ultrastructure. Proc. Natl. Acad. Sci. USA 106, 3125–3130 (2009)

Assessing Biological Samples with Scanning Probes

A. Engel

Summary. Scanning probe microscopes raster-scan an atomic scale sensor across an object. The scanning transmission electron microscope (STEM) uses an electron beam focused on a few Å spot, and measures the electron scattering power of the irradiated column of sample matter. Not only does the STEM create dark-filed images of superb clarity, but it also delivers the mass of single protein complexes within a range of 100 kDa to 100 MDa. The STEM appears to be the tool of choice to achieve high-throughput visual proteomics of single cells. In contrast, atomically sharp tips sample the object surface in the scanning tunneling microscope as well as the atomic force microscopes (AFM). Because the AFM can be operated on samples submerged in a physiological salt solution, biomacromolecules can be observed at work. Recent experiments provided new insights into the organization of different native biological membranes, and allowed molecular interaction forces, that determine protein folds and ligand binding, to be measured.

21.1 Introduction

To raster-scan a small probe over an object for mapping its local properties is an old idea invented in the early days of electron microscopy [1]. Three decades later, Albert Crewe and collaborators took up this approach and by using a field emission electron source they produced an electron probe of a few Å in diameter and were thus able to visualize single heavy atoms with superb contrast [2]. The scanning transmission electron microscope (STEM) thus promoted attempts to sequence DNA, but electron beam-induced motion of heavy atom labels attached to specific nucleotides prevented a breakthrough of this approach [3]. Nevertheless, the STEM found its application in biology not only for the clarity of images it produced but also for its analytical power. The STEM turned out to be an ideal instrument for mass measurements of single biomacromolecules based on electron scattering [4].

Another decade later, Gerd Binnig and Heinrich Rohrer took the tungsten wire used in the STEM to extract electrons by field emission, and mounted it on a piezo-driven scanner. Instead of deflecting the electron beam by a magnetic field, they moved the atomically sharp tip over a conducting surface and simply measured the current of electron tunneling from tip to sample. The short decay length of this process allowed the tip to be guided with sub-Å accuracy over the sample, thus mapping its corrugations at atomic scale resolution [5]. The amazing simplicity of the scanning tunneling microscope (STM) made it a big success, and its invention may be considered as the hour of birth of nanoscale sciences. Silicon atoms were imaged with this new microscope with unprecedented clarity. Since proteins, nucleic acids, and lipids are notorious insulators, the scanning tunneling microscope had little impact in biology, although physicists were again tempted to use it as a tool for reading the DNA sequence, unfortunately without much success. Nevertheless, the ice was broken and soon Gerd Binnig and colleagues produced the atomic force microscope (AFM), which allowed insulators to be imaged at nanoscale resolution [6]. This microscope operates in vacuum, air, and liquids and opened avenues to observe proteins, DNA, lipids, and biological membranes in their native environment [7], allowing single molecules to be addressed and studied at work [8]. Other types of scanning probe microscopes emerged from the hands of creative physicists, but none of them had the same impact in biology as did STEM and AFM.

21.2 The Scanning Transmission Electron Microscope

In the STEM, the scanning electron beam focused to a diameter of a few Å irradiates a roughly cylindrical volume of a thin sample. All atoms of this cylinder scatter electrons elastically and inelastically according to their scattering cross-section. Since no post-specimen optical system is required, essentially all scattered electrons can be captured. Knowledge about the chemical composition of the sample allows the mass of the irradiated cylinder to be calculated from the number of electrons scattered into a given collection angle and the number of impinging electrons. Thus, the STEM allows quantitative information about each cylindrical volume element to be collected by raster-scanning the beam across the sample, acquiring in this way projection maps of single protein complexes adsorbed to a thin carbon film (Fig. 21.1a). Their mass is calculated by integrating all electrons scattered from an area that includes the particle projection and subtracting the counts resulting from the corresponding piece of carbon film, normalizing this difference with the recording dose and multiplying it with a calibration factor (Figs. 21.1b and c). Heterogeneous samples can be fully analyzed in this manner, and particle projections can be sorted according to their mass thus providing the link between mass and particle shape (Fig. 21.1d). The precision and reproducibility of this method compares favorably to that of the analytical ultracentrifuge, and it allows particles ranging from about 100 kDa to 100 MDa to be analyzed routinely.

Single unstained protein complexes must be prepared by freeze-drying to prevent structural collapse during dehydration [10]. STEM dark-field images

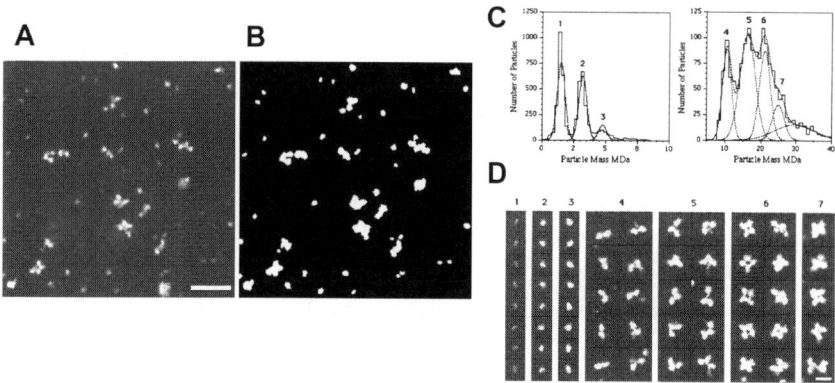

Fig. 21.1. STEM-automated mass measurements provide the link from mass-to-shape. (a) Low dose dark-field image of unstained freeze-dried supraspliceosome complexes [9]. Scale bar: 100 nm. (b) Same as A after segmentation. The black area serves to calculate the average background. Electron counts are integrated for all individual white areas, background counts are subtracted and the difference is divided by the recording dose and multiplied by the calibration factor to get the mass of the respective particle. (c) Mass values are binned into histograms. (d) Galleries of particle projections having masses within a narrow range are assembled from histogram peaks

taken at a dose of typically 300 electrons/nm^2 are impaired by statistical noise, preventing identification of fine structural details. For imaging single complexes in pursuit of the goal to visualize their structure, negative staining is of advantage in spite of its possible adversary effect on the protein preparation. Since heavy atom stains scatter electrons elastically, approximately two orders of magnitude more than the light atoms of biological matter, the stain surrounding the single molecules and filling their hydrophilic cavities provides the contrast, which is particularly clear in the STEM dark-field images (Fig. 21.2). Such images recorded over the years from many different samples have given tremendous insight into biological questions of interest.

21.3 High-Throughput Visual Proteomics

Offering the possibility to determine the mass of single protein complexes, the STEM appears to be an ideal instrument for visual proteomics. As this terminology coined by the Baumeister laboratory promises, the proteome of a single cell is being visualized – providing access to individual protein complexes, as they exist in the cell by cryo-electron tomography [15]. Since eukaryotic cells are in general larger than about 10–50 μm, they are too thick for imaging their structure at 2–3 nm resolution as would be required to identify the complexes and their subunits. Therefore, we propose to use microfluidic circuits to grow and lyse single cells [16–18], to fractionate their contents [19, 20],

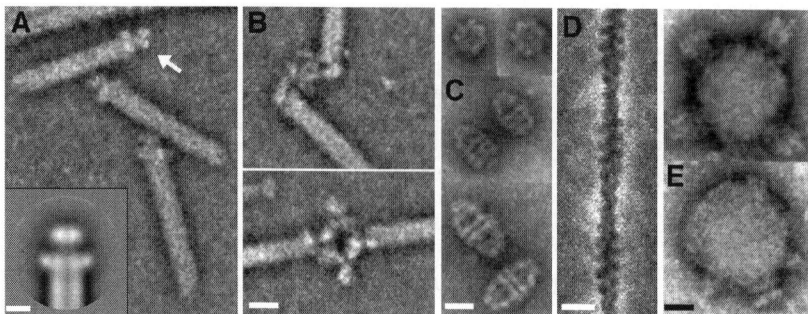

Fig. 21.2. Negatively stained single protein complexes and organelles visualized by STEM dark-field mode. (**a**) The tip structure of the injectisome needle. On single images, the essential features are directly seen [11]. The average in the inset is from only a few particles, and it shows the central channel (Scale bar: 5 nm). (**b**). Antibodies grown against protein LcrV attach to the tip structure. (**c**) Various conformations of the bacterial chaperon GroEL have been observed [12]. Top views are frequently found in preparations of GroEL – they exhibit the 7-fold symmetry of the complex (*top*). The most frequent conformation of GroEL–GroES complexes exhibit a 1:1 stoichiometry. In one of the two complexes, the substrate bound to the rim of the cylinder is distinct (*middle*). GroEL-GroES complexes having a stoichiometry of 1:2 are also seen. These complexes are referred to as 'footballs' (*bottom*). (**d**) Actin filament revealing the helical arrangement of single actin molecules [13]. E) Synaptic vesicles are densely packed with different enzymes. Large complexes, such as the V-type ATPase, are clearly seen [14]. Scale bars: 10 nm

and to deposit the fractions by spotting 10 pL drops per square of an electron microscopy grid. In this way, the final sample will be thin enough for high-resolution imaging, sample deposition devices can be tuned to desalt the sample, to add negative stain, or to vitrify the grid once all grid squares are populated.

Preliminary data illustrate that samples can be prepared by 10^6 smaller sample volumes compared to current procedures, i.e. 10 pL per sample rather than 10 µL. The ambitious expectation is that the state-of-the-art microfluidic circuits will produce cell fractions in a highly reproducible manner, that this protocol will allow on the fly cross-linking, that fractionation will be done within few minutes, that the output stream can be divided into samples for STEM mass measurements as well as for negative staining. Taking a typical cell of about 10 µm diameter, assuming its cytosol to have a concentration of 100 mg ml^{-1}, and the dilution factor required to achieve an appropriate particle density on the grid to be 10^4, we find that about 10^3 droplets of 10 pL can be deposited. In this way, the contents of a single cell would be deposited on two grids, and grid squares could be visually inspected like the pages of a book (Fig. 21.3).

The readout by STEM will be achieved in either of the two ways. First, hierarchical automated particle selection and segmentation algorithms will

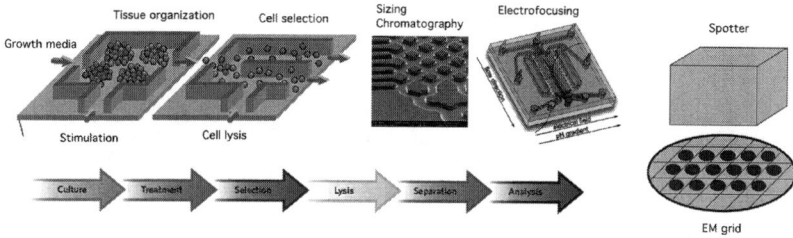

Fig. 21.3. Visual proteomics require samples to be prepared by a microfluidic circuit. Cells grown in microreactors are released and sorted to reach a device that lyses them. Various well-known separation circuits are used in series for generating suitable fractions of the lysate. Importantly, a spotting device akin to those used for DNA array production produces 10 pL droplets that are deposited on an EM grid. One cell will provide enough material to cover 1 to a few grids, when one droplet is deposited per grid square

evaluate the mass of all visible particles, and particle projections will be sorted according to their mass (see Fig. 21.1). Mass-related projections can be classified to obtain mass-maps of complexes viewed from different directions. This protocol establishes the mass-to-shape relation of complexes in inspected fractions, which will complement much more accurate mass determination by mass spectrometry. Secondly, hierarchical automated particle selection will be used to select projections of negatively stained particles, which will be classified based on shapes obtained from mass-mapping. Refined classification and pattern recognition protocols will then be used to identify complexes by comparison with look-up tables composed of projections calculated from atomic scale structural models of specific complexes. In this way variations between single cells submitted to specific perturbations can be efficiently assessed at the level of single complexes.

In addition, initial 3D maps of yet uncharacterized, but reproducibly observed complexes can be obtained from projections of negatively stained samples. Ultimately, the visual proteomics chain will be completed by the inspection of vitrified cell fractions, using cryo-electron microscopy. By sorting out particle projections based on all information established with mass-mapping and 3D reconstruction of negatively stained complexes, high-resolution 3D maps will be obtained. Combined with mass spectrometry data from the respective fractions, these 3D maps will provide a solid foundation for creating atomic scale models of all complexes identified.

21.4 The Atomic Force Microscope

In the AFM, an atomically sharp probe, the tip of a pyramidal structure at the end of a thin cantilever is raster-scanned over the sample surface [6]. The highest resolution images have been taken in buffer solution while operating

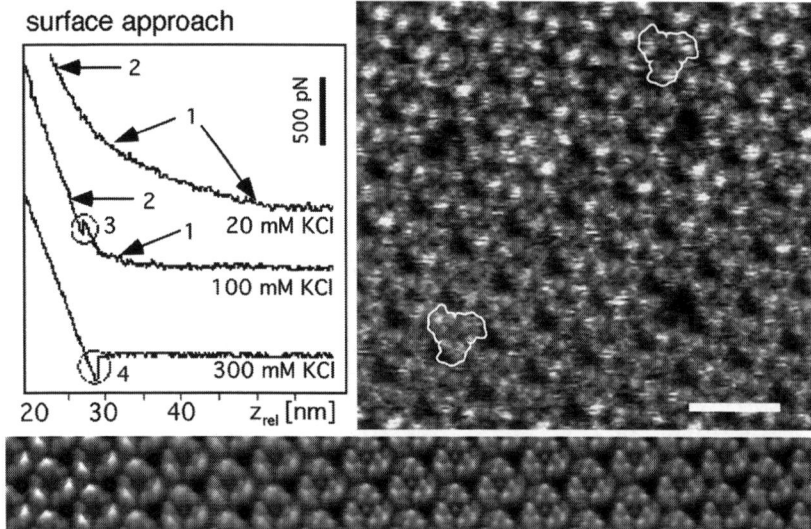

Fig. 21.4. High-resolution imaging of a native membrane by AFM. Top left: The adjustment of the buffer's ionic strength is critical [21]. Force curves show that in 100 mM KCl forces around 50–100 pN deflect the cantilever when there is less than 5 nm to go before contact. *Top right*: Bacteriorhodopsin trimers are distinct as they adopt different conformations. The latter variation is the result of tip force variations [26]. Loops may be in their most extended conformation when the AFM is operated at minimum force. Pressing the cantilever down onto the membrane by an additional 50–100 pN bends the loops away. *Bottom*: The conformational transition is shown in the morphed averages of different states

the AFM in contact mode. Here the forces acting on the sample are optimized in two ways. First, the servo system that measures the cantilever deflection induced by the sample surface corrugations is tuned to react quickly to deflection changes and to control the piezo motor moving the sample vertically with highest possible sensitivity. In this way, the vertical forces between the tip and sample are minimized, while the latter is raster-scanned below the tip. Secondly, pH and ionic strength of the buffer are adjusted to induce a negatively charged sample surface, and to tune the decay length of electrostatic repulsion so that the vertical force of the tip acting on the sample is distributed over an area of some 10 nm diameter [21]. Hence, the force that must be applied to the cantilever to obtain stable operation is balanced by the electrostatic repulsion, which also balances the van der Waals attraction of the tip that is in contact with the sample surface (Fig. 21.4a). Force-deflection curves acquired by extending the tip to the sample surface and retracting it from the same allow ideal imaging conditions to be identified. Tip structure and corrugation of the sample dictate the achievable resolution. Protrusions beyond some 10 nm in height cannot be imaged at high resolution as a result of (1) the conical tip and (2) the flexibility of the protrusion. Nevertheless, under ideal conditions a lateral resolution of 1 nm or even better can be achieved, and force-induced conformational changes can be observed (Fig. 21.4). Depending on the mechanical nature of the sample, the vertical resolution of an AFM operated under optimized conditions can reach 0.1 nm.

Although the performance of modern AFMs is impressive, there is room for improvement for biological applications. Sample screening is slow because AFMs acquire usually not much more than 1–2 frames per minute. Combining an AFM with a high-power light microscope could be a practical solution to this problem, but then compromises need to be made concerning the mechanical stability of the system. The mechanics of the cantilever and its hydrodynamic properties set the limits for the scan speed and for the thermal noise when the AFM is operated in liquid [22]. In general, short and relatively stiff cantilevers are better suited for scanning at elevated speed (Table 21.1). But the stiffer a lever is the more sensitive the deflection detector needs to be to prevent the force-induced sample damage.

Fabry Perot-based interferometer deflection detection is currently the most sensitive method [23]. Friction forces are inevitable in contact mode AFM, and they can be detrimental to fragile biological samples. Tapping mode AFM has been introduced many years ago and was demonstrated to prevent friction forces and hence, the lateral displacement of weakly immobilized samples, such as single protein complexes adsorbed to mica [24]. Yet even the small fraction of sample contact in tapping suffices to induce conformational changes. Hence, a further improvement appears to be obtained by measuring the force-induced detuning of the cantilever oscillating at its resonance frequency [25]. This frequency-modulated (FM) AFM operation mode ideally combines stiff and short cantilever oscillation even in buffer solution at 0.2–0.8 MHz, with the ultimate sensitivity of a Fabry Perot interferometer [23]. Unfortunately, such instruments are not as yet commercially available and have therefore not been widely used.

21.5 Imaging Native Membranes in Buffer Solution

High resolution cannot be attained on surfaces of living cells simply because these are soft, dynamic structures that retract upon contact with a scanning tip. Native membranes or reconstituted arrays of densely packed membrane proteins, however, can be adsorbed to freshly cleaved mica and imaged at sufficiently high resolution to build atomic models of native membrane protein arrangements. This is because membrane proteins often exhibit loops that protrude only by a few nm out of the bilayer, and the surface of such proteins can often be mapped at lateral resolutions better than 1 nm (Fig. 21.4). Imaging of native membranes by AFM has given significant new information about functionally relevant native protein–protein interactions.

Table 21.1. Physical properties of AFM cantilevers and thermal noise according to [22]. It is evident that stiff cantilevers exhibit better performance than soft cantilevers. However, the practical limitation is then imposed by the sensitivity of the deflection detector

k_L	0.1				10				100			
l	20		100		20		100		20		100	
w	5	10	10	20	5	10	10	20	5	10	10	20
t	0.15	0.12	0.60	0.48	0.70	0.56	2.78	2.21	1.51	1.20	5.98	4.75
f_{ov}	550	437	87	69	2553	2021	404	321	5,488	4355	871	692
f_{ol}	104	58	22	13	1,198	670	240	142	3,505	2,076	658	418
Q	1.5	1.9	1.5	1.9	4.3	5.1	4.8	5.3	8.5	9.2	10.6	10.2
FTN	41.2	48.6	89.3	105.5	71	87	151	187	94	117	194	248
z_{th}	412	486	893	1055	7.1	8.7	15.1	18.7	0.9	1.2	1.9	2.5

k_L: Spring constant (N m^{-1})

l, w, t: Cantilever length, width, and thickness (in mm)

f_{ov} and f_{ol}: Resonance frequency in vacuum and water, respectively in kHz [22]

Q: Quality factor in water

$FTN = F_{th}/B^{1/2}$: Force thermal noise limit in fN Hz$^{-1/2}$

z_{th}: Metric thermal noise limit $= FTN/k_L$ in fm Hz$^{-1/2}$

Rows of rhodopsin dimers found in bovine and murine disc membranes support the view that G-protein coupled receptors (GPCRs) may operate as dimers or higher oligomers (Fig. 21.5a; [27]). Modeling based on topographs of disc membranes allowed the most prominent interface in rhodopsin oligomers involving helices IV and V to be identified (Fig. 21.5b; [28]). Interestingly, a cysteine scan in the dopamine 2 receptor D2R has revealed that the same helices are involved in D2R dimerization (Fig. 21.5c; [30]). The light-dependent rearrangement of photosynthetic complexes in native membranes of *Rsp. photometricum* in response to different light intensities during cell growth has been unveiled by AFM of native photosynthetic membranes [31]. Realistic atomic models of the supramolecular assembly of LH2 and core complexes in high-light adapted or LH2-only antenna domains in low-light adapted membranes were assembled based on high-resolution topographs acquired by AFM [32]. High-resolution images of native lens fiber cell membranes confirmed the notion that the conformation of AQP0 resolved to 1.9 Å resolution corresponds to the native state of this water channel, whose second biological function is to form fiber cell junctions [33]. Native arrangements of the mitochondrial outer membrane channel VDAC has provided solid evidence for the existence of single VDAC proteins diffusing in the membrane, an observation that was in contrast to previous data that suggested these channels to exist as trimers or hexamers [34].

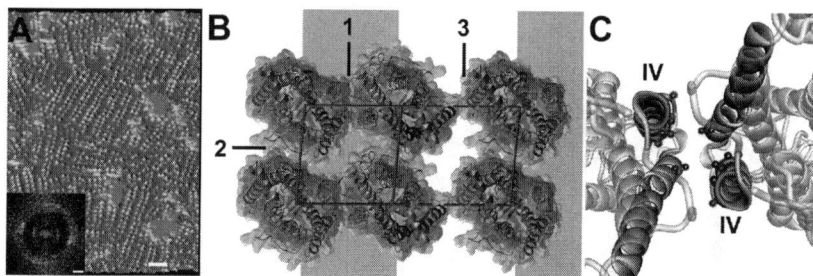

Fig. 21.5. Rhodopsin forms rows of dimers [27]. (a) An AFM topograph of native disc membranes from mouse retina reveals the packing of rhodopsin (scale bar: 10 nm). Paracrystallinity of these arrays is documented by the power spectrum of this topograph (inset, scale bar: $(10\,nm)^{-1}$). (b) Lattice vectors of the most densely-packed areas set stringent limitations on the packing possibilities [28]. The best model based on the rhodopsin structure [29] exhibits strong contacts formed by helices 4 and 5 that are likely to be the major dimeric interface (1). Dimer rows are formed as a result of weak interactions (2). Side-by-side packing of row leading to paracrystalline arrays is the result of a small interface at the end of helix 1(3). (c) Amazingly, a Cys scan done on dopamine 2 receptor reveals the residues at the dimeric interface to be on helices IV and V [30], confirming the rhodopsin dimer model shown in **b**. (Cysteines inducing cross-linking are indicated by small spheres).

21.6 Assessing Forces that Determine Stability and Interactions of Membrane Proteins

Advances in single-molecule force spectroscopy of soluble proteins [35] and cells tethered between a support and a cantilever [36,37] has stimulated experiments on bacterial surface layers [38] and on bacteriorhodopsin (bR) [39]. After imaging a membrane, densely packed with proteins at high resolution, stopping the scan and pressing the tip down into the sample with about 200 pN for 0.5 s attaches individual proteins to the tip. Upon tip retraction, the force-distance trace is recorded to document the protein unfolding and extraction process. Subsequent imaging of the sample relates the force-distance traces to the structural damage created by the protein extraction (Fig. 21.6). The short contact force pulse initiates physisorption between the tip and protein, most likely by the breakdown of the hydration shells of both the protein and the tip. Physisorbed molecules may withstand pulling forces of several hundred pN before they detach, indicating that multiple local interactions such as van der Waals forces, charge interactions, and hydrogen bonds stabilize this contact [40].

To interpret the force-distance traces, many need to be recorded and sorted according to their length. Since imaging is not required once a suitable membrane patch is identified, acquisition and processing can be automated, and many full-length force-distance traces averaged. Depending on the termini exposed to the tip, different unfolding characteristics can be expected. Comparing engineered isoforms of the membrane protein that differ in their termini can facilitate the interpretation of such force curves [41]. How the load acting on the protein is distributed in the membrane has been an open question. Indeed, whether parts of the protein on the opposing side of the membrane locally interact with the support or whether the membrane stiffness suffices to distribute the force over a larger area was not known. Experiments on multiple membrane stacks unambiguously showed that the stiffness of the membrane is sufficient, and that under suitable conditions the adhesion between the lower loops and the support can be reduced below the force resolution limit [26]. A multitude of membrane proteins has been investigated by single-molecule force spectroscopy and the force-distance traces provided a richness of novel information that would not have been accessible otherwise [43–48].

During protein unfolding, the force that the cantilever exerts on the protein is transduced along the peptide backbone into the membrane protein. The protein is stabilized in its conformation by the interplay between local forces, which can be represented in their full complexity by the potential energy landscape. The protein being unfolded is dragged in the 3N-dimensional energy landscape along a given direction and encounters a barrier, which needs to be conquered to continue the unfolding process. Two contributions act together in this process: the external force transmitted along the backbone of the already unfolded part of the protein and the thermal fluctuations. If the pulling force increases slowly, it will be more likely that the thermal fluctuations help to

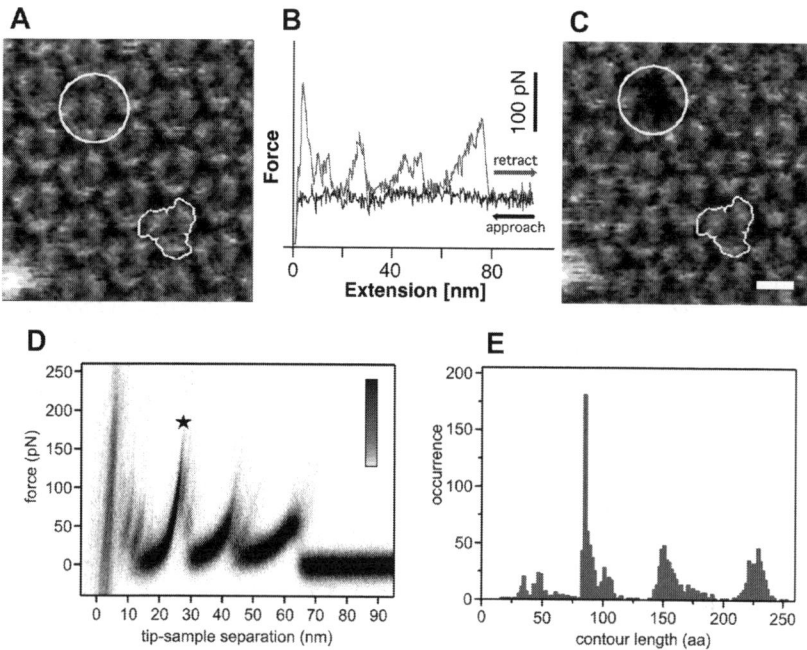

Fig. 21.6. Single-molecule force spectroscopy on purple membrane [39]. (a) Trimers are arranged in a trigonal lattice with 6.1 nm unit cell length. Pressing the tip down onto the bR molecule, marked by circle, by enhancing the force to about 200 pN for 0.5 s attaches the C-terminus. (b) Upon tip retraction, the force required to unfold the bR molecule is manifested by the cantilever deflection trace. The contour length of the polypeptide is longer than 80 nm. (c) The high-resolution topographs subsequently recorded show the vacancy left by the unfolded bR (scale bar: 5 nm). (d) Pulling bR molecules out of the membrane can be automated to acquire a significant number of force-retraction curves [41]. (e) Determining the contour length of individual unfolded peptide stretches by fitting the worm-like-chain model allows a contour length histogram to be established. From this the barriers related to the respective unfolding segments can be mapped onto the secondary structure model of a membrane protein [42]

overcome the remaining barrier than if the external force increases sharply. The point where the barrier is conquered will, therefore, depend on the rate at which the force is increased [40, 49].

At a given force, the elasticity of covalent bonds of the amino acid backbone gives rise to a length increase. But thermal fluctuations act on the backbone, which on an average pulls the cantilever closer to the membrane, a phenomenon referred to as entropic elasticity of linear polymers. The worm-like chain model [50] describes the polymer as an elastic rod with bending stiffness submitted to thermal fluctuations that decrease the end-to-end distance of the rod. Alternatively, the freely jointed chain model calculates the conformational freedom of jointed segments with random orientation as a function of the end-to-end distance [35]. Both models are equivalent descriptions, and have been used widely to interpret force-distance profiles. Importantly, once the a persistence length of the entropic spring is chosen, which has been determined to be 0.4 nm for typical polypeptides [51], the only free parameter for fitting analytical curves to the experimental ones is the number of residues in the chain. Therefore, the averaged force curves provide a solid basis to localize all barriers occurring along the unfolding path with an accuracy of a few residues. Instead of fitting a single curve to the force profile average, curves can be fitted to individual experimental force-distance profiles to establish a histogram of the corresponding residue numbers obtained [41].

Energy barriers may not only be signatures of the intramolecular bonds, but they may also be modulated by the interaction of ligands with the protein. With the option to localize such barriers with the precision of a few amino acids, an attempt to screen the potential landscape of the protein for signatures of ligand interaction seems not only feasible but also attractive for assessing the interaction of drugs targeting membrane proteins. Müller and coworkers have unfolded the sodium proton anti-porter in its active and inactive state, and attributed a barrier around aa 225 to binding of Na^+ [45]. They further found that functional inhibition by 2-aminoperimidine results in the formation of yet another barrier at around aa 85, which was interpreted as a ligand-induced stabilization of the loop close to aa 85, which together with other parts of the protein may form the ligand-binding pocket. This pioneering experiment opens the path for more ligands and binding pockets to be identified in the long list of health relevant membrane proteins where no high-resolution structural information is otherwise available [44].

21.7 Perspectives

A wealth of information has been obtained by the observation of single proteins using scanning probe microscopes. The requirements of large-scale analyses of tissues and cells down to the molecular level, currently achieved in genomics and proteomics projects, can also be fulfilled by automated high-throughput sample preparation and imaging to be developed in the future. Such approaches will make visual proteomics a routine procedure that will not only be be be used in the basic research, but will also serve as an advanced diagnostic tool in clinical applications.

High-throughput single-molecule force spectroscopy has the potential to map binding sites of ligands at the single residue level and determine their binding constant while the respective membrane bound receptor is in its native environment, i.e. the lipid bilayer. Thus, the use of AFM will not only mature in basic research fields, but will also find new applications in the life science industry.

Acknowledgements

This work has been supported by the Swiss National Science Foundation (SNF), the National Center of Competence in Research (NCCR) of Structural Biology, the NCCR of Nanoscale Sciences, the SNF grant 3100A0–108299 to AE, EU grant 035995-2, the University of Basel, and the Maurice E. Müller Foundation of Switzerland. The author is indebted to Ansgar Philippsen for his help with Fig. 21.5c, and thanks Dimitrios Fotiads, Hermann Gaub, Daniel Müller, Shirely Müller, Krzysztof Palczewski, and Simon Scheuring for numerous stimulating discussions.

References

1. M.V. Ardenne, Z. Physik. **109**, 553 (1938)
2. A.V. Crewe, J. Wall, J. Langmore, Science **168**, 1338–1340 (1970)
3. M.D. Cole, J.W. Wiggins, M. Beer, J. Mol. Biol. **117**, 387–400 (1977)
4. A. Engel, Ultramicroscopy **3**, 273–281 (1978)
5. G. Binnig, H. Rohrer, Helv. Phys. Acta **55**, 726–735 (1982)
6. G. Binnig, C.F. Quate, C. Gerber, Phys. Rev. Lett. **56**, 930–933 (1986)
7. B. Drake, C.B. Prater, A.L. Weisenhorn, S.A. Gould, T.R. Albrecht, C.F. Quate, D.S. Cannell, H.G. Hansma, P.K. Hansma, Science **243**, 1586–1589 (1989)
8. A. Engel, D.J. Müller, Nat. Struct. Biol. **7**, 715–718 (2000)
9. S. Müller, B. Wolpensinger, M. Angenitzki, A. Engel, J. Sperling, R. Sperling, J. Mol. Biol. **283**, 383–394 (1998)
10. S.A. Müller, K.N. Goldie, R. Buerki, R. Haering, A. Engel, Ultramicroscopy **46**, 317–334 (1992)
11. C.A. Mueller, P. Broz, S.A. Muller, P. Ringler, F. Erne-Brand, I. Sorg, M. Kuhn, A. Engel, G.R. Cornelis, Science **310**, 674–676 (2005)
12. A. Engel, M.K. Hayer-Hartl, K.N. Goldie, G. Pfeifer, R. Hegerl, S. Müller, A. da Silva, W. Baumeister, F.U. Hartl, Science **269**, 832–836 (1995)
13. A. Bremer, C. Henn, K.N. Goldie, A. Engel, P.R. Smith, U. Aebi, J. Mol. Biol. **242**, 683–700 (1994)
14. S. Takamori, M. Holt, K. Stenius, E.A. Lemke, M. Gronborg, D. Riedel, H. Urlaub, S. Schenck, B. Brugger, Cell **127**, 831–846 (2006)
15. S. Nickell, C. Kofler, A.P. Leis, W. Baumeister, Nat. Rev. Mol. Cell Biol. **7**, 225–230 (2006)
16. P.J. Hung, P.J. Lee, P. Sabounchi, N. Aghdam, R. Lin, L.P. Lee, Lab Chip **5**, 44–48 (2005)
17. D. Di Carlo, C. Ionescu-Zanetti, Y. Zhang, P. Hung, L.P. Lee, Lab Chip **5**, 171–178 (2005)
18. J.T. Nevill, R. Cooper, M. Dueck, D.N. Breslauer, L.P. Lee, Lab Chip **7**, 1689–1695 (2007)
19. D.N. Breslauer, P.J. Lee, L.P. Lee, Mol. Biosyst. **2**, 97–112 (2006)
20. N. Pamme, Lab Chip **7**, 1644–1659 (2007)
21. D.J. Müller, D. Fotiadis, S. Scheuring, S.A. Müller, A. Engel, Biophys. J. **76**, 1101–1111 (1999)

22. J.E. Sader, J. Appl. Phys. **84**, 64–76 (1998)
23. B.W. Hoogenboom, P.L.T.M. Frederix, D. Fotiadis, H.J. Hug, A. Engel, Nanotechnology **19**, 384019 (2008)
24. C.A.J. Putman, K.O. Vanderwerf, B.G. Degrooth, N.F. Vanhulst, J. Greve, Appl. Phys. Lett. **64**, 2454–2456 (1994)
25. B.W. Hoogenboom, H.J. Hug, Y. Pellmont, S. Martin, P.L.T.M. Frederix, D. Fotiadis, A. Engel, Appl. Phys. Lett. **88**, 193109 (2006)
26. D.J. Müller, J.B. Heymann, F. Oesterhelt, C. Möller, H. Gaub, G. Büldt, A. Engel, Biochim. Biophys. Acta **1460**, 27–38 (2000)
27. D. Fotiadis, Y. Liang, S. Filipek, D.A. Saperstein, A. Engel, K. Palczewski, Nature **421**, 127–128 (2003)
28. Y. Liang, D. Fotiadis, S. Filipek, D.A. Saperstein, K. Palczewski, A. Engel, J. Biol. Chem. **278**, 21655–21662 (2003)
29. K. Palczewski, T. Kumasaka, T. Hori, C.A. Behnke, H. Motoshima, B.A. Fox, I. Le Trong, D.C. Teller, T. Okada, Science **289**, 739–745 (2000)
30. W. Guo, L. Shi, M. Filizola, H. Weinstein, J.A. Javitch, Proc. Natl. Acad. Sci. U S A. **102**, 17495–17500 (2005)
31. S. Scheuring, J.N. Sturgis, Science **309**, 484–487 (2005)
32. S. Scheuring, T. Boudier, J.N. Sturgis, J. Struct. Biol. **159**, 268–276 (2007)
33. S. Scheuring, N. Buzhynskyy, S. Jaroslawski, R.P. Goncalves, R.K. Hite, T. Walz, J. Struct. Biol. **160**, 385–394 (2007)
34. B.W. Hoogenboom, K. Suda, A. Engel, D. Fotiadis, J. Mol. Biol. **370**, 246–255 (2007)
35. M. Rief, M. Gautel, F. Oesterhelt, J.M. Fernandez, H.E. Gaub, Science **276**, 1109–1112 (1997)
36. M. Grandbois, W. Dettmann, M. Benoit, H.E. Gaub, J. Histochem. Cytochem. **48**, 719–724 (2000)
37. H. Clausen-Schaumann, M. Seitz, R. Krautbauer, H.E. Gaub, Curr. Opin. Chem. Biol. **4**, 524–530 (2000)
38. D.J. Müller, W. Baumeister, A. Engel, Proc. Natl. Acad. Sci. USA **96**, 13170–13174 (1999)
39. F. Oesterhelt, D. Oesterhelt, M. Pfeiffer, A. Engel, H.E. Gaub, D.J. Müller, Science **288**, 143–146 (2000)
40. M. Seitz, C. Friedsam, W. Jostl, T. Hugel, H.E. Gaub, Chemphyschem **4**, 986–990 (2003)
41. P. Bosshart, F. Casagrande, P. Frederix, M. Ratera, C. Bippes, D. Müller, M. Palacin, A. Engel, D. Fotiadis, Nanotechnology **19**, 384014 (2008)
42. A. Engel, H.E. Gaub, Annu. Rev. Biochem. **77**, 127–148 (2008)
43. H. Janovjak, H. Knaus, D.J. Muller, J. Am. Chem. Soc. **129**, 246–247 (2007)
44. A. Kedrov, H. Janovjak, K.T. Sapra, D.J. Muller, Annu. Rev. Biophys. Biomol. Struct. **36**, 233–260 (2007)
45. A. Kedrov, M. Krieg, C. Ziegler, W. Kuhlbrandt, D.J. Muller, EMBO Rep. **6**, 668–674 (2005)
46. A. Kedrov, S. Wegmann, S.H. Smits, P. Goswami, H. Baumann, D.J. Muller, J. Struct. Biol. **159**, 290–301 (2007)
47. C. Möller, D. Fotiadis, K. Suda, A. Engel, M. Kessler, D.J. Müller, J. Struct. Biol. **142**, 369–378 (2003)
48. J. Preiner, H. Janovjak, C. Rankl, H. Knaus, D.A. Cisneros, A. Kedrov, F. Kienberger, D.J. Muller, P. Hinterdorfer, Biophys. J. **93**, 930–937 (2007)

49. E. Evans, F. Ludwig, J. Phys. Condens. Matter. **11**, 1–6 (1999)
50. M.S. Kellermayer, S.B. Smith, H.L. Granzier, C. Bustamante, Science **276**, 1112–1116 (1997)
51. H. Dietz, F. Berkemeier, M. Bertz, M. Rief, Proc. Natl. Acad. Sci. USA **103**, 12724–12728 (2006)

Part VIII

Single Molecule Microscopy in Individual Cells

Enzymology and Life at the Single Molecule Level

X. Sunney Xie

Summary. The advent of room-temperature single-molecule imaging and spectroscopy in the early 1990s made it possible to follow biochemical reactions and conformational dynamics of an individual enzyme molecule in real time, yielding new information about the working of enzymes *in vitro*. This eventually led to the recent success of probing single-molecule biochemical reactions in a living cell with high specificity, millisecond time resolution, and nanometer spatial precision. We have studied how gene expression and regulation occur at the single-molecule level in living bacterial cells. The examples herein illustrate the impact of the single-molecule approach on biological discovery, as well as prospects for medicine.

22.1 Introduction

Advances in life sciences in the last half-century, from the discovery of DNA structure to the crystal structure of enzymes, have been facilitated by the development of physical tools, such as X-ray crystallography. We face new challenges in understanding how the enzymes work in real time, how they work individually, how they work together in a living cell, and how the different genes get turned on and off in a living cell.

At the start of my independent career in the early 1990s, I was very fortunate to be able to participate in the development of single-molecule imaging and spectroscopy at room temperature, which was prompted by prior success under cryogenic conditions [1, 2]. The room-temperature work was initially accomplished with near-field microscopes [3–5], then much more easily with far-field fluorescence microscopes [6, 7]. As recognized in 1976 [8], the key idea behind single-molecule detection at room temperature was to use a microscope to reduce an excitation volume in order to suppress the background signal [9, 10]. In my opinion, these developments on room-temperature detection, imaging and spectroscopy of single-molecules with a far-field microscope, together with the then emerging single-molecule manipulation techniques [11–13], opened the exciting possibility of studying a large variety of single-molecule behaviors in biology, beyond single ion channel recording [14].

As the methodologies were being developed and refined, the challenge, both technical and intellectual, was how to create a new science with new tools at the boundaries of physics, chemistry and biology. The application of single-molecule imaging, spectroscopy and manipulation to biology has led to widespread research activities around the world, which have made an impact on biochemistry and molecular biology. This is primarily because many compelling problems in biology can best be addressed with the single-molecule approach.

My group's first step in biochemistry was to monitor the biochemical reactions and conformational dynamics of a single enzyme molecule in real time by fluorescence detection [15]. Thanks to the development of *in vitro* assays, single-molecule enzymology has since provided mechanistic understanding of specific systems, as well as fundamental insights into enzymatic catalysis, and even the prospect of human genome sequencing by single-molecule methods.

The utility of single-molecule studies in biology is best illustrated by the fact that in a living cell there are only one or two copies of a particular gene on the chromosome DNA, at which fundamental biological processes such as gene expression and regulation take place. Consequently, these processes occur stochastically, and are intrinsically asynchronous among different cells. Hence, understanding of these processes requires real-time observations with single-molecule sensitivity. In 2006, we developed strategies to study these processes at the single-molecule level in living cells [16, 17]. Since then, it has been a revelation to see quantitative understanding of these processes emerging from these studies.

Below I summarize my group's work in single-molecule enzymology and gene expression and regulation in living bacteria, with a commentary on the future of single-molecule studies in biology and medicine.

22.2 Single Molecule Enzymology: The Fluctuating Enzyme

Essential to all life processes, enzymes are biological catalysts that accelerate biochemical reactions with high efficiency and high fidelity, unmatched by artificial systems. Despite more and more structures and interaction networks of enzymes becoming known, the quest for understanding how an enzyme works in real time at the molecular level remains a vital question.

In 1998, we reported the real-time observation of enzymatic turnovers of a single-molecule cholesterol oxidase, a flavoenzyme that catalyzes oxidation of cholesterol by oxygen [15] (Fig. 22.1A). The active site of the enzyme, flavin adenine dinucleotide (FAD), (Fig. 22.1B), is naturally fluorescent in its oxidized form but not in its reduced form. With excess amounts of cholesterol

A   **B**

E-FAD E-FADH⁻

H₂O₂ ⟵ O₂

C

Time (ms)

D

$$E + S \xrightarrow{k_1} ES \xrightarrow{k_2} Product$$

On Time (ms)

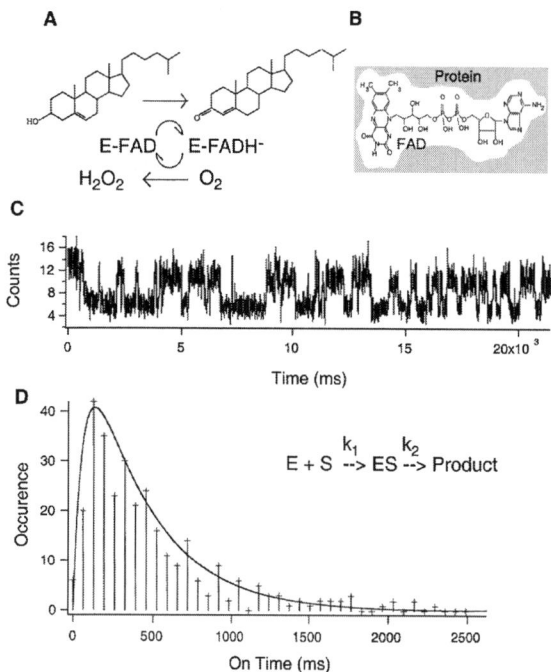

Fig. 22.1. (**A**) Enzymatic cycle of cholesterol oxidase which catalyzes the oxidation of cholesterol by oxygen. The enzyme's naturally fluorescent FAD active site is first reduced by a cholesterol substrate molecule, generating a non-fluorescent FADH₂, which is then oxidized by oxygen. (**B**) Structure of FAD, the active site of cholesterol oxidase. (**C**) A portion of the fluorescence intensity time trace of a single cholesterol oxidase molecule. Each on-off cycle of emission corresponds to an enzymatic turnover. (**D**) Distribution of emission on-times derived from (C). The solid line is the convolution of two exponential functions with rate constants $k_1 [S] = 2.5\,s^{-1}$ and $k_2 = 15.3\,s^{-1}$, reflecting the existence of an intermediate, ES, the enzyme–substrate complex, as shown in the kinetic scheme in the inset. From ref. [15]

and oxygen, the emission from a single enzyme molecule confined in an agarose gel exhibits an on-off behavior (Fig. 22.1C), each on-off cycle corresponding to an enzymatic turnover. The exponential rise and decay of the histogram of the waiting times indicate the existence of the enzyme–substrate complex (Fig. 22.1D).

By conducting statistical analyses of the data, we found that a single enzyme molecule exhibits fluctuations of catalytic rates – a single enzyme molecule does not have a rate constant! This phenomenon, which had been hidden in the conventional experiments, turned out to be general.

We attributed the rate constant fluctuation to conformational interconversion, which had been inferred by Frauenfelder and co-workers in their earlier work on photolysis of heme proteins [18]. We developed a method that made

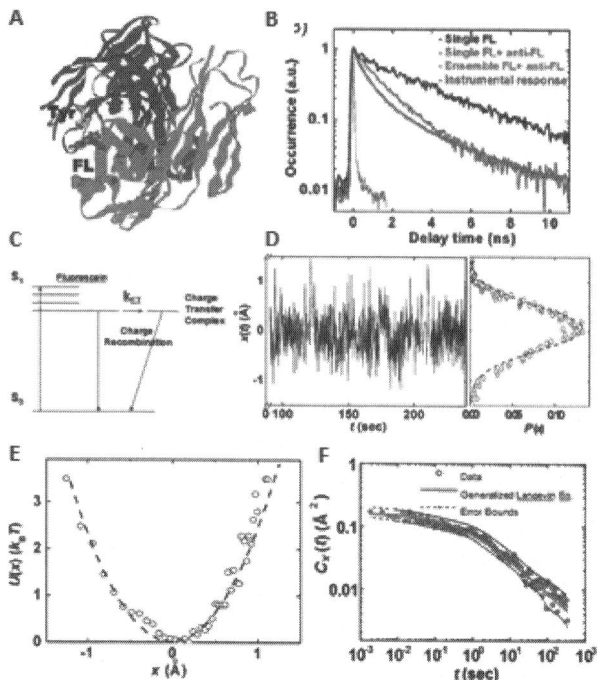

Fig. 22.2. (**A**) Fluorescein (FL) and anti-FL complex with the electron transfer donor and acceptor, Tyr37 and FL, respectively. (**B**) Monoexponential fluorescence lifetime decay for a single FL molecule (*black curve*), multiexponential fluorescence decay for the FL/anti-FL complex at both ensemble (*green curve*) and single-molecule (*red curve*) levels, and the instrumental response function (*blue curve*). (**C**) Energy diagram of the ground and excited states of fluorescein and the charge transfer state. (**D**) A segment of the donor–acceptor distance, $x(t)$ trajectory with the corresponding probability density function $P(x)$ with a Gaussian fit. (**E**) Harmonic potential of mean force, $U(x) = -k_\mathrm{B}T \ln[P(x)]$. (**F**) Autocorrelation of the donor–acceptor distance in (E) that shows the broad range of time scale distance fluctuation. From ref. [20]

it possible to directly observe conformational changes of an individual protein (e.g., Fig. 22.2A and B). [19,20] This was accomplished by using electron transfer as a distance-dependent probe (Fig. 22.2C) [19], which is complementary to fluorescent resonant energy transfer (FRET) [21], but is capable of probing angstrom scale distance fluctuation in an intact protein (Fig. 22.2A). The distance between an electron transfer donor and an acceptor was found to take place at a broad range of time scales, ranging from 10 ms to 10 s (Fig. 22.2D–F)

This equilibrium conformational fluctuation is again a general phenomenon that occurs in every system that we have studied, taking place at the same time scales at which we observed the enzymatic rate fluctuations. Such conformational fluctuations at a broad range of time scales result from the flexible but glassy nature of enzymes as biopolymers, yet interestingly exhibit a large effect in the rates of enzymatic reactions.

Like enzymology in general, single-molecule enzymology is primarily limited by assay developments. Here, I list the single-molecule enzymatic turnover assays by fluorescence detection using

1. A fluorescent active site, as described for cholesterol oxidase
2. Fluorescently labeled substrate molecules, such as ATP or dNTP, which require either low substrate concentrations [22] or zero-mode waveguide to suppress [23]
3. FRET pair that reports conformational changes triggered by enzymes, such as staphylococcal nuclease [24], a ribozyme [25], T4 Lysozyme [26]
4. Intensity change of a fluorophore that is associated with conformational change induced by a DNA polymerase [27]
5. Fluorogenic substrate molecules that produce fluorescent product molecules by enzymatic reactions, as proposed for horseradish peroxidase [28]. Because fluorophores are continuously replenished and diffuse away from the excitation volume, this assay circumvents the photobleaching problem that hampers all the other assays, and provides extremely long time traces [29, 30]

Most ensemble enzymatic kinetics have been satisfactorily described by the classic Michaelis-Menten equation [31]. The observation of conformational dynamics occurring on multiple time scales raises an intriguing question: why does the Michaelis-Menten equation work so well despite the broad distributions and dynamic fluctuations at the single-molecule level?

We conducted a single-molecule experiment on β-galactosidase with a fluorogenic substrate that offers superb statistics (Fig. 22.3). This study allowed us to confirm the dispersed kinetics (Fig. 22.3C) and fluctuations of the catalytic rate at multiple time scales, from 10 ms to 10 s (Fig. 22.3D). Interestingly, despite the fluctuations, the Michaelis-Menten equation still holds on a single enzyme basis, except k_{cat} and K_m bear different microscopic interpretations, i.e., the weighted average of different conformers [29].

This explains why the Michaelis-Menten equation works so well even in the presence of large fluctuations at a broad range of time scales. This intrinsic fluctuation is not significant for a system that comprises a large number of enzyme molecules; knowing the ensemble, the average result would be sufficient. However, if, in a living cell, there is only one or a few copies of a particular enzyme, these fluctuations may result in a large physiological effect.

It is important to stress that single-molecule enzyme turnover experiments are distinctly different from conventional *in vitro* enzymatic assays in that they are conducted under a nonequilibrium steady state condition with invariant substrate concentrations [32, 33], which is the usual condition in a living cell.

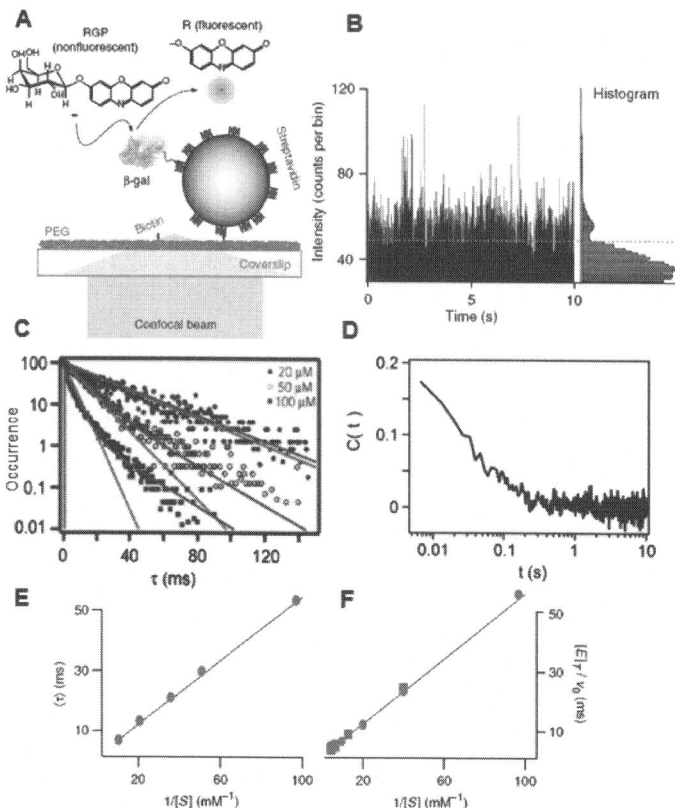

Fig. 22.3. (A) One β-galactosidase enzyme molecule is linked to a streptavidin-coated polystyrene bead via a flexible PEG linker. The bead binds to the biotin-PEG surface of a cover slip. The hydrolysis of the photogenic substrate RGP is catalyzed by an enzyme, and the fluorescent product resorufin (R) is monitored in the diffraction-limited confocal volume. (B) A segment of the chronological waiting time trajectory as a function of time. (C) Histograms of the waiting time distributions at different substrate concentrations [S]. (D) Normalized autocorrelation functions of waiting times in the time trace in (B). This highlights the broad range of time scales of turnover rate fluctuations, spanning at least four decades from 10^{-3} to 10 s. (E) Single-molecule Lineweaver-Burke plot of mean waiting times vs. 1/[S]. (F) Ensemble averaged Lineweaver-Burke plot agrees well with (E), indicating that the Michaelis-Menten equation holds on a single-molecule basis despite the fluctuation. From ref. [29].

22.3 Gene Expression and Regulation in Bacteria: Life at the Single-Molecule Level

A compelling challenge is conducting single-molecule experiments in a living cell. We have developed two strategies for studying DNA protein interactions at the single-molecule level: detection by localization and stroboscopic illumination, which allow probing of individual fluorescent protein molecules (FP) with specific labels in millisecond time resolution and nanometer spatial precision [34]. Prior to our work, tandem repeats of FPs had been used to monitor single mRNA molecules in live cells [35, 36]. We have used our single FP sensitivity to study a variety of fundamental processes in molecular biology.

Gene expression is a single-molecule problem because most genes often exist in one or two copies per cell. In 2008, we reported the first direct observation of the production of protein molecules as they are generated one at a time, in a single live *E. coli* cell, yielding quantitative information about gene expression [16, 17] (Fig. 22.4). Under the repressed condition, protein molecules are produced in bursts (Fig. 22.4B), with each burst originating from a stochastically-transcribed single messenger RNA (mRNA) molecule; the protein copy numbers in a burst follow an exponential distribution (Fig. 22.4D), as was previously predicted theoretically [37].

Gene expression is regulated by transcription factors (TFs), which bind to specific sequences on chromosomal DNA, called operators, in controlling the transcription by RNA polymerase. We made the first real-time observation of binding and dissociation of a TF on chromosome DNA in a live cell with the method of detection by localization [38]. The *lac* repressor was fused with a yellow fluorescent protein, and monitored in response to the addition and dilution of an inducer, a lactose analog [34]. At high inducer concentration, upon inducer binding within a few seconds, the *lac* repressor dissociates from the *lac* operators (Fig. 22.5A). After a sudden dilution of extracellular inducer concentration by a factor of 50, the rebinding of a *lac* repressor to the operators takes 60 s (Fig. 22.5B). It had been previously deduced from indirect experiments that a *lac* repressor molecule finds an operator through a combined 1D diffusion along DNA and 3D diffusion through cytoplasm (Fig. 22.6A) [39]. Using the method of stroboscopic illumination (Fig. 22.6B and C), we proved quantitatively that during the $\sim$60 s search time for the operator in the *E. coli* genome, a *lac* repressor spends $\sim$90% of its time non-specifically bound to and diffusing along DNA with a residence time of $\sim$ 5 ms (Fig. 22.6D).

We then studied the gene regulation of the lactose metabolism in *E. coli* using the classic *lac* operon [40] which includes the *lacZ* and *lacY* genes encoding β-galactosidase and lactose permease, respectively [41] (Fig. 22.7A). β-galactosidase catalyzes the hydrolysis of lactose, whereas lactose permease is a membrane channel for lactose to enter the cell. Expression of the *lac* operon is regulated by the *lac* repressor. Under high extracellular concentrations of inducers, e.g., methyl-β-D-thiogalactoside (TMG), the dissociation of the *lac* repressor from the operators allows transcription; the *lac* genes are fully expressed for every cell in a population. However, at moderate inducer concentrations, the *lac* genes are highly expressed in only a fraction of a population, which allows the entire *E. coli* population to conserve resources in the environment. Thus, genetically identical cells in the same environment can exhibit different phenotypes. A single cell's decision on the phenotype

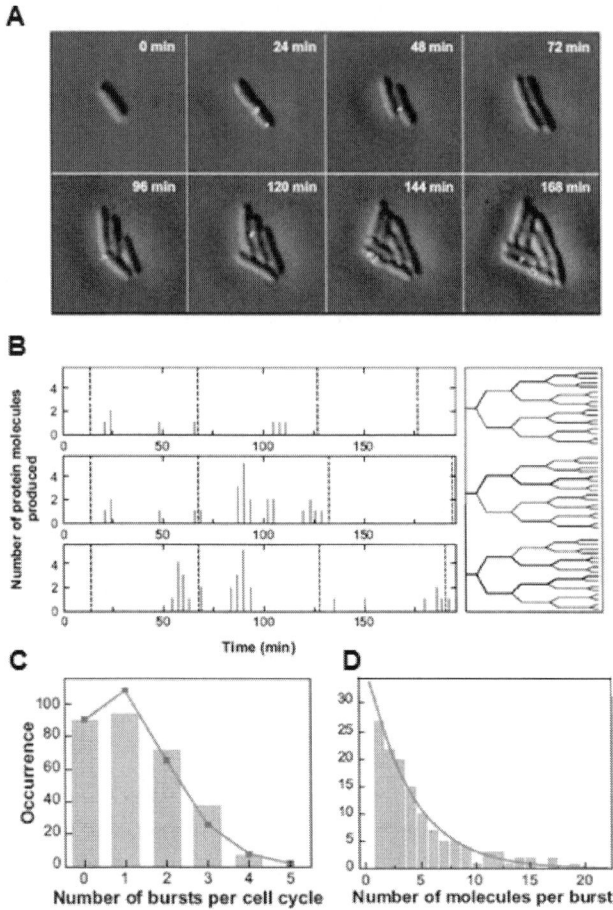

Fig. 22.4. (**A**) Time-lapse movie of fluorescence images (*yellow*) overlaid with simultaneous DIC images (*gray*) of *E. coli* cells expressing a membrane protein fused with YFP under the repressed condition. Each yellow spot is due to one YFP generated by gene expression. (**B**) Time traces of the expression of YFP molecules (*left*) along three particular cell lineages (*right*). The vertical axis is the number of protein molecules newly synthesized during the last 3 min. The dotted lines mark the cell division times. Protein production occurs in stochastic bursts, each due to one copy of mRNA and generates variable numbers of YFP molecules. (**C**) Histogram of the number of expression bursts per cell cycle. The fit is a Poisson distribution of an average of 1.2 mRNA per cell cycle. (**D**) Distribution of the number of YFPs in each gene expression burst, which follows an exponential distribution with an average of four molecules per burst. From ref. [16]

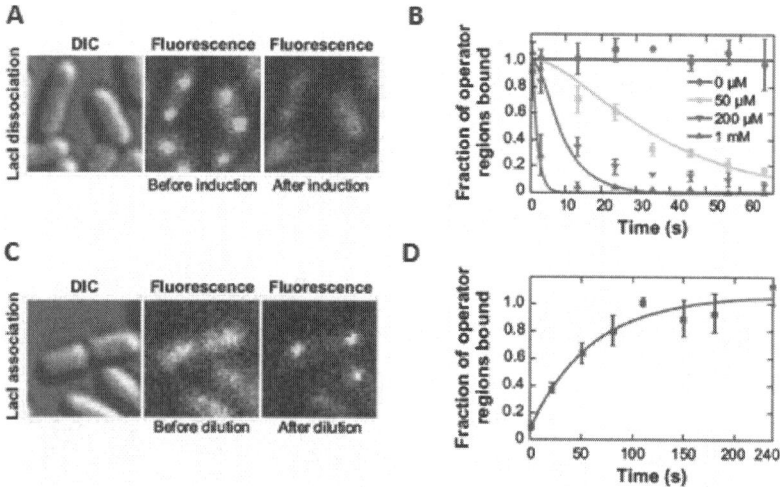

Fig. 22.5. (A) *E. coli* cells with YFP-labeled *lac* repressor before and 40 s after addition of inducer IPTG to a final concentration of 1 mM. (B) Fraction of *lac* operons in a cell population bound to repressors is plotted as a function of time after adding various concentrations of IPTG. (C) *E. coli* cells with YFP-labeled *lac* repressor before and 1 min after rapid dilution of IPTG from 100 to 2 mM. (D) Fraction of the operator region that is bound to the repressor as a function of time after the rapid dilution of inducer concentration by factor of 50. The rebinding times are exponentially fitted with a time constant of 60 s. From ref. [38]

to be exhibited is made by the bistable genetic switch, the *lac* operon. We investigated the molecular mechanism that controls the phenotype switching of a single cell.

We labeled the lactose permease with a yellow fluorescent protein. The two specific phenotypes under discussion are: first, above a certain threshold number of the permease, a cell has a fluorescent membrane and is capable of lactose metabolism; second below this threshold, a cell is non-fluorescent and is incapable of lactose metabolism. The two phenotypes coexist in a population of cells (Fig. 22.7B) and show a bimodal distribution (Fig. 22.7C). This threshold is determined to be ~300, corresponding to a big burst of permease production.

The *lac* repressor is a tetramer that can simultaneously bind to two operators to form a DNA loop and dissociates from DNA under a high inducer concentration [38] (Fig. 22.8A). Under low inducer concentrations (Fig. 22.8B), the repressor cannot be pulled off the DNA by the inducer. Rather, spontaneously, partial dissociations of the repressor result in transcription of one mRNA and a small burst of proteins, as seen in Fig. 22.4B. However, infrequent events of complete dissociation of the repressor result in large bursts of permease expression that trigger induction of the *lac* operon

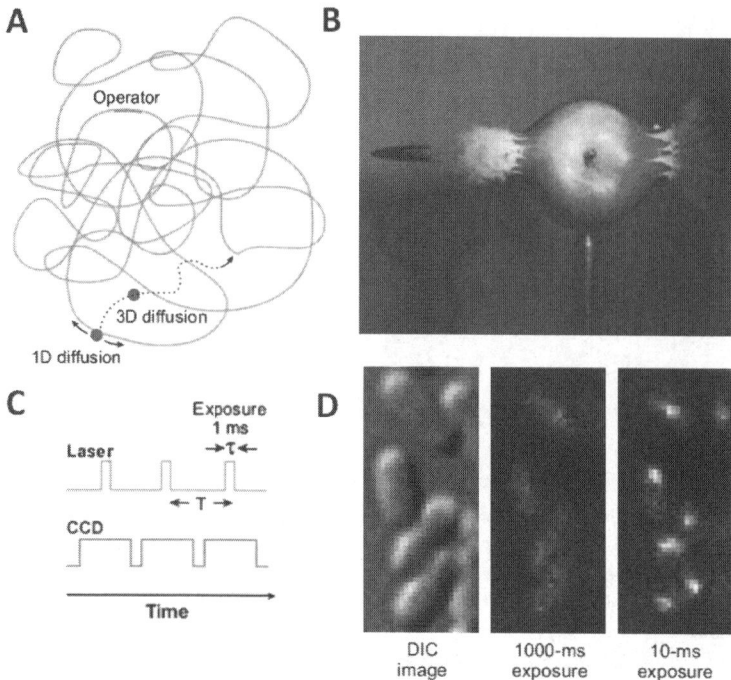

Fig. 22.6. (**A**) In searching for a target DNA sequence, a DNA repressor first non-specifically binds to DNA and undergoes 1D diffusion along a short segment of DNA before dissociating from DNA, diffusing in 3D through the cytoplasm, and rebinding to a different DNA segment. (**B**) Image of a bullet passing through an apple. From the Harold and Esther Edgerton Foundation. (**C**) Timing diagram for stroboscopic illumination. Each laser pulse is synchronized to a CCD frame that lasts for time T. (**D**) Two fluorescence images with different exposure times and the corresponding DIC image of the IPTG-induced *E. coli* cells. At 1 s, individual *lac* repressor-Venus molecules appear as diffuse fluorescence background. At 10 ms, they are clearly visible as nearly diffraction-limited spots. This indicates that the residence time of repressor is ∼10 ms. From ref. [38]

(Fig. 22.8C). We now understand the working of the bistable genetic switch at the molecular level [41].

This study proves that the stochastic single-molecule event of complete dissociation of the tetrameric *lac* repressor from DNA is solely responsible for the life changing decision of the cell, switching from one phenotype to another. This finding highlights the importance of single-molecule behaviors in biology [41].

22.4 In the Future

Single-molecule enzymology has yielded new information about how macro-molecular machines work *in vitro*. No doubt we have so much left to learn; the field will be active for many years to come.

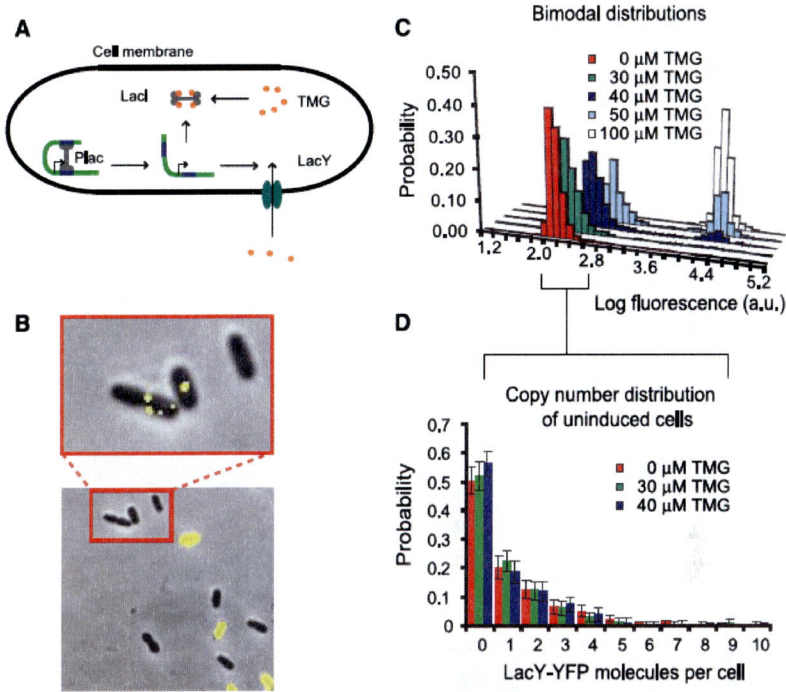

Fig. 22.7. (A) Gene expression of lactose permease in *E. coli*. Expression of permease *Lac* I increases the intracellular concentration of the inducer TMG, which causes the dissociation of *Lac*I from the promoter, leading to even more expression of permeases. **(B)** Cells expressing a *Lac*Y-YFP fusion exhibit all-or-none fluorescence in a fluorescence-phase contrast overlay (bottom, image dimensions 31 μm × 31 μm). Cells with a sufficient number of permeases are fully induced, whereas cells with too few permeases will stay uninduced. Fluorescence imaging with high sensitivity reveals single molecules of permease in the uninduced cells (top, image dimensions 8 μm × 13 μm). **(C)** Bimodal fluorescence distributions show that a fraction of the population exists either in an uninduced or induced state, with the relative fractions depending on the TMG concentration. **(D)** The distributions of *Lac*Y-YFP molecules in the uninduced fraction of the bimodal population at different TMG concentrations, measured with single-molecule sensitivity, indicate that one permease molecule is not enough to induce the *lac* operon. From ref. [41]

Meanwhile, an ultimate application of single-molecule enzymology that might have a societal impact is DNA sequencing, i.e., the real-time monitoring of a DNA polymerase that continuously replicates a single strand DNA template by incorporating different dye-labeled dNTPs. Currently, two commercial single-molecule sequencers have been developed [42, 43]. While others, including my group, are exploring complementary technologies, single-molecule sequencing offers the prospect of long read lengths and cost reduction, which will facilitate personalized medicine.

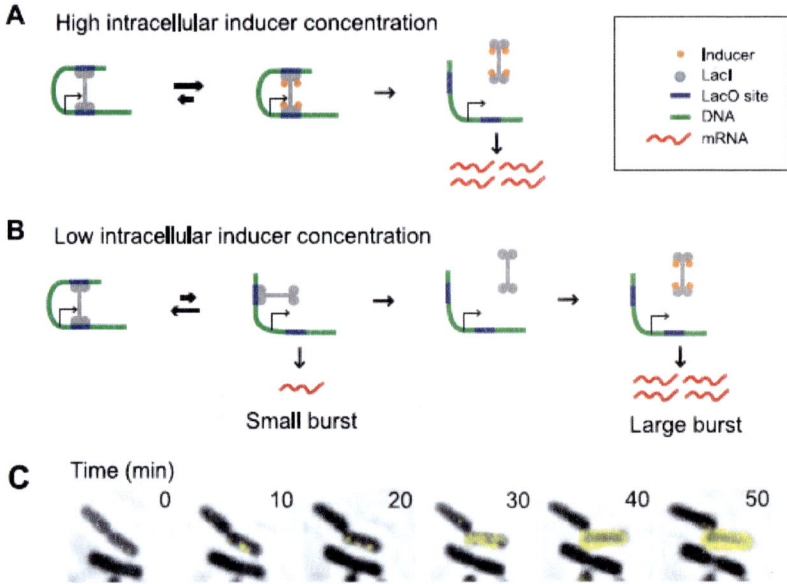

Fig. 22.8. (A) At high concentration of intracellular inducer, the repressor dissociates from its operators, as described by Monod and Jacob [40]. (B) At low concentrations of intracellular inducer, partial dissociation from one operator by the tetrameric LacI repressor is followed by a fast rebinding. Consequently, no more than one transcript is generated during such a brief dissociation event. However, the tetrameric repressor can dissociate from both operators stochastically and then be sequestered by the inducer so that it cannot rebind, leading to a large burst of expression. (C) A time-lapse sequence captures a phenotype-switching event. In the presence of 50 mM TMG, one such cell switches phenotype to express many LacY-YFP molecules (yellow fluorescence overlay) whereas the other daughter cell does not. From ref. [41]

In our live cell work, we have chosen to study the *lac* operon, an extremely well-characterized system, in order to validate our new experimental approach, from which we have already gained new knowledge. Our goal is to apply this methodology to less well-investigated systems. To achieve this, we have constructed an *E. coli* library with each gene tagged with a yellow fluorescent protein. We have found that a large number of genes are expressed in less than a few protein molecules per cell. This further justifies why single-molecule measurements are essential. This library will serve as a basis for new discoveries.

Can single-molecule studies help medicine? I believe the answer is yes. For example, the single-molecule question regarding genetic switches as discussed above is pertinent to research on tuberculosis, a deadly bacterial disease that kills two million people each year. There is a general phenomenon in bacteria: a small population of abnormal cells, called persisters, is drug resistant.

They have the same genes as normal cells, but a drug-resistant phenotype. The biology of persisters is not understood. With the bacterial library, we are in a position to study the gene expression of persisters with single-molecule sensitivity, which could provide clues for developing drugs against tuberculosis.

In addition to bacterial research, we are attempting to conduct similar single-molecule experiments on mammalian cells, the study of which is in high demand. The question about how cells or identical genes develop different phenotypes for the *lac* operon and persisters is also pertinent to stem cells, which again emphasizes that single-molecule behaviors are important to biology.

I hope the few examples highlighted above help to illustrate that the single-molecule approach has matured as a powerful tool and offers exciting opportunities for biological discoveries. I believe the best that single-molecule science and technology can offer to biology and medicine is yet to come.

Acknowledgments

It is my pleasure to express my gratitude to current and former members of my group for their contributions summarized herein, especially Bob Dunn, Peter Lu, Haw Yang, Hongye Sun, Antoine van Oijen, Guobin Luo, Brian English, Wei Min, Ji Xiao, Ji Yu, Long Cai, Nir Friedman, Johan Elf, Gene-Wei Li, Paul Choi, Kirsten Frieda, Huiyi Chen, Yuichi Taniguchi, Peter Sims, Will Greenleaf, Sangjin Kim, Rahul Roy, and Srinjan Basu. I also acknowledge fruitful collaborations with Luying Xun, Greg Schenter, Sam Kou, Binny Cherayil, Qian Hong, Greg Verdine, Eric Rubin, Martin Karplus, Attila Szabo, Biman Bagchi and Andrew Emili. I am grateful to the DOE, NIH, NSF and the Bill and Melinda Gates Foundation for supporting our ventures.

References

1. W.E. Moerner, L. Kador, Phys. Rev. Lett. **62**, 2535–2538 (1989)
2. M. Orrit, J. Bernard, Phys. Rev. Lett. **65**, 2716–2719 (1990)
3. E. Betzig, R.J. Chichester, Science **262**, 1422–1425 (1993)
4. X.S. Xie, R.C. Duun, Science **265**, 361–364 (1994)
5. W.P. Ambrose et al., Science **265**, 364–367 (1994)
6. J.J. Macklin, J.K. Trautman, T.D. Harris, L.E. Brus, Science **272**, 255–258 (1996)
7. X.S. Xie, J.K. Trautman, Ann. Rev. Phys. Chem. **49**, 441 (1998)
8. T. Hirschfeld, Appl. Opt. **15**, 2965–2966 (1976)
9. E.B. Shera, et al., Chem. Phys. Lett. **174**, 553–557 (1990)
10. R. Rigler, J. Widengren, BioScience **3**, 180–183 (1990)
11. A. Ashkin, J.M. Dziedzic, J.E. Bjorkholm, S. Chu, Opt. Lett. **11**, 288–290 (1986)
12. S.M. Block et al., Nature **348**, 348–352 (1990)
13. S.B. Smith, Y. Cui, C. Bustamante, Science. **271**, 795–799 (1996)

14. B. Sackman, E. Neher, *Single-channel recording*, 2nd edn. (Plenum, New York and London, 1995)
15. H.P. Lu et al., Science **282**, 1877 (1998)
16. J. Yu et al., Science **311**, 1600 (2006)
17. L. Cai, N. Friedman, X.S. Xie, Nature **440**, 358 (2006)
18. R.H. Austin et al., Biochem. **14**, 5355–5373 (1975)
19. H. Yang et al., Science **302**, 262 (2003)
20. W. Min et al., Phys. Rev. Lett. **94**, 198302 (2005)
21. S. Weiss, Science **283**, 1676–1683 (1999)
22. Funatsu T et al., Nature **374**, 555 (1995)
23. M.J. Levene, J. Korlach, S.W. Turner, M. Foquet, H.G. Graighead, W.W. Webb, Science **299**, 682–686 (2003)
24. T. Ha et al., PNAS **96**, 893–898 (1999)
25. X. Zhuang et al., Science **288**, 2048–2051 (2000)
26. C. Yu, D. Hu, E.R. Vorpagel, H.P. Lu, J. Phys. Chem. B. **107**, 7947 (2003)
27. G. Luo et al., PNAS **104**, 12610–12615 (2007)
28. L. Edman et al., Chem. Phys. **247**, 11–22 (1999)
29. B. English et al., Nat. Chem. Biol. **2**, 87–94 (2006)
30. K. Velonia, O. Flomenbom, D. Loos, S. Masuo, M. Cotlet, Y. Engelborghs, J. Hofkens, A.E. Rowan, J. Klafter, R.J.M. Nolte, F.C. de Schryver, Angew. Chem. Int. Ed. (2004) **43**, 2–6 (2005)
31. L. Michaelis, M.L. Menten, Biochem. Z. **49**, 333–369 (1913)
32. W. Min et al., Nano Lett. **5**, 23773–2378 (2005)
33. W. Min, X.S. Xie, B. Bagchi, J. Phys. Chem. **112**, 454–466 (2008)
34. X.S. Xie et al., Ann. Rev. Biophys. **37**, 417–444 (2008)
35. E. Bertrand, P. Chartrand, M. Schaefer, S.M. Shenoy, R.H. Singer, R.M. Long, Mol. Cell. **2**, 437–445 (1998)
36. I. Golding, J. Paulsson, S.M. Zawilski, E.C. Cox, Cell **123**, 1026–36 (2005)
37. O.G. Berg, J. Theor. Biol. **71**, 587–603 (1975)
38. J. Elf, G. Li, X.S. Xie, Science **316**, 1191 (2007)
39. P.H. Von Hippel, O.G. Berg, J. Biol. Chem. **164**, 675–78 (1989)
40. F. Jacob, J. Monod, J. Mol. Biol. **3**, 318 (1961)
41. P. Choi, L. Cai, X.S. Xie, Science **322**, 442–446 (2008)
42. T.D. Harris et al., Science **320**, 106–109 (2008)
43. J. Eid et al., Science **323**, 133–138 (2009)

Controlling Chemistry in Dynamic Nanoscale Systems

Aldo Jesorka, Ludvig Lizana, Zoran Konkoli, Ilja Czolkos, and Owe Orwar

23.1 Introduction

The biological cell, the fundamental building block of the living world, is a complex maze of compartmentalized biochemical reactors that embed tens of thousands of chemical reactions running in parallel. Several, if not all, reactors are systematically interconnected by a web of nanofluidic transporters, such as nanotubes, vesicles, and membrane pores with ever-changing shapes and structures [1]. To initiate, terminate, or control chemical reactions, small-scale poly-/pleiomorphic systems undergo rapid and violent shape changes with energy barriers close to k_BT, where, due to the small dimensions, diffusional mixing of reactants is rapid. The geometry, i.e. volume, and shape changes can be utilized to control both kinetic and thermodynamic properties of the system. This is in sharp contrast to the man-made macroscopic bioreactors, in which mixing of reactants is aided by mechanical means, such as stirring or sonication, under the assumption that reactions take place in volumes that do not change over time. Such reaction volumes are compact, like a sphere, a cube, or a cylinder, and do not provide for variation of shape. Ordinarily, reaction rates, mechanisms, and thermodynamic properties of chemical reactions in condensed media are based on these assumptions. A number of important questions and challenges arise from these facts. For example, how will we achieve fundamental understanding of how reactor shape affects chemistry on the nanoscale, how do we develop appropriate and powerful experimental model systems, and last but not least what impact will this knowledge have on the design and function of nanotechnological devices with new operation modes derived from natural principles.

A number of instances are already known that provide knowledge about shape and volume dependence of cellular physiological phenomena. For instance, optimal pH-levels are maintained by cell volume regulation [2]. Cdc42 (a member of the Rho family of small GTPases) is regulated by cell shape [3]. Cell volume control is crucial for downregulation of the HOG pathway. Swelling and shrinkage of mitochondria affect ADP/ATP levels [4].

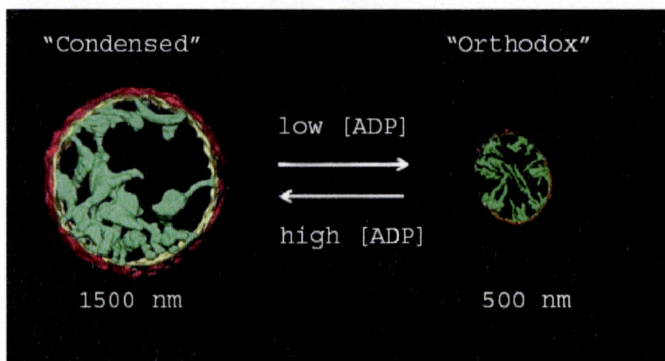

Fig. 23.1. ADP-driven volume and structural changes of mitochondria. Depending on the ADP concentration, the morphology of the interior changes as well as the volume of the mitochondrion, effectively regulating ATP production. Reproduced with permission [6]

Rac- and Rho-type GTPases can regulate the formation of protrusive lamellipodia, and cell retraction as well as cortical tension. Several local signaling networks have been identified that in a hierarchical way control cell shape and migration [5].

Figure 23.1 shows an example of a complex shape and volume change in a mitochondrion, visualized as a 3D-enhanced model obtained by transmission electron microscopy. The change is driven by a shift in ADP concentration. At low ADP levels, an "orthodox" shape with a small diameter and less voluminous interior is assumed, while at high ADP levels, a considerably larger diameter with a larger-volume interior is prevalent. The system undergoes a chemically self-regulated structural transition, that is associated with a functional transition depending on the physiological need of the cell, in this case the availability of ATP [6–8].

Without doubt, the extent and the rates of shape and volume changes in biological reactors, as well as certain structured geometries along with their dynamic behavior need to be understood in the context of biology. Also, it is of interest to understand how shape dynamics can be directly or indirectly used to introduce and exploit certain functionalities in nanofluidic devices, for instance to control the diffusional transport of chemical reactants in an artificial reactor/conduit network. Moreover, parallels can directly be drawn from living or artificial, biology-derived systems to fully artificial models. For example, novel nanotechnological devices that exploit key aspects of shape and volume changes can be used to control both the kinetics and thermodynamic properties of chemical reactions utilized in e.g. sensors, separations, and nanoelectronics.

Living biological systems are complex, and their functionality needs to be reduced to the relevant key fundamental properties in order to be described and modeled. Key features that need to be included in such experimental models include small scale (nm-to-low µm), an aqueous environment (defined pH, biological compatibility, etc.), controlled chemistry (concentration, equilibrium constants, rate constants, etc.), controlled geometry, shape (i.e. these parameters should possibly change as a chemical reaction proceeds within the reactor), and finally the possibility to follow reactions by sensitive and selective means of chemical and physical analysis.

Transitions in reactor shape from 3D to 2D projections or functional surfaces are interesting not the least from a technological point of view. It has to be mentioned that the reduction to two-dimensional structures or surfaces is easily possible due to the unique material properties of the basic building blocks, namely self-assembly and self-organization in a fluid-crystalline matrix.

Figure 23.2 depicts four models of reactor shape changes in order to study and evaluate the chemical and physical peculiarities of volume- and shape-change modulated kinetics as well as thermodynamics of bio-related (enzymatic) reactions.

An attractive method to construct model geometries that can undergo dynamic shape and volume changes involve the use of molecules that aggregate or self-organize through noncovalent interactions [9], forming supramolecular structures. Examples of such materials include microemulsions, surfactant membranes, and some amphiphilic polymer systems that can self-assemble into mesoscale structures, e.g. nanotubes or molecularly thin films. Biological systems successfully employ self-assembled surfactant membranes composed mainly of phospholipids to provide a universal solution to e.g. transport, reaction system enclosure, and sorting of molecular species. A variety of possibilities for using self-assembled lipid membranes for construction of artificial nano- and microdevices intended for storage, transportation, and mixing of extremely small volumes (10^{-12}–10^{-21} L) of aqueous solutions of chemically active entities have been previously reported. In the following chapter, shape and volume changes of lipid membrane constructs as a means to control chemical reaction parameters, such as reaction rates, are summarized. The different model strategies presented in Fig. 23.2 are discussed together with their respective theoretical frameworks, and examples from biology and state-of-the-art nanotechnology are provided. We begin with a short introduction to the model systems used and their relevant physical properties. A separate section is dedicated to two-dimensional reactors, and associated diffusive and tension-driven transport phenomena.

23.2 Liposome Networks and Nanochemistry

Chemical reactions occurring in geometries resembling those found in a cell can be investigated with the help of nanotube-vesicle networks (NVNs, Fig. 23.3 [10]). NVNs are highly flexible lipid bilayer structures in which the main building blocks are surface-immobilized vesicles connected by nanotubes.

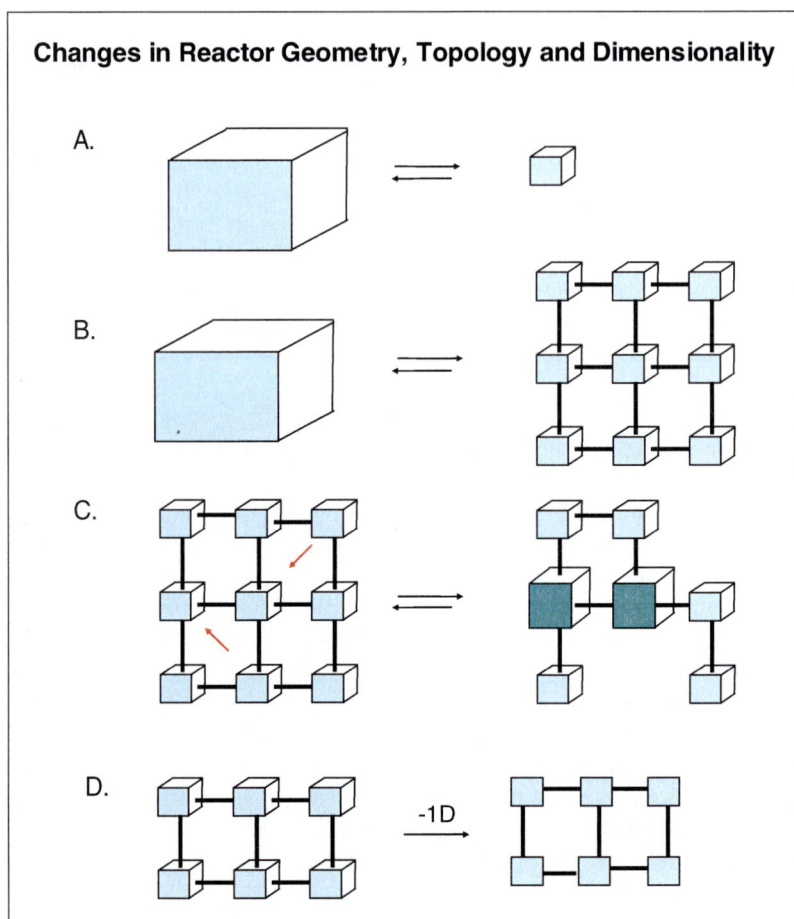

Fig. 23.2. Schematic representation of model systems for volume and shape changes in reactor geometry, topology, and dimensionality, using simple cube geometries and interconnecting conduits. (a) Change in reactor volume; (b) Change in topology, from compact-to-structured geometries; (c) Dynamic changes in reactor topology and volume. (d) Change in dimensionality, from a three-dimensional network to a surfaces of isolated areas interconnected by lanes

The physical properties, especially the bending flexibility of the lipid assemblies, facilitate the formation of highly complex structures in which topology, vesicle volumes, and tube lengths as well as radii can be constructed with high precision and modified at will (Fig. 23.3).

We have developed a variety of experimental techniques to control the geometry, dimensionality, topology, and functionality in surfactant membranes that can be directly useful in nanoscale network and device design [11–13]. Methods based on self-assembly, self-organization, and forced shape

transformations using modern micromanipulation tools have been developed
to form synthetic or semi-synthetic enclosed lipid bilayer structures with a
range of properties similar to biological nanocompartments. There is also
the possibility to load different vesicles with different chemical components
(Fig. 23.4).

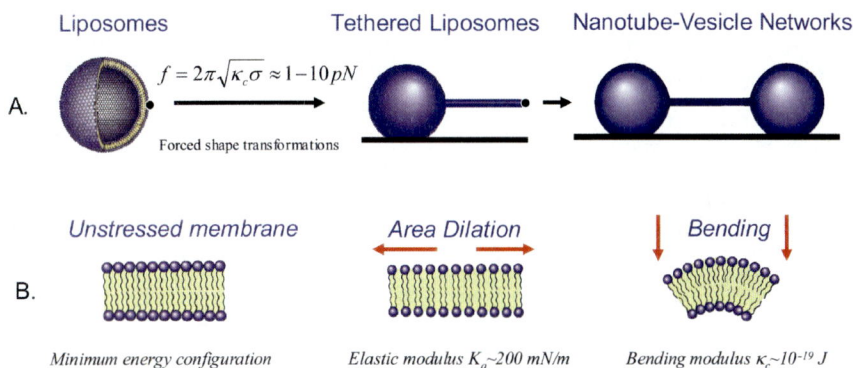

Fig. 23.3. (a) Schematic depiction of the creation of a lipid nanotube by application
of a point force, allowing interconnection of vesicles. (b) Mechanical stresses and
physical properties of planar phospholipid assemblies

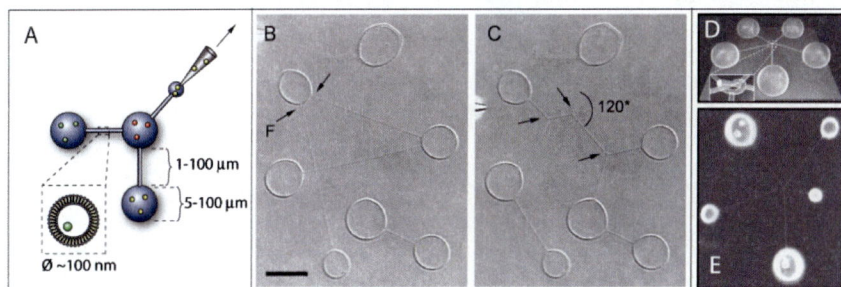

Fig. 23.4. (a) Construction strategy for vesicle-nanotube networks with functionalized interiors. Individual vesicles are constructed by the use of a pipet
micromanipulation technique, where each individual vesicle is generated with a
different back-filled needle, leading to differentiated interior chemical composition.
Size ranges for typical vesicle and nanotube dimensions are indicated as well as
the approximate size relations between a protein molecule and a nanotube cross
section. (b, c) Brightfiled micrographs of a large network of nanotube interconnected vesicles, before (b) and after (c) self-optimization of the nanotube links. The
arrows in (b) indicate the initial force triggering the self-optimization. The arrows
in (c) indicate the three-way junctions that have formed with an optimum angle
between the nanotubes. Nanotubes can be constructed in a manner that inhibits
self-optimization, e.g. by introduction of knots, as shown in the schematic drawing
(d) and the corresponding brigthfield micrograph (e). Pictures are reproduced with
permission

Due to the small length-scales of these systems, chemical gradients equilibrate on the order of subseconds to minutes by diffusion, which by appropriate network design can be harnessed to work as a controlled means of reactant mixing and transport. These features make NVNs ideal for studying chemical reactions under biologically relevant conditions: reacting species, e.g. enzymes and substrates, can be delocalized in space into nanotube-connected lipid containers, and the mixing rate is determined by network structure which can, if desirable, be reconfigured as the reactions progress. This contrasts macroscopic systems in which diffusion is a very inefficient way of transporting solutes.

The notion of a network is very general, and can describe any system where certain components (the network nodes) interact with each other. Each interaction between two nodes results in a link between them [14]. The study of networks is relevant in many scientific fields, in which a huge number of networks spanning over many disciplines share common features [15,16], such as the small-world and the scale-free character. Network structures have been shown to exist in, for instance, social contacts and neural networks [17]. The scale-free character reveals that the network is extremely heterogeneous in the sense that the topology is dominated by a few highly connected nodes (hubs), linking together the rest of the less connected ones. Two representatives exhibiting scale-free properties are the connectivity of the Internet and the signaling pathways in which biomolecules are linked via chemical reactions [18].

23.3 Diffusion as an Efficient Means of Transport in Micro- and Nanoscale Chemical Reactors: k_BT–Driven Fluidic Devices

Diffusion is the process by which molecules and colloidal particles undergo thermally induced motion through space. When length-scales approach micro- and nanometer dimensions, diffusion becomes an efficient means of transport as well as of mixing. For example, in compartments only a few times larger than the molecules themselves, diffusive mixing may be faster than the duration of single molecular reaction events e.g. catalytic cycles [19]. This implies that the need to redistribute reactants by e.g. stirring or external fields in small-scale systems is unnecessary. Instead, one can rely on diffusion as means of transport, the rate of which can be controlled by a proper reactor design.

The dynamic behavior of molecules diffusing within interconnected subvolumes is, in principle, contained in the diffusion equation. However, obtaining the solution for a system of arbitrary geometry and topology is not straightforward, which makes transport properties very demanding to acquire. For example, small tube openings imply that the containers to a good approximation can be treated as being homogeneously mixed, reducing all intracontainer dynamics to that of a well-stirred tank. Lizana and Konkoli [20] developed a transport model describing how the concentration $c_k(t)$ in each container k in a network changes over time t

$$\frac{dc_k(t)}{dt} = \sum_{l=1}^{m} \Lambda_{kl} \kappa_{kl} [c_l(t) - c_k(t)], \quad k = 1, \ldots, m \qquad (23.1)$$

where m is the number of containers in the network and Λ is a connectivity matrix. $\Lambda_{kl} = 1(0)$ if containers k and l are (not) linked (note: $\Lambda_{kl} = \Lambda_{lk}$ and $\Lambda_{kk} = 0$). The transport rate constant

$$\kappa_{kl} = \frac{\pi a_{kl} D_{kl}}{V_k \ell_{kl}} \qquad (23.2)$$

contains the geometrical parameters of the network, where V_l denotes the container volume, D_{kl} is the diffusion constant in the tube, and a_{kl} and ℓ_{kl} stands for the tube radius and length, respectively [21].

The solution to Eq. (23.1) is in general a sum of exponentials where the most elementary case is two containers interconnected via a tube, to which the solution is

$$c_{1,2}(t) = c_{1,2}(\infty) + [c_{1,2}(o) - c_{1,2}(\infty)] \exp\left(-\frac{\pi a_{12}^2 D_{12} t}{\ell_{12}} \frac{V_1 + V_2}{V_1 V_2}\right) \qquad (23.3)$$

When the tubes are very long, the molecules reside in the tubes for a substantial amount of time, leading to a time-dependent transport rate constant κ_{kl}. It can be shown that the decay in concentration is given by $c_k(t) \propto (D_{kl}t)^{-1/2}$ for short times, i.e. $t \ll \tau_{\text{transp}}$. [20].

The dynamics of reaction-diffusion processes in complex networks is also interesting outside a biological context. The main reasons are the intriguing features which differ substantially from bulk systems [22]. A number of studies have shown anomalous properties in equilibrium densities of reactant molecules [23] and reaction order [24].

Chemical reactions in small-scale systems are often associated with diffusion of the reactant molecules. Since transport properties set a fundamental limit to how fast reactants can actually meet, it is possible to regulate reaction rates by designing size, shape, and dimensionality of the reaction chamber in a proper way. For studying the impact of geometry and topology on enzyme-catalyzed reactions, NVNs are well suited. A technique to initiate chemical reactions involving few reactants inside NVNs has been described [25]. The shape of these networks is under dynamic control, allowing for transfer and mixing of two or several reactants at will. Often, reaction-diffusion experiments are conducted when a substrate is first distributed into specific network nodes, and then an enzyme is introduced to the network from a nanotube to a newly formed vesicle (Fig. 23.5). The enzymatic reaction commonly employed in these studies is a two-step process in which the enzyme, alkaline phosphatase (ALP), cleaves off two phosphate groups from fluorescein diphosphate

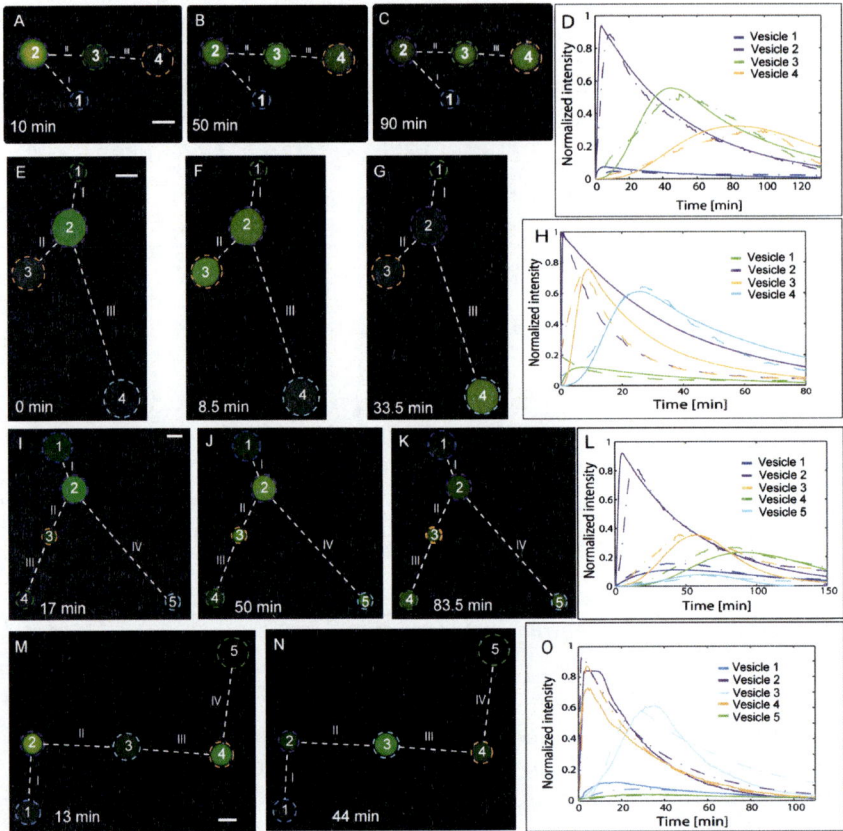

Fig. 23.5. Controlling enzymatic reactions by network geometry. Fluorescent micrographs (A–C, E–G, I–K, M–N) and normalized fluorescence intensities vs. time (D, H, L, and O), representing product formation from an enzymatic reaction in networks with different geometries. Initially, the enzyme-filled vesicles are vesicle 1 in images A–C, E–G, and I–K and vesicles 1 and 5 in M–N. All other vesicles are substrate filled. The broken lines in graphs D, H, L, and O show the theoretical fit to the experimentally measured product formation. The scale bars in the images represent 10 μm. Pictures are reproduced with permission [26]

(FDP), forming fluorescein (F). Fluorescein is a highly fluorescent molecule that can be detected under a LIF confocal microscope with great sensitivity.

The enzyme will diffuse throughout the network down its concentration gradient and explore each substrate-filled node where it exerts its catalytic activity turning substrate into product. Since reactant transport is highly affected by network morphology, e.g. size and connectivity, different locations can be programmed to experience different concentrations of reactants over time, yielding a means of controlling reaction rates in given nodes within a network [26]. Initially, reaction-diffusion systems have been studied using static network geometries. For example, it has been shown that an ordinary enzyme-catalyzed reaction can display front propagating properties, and that the reaction dynamics can be controlled directly by the geometry of the network [26]. Furthermore, filtering of chemical signals has been proposed in static networks having immobilized enzymes [27].

23.4 Chemical Reactions in Dynamic Nanoscale Volumes

The rate of a chemical reaction taking place inside a solitary giant liposome (or any other reactor) can be controlled by decreasing and increasing the vesicle volume, where a larger (smaller) volume leads to a lower (higher) reactant concentration (given a constant number of molecules) and hence, reduced (increased) reaction rates. As an example, consider the second-order equilibrium reaction

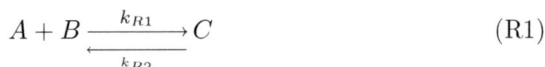

$$A + B \underset{k_{R2}}{\overset{k_{R1}}{\rightleftharpoons}} C \tag{R1}$$

The rate of the forward reaction in (i.e. the number of C molecules formed per unit of time) is favored by a small volume since the probability of reactants A and B meeting is higher. However, the backward reaction is determined by the rate of dissociation of C, and is independent of volume. According to the *Van't-Hoff–Le Chatelier* principle, it is thus possible to shift the equilibrium concentrations of A, B, and C by changing the container volume. The formation of C is favored by a small volume, and vice versa. If the volume increase or decrease is made at a constant rate, $dV/dt = k_V$, the concentration of A, B, and C change over time as:

$$\frac{dc_A(t)}{dt} = -k_{R1}c_A(t)c_B(t) - k_{R2}c_C(t) - c_A(t)\frac{k_V}{V(t)}$$
$$\frac{dc_B(t)}{dt} = -k_{R1}c_A(t)c_B(t) - k_{R2}c_C(t) - c_B(t)\frac{k_V}{V(t)} \tag{23.4}$$
$$\frac{dc_C(t)}{dt} = k_{R1}c_A(t)c_B(t) + k_{R2}c_C(t) - c_C(t)\frac{k_V}{V(t)}$$

The first two terms on the right hand side accounts for changes in reaction rate, while the third term accounts for the change in concentration due to a change in volume. The extent of the reaction is obtained as $n_q(t) = c_q(t)V(t)$, $q = A$, B, C. For a Michaelis-Menten (MM) enzymatic reaction:

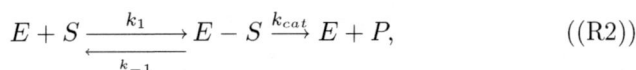

$$E + S \underset{k_{-1}}{\overset{k_1}{\rightleftharpoons}} E - S \overset{k_{cat}}{\rightarrow} E + P, \tag{(R2)}$$

the change in concentration of product $c_p(t)$, substrate $c_S(t)$, and enzyme $c_E(t)$ *is given by:*

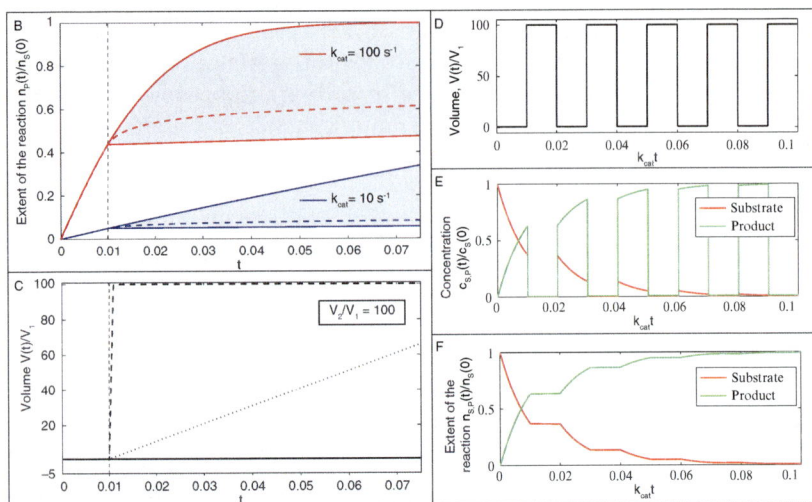

Fig. 23.6. Effects of volume changes on an enzymatic reaction in a solitary lipid vesicle (numerical solutions). (**a**) Extent of reaction in a solitary vesicle as a function of $k_{cat}t$, where the volume change is 100-fold. The red and blue lines indicate a "fast" and a "slow" reaction, respectively. The upper solid lines indicate no volume change, while the lower solid lines indicate very rapid volume change. Dashed lines represent an intermediate rate of volume change that significantly affects the reaction rate. (**b**) Solitary vesicle volume vs. time. The dashed, dotted, and solid lines represent a rapid, intermediate, and no volume change, respectively. The time at which the volume change is initiated ($k_{cat}t = 0.01$) is indicated by a vertical dashed line. (**c–e**) Dynamics of an enzymatic reaction taking place in a solitary vesicle undergoing periodic volume contraction–dilation transitions. Shown are the transformation between V_1 and V_2 (**c**), the behavior of substrate and product concentration (**d**), and the extent of the reaction (moles) (**e**). ([28], reproduced with permission)

$$\frac{dc_P(t)}{dt} = k_{cat}c_E(t)\frac{c_S(t)}{K_M + c_S(t)} - c_P(t)\frac{k_V}{V(t)}$$

$$\frac{dc_S(t)}{dt} = -k_{cat}c_E(t)\frac{c_S(t)}{K_M + c_S(t)} - c_S(t)\frac{k_V}{V(t)}, \qquad (23.5)$$

$$\frac{dc_E(t)}{dt} = -c_E(t)\frac{k_V}{V(t)}$$

where $K_M = (k_{cat} + k_{-1})/k_1$. Note that E–S is an enzyme–substrate complex. In order to calculate the extent of the reaction as a function of time at different rates of volume change, we have introduced the parameter $k_V' = (k_{cat}/K_M) \times n_E$, which is a measure of enzymatic reaction efficiency multiplied by the number of enzyme molecules n_E (assumed constant) that participate in the reaction [28].

Figure 23.6 shows the extent of an enzymatic reaction $n_P(t)/n_S(0)$ as a function of $k_{cat}\, t$, when the container volume is increased 100-fold (~ 4.6 fold increase in radius) at three volume expansion rates ($k_V = 0$, 10^3, $10^5\,\mathrm{dm}^3\,\mathrm{s}^{-1}$, going from V_1 to V_2) and for two different values of k_{cat}, representing a "fast" ($k_{\text{cat}} = 100\,\text{s}^{-1}$, red) and a "slow" ($k_{\text{cat}} = 10\,\text{s}^{-1}$, blue) reaction, respectively, under conditions $c_S(0) = 100\,\mu\text{M}$, $K_{\text{M}} = c_E(0) = 1\,\mu\text{M}$ and $V_1 = 10^{-12}\,\text{dm}^3$. These results indicate that reactions can be turned on and off, and be tuned, respectively, by volume changes given that the rate of volume change outcompetes the overall rate of product formation, and, in addition, that the actual size of the volume resulting from the expansion becomes so large that the substrate is diluted well below K_{M}. In order to see if volume changes could affect the enzyme activity in the Krebs cycle, we applied the solitary vesicle model to a mitochondrial-sized volume (diameter $= 0.5\,\mu\text{m}$).

Even though this is a very simplistic model, the calculations suggest that mitochondria and small biological compartments in general, can employ volume swelling and shrinkage as a means of regulating rates of biochemical reactions [28]. It also suggests that normal volume fluctuations in mitochondria does not alter substantially the overall performance (yield) of the Krebs cycle.

23.5 Shape and Volume Changes as Control Mode for Reaction Rates

Some biological structures, such as the Golgi-ER system, are of a more complex geometry and topology than solitary spheres. We have, therefore, investigated if a chemical reaction could be initiated or boosted in certain nodes (i.e. change of local rate of product formation) within a reactor network as a function of shape factors, such as network connectivity. Specifically, we investigated a linear-to-circular transformation by end-point conjugation, i.e. a topology transformation, in a four-container network in two different systems: (a) an enzyme-catalyzed reaction-diffusion model and (b) an enzyme-catalyzed reaction-diffusion model with self-inhibition. In these experiments, the volume remains constant and the observed changes in the spatiotemporal profiles of product formation are solely due to a change in container connectivity. A similar situation was reported in [29, 30], however, the structural changes considered here are not smooth.

We studied, *vis-a-vis* a linear reference case (Fig. 23.7a), how modification of network topology (linear-to-circular transformation) affects the reaction rate in individual network nodes. Specifically, containers γ_1 and γ_4 were connected by a nanotube ~ 50 min after the reaction was started using microelectrofusion. The temporal intensity profile in each vesicle is shown in Fig. 23.7b. The new tubular connection results in an increased rate of product formation in γ_4, indicating that enzymes access this location more easily through the new link rather than diffusing through the whole network. The effects of product inhibition were also investigated (Fig. 23.7c). The experimental findings in all cases are well reproduced by theory (Fig. 23.7, bottom panels) [28].

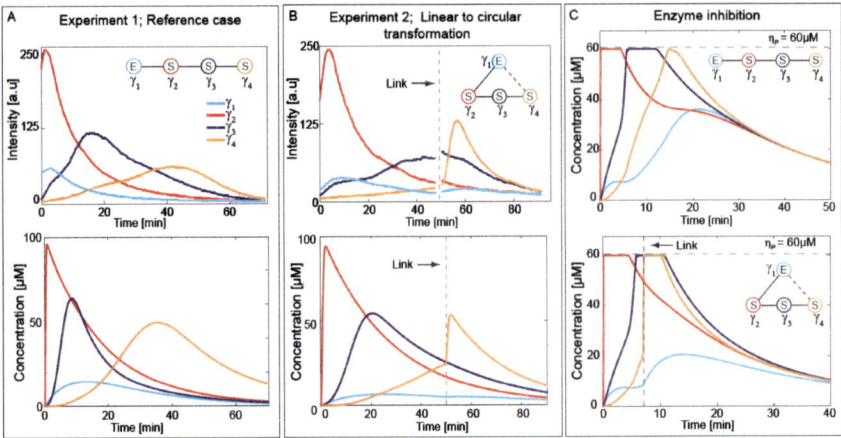

Fig. 23.7. Dynamics of an enzymatic reaction in lipid nanotube networks with variable topology: numeric calculations (bottom)/fluorescence intensity of the reaction product (top) vs. time for three differently chosen network geometries. (**a**) Reference experiment: a static four-vesicle network. The product concentration displays a cascade-like behavior in time and space. (**b**) Linear-to-circular topology change in the four-vesicle network (**c**) A model study of the effect of product inhibition as the linear four-vesicle network (*top panel*) undergoes the same change in structure (*bottom panel*) as the network in the reference experiment ([28], reprinted with permission)

The cases discussed above indicate that nanotubes rapidly connecting distant parts of the network can either initiate or boost ongoing reactions by locally increasing the reactant concentration, which may be exploited to control the overall network dynamics. Note that the local amplification of reactant density in these three experiments are induced only by a change of connectivity, primarily facilitating enzyme passage, the number of enzyme molecules remains constant. Consequently, tubular connections found, for instance, inside Golgi stacks might dramatically change the reaction-diffusion dynamics of the system and may therefore serve as important instruments for controlling local synthesis of biochemical compounds [28].

23.6 2D-Nanofluidic Surface Reactors: A New Technological Concept

Biomimetic membrane-enclosed compartments have numerous advantages when it comes to model biological systems. However, at shorter length scales, surface properties become increasingly important. In applications where the length scales are of the order of micrometers to submicrometers, the chemistry can be completely dominated by surface interactions [31]. This fact and the reasoning that the micro/nano fabrication of three-dimensional reactor-conduit assemblies is not yet optimized for larger scale fabrication efforts towards e.g. bioanalytical or similar devices, triggered the interest in interfacial phenomena between self-assembled lipid films and hydrophobic surfaces. Such structures in the two-dimensional domain would offer a number of benefits, such as the availability of networks in which the connectivity (or number of available pathways), the connectivity strength, the reactant concentration, material flux, mixing ratios between reactants, and product yield can be far more accurately controlled. Applications are immediately within reach, for example, non-linear dilution networks or networks for parallel or combinatorial synthesis. The capacity for single-molecule studies is present, due to the nature of the monomolecularly thin fluid films which can either bind or temporarily accommodate and release other molecules of comparable hydrophobicity within the film structure. Most interestingly, to change a solvent composition on demand [32, 33], through release of materials from surfaces, is thus very attractive because it does not induce volume flow or pressure gradients, i.e. the rest of the system remains undisturbed.

We have introduced a novel technique for the controlled spreading and mixing of lipid monolayers from multilamellar precursors on surfaces covered by the hydrophobic epoxy resin SU-8, or similar coatings. The lipid spreads as a monolayer due to the high interface tension between SU-8 and the aqueous environment. A micropatterned device with SU-8 lanes, injection ports, and mixing regions, surrounded by hydrophilic Au was constructed to allow handling of lipid films and to achieve their mixing at controlled stoichiometry. Our findings offer a new approach to dynamic surface fictionalization and decoration as well as surface-based catalysis and self-assembly [34–36]. Current application examples are multicomponent surface mixers of well-defined stoichiometry and devices for DNA manipulation and release.

When multilamellar lipid vesicles are placed on a SU-8 substrate, the contained lipid rapidly spreads out over time as a monolayer on the entire available surface (Fig. 23.8a). The contact angle of water on SU-8 was determined to be $91.4° \pm 1.5°$. The formed lipid patches are perfectly circular as shown in Fig. 23.8b. The multilamellar vesicles are eventually entirely depleted in the process. The tension induced by SU-8 is sufficiently large to disrupt the structure of the multilamellar vesicle, and the surface adhesion energy of lipids on SU-8, Σ, equals the tension at the spreading edge, and is therefore also equal to the lipid/SU-8 adhesion energy.

SU-8 is a chemically amplified bisphenol A-based epoxy photoresist and it is therefore reasonable to assume that the surface tension SU-8/water could be as high as 47 mN/m [37]. In refs. [38] and [39], the dynamics of spreading was modeled by balancing the lipid film Marangoni stress with the sliding friction force between the lipid film and surface (per unit area). Estimated spreading coefficients on SU-8 are in the range of $1–3\mu m^2/s$.

We conceptualized and microfabricated the SU-8 patterns on Au as base layer, which in contrast to SU-8 is hydrophilic and does not promote lipid

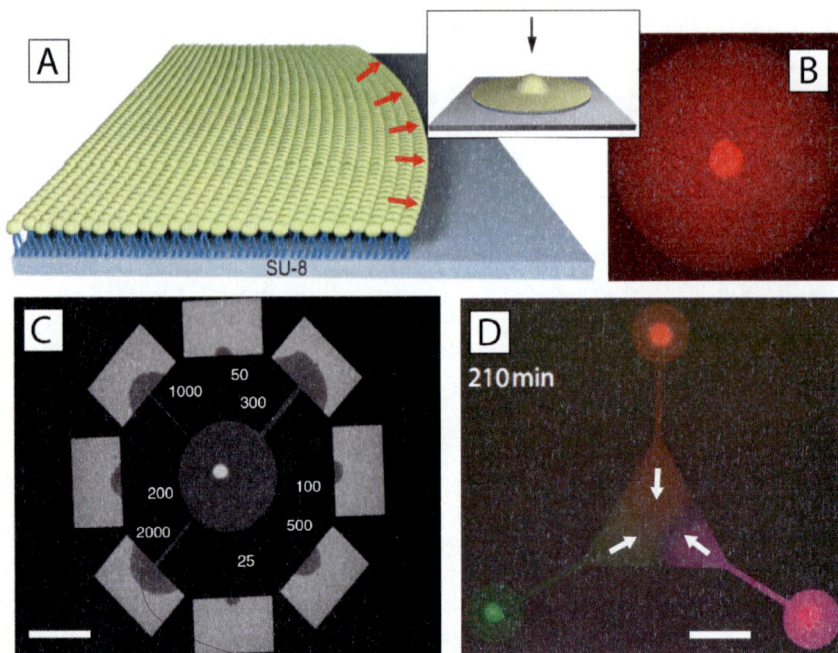

Fig. 23.8. Lipid spreading and mixing on microfabricated hydrophobic SU-8 structures. (**a**) Schematic of the spreading lipid film. The red arrows indicate the spreading film front. (**b**) Confocal micrograph of a rhodamine phosphatidylethanolamine-stained circular lipid patch (*right*). Inset: Perspective view of the lipid patch as it draws material from a liposome that has been deposited as initial source of material. The arrow serves to illustrate the reduction of the liposome size as the spreading progresses. (**c**) Lipid spreading across nano/microsized SU-8 lanes of different widths. The light areas are SU-8 surface patches. Situated in the center is a liposome as source of lipid. The dark semi-circular grey areas are the lipid film fractions that reach the target areas through the individual lanes. The size of each spread patch is proportional to the lane width, which ranges from 25 to 3,000 nm. (**d**) Ternary lipid mixing device. Three different lipid fractions were applied to the three circular source patches as individually labeled multilamellar liposomes. After approximately 4 min of spreading in the direction indicated by the white arrows, the films merge in the triangular central area in a stoichiometrically exactly defined composition. Scale bars in the lower panels: 20 μm. The images are reproduced with permission

spreading. Since SU-8 is a negative photoresist, it offers the opportunity to easily generate structures on the micrometer scale [40] whose shape was designed to support the lipid film formation and controlled stoichiometric mixing. We further explored the spreading of lipid films on nanosized lanes (Fig. 23.8c), where the expected uniformity of the spreading velocity could be confirmed by the lane-width-dependent size of the lipid patch in the target area, and created binary and ternary mixers having two and three injection pads for multilamellar vesicles, respectively, and one centrally placed mixing region (Fig. 23.8d). These mixers are examples of true 2D chemical reactors, since the composition of reactants meeting in its mixing zone, either by diffusion within the monolayer or coupled to diffusion of the lipids they are covalently conjugated to, can be tightly controlled through the mixing parameters [34]. This work represents the starting point for the controlled modification of two-dimensional liquids, and is a solid and easy-to-obtain platform for applications in the areas of surface functionalization and decoration as well as surface-based catalysis and self-assembly including supramolecular and nanoscale assembly.

Another model system that demonstrates the applicability of the concept to more complex biomolecules, namely DNA, has also been investigated [35, 36]. DNA as well as other polynucleotides represent common classes of compounds whose specific adsorption, desorption, and binding is of great interest to control in small-scale systems [41–43], mainly in processing and analyzing biological samples. Many applications in biotechnology and bioanalysis are based on surface-assisted DNA hybridization. In the final section, a methodology is presented for specific adsorption of hydrophobic ssDNA conjugates to SU-8 surfaces, hybridization with complementary ssDNA (Fig. 23.9a–d) and, subsequently, for re-release of the bound DNA (ss or ds) by initializing competition with a spreading lipid film on the same surface (Fig. 23.9e–f).

Cholesteryl-TEG-modified oligonucleotides adsorb efficiently on highly hydrophobic SU-8 surfaces whereas non-modified oligonucleotides stay in solution, and can subsequently be hybridized by their complementary ssDNA. The coupling of ss-chol-DNA to SU-8 involves strong hydrophobic interactions; no evidence is present that covalent bonds are formed. Fluorescence resonant energy transfer (FRET) is applied to visualize both the binding of chol-DNA and the hybridization (Fig. 23.9a–c). The stepwise deposition/rinsing immobilization route (Fig. 23.9d) for deposition, binding, and hybridization grants an advantage over other methods that involve functionalized surfaces by eliminating the need for surface activation. High and reproducible yields of hybridization of complementary strands to immobilized chol-DNA are obtained.

After immobilization of conjugated DNA and, if desired, reaction with complementary strands, surface-bound DNA can be released from the SU-8 patches by competition with a spreading lipid monolayer. As discussed above, a lipid reservoir spreads readily due to screening of the hydrophobic surface energy between the SU-8 and water by the forming lipid film, resulting in a free energy gain. This process can be initiated by placing a multilamellar vesicle onto a DNA-covered SU-8 structure, schematically depicted in Fig. 23.9e. Both ssDNA and DS DNA are affected by the spreading lipid film and lift off to near completeness.

Almost immediately when the multilamellar vesicle gets into contact with the chol-DNA-covered SU-8 surface, chol-DNA detaches. As seen on the spreading lane in Fig. 23.9f, the change in fluorescence intensity is abrupt

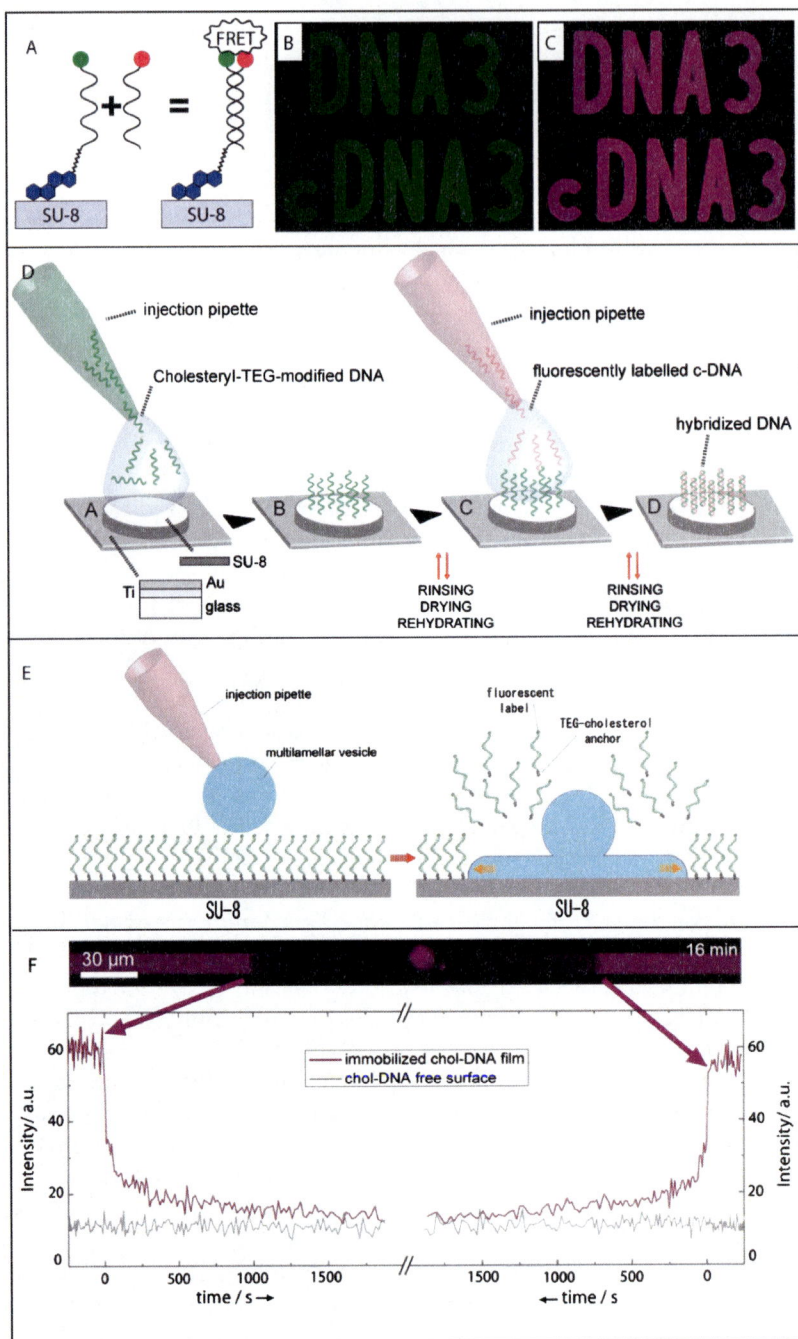

Fig. 23.9. Controlled deposition and hybridization of DNA on hydrophobic SU-8 surface patches (**a–d**) and defined displacement of DNA from such areas (**e–f**). (**a**) Schematic drawing of the individually labeled and cholesterol-modified DNA

and follows nearly a step function where the base levels correspond to those observed for DNA-free surfaces and from the slope it is evident that the release is rapid. Two conclusions can be drawn from these experiments: On chol-TEG-DNA coated SU-8, the lipid spreads at the same velocity as on bare SU-8 surfaces, and the majority of the DNA adducts are released from the surface immediately upon contact with the lipid film. The spreading power S represents the difference in free energy between lipids on the surface and in the reservoir (interfacial tension, σ_{Lipid}). When chol-DNA is immobilized on the surface, the lipid spreading power is reduced by the free energy difference for chol-DNA (interfacial tension, σ_{DNA}) in solution and on the surface.

The geometry of the hydrophobic structures which support lipid film formation can be arbitrarily controlled within the limits of photolithographic microfabrication in SU-8 and extended by applying direct writing ebeam lithography. This gives opportunities to create arrays of arbitrary patterns of immobilized DNA. It, furthermore, allows the generation of complex concentration profiles of DNA in solution, as the DNA is released by a spreading lipid film. We successfully microfabricated SU-8 patterns of spiral, triangular, comb, meander, and straight lanes where we show the spreading of a lipid monolayer and subsequent chol-DNA release into the solution.

This second example demonstrates that the SU-8-based micro/nanostructured surfaces allow for rapid binding and manipulation of hydrophobically modified DNA at SU-8 surfaces and its controlled release by lipid film spreading. Using a combination of nano- and micro-meter-sized SU-8 patterns, DNA concentration gradients in solution can be created dynamically. Moreover, other molecules with different hydrophobicites can likely be used such that the hydrophobicity of the molecules can be exploited as a sorting tool to modify surface properties of SU-8. It is, furthermore, possible to heat

---

Fig. 23.9 (continued). molecules used in the FRET experiment. (**b**) Confocal micrograph showing the fluorescence emission of the donor molecules adsorbed to the SU-8 surface when the donor was specifically irradiated. (**c**) Confocal micrograph showing the fluorescence emission of the acceptor molecules after hybridization when the donor was specifically irradiated. The emission is due to resonant energy transfer from the irradiated donor, indicating successful hybridization. (**d**) Procedure of DNA immobilization and hybridization on microfabricated SU-8 surface patches. The sequential steps are indicted by black arrows, with intermediate rinsing and drying steps. (**e**) Schematic drawing of selective displacement of fluorescently labeled DNA from a SU-8 surface by lipid molecules in a spreading film originating from a multilamellar vesicle. (**f**) Confocal micrograph of a lane initially covered with labeled DNA (magenta), where a part of the DNA has been displaced by a spreading lipid film (uncolored). Part of the DNA migrates into the hydrophobic center liposome, rendering it fluorescent as well. The bottom panel shows the corresponding chart intensity vs. time chart of the fluorescence intensity at the border between lipid and DNA. Arrows indicate the points where the intensities were obtained. Pictures are reproduced with permission

the surface by embedding, e.g. surface-printed micro heaters below the SU-8 substrate. Hence, by using lipids with well-defined transition temperatures, spreading and thus DNA release can potentially be started or stopped on demand. The techniques presented here for ssDNA and dsDNA immobilization and release represent simple and highly efficient small-scale procedures with potential applications in DNA microarrays, microfluidic devices, and functionalized surfaces i.e. in cantilevers, Quartz Crystal Microbalance (QCM), and Surface Plasmon Resonance (SPR).

23.7 Summary

Spatial organization and shape dynamics are inherent properties of biological cells and cell interiors. There are strong indications that these features are important for the in vivo control of reaction parameters in biochemical transformations. Nanofluidic model devices founded on surfactant systems, such as phospholipids or phospholipid mixtures, that can be assembled and manipulated through a combination of self-assembly and forced shape transformations, offer numerous practical benefits since they closely resemble their biological counterparts both in function and in structure. To date, these systems belong to a rare group of techniques that are used to model shape and volume changes on the micrometer and nanometer-scale on relevant time scales.

Diffusion is an efficient means of materials transport in natural and artificial nanoscale systems and can be readily employed in the study of enzymatic reactions in fluid membrane reactors of static or of changing geometries and morphologies. Other means of transport, e.g. electrophoretic or tension-driven modes are also available.

Most importantly, reaction rates in nanofluidic systems can be controlled both by shape and volume changes. The important interplay between chemical reactions and geometry has been conceptualized within a theoretical framework for ultra-small volumes and tested on a number of experimental systems, opening pathways to more complex, dynamically compartmentalized ultra-small volume reactors, or artificial model cells, that offer more detailed understanding of cellular kinetics and biophysical phenomena, such as macromolecular crowding.

A projection of nanotube vesicle networks onto surfaces is a viable strategy to overcome challenging difficulties with respect to stability, portability, and ease of fabrication.

The negative photoresist SU-8 has been utilized as a hydrophobic, structured support with feature sizes in the "μm" and "nm" range, accommodating hydrophobic or hydrophobized molecularly thin films. A unique feature of such structures is the controlled and stoichiometrically well-defined mixing of dynamically flowing surface coatings, for example, through the formation of spreading and mixing lipid monolayers. Moreover, other immobilization strategies based on hydrophobic interactions have been explored and established, such as the surface attachment of cholesterol-modified DNA, serving as anchors for complementary DNA recognition. Immobilized chol-TEG-DNA shows robust and efficient attachment, high surface coverage, and is well accessible for complementary strands. dsDNA disassociation and hybridization on chip via surface printed thin film heaters or infrared laser light is a possible extension of the concept. Controlled release of chol-DNA molecules from SU-8 surfaces gives the possibility to dynamically change surface and/or solution properties in micro and nanoreactor applications, opening access to stable 2D chemistry on surface-based devices with potential for easy interfacing with conventional microfluidic devices.

Acknowledgments

The work described herein was supported by the Swedish Foundation for Strategic Research, the European Research Council, the Wallenberg Foundation, and the Swedish Research Council.

References

1. X.S. Xie, P.J. Choi, G.W. Li, N.K. Lee, G. Lia, Ann. Rev. Biophys. **37**, 417–444 (2008)
2. H. Volkl, G.L. Busch, D. Haussinger, F. Lang, FEBS Lett. **338**, 27–30 (1994)
3. P. Nalbant, L. Hodgson, V. Kraynov, A. Toutchkine, K.M. Hahn, Science **305**, 1615–1619 (2004)
4. L. Packer, J. Cell Biol. **18**, 487 (1963)
5. A. Hall, Biochem. Soc. Trans. **33**, 891–895 (2005)
6. C.A. Mannella, Biochim. Biophys. Acta (BBA) – Mol. Cell Res. **1763**, 542–548 (2006)
7. C.R. Hackenbrock, J. Cell Biol. **30**, 269–297 (1966)
8. C.R. Hackenbrock, J. Cell Biol. **37**, 345–369 (1968)
9. J.M. Lehn, Proc. Natl. Acad. Sci. U S A. **99**, 4763–4768 (2002)
10. R. Karlsson, A. Karlsson, A. Ewing, P. Dommersnes, J.F. Joanny, A. Jesorka, O. Orwar, Anal. Chem. **78**, 5960–5968 (2006)
11. M. Karlsson, K. Sott, A.S. Cans, A. Karlsson, R. Karlsson, O. Orwar, Langmuir **17**, 6754–6758 (2001)
12. R. Karlsson, A. Karlsson, O. Orwar, J. Phys. Chem. B. **107**, 11201–11207 (2003)
13. R. Karlsson, A. Karlsson, O. Orwar, J. Am. Chem. Soc. **125**, 8442–8443 (2003)
14. A.L. Barabási, Linked: The New Science of Networks. (Perseus Publishing, Oxford, 2002)
15. R. Albert, A.L. Barabasi, Rev. Mod. Phys. **74**, 47–97 (2002)
16. S.H. Strogatz, Nature **410**, 268–276 (2001)
17. D.J. Watts, S.H. Strogatz, Nature **393**, 440–442 (1998)
18. H. Jeong, B. Tombor, R. Albert, Z.N. Oltvai, A.L. Barabasi, Nature **407**, 651–654 (2000)
19. P. Stange, A.S. Mikhailov, B. Hess, J. Phys. Chem. B. **104**, 1844–1853 (2000)

20. L. Lizana, Z. Konkoli, Phys. Rev. E. **72**, (2005)
21. L. Dagdug, A.M. Berezhkovskii, S.Y. Shvartsman, G.H. Weiss, J. Chem. Phys. **119**, 12473–12478 (2003)
22. L.K. Gallos, P. Argyrakis, J. Phys.-Condens Matter **19**, 10 (2007)
23. V. Colizza, R. Pastor-Satorras, A. Vespignani, Nat. Phys. **3**, 276–282 (2007)
24. L.K. Gallos, P. Argyrakis, Phys. Rev. Lett. **92**, 4 (2004)
25. A. Karlsson, K. Sott, M. Markstrom, M. Davidson, Z. Konkoli, O. Orwar, J. Phys. Chem. B. **109**, 1609–1617 (2005)
26. K. Sott, T. Lobovkina, L. Lizana, M. Tokarz, B. Bauer, Z. Konkoli, O. Orwar, Nano Lett. **6**, 209–214 (2006)
27. L. Lizana, Z. Konkoli, O. Orwar, J. Phys. Chem. B. **111**, 6214–6219 (2007)
28. L. Lizana, B. Bauer, O. Orwart, Proc. Natl. Acad. Sci. USA. **105**, 4099–4104 (2008)
29. Z. Konkoli, Phys. Rev. E. **72**, 011917 (2005)
30. Z. Konkoli, J. Phys.-Condens. Matter **19**, 065149 (2007)
31. D.T. Chiu, C.F. Wilson, A. Karlsson, A. Danielsson, A. Lundqvist, A. Stromberg, F. Ryttsen, M. Davidson, S. Nordholm, O. Orwar, R.N. Zare, Chem. Phys. **247**, 133–139 (1999)
32. B.A. Grzybowski, K.J.M. Bishop, C.J. Campbell, M. Fialkowski, S.K. Smoukov, Soft Matter **1**, 114–128 (2005)
33. C.J. Kastrup, M.K. Runyon, F. Shen, R.F. Ismagilov, Proc. Natl. Acad. Sci. USA. **103**, 15747–15752 (2006)
34. I. Czolkos, Y. Erkan, P. Dommersnes, A. Jesorka, O. Orwar, Nano Lett. **7**, 1980–1984 (2007)
35. Y. Erkan, I. Czolkos, A. Jesorka, L.M. Wilhelmsson, O. Orwar, Langmuir **23**, 5259–5263 (2007)
36. Y. Erkan, K. Halma, I. Czolkos, A. Jesorka, P. Dommersnes, R. Kumar, T. Brown, O. Orwar, Nano Lett. **8**, 227–231 (2008)
37. C.A. Harper, *Plastics Materials and Processes: A Concise Encyclopedia* (Wiley Interscience, 2003)
38. J.O. Radler, T.J. Feder, H.H. Strey, E. Sackmann, Phys. Rev. E. **51**, 4526–4536 (1995)
39. J. Nissen, S. Gritsch, G. Wiegand, J.O. Radler, Eur. Phys. J. B. **10**, 335–344 (1999)
40. J. Zhang, K.L. Tan, G.D. Hong, L.J. Yang, H.Q. Gong, J. Micromech. Microeng. **11**, 20–26 (2001)
41. A.K. Pannier, L.D. Shea, Mol. Therapy **10**, 19–26 (2004)
42. A.L. Hook, H. Thissen, J.P. Hayes, N.H. Voelcker, Biosens. Bioelectron. **21**, 2137–2145 (2006)
43. A.L. Hook, H. Thissen, N.H. Voelcker, Trends Biotechnol. **24**, 471–477 (2006)

Part IX

Catalysis of Single Enzyme Molecules

Single-Molecule Protein Conformational Dynamics in Enzymatic Reactions

H. Peter Lu

Summary. Enzymes involve many critical biological processes, and for some extends, the biological clock of a living cell is often regulated by enzymatic reactions. An enzymatic reaction involves active substrate–enzyme complex formation, chemical transformation, and product releasing, as we know of the Mechalis–Menten mechanism. Enzymes can change the biological reaction pathways and accelerate the reaction rate by thousands and even millions of times. It is the enzyme–substrate interaction and complex formation that play a critical role in defining the enzymatic reaction landscape, including reaction potential surface, transition states of chemical transformation, and oscillatory reaction pathways. Subtle conformational changes play a crucial role in enzyme functions, and these protein conformations are highly dynamic rather than being static. Using only a static structural characterization, from an ensemble-averaged measurement at equilibrium is often inadequate in predicting dynamic conformations and understanding correlated enzyme functions in real time involving in nonequilibrium, multiple-step, multiple-conformation complex chemical interactions and transformations.

Single-molecule assays have revealed static [1–5] and dynamic [3–5] disorders in enzymatic reactions by probing co-enzyme redox state turnovers [3] and enzymatic reaction product formation in real time [4, 5]. Static and dynamic disorders [6–10] are, respectively, the static rate inhomogeneities between molecules and the dynamic rate fluctuations for individual molecules. Dynamic disorder, which is not distinguishable from static disorder in an ensemble-averaged measurement, has been attributed to protein conformational fluctuations [3–5, 11]. The protein conformational motions at the enzyme active site, which include enzyme–substrate complex formation, enzymatic reaction turnovers, and product releasing, are mostly responsible for the inhomogeneities in enzymatic reactions [3–5]. Consequently, direct observations of conformational changes along enzymatic reaction coordinates are often crucial for understanding inhomogeneities in enzymatic reaction systems [12].

We have applied single-molecule spectroscopy and imaging to study complex enzymatic reaction dynamics and the enzyme conformational changes, focusing on the T4 lysozyme enzymatic hydrolyzation of the polysaccharide walls of *Escherichia coli B* cells. By attaching a donor–acceptor pair of dye molecules site-specifically to noninterfering sites on the enzyme, we were able to measure the hinge-bending conformational motions of the active enzyme by monitoring the donor–acceptor emission intensity as a function of time. We have also explored a combined approach, applying molecular dynamics (MD) simulation and a random-walk model based on the single-molecule experimental data. Using this approach, we analyzed enzyme–substrate complex formation dynamics to reveal (1) multiple intermediate conformational states, (2) oscillatory conformational motions, and (3) a conformational memory effect in the chemical reaction process [13]. Moving forward to study enzymatic dynamics and enzyme conformational dynamics in living cells, we have developed a single-molecule enzyme delivery approach to place an enzyme specifically to an enzymatic reaction site on a cell membrane.

24.1 T4 Lysozyme Conformational Changes Under Enzymatic Reaction Turnovers

The wild-type T4 lysozyme has two domains connected by an α-helix (Fig. 24.1a) [14–17] that adopts different conformations as the two domains undergo a hinge-bending motion [12, 14–17]. Although there have been no direct observations of how much the active site opens to initiate the formation of an enzyme–substrate complex under physiological conditions, literature shows that the scale of the motion can be between 4 and 8 Å [12, 14–17]. The difficulties for ensemble-averaged and static structure measurements are that the dynamic enzyme–substrate intermediate states are extremely difficult, if not impossible, to measure. So far, a sufficient understanding of dynamic domain hinge-bending motions of the T4 lysozyme has not been obtained from direct measurements of structure prior to this single-molecule experiment. The domain motion of T4 lysozyme is rather complex and contains motions besides hinge-bending. It is reasonable to assume that the hinge-bending motion in nature involves multiple coupled nuclear coordinates that can be projected to a nuclear coordinate associated with the α-helix. Based on our study [12] and a previous [18] MD simulation of wild-type T4 lysozyme in solution, the distance change between the two dye-tethered cysteine residues is from 30.5 to 35 Å, i.e., the donor–acceptor distance change is about 4.5 Å.

24.2 Probing Protein Conformational Changes by Intramolecular Single-Pair FRET

By attaching a donor–acceptor TMR/Texas Red pair of dye molecules site-specifically to noninterfering sites on the enzyme (Fig. 24.1a) [12], we were able to measure the hinge-bending motions of the enzyme by monitoring the donor–acceptor emission intensities (Fig. 24.1b) as a function of time. We estimated the Förster distance R_0 [19] of a TMR/Texas Red pair to be about

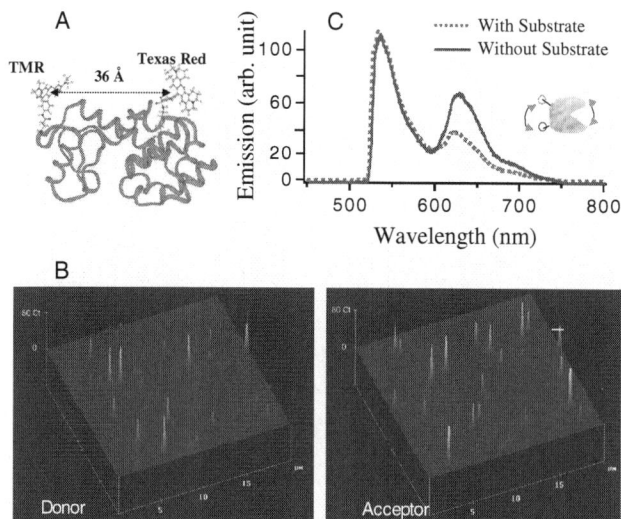

Fig. 24.1. Single-molecule recording of T4 lysozyme conformational motions and enzymatic reaction turnovers of hydrolysis of an *E. coli B* cell wall in real time. **(a)** Crystal structure of wild-type T4 lysozyme. The protein was labeled with Texas Red maleimide and tetramethylrhodamine iodoacetamide by thiolation to Cys-54 and Cys-97. The significant advantage of our site-specific covalent dye labeling is that the attached donor–acceptor pair can sense the relative motion of the two domains in T4 lysozyme without perturbation of the enzymatic activity [12]. **(b)** Fluorescence image (20 × 20 μm) of single-T4 lysozyme molecules tethered to the hydrocarbon-modified glass surface of a cover slip under pH 7.2 buffer aqueous solution. The fluorescence emission was split by a dichroic beamsplitter (595-nm long-pass). The donor (*left panel*) and acceptor (*right panel*) emissions were detected separately by a pair of avalanche photodiode detectors after passing through a 570-nm band-pass filter (20-nm bandwidth) and a 615-nm long-pass filter, respectively. The two images were taken from the same area of a sample with an inverted fluorescence microscope by raster-scanning the sample with a focused laser beam of 100 nW at 532 nm. Each individual peak is attributed to a single T4 lysozyme molecule. The intensity variation among the molecules is predominately due to spFRET. **(c)** The fluorescence spectra of Alexa 488/Alexa 594 labeled T4 lysozyme mutant (E11A) excited at 488 nm. The blue and red lines are the fluorescence spectra of the enzyme in solution without and with substrate *E. coli B* cell walls present, respectively. The two spectra are normalized to the same maximum point, to aid the view

50 ± 5 Å. Although the change of the acceptor fluorescence intensity is less than 10%, the lengths of the two covalent linking groups of the donor and acceptor dipoles, which have an actual donor–acceptor distance change of 5.5 Å [12], can cause a change of 2 ∼ 3 times in donor fluorescence intensity [12, 19].

24.3 Ensemble-Averaged and Single-Molecule Control Experiments

To demonstrate the feasibility of applying FRET to probe the conformational changes of T4 lysozyme proteins under binding and unbinding hinge-bending motions, we measured the ensemble-averaged FRET emission spectra from the donor–acceptor-labeled T4 lysozyme, both with and without a substrate (Fig. 24.1c). Although the wild type is the best candidate for such a measurement, the enzymatic reaction is too fast even for the concentration of the dye-labeled enzyme at the detection limit of a fluorometer, and the substrate concentration cannot be maintained at constant level within seconds of the spectrum-collection time. Consequently, we used the mutant E11A instead, which has binding and unbinding activity but does not catalyze the hydrolysis of the substrate. Figure 24.1c shows the emission spectra of the donor–acceptor labeled proteins in a buffer solution without (solid line) and with (dashed line) the addition of substrate *E. coli B* cell walls. The ratio of spectral intensities of the donor emission vs. acceptor emission increases on attaching to the substrate. Modeling the expected motions during opening of the enzyme active-site region shows that the donor–acceptor distance will increase, decreasing the efficiency of donor–acceptor energy transfer and yielding the observed increase in the donor–acceptor spectral intensity ratio.

The contribution to the donor–acceptor spectral intensity change of the ensemble-averaged measurement has to include additional factors besides FRET. The apparent donor–acceptor intensity ratio change in this ensemble-averaged measurement may not be equal to the ratio change in the single-molecule FRET measurement. The T4 lysozyme mutant E11A sample contains donor–donor labeled product, for which the distance change is associated with the self-quenching effect [12]. The distance increase between the two donors reduces the self-quenching and consequently increases the donor fluorescence intensity. The fraction of time of an "open" state during the enzymatic reaction has also an important influence on the donor–acceptor intensity ratio change. This donor–donor self-quenching effect enhances the measured donor–acceptor emission intensity ratio difference with and without the substrate (Fig. 24.1c). Nevertheless, the ensemble fluorescence intensity measurement indicates the conformation change of T4 lysozyme upon binding to the substrate although it is essentially a qualitative confirmation of FRET existence under the interaction between the enzyme and substrate.

Figure 24.2a shows dual fluorescence intensity trajectories simultaneously recorded from a donor–acceptor labeled T4 lysozyme in the presence of substrate at pH 7.2. The anticorrelated fluctuations (Fig. 24.2a and b) are due to spFRET, reporting the donor–acceptor distance change associated with the protein conformational motion. Likewise, fluorescence trajectories of donor–acceptor labeled T4 lysozyme without substrates did not show anticorrelated behavior (Fig. 24.2c and d). We attribute this conformational motion to an enzymatic-related motion, most likely the open-closed hinge-bending motion

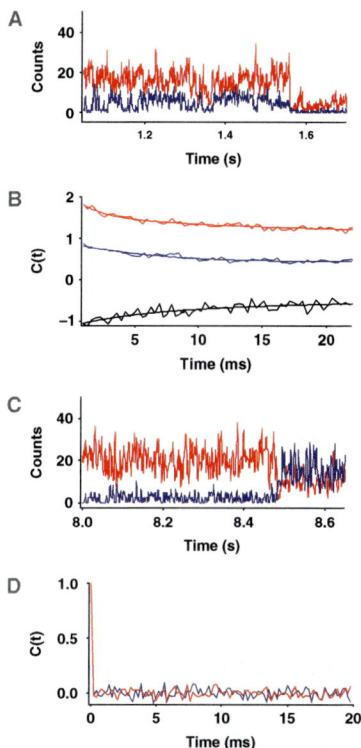

Fig. 24.2. Single-molecule recording of T4 lysozyme conformational motions and enzymatic reaction turnovers of hydrolysis of an *E. coli* B cell wall in real time. **(a)** This panel shows a pair of trajectories from a fluorescence donor tetramethyl-rhodamine (*blue*) and acceptor Texas Red (*red*) pair in a single-T4 lysozyme in the presence of *E. coli* cells of 2.5 mg/mL at pH 7.2 buffer. Anticorrelated fluctuation features are evident. **(b)** The correlation functions ($C(t)$) of donor ($\langle \Delta I_d (0) \, \Delta I_d (t) \rangle$, blue), acceptor ($\langle \Delta I_a (0) \, \Delta I_a (t) \rangle$, red), and donor–acceptor cross-correlation function ($\langle \Delta I_d (0) \, \Delta I_d (t) \rangle$, black), deduced from the single-molecule trajectories in (**a**). They are fitted with the same decay rate constant of $180 \pm 40 \, \mathrm{s}^{-1}$. A long decay component of $10 \pm 2 \, \mathrm{s}^{-1}$ is also evident in each autocorrelation function. The first data point (not shown) of each correlation function contains the contribution from the measurement noise and fluctuations faster than the time resolution. The correlation functions are normalized, and the $\langle \Delta I_a (0) \, \Delta I_a (t) \rangle$ is presented with a shift on the y axis to enhance the view. **(c)** A pair of fluorescence trajectories from a donor (*blue*) and acceptor (*red*) pair in a T4 lysozyme protein without substrates present. The acceptor was photo-bleached at about 8.5 s. **(d)** The correlation functions ($C(t)$) of donor ($\langle \Delta I_d (0) \, \Delta I_d (t) \rangle$, *blue*), acceptor ($\langle \Delta I_a (0) \, \Delta I_a (t) \rangle$, *red*) derived from the trajectories in (**c**). The autocorrelation function only shows a spike at $t = 0$ and drops to zero at $t > 0$, which indicates that only uncorrelated measurement noise and fluctuation faster than the time resolution recorded (Adapted with permission from [12]. Copyright 2003 American Chemical Society)

of the T4 lysozyme under enzymatic reaction conditions. This attribution is consistent with the results of the ensemble-averaged FRET measurements made with and without substrate (Fig. 24.1c). Further evidence that we are measuring the hinge-bending motion comes from evaluating autocorrelation functions calculated from the single-molecule trajectories when changing the laser excitation intensity and the physiological conditions of pH and substrate concentration.

The fluorescence intensity trajectories of the donor $(I_d(t))$ and acceptor $(I_a(t))$ give autocorrelation times (Fig. 24.2b) indistinguishable from fitting an exponential decay to the autocorrelation functions, $\langle \Delta I_d (0) \Delta I_d (t) \rangle$ and $\langle \Delta I_a (0) \Delta I_a (t) \rangle$, where $\Delta I_d(t)$ is $I_d(t) - \langle I_d \rangle$, $\langle I_d \rangle$ is the mean intensity of the overall trajectory of a donor, and $\Delta I_a(t)$ has the same definition for an intensity trajectory of an acceptor. In contrast, the cross-correlation function between the donor and acceptor trajectories, $\langle \Delta I_d (0) \Delta I_d (t) \rangle$, is anticorrelated with the same decay time (Fig. 24.2b) which supports our assignment of anticorrelated fluctuations of the fluorescence intensities of the donor and acceptor to the spFRET process.

24.4 Single-Molecule Photon-Stamping Spectroscopy Probing spFRET and Nanosecond Anisotropy

Our further control experiment proved that the tethered enzyme can freely rotate and has a minimum perturbation from the hydrocarbon-modified glass surface. We developed and applied a new approach using a multi-channel photon time-stamping TCSPC apparatus, based on techniques previously reported [20, 21]. It combines the advantages of both the sub-nanosecond time resolution of time-correlated single-photon counting (TCSPC) and single-molecule time trajectory recording [22]. For each recorded photon, the detector identification, arrival time, and the delay time between laser excitation pulse and detected photon in nanoseconds or picoseconds were obtained (Fig. 24.3a). The photon time-stamping TCSPC data contain nonscrambled and detailed information, adding a delay-time trajectory (the consecutive delay times between the excitation pulse and each emitted photon) and an arrival-time trajectory (the consecutive arrival time of each detected photon). The data can be used to analyze the single-molecule dynamics from nanosecond to second time-scales (Fig. 24.3a).

Using this photon-stamping technique, we were able to study single-molecule FRET and anisotropy dynamics (Fig. 24.3b) over a wide range of time-scales, both the nanosecond fast dynamics and the sub-millisecond or longer slow dynamics [22]. In our single-molecule enzymatic reaction study, the dye-labeled T4 lysozyme proteins were covalently tethered on a hydrocarbon-modified glass surface (Fig. 24.1b and c) [12,22]. Measuring the single-molecule anisotropy, we found that the tethered T4 lysozyme proteins were fully mobile and that there was no other perturbation on the activity other than the spatial

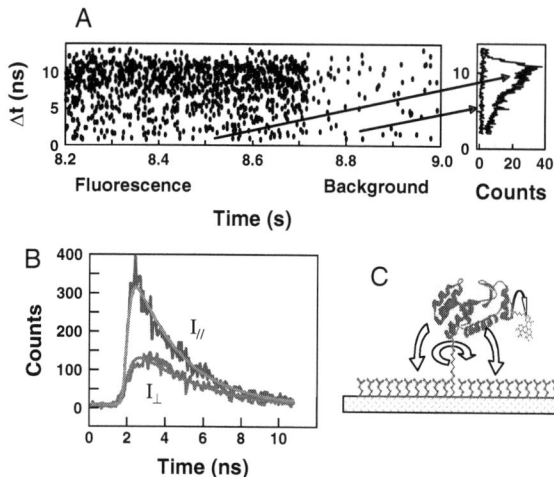

Fig. 24.3. (a) Schematic of a single-molecule confocal microscope with two-channel photon time-stamping TCSPC. It allows us to study the fluorescence intensity, lifetime, and anisotropy of a single molecule simultaneously by recording the arrival time and delay time of each fluorescence photon. The top-left plot is an example of the raw data of the photon time-stamping TCSPC of a detector channel. Each dot corresponds to a photon detected plotted by its arrival time (t) and delay time (Δt) (raw output from TAC in reverse timing). The fluorescence intensity trajectory (not shown) can be calculated from the histogram of arrival time (t) with a given time-bin resolution. The molecule was photo-bleached at 8.71 s. The nanosecond fluorescence decay curves (top-right plot) are the histograms of the delay time of the fluorescence photons ($t < 8.1$ s) and background photons ($t > 8.1$ s). (b) The parallel and perpendicular fluorescence decay and single exponential fits of a single T4 lysozyme/Alexa 488 molecule covalently linked to surface. The decay curves are integrated from all detected photons of parallel and perpendicular channels before photo-bleaching. (c) A cartoon shows that the T4lysozyme protein tethered to hydrocarbon modified glass surface through bi-function linker. The enzyme can freely move in buffer solution and its enzymatic activity is not perturbed by the surface

confinement due to the tethering to the surface [12, 22]. The tethered proteins stayed predominately in solution, rather than being fixed on a modified surface (Fig. 24.3c).

24.5 Probing Single-Molecule Conformational Motion Dynamics under Enzymatic Reaction

The substrate for the T4 lysozyme enzymatic reaction is peptidoglycan from the *E. coli B* cell wall, a cross-linked polymer containing polysaccharide and amino acids. Due to the relatively lower density of the binding sites at the substrate, a hydrolysis reaction often has an "incubation time" when the T4

lysozyme interacts with the substrate at a slow rate or less efficiently. In the
single-molecule reaction, we introduced nanomolar unlabeled T4 lysozyme
into the substrate solution for the purpose of "incubation" to maintain the
normal reaction condition of the surface-tethered dye-labeled T4 lysozyme
enzyme.

Single-molecule spFRET fluorescence trajectories contain detailed information about the conformational motion associated with the enzymatic
turnovers. The upper panel in Fig. 24.4 shows an expanded portion of a
trajectory (middle panel) recorded from donor fluorescence of a single-pair
donor–acceptor labeled protein with substrate present. By comparison, the
lower panel shows a portion of a donor-fluorescence trajectory recorded
from a donor-only labeled T4 lysozyme under the same conditions. The

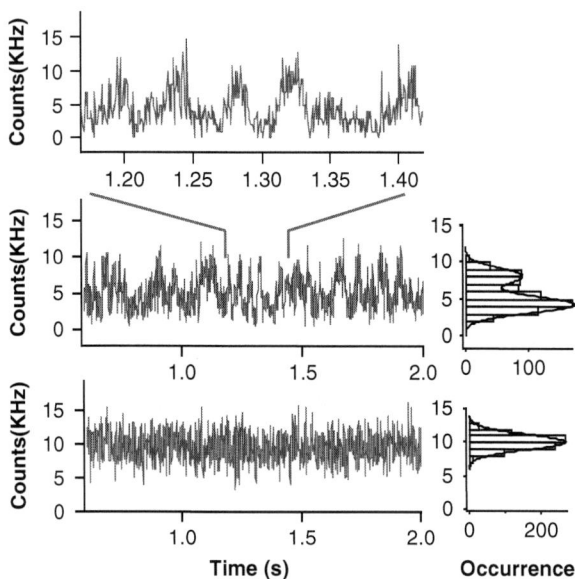

Fig. 24.4. Simultaneous probing of a single T4 lysozyme enzymatic reaction
turnover trajectory with correlated hinge-bending conformational motions of the
enzyme under hydrolysis of polysaccharide of a cell wall. The data in the three panels were recorded in 0.65 ms per channel at the same enzymatic reaction condition.
The upper panel shows an expanded portion of a trajectory (middle panel) from
donor fluorescence of a donor–acceptor labeled single-T4 lysozyme. Intensity wig-
gles in the trajectory are evident beyond the measurement shot noise. The lower
panel shows a portion of a trajectory recorded from a donor-alone-labeled enzyme.
The fluorescence intensity distributions derived from the two trajectories are shown
in the insets of the middle and lower panels. The solid lines are fit using bimodal
and Gaussian functions, respectively. The T4 lysozyme was covalently linked to a
hydrocarbon-modified glass cover slip by the bi-functional linker SIAXX (Molecular Probes, Inc.) (Adapted with permission from [12]. Copyright 2003 American
Chemical Society)

large-amplitude, lower-frequency wiggling of the donor fluorescence intensity in the upper panel is largely absent from the trajectory in the lower panel. The inset in Fig. 24.4 (middle panel) shows a bimodal fluorescence intensity distribution that reflects the open and closed conformational states of a T4 lysozyme. By contrast, only a Poisson-shaped distribution was deduced from the trajectory of a donor-alone labeled single enzyme molecule. This indicates that only fast fluctuations and uncorrelated noise were recorded, since the donor dye alone is not sensitive to the enzyme open–close hinge-bending motions (Fig. 24.4, lower panel).

We have attributed each wiggle to a hinge-bending motion involved in either an enzymatic reaction or a nonproductive binding and releasing of the substrate [12]. The donor fluorescence intensity increases as the active site opens due to substrate inserting and decreases as the active site closes to form an active enzyme–substrate complex. This attribution is consistent with the control results of the ensemble-averaged FRET measurements (Fig. 24.1c) and the single-molecule FRET intensity fluctuation dynamics analysis (Fig. 24.2) made with and without a substrate [12]. Further evidence that we are measuring the hinge-bending motion comes from evaluating autocorrelation functions calculated from the single-molecule donor and acceptor trajectories when changing the laser excitation intensity, the pH, and the substrate concentration. We did not observe a dependence of fluctuation correlation time on the excitation rate, indicating that the fluctuations were spontaneous rather than laser-driven. However, we did observe autocorrelation rate constants that differed by a factor of 2 over the pH range from 7.2 to 6.0 [12]. This twofold decrease in decay rate constant is consistent with the enzymatic activity decrease measured by ensemble-averaged assays at pH 7.2 and 6.0 [23].

From these simultaneous observations of protein conformational changes and enzymatic reactions in real time, we propose a mechanism similar to the Mechalis–Menton mechanism (24.1), where E, S, and P represent the enzyme, substrate, and product, respectively, and ES and ES* represent the enzymatic reaction complex in nonactivated and activated states, respectively:

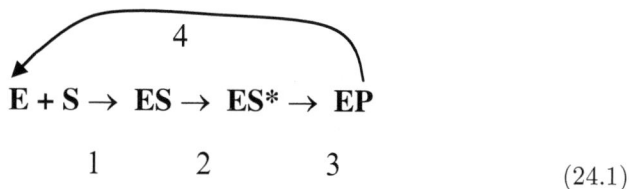

$$
\underset{1\qquad\quad 2\qquad\quad 3}{E + S \rightarrow ES \rightarrow ES^* \rightarrow EP} \quad \overset{4}{\curvearrowleft}
$$

$$(24.1)$$

The enzyme and polysaccharide chains engage in a nonspecific interaction through diffusion to form the nonactivated complex state, ES. Then, the active-site cleft binds to a six-glycoside segment of the polysaccharide chain to form specific hydrogen bonds in the activated complex state, ES* [17]. EP represents the enzyme–product complex immediately after completion of the hydrolysis reaction. The subsequent release of product (step 4 in (24.1)) and the enzyme searching for the next reactive site in the substrate complete the enzymatic turnover cycle.

24.6 Mechanistic Understanding of the Conformational Dynamics in T4 Lysozyme Enzymatic Reaction

Based on our single-molecule experimental trajectory data (Figs. 24.4 and 24.5), the donor–acceptor distance and the donor fluorescence intensity increase when the active site opens up to form a nonspecific binding complex (ES) with the substrate, corresponding to the process of $E + S \rightarrow ES$ (Fig. 24.5b). The donor–acceptor distance and the donor fluorescence intensity decrease when the active site closes to form an active complex (ES*), corresponding to the process of $ES \rightarrow ES^*$ (Fig. 24.5b). There are no measurable spFRET changes, implying no significant conformational motions in the process $ES^* \rightarrow EP$ or in the product-releasing process of $EP \rightarrow E + P$. Therefore, the donor fluorescence intensity in a single-molecule time trajectory increases in step 1, decreases in step 2, and remains low in steps 3 and 4 (24.1). The hydrolysis reaction occurs in step 3 (Fig. 24.5b and c).

The formation of ES and ES* involves significant domain breathing-type hinge-bending motions along the α-helix and is probed in real time by recording single-molecule spFRET trajectories that record the formation times, t_{open}, of enzymatic intermediate ES and ES* states from the single-molecule enzymatic turnover trajectories (Fig. 24.5b and c) [12]. Figure 24.5a shows a Gaussian-shaped distribution of the open-time (t_{open}) deduced from a single-molecule trajectory. The first moment of the distribution, $\langle t_{open} \rangle = 19.5 \pm 2$ ms, corresponds to the mean time of the processes of $E+S \rightarrow ES \rightarrow ES^*$. The standard deviation of the distribution, $\sqrt{\langle \Delta t_{open}^2 \rangle} = 8.3 \pm 2$ ms, reflects the distribution bandwidth. For the individual T4 lysozyme molecules examined under the same enzymatic reaction conditions, we found that the first and second moments of the single-molecule t_{open} distributions are homogeneous, within the error bars. The hinge-bending motion allows sufficient structural flexibility for the enzyme to optimize its domain conformation: the donor fluorescence essentially reaches the same intensity in each turnover, reflecting the domain conformation reoccurrence. The distribution with a defined first moment and second moment shows typical oscillatory conformational motions. The nonequilibrium conformational motions in forming the active enzymatic reaction intermediate states intrinsically define a recurrence of the essentially similar potential surface for the enzymatic reaction to occur, which represents a memory effect in the enzymatic reaction conformational dynamics [12, 41, 42].

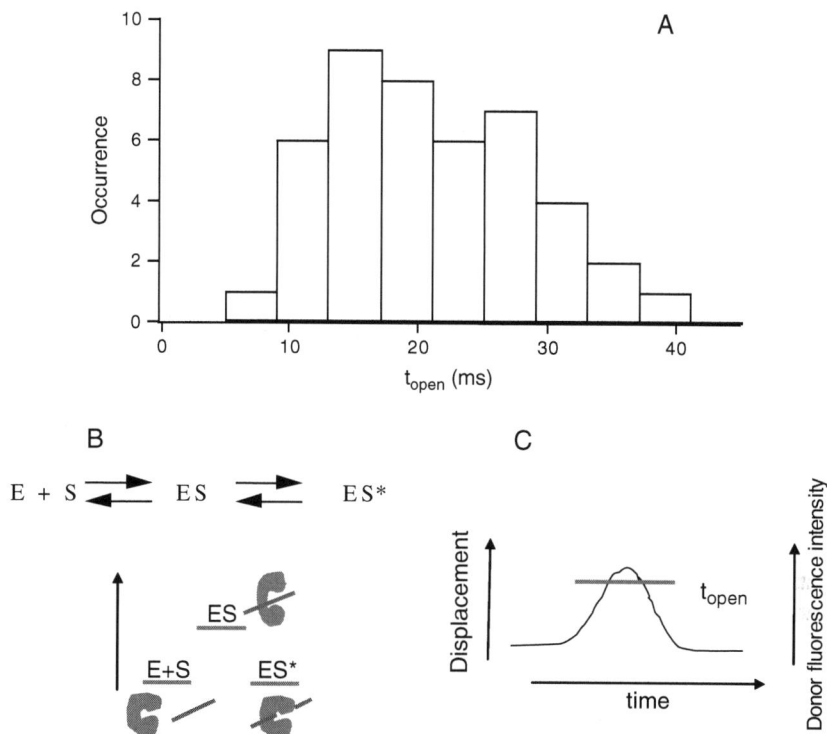

Fig. 24.5. (a) Active-complex, ES*, formation time (t_{open}) distribution deduced from a single T4 lysozyme fluorescence trajectory under enzymatic reaction. The t_{open} is the duration time of each wiggling of the intensity trajectory above a threshold. The threshold is determined by the 50% of the bimodal intensity distribution (see Fig. 24.4, inset). The mean open time, $\langle t_{open}\rangle$, is 19.5 ± 2 ms and the standard deviation of the open time is 8.3 ± 2 ms. (b) The measured t_{open} reflects the time for the formation of active enzyme–substrate complex (ES*). The enzyme active site opens up to take substrate in and form the nonspecific binding complex (ES) and close down to form the active complex (ES*). The single-molecule open–close hinge-bending motions are measured by single-molecule FRET spectroscopy and recorded in the FRET time trajectories.(c) An illustration of the t_{open} measured from the donor intensity changes

24.7 Molecular Dynamics (MD) Simulation of the Active-Site Conformational Motions in Forming an Active Enzyme–Substrate Complex of the Enzymatic Reaction

MD simulation is intrinsically a single-molecule computational experiment of obtaining picosecond to nanosecond short-time trajectories as "snapshots" of the single-molecule experimental trajectories, thus being complementary to

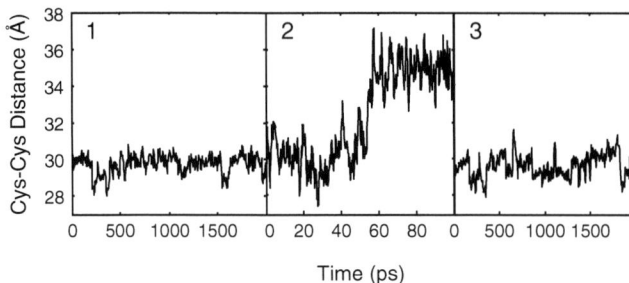

Fig. 24.6. Molecular dynamics (MD) simulation of T4 lysozyme hinge-bending motion. Three conformation trajectories of T4 lysozyme in solution at room temperature are presented: free enzyme without substrate (E), active enzyme–polysaccharide complex (ES*), and nonspecific binding of polysaccharide (E + S → ES) (*left panel*). The time trajectory of distance between two −SH of Cys-54 and Cys-97 of free enzyme (*middle panel*). The time trajectory of nonactive enzyme–substrate formation dynamics as the polysaccharide moves into the active site of the T4 lysozyme (*right panel*). The time trajectory of the distance between two −SH of Cys-54 and Cys-97 of T4 lysozyme–polysaccharide active complex. Enzyme structure (E) was taken from the 1.7-Å X-ray crystallographic structure (PDB entry 3LZM) and included 152 water molecules. The system was placed in a periodic cube of 73.34 Å per side and filled with 11,948 SPCE water molecules, including a section of the substrate (ES*). A six-unit oligosaccharide consisting of alternating N-acetylmuramic acid (NAM) and N-acetylglucosamine (NAG) was positioned in the active site with the aid of superimposing the lysozyme mutant adducted with substrate cleaved from the cell wall of *E. coli* (PDB entry 148 L) (Adapted with permission from [12]. Copyright 2003 American Chemical Society)

single-molecule spectroscopy. We applied MD simulation to further explore the conformations of the intermediate states and the essential conformational change coordinates of the enzymatic reaction in combination with the spFRET data [12]. Using MD simulation, we sampled the T4 lysozyme conformation states of the enzyme alone (E) (Fig. 24.6, left panel), before its interaction with substrate, the substrate–enzyme complex formation (E + S → ES) (Fig. 24.6, middle panel), the enzyme with the polysaccharide substrate in the active site (*ES**) (Fig. 24.6, right panel), and the product release (*EP* → E + P). Our MD simulation showed a narrow gateway to the active site, blocked by an arginine–glutamic acid salt bridge, which has to be opened before the substrate molecule can diffuse into the active site to form the substrate–enzyme complex (E + S → ES → ES*). When the substrate goes through the gateway to enter the active site, forming the nonactive complex (ES) (Fig. 24.6), the two domains of the enzyme exhibit a breathing-type hinge-bending motion along the α-helix, causing a 4.5 ± 0.2 Å distance increase between the two cysteines. This motion propagates to a 5.5 ± 0.2 Å increase of the average distance between the donor–acceptor dipoles. This result is consistent with that of the donor and acceptor emission intensity changes measured in the spFRET trajectories, in which more than two times of the donor emission intensities can be observed (Fig. 24.4).

24.8 The Conformational Change Energy Landscape

The Gaussian-shaped distribution of the t_{open} and the ramping changes of intensity at a millisecond time-scale in the single-molecule fluorescence time trajectories suggest that the protein hinge-bending conformational changes involve multiple intermediate conformational states and memory effect [3–5, 12, 41, 42]. The memory effect is intrinsic, due to which the dominant nuclear coordinate is essentially the same for the bending motion that opens and closes the active site and, most likely, is associated with the α-helix hinge of the T4 lysozyme and relatively identical substrate–enzyme electrostatic force of the induced conformational changes. The results of the MD simulation suggest that the dominant driving force for $E + S \rightarrow ES$ is the positive surface charge of the enzyme from surface amino acid residues (arginine and lysine) interacting with the negatively charged polysaccharide substrate. The driving force for $ES \rightarrow ES^*$ includes the formation of six hydrogen bonds in the active site of ES^*.

We model the hinge-bending motion associated with interactions between the enzyme and substrate as a classical particle one-dimensional multiple-step random walk in the presence of a force field [12, 24], $\{n(t)\}$, where $n(t)$ is the step index (Fig. 24.7). The position distribution density function $P_n(t)$ can be calculated [12] by

$$\frac{dP_n(t)}{dt} = k_f P_{n-1}(t) + k_b P_{n+1}(t) - (k_f + k_b)P_n(t), \qquad (24.2)$$

where k_f and k_b are the forward and backward step rate constants, respectively. The driving force of the electrostatic interaction between the T4 lysozyme and the substrate [12] would tend to make $k_f > k_b$.

The first moment $\langle n(t) \rangle = (k_f - k_b)t$ and standard deviation $\sqrt{\langle \Delta n(t)^2 \rangle} = \sqrt{(k_f + k_b)t}$ reflect the drifting and the spreading of the probability $P_n(t)$, respectively. Assuming the random-walk step-size of L and a drifting distance of $X_n = nL$, we have

$$\langle DX_n(t)^2 \rangle = L^2(k_f + k_b)t \text{ and } \langle X_n(t) \rangle = L(k_f + k_b)t. \qquad (24.3)$$

Considering the one-dimensional random walk and approximate Gaussian-shape of $P_n(t)$ [12], we have

$$D = \langle DX_n(t)^2 \rangle / 2t = L^2(k_f + k_b)/2 \qquad (24.4)$$

and

$$\langle v \rangle = \langle X_n(t) \rangle / t = L(k_f - k_b) \qquad (24.5)$$

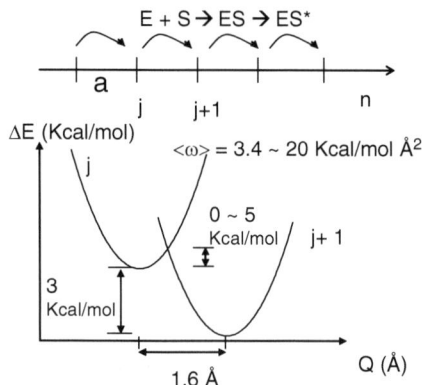

Fig. 24.7. Based on the results of the single-molecule spectroscopy measurements, an attempt to estimate the energy potential surface of T4 lysozyme–substrate complex formation process. The conformational change dynamics involving multiple intermediate states is analyzed based on a one-dimensional random walk model coupled with the parameters obtained from our single-molecule experimental spectroscopy and MD simulation. The conformational motion in each enzymatic turnover cycle involves six intermediate states based on the model analysis. Since the total energy change between E and ES* states is about $-18\,$kcal/mol, assuming six intermediate states would result in an average energy difference of $3\,$kcal/mol for each associated conformational state along the α-helix coordinate during the hinge-bending motion. We postulate that the activation energy associated with the k_f of the forward step is about $0 \sim 5\,$kcal/mol, considering the slow forward rate and the entropy decrease in the complex formation process. It is reasonable to assume the energy potential surface along the conformational change nuclear coordinate to be parabolic for each intermediate state. Therefore, the averaged force constant of the potential surface for each intermediate state is calculated to be $3.4 \sim 20\,$kcal/mol/Å^2 (Adapted with permission from [12]. Copyright 2003 American Chemical Society)

where $\langle v \rangle$ is the mean drifting velocity of the conformational change along the α-helix coordinate. With the approximation of

$$\frac{\sqrt{\langle \Delta X_N(t)^2 \rangle}}{\langle X_N(t) \rangle} = \frac{\sqrt{\langle \Delta t_{\text{open}}^2 \rangle}}{\langle t_{\text{open}} \rangle} \tag{24.6}$$

at a long-time limit, where N is the index of the final state, we have

$$D = \frac{\left(\sqrt{\langle \Delta t_{\text{open}}^2 \rangle} \langle X_N(t) \rangle \right)^2}{2 \langle t_{\text{open}} \rangle^3}, \tag{24.7}$$

where D is the diffusion coefficient.

The total drifting distance of the conformational open–close motion in one enzymatic reaction turnover, $\langle X_N(t) \rangle$, is about $9\,$Å, based on our MD simulation (Fig. 24.6). The mean open-time, $\langle t_{\text{open}} \rangle$ and the standard deviation of the open-time distribution, $\sqrt{\langle\Delta t_{\text{open}}{}^2\rangle}$, were measured to be 19.5 ± 2 and 8.3 ± 2 ms, respectively. Therefore, the mean drifting velocity, $\langle v\rangle$, of the conformational change in $E + S \rightarrow ES \rightarrow ES^*$ is $\langle v\rangle = 4.6 \times 10^{-6}$ cm/s, and the diffusion coefficient is, therefore, $D = 3.8 \times 10^{-14}$ cm^2/s.

We further characterized the energy landscape of the hinge-bending conformational change dynamics by estimating the minimum number of the intermediate conformational states involved in the complex formation and by calculating the averaged rates of forming these conformational states. From (24.4) and (24.5), we have

$$D/\langle v\rangle = L\left[(k_{\text{f}} + k_{\text{b}})/2(k_{\text{f}} - k_{\text{b}})\right] \text{ and } L = 2D/\langle v\rangle = 1.6\,\text{Å, when } k_{\text{b}} \rightarrow 0 \tag{24.8}$$

Therefore, the minimum number of conformations is $m = \langle X_N(t)\rangle/L = 5.6$. The average rate for each step is $m/\langle t_{\text{open}}\rangle = 280\,\text{s}^{-1}$. This result of $m > 2$ at the limit of $k_{\text{b}} \rightarrow 0$ suggests that there are more than two conformational intermediate states in addition to ES and ES*. With the assumption of $k_{\text{b}} \rightarrow 0$, the friction coefficient can be estimated using the Einstein relationship, $\xi = kT/D$, giving $\xi = 1.1\,\text{erg s/cm}^2$. The energy consumed by friction in the drifting process is $E_{\text{f}} = \xi\langle v\rangle\langle X_N(t)\rangle = kT\langle v\rangle\langle X_N(t)\rangle/D = 4.5 \times 10^{-20}$ J. From MD simulation of the E state and the ES* state energies, our best estimate of the total energy change between the two states is -18 kcal/mol (-1.25×10^{-19} J per molecule) that constitutes the energy gains from electrostatic, van der Waals, and hydrogen-bond formation terms. Therefore, we estimate that 36% of the total energy change between E and ES* is spent on the friction along the reaction coordinate. Since the total energy change between the E state and the ES* state is about -18 kcal/mol, assuming six intermediate states would result in an average energy difference of 3 kcal/mol for each associated conformational state along the α-helix coordinate during the hinge-bending motion. We postulate that the activation energy associated with the k_{f} of the forward step is about $0 \sim 5$ kcal/mol, considering the slow forward rate and the entropy decrease in the complex formation process [14]. It is reasonable to assume the energy potential surface along the conformational change nuclear coordinate to be parabolic for each intermediate state. Therefore, the averaged force constant of the potential surface for each intermediate state is calculated to be $3.4 \sim 20$ kcal/mol/Å (Fig. 24.7) [12].

The information about the existence of the multiple intermediate conformational states involving the enzymatic active complex formation and a detailed characterization of the energy landscape (Fig. 24.7) of the complex formation process cannot be obtained either by only an ensemble-averaged experiment, only a single-molecule experiment, or a solely computational approach. The combined approach demonstrated here is essential to achieve the potential of both single-molecule spectroscopy and MD simulations for studies of slow enzymatic reactions and protein conformational change dynamics.

The single-molecule FRET donor–acceptor fluorescence intensity trajectories can also be analyzed by using a Generalized Langevian Equation (GLE) approach [25, 26]. In this approach, the FRET efficiency time trajectories are converted to donor–acceptor distance time trajectories and fluctuation velocity trajectories. A memory kernel and time-dependent friction, $\zeta(t)$, can then be deduced from analyzing the autocorrelation function of the velocity time trajectories and convolution fitting using the GLE analysis [25, 26]. Based on the fluctuation–dissipation Theorem, the environmental force fluctuation autocorrelation function is defined by the time dependent friction, $k_B T \zeta(t) = \langle F(0) F(t) \rangle$, where k_B is the Boltzman constant, T is the temperature, and $F(t)$ is the fluctuating force [25–27]. Analyzing the force fluctuation dynamics, we observed a slow force fluctuation at the rate of comparable to the hinge-bending open–close conformational motion rate. The force fluctuation around the enzymatic reaction conformational coordinate is not completely random, and it is most likely that the other domains of the enzyme also involve in the conformational fluctuation at a similar rate of hinge-bending motions and the overall fluctuations are coupled to the enzymatic hinge-bending motion dynamics [42].

To further analyze the memory effect and the identification of multiple conformational intermediate states involved in the enzymatic conformational motions, we have simulated the t_{open} time trajectories based on multi-step Poisson rate process model by calculating convolutional random walk step time at 3.25 ms in average [42]. Our computational simulation shows that the multiple (index number of m) consecutive Poisson rate processes give a Gaussian-like t_{open} time distribution at mean of 19 ms when $m = 6$ (Fig. 24.8a), which is consistent with our Random model analytical results discussed earlier. We construct a two-dimensional joint probability distributions, $\delta(\tau_i, \tau_{i+1})$, for the t_{open} times separated by 1 index number, i.e., t_{i+1} and t_i; a joint probability distribution of t_{open} and the adjacent t_{open} in a simulated single-molecule conformational motion time trajectory. Figure 24.8b shows the two-dimensional joint probability distribution calculated from the conformational motion time trajectory when assuming $m = 6$; i.e., assuming there are 5–6 intermediate conformational states (random walk steps) in an enzymatic reaction active complex formation cycle of $E + S \rightarrow ES \rightarrow ES^*$. The distribution $\delta(\tau_i, \tau_{i+1})$ shows a characteristic diagonal feature of memory effect in the t_{open}, reflecting that a long t_{open} time tends to be followed by a long one and a short t_{open} time tends to be followed by a short one. Furthermore, Fig. 24.8b also revealed a typical oscillatory conformational motion dynamics at nonequilibrium that there are clearly defined first and second moments of the conformational motion times [43, 44].

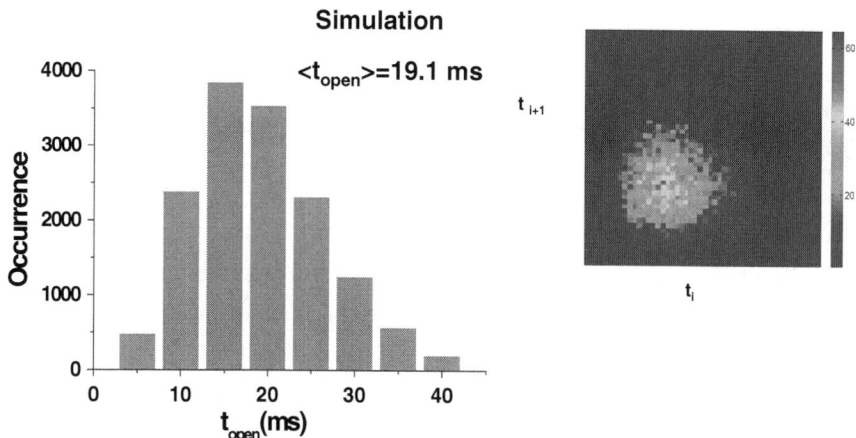

Fig. 24.8. Computational simulation analysis of conformational dynamics in T4 lysozyme enzymatic reaction. **(a)** Histograms of t_{open} calculated from a simulated single-molecule conformational change trajectory, assuming a multiple consecutive Poisson rate processes representing multiple ramdom walk steps. **(b)** Two-dimensional joint probability distributions $\delta\ (\tau_i,\ \tau_{i+1})$ of adjacent pair t_{open} times. The distribution $\delta(\tau_i,\ \tau_{i+1})$ shows clearly a characteristic diagonal feature of memory effect in the t_{open}, reflecting that a long t_{open} time tends to be followed by a long one and a short t_{open} time tends to be followed by a short one

24.9 Placing Single-Molecule T4 Lysozyme Enzymes on a Bacterial Cell Surface: Toward Studying Single Molecule Enzymatic Reaction Dynamics in Living Cells

To directly probe the T4 lysozyme enzymatic reaction on a living bacterial cell wall, we have developed and used a combined single-molecule placement approach and spectroscopy analyses (Fig. 24.9) [28]. Placing an FRET donor–acceptor dye-labeled single T4 lysozyme molecule on a targeted bacterial cell wall by using a nano-liter hydrodynamic injection approach (Fig. 24.9a and b), we monitored single-molecule rotational motions during binding, attachment to, and dissociation from the cell wall by tracing single-molecule fluorescence intensity time trajectories and polarization (Fig. 24.9c). Applying single-molecule fluorescence polarization measurements to characterize binding and motions of the T4 lysozyme molecules, we observed that the motions of wild-type and mutant T4 lysozyme proteins are essentially the same whether under enzymatic reaction or not. The changing of the fluorescence polarization suggests that the motions of the T4 lysozyme are associated with orientational rotations. This observation also suggests that the T4 lysozyme binding–unbinding motions on cell walls involve a complex mechanism beyond a single-step first-order rate process.

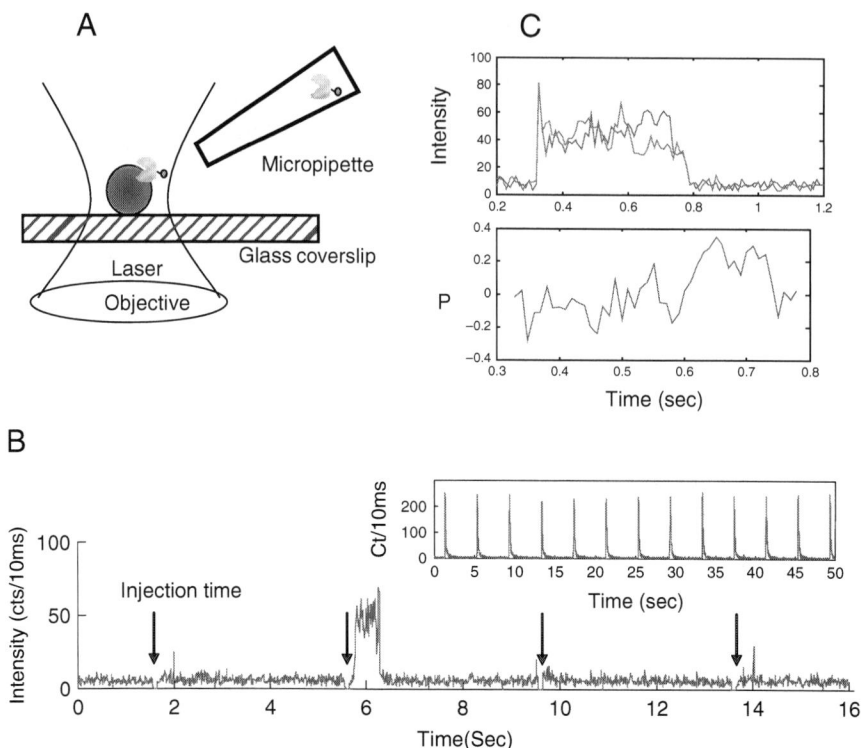

Fig. 24.9. Placing Single-Molecule T4 Lysozyme Enzymes on a Bacterial Cell Surface. (a) Experiment setup. The *E. coli* cells were immobilized on a clean coverslip. The excitation laser was focused on the cell. A glass micropipette filled with enzyme solution was placed near the cell on the focal point by a micromanipulator. The solution in the micropipette was injected by a picoliter injector. We used a hydrodynamic nanoliter liquid injection technique using a micropipette. A picoliter injector (Harvard/Medical Systems PLI-100) was used to inject a controlled volume of solution to a cell wall substrate on a glass surface. The tip was placed $2 \sim 3\,\mu$m from the laser focal point where a cell wall piece was imaged. (b) A segment of a typical fluorescence time trajectory of injecting of 10^{-8} M WT-Alexa 488 to a cell wall. Injection occurred every 4 s. We attribute the fluorescence intensity peaks to single molecules because they are quantized and drop to the background level in one step and because their intensity levels are similar to the intensity of immobilized single T4 lysozymes on a glass surface. Furthermore, the peaks are more likely to be observed immediately after injection as opposed to a random time after injection. (*Inset*) The fluorescence time trajectory of injecting 10^{-5} M Alexa 488 into the laser focal point for a duration of 20 ms. The counting dwell time of the trajectory was 10 ms. (c) (*upper panel*) The fluorescence intensity trajectory of the two polarization components and (*lower panel*) the polarization trajectory. Data are from one *E. coli* cell wall and many molecules placed on the cell

The major challenge in conducting studies of single-molecule complex enzymatic reactions in living cells is the autofluorescence background generated by a cell under laser excitation. Nevertheless, single-molecule spectroscopy is useful and promising for uncovering enzymatic reactions on the cell walls or membranes where the density of the fluorescent proteins is low. T4 lysozyme can attach and bind to cell walls and degradate the cell wall by catalyzing the hydrolysis of cell wall peptidoglycan. Much is still largely unknown about the complex mechanism and inhomogeneous dynamics of enzyme–cell wall binding interactions, association and dissociation, and enzyme diffusion motions in the enzymatic reaction process. Overall, it is still a mystery how the T4 lysozyme efficiently hydrolyzing a cell wall, which is a covalently bonded polymer network with a heterogeneous structure and inhomogeneous electrostatic distribution.

By controlling the injection volume and concentration of the solution, we were able to deliver one or less than one enzyme molecule to the cell wall resulted from each injection pulse. Figure 24.9b show that the fluorescence intensity jumped to a higher level once a T4 lysozyme was delivered and bound to the cell wall after the injection pulse, and that the intensity dropped back to the background level after the molecule was photo-bleached or detached from the cell wall. In our single-molecule injection imaging experiments, autofluorescence from the cell was minimal because before the injection sequence began, the cell wall was pre-photo-bleached by laser with 100-times stronger power over a period of minutes. However, most molecules in the injection pulse flow away so that they would not be able to bind to or even collide with the cell wall. Typically, the ratio of the single-molecule placing is 1 out of $3 \sim 10$ injection pulses (Fig. 24.9b) [28].

Two types of T4 lysozyme bindings to cell walls are possible: (1) nonspecific attachment and (2) chemical binding associated with the hydrolysis reaction. Events indicated by the fluorescence intensity dropping to background level are associated with T4 lysozyme diffusing away from the cell wall. When the T4 lysozyme attached to the cell wall, many enzymatic reaction turnovers likely occurred; in experiments, we have observed that the cell wall typically shrinks and eventually disappears from the imaging field of view [12].

To characterize the binding and the motions of the placed single T4 lysozyme molecules on cell walls, we used single-molecule fluorescence polarization measurements. The orientation of the single-molecule transition dipole can be probed by either linear polarized excitation or linear polarized emission. In this work, the excitation light was unpolarized. The emission was split into orthogonal (s polarized I_1 and p polarized I_2) polarizations and detected by two photon detectors [22]. The intensity trajectories probed at the two orthogonal polarizations are shown in Fig. 24.9c (upper panel). The polarization P is defined as:

$$P = \frac{I_1 - I_2}{I_1 + I_2}$$

If the rotation time is much longer than the fluorescence intensity averaging time, P can be used to determine the orientation of the transition dipole within an accessible space.

The possibility of the T4 lysozyme orientation rotation being faster than 5 ms of bin time in fluorescence intensity collection was ruled out, suggesting that T4 lysozyme did not freely rotate on the cell wall. Furthermore, the trajectory of P in Fig. 24.9c showed a slow change with time, which corresponds to a slow rotation of the single-molecule orientation. The changing of P during high levels of fluorescence intensity (Fig. 24.9c, lower panel) suggests that the motions of the T4 lysozyme are associated with orientational rotations. It is likely that within one apparent binding event, the enzyme has done multiple reactions in multiple sites on the cell wall. This observation is consistent with our findings that the dynamics of enzyme–substrate interactions that form complexes with polysaccharides in cell walls involves both the attachment of the T4 lysozyme and the binding for enzymatic reaction [12].

24.10 The Technical Limitations and Possible Improvements of Single-Molecule Enzymology Analyses

Single-molecule enzymology has been under a rapid development, which presents one of the most exciting trends of the new biophysical methodologies. A few different single-molecule assays have been demonstrated by probing: (1) single molecule products [4, 5], (2) individual electrophoresis zones of products from a single enzyme [1, 2], (3) single substrate molecules [29], (4) a fluorescent enzyme active site [3], (5) intermolecular spFRET [30], and (6) intramolecular spFRET [12]. Each approach has advantages and shortcomings. Probing single molecules of a product or substrate directly measures the individual enzymatic reaction turnover in real time. Probing the zones of product molecules obtains the overall single-molecule enzymatic reaction rates, but provides little or no information on the reaction-associated enzyme conformational changes. Probing fluorescent active-site fluorescence, however, can yield information on both single-molecule enzymatic reaction dynamics and the collective active-site conformational fluctuation dynamics. For example, probing the redox state of a co-enzyme, flavin adenine dinucleotide toggles between oxidized and reduced states in cholesterol oxidase enzymatic reaction turnovers [3]. However, such approach makes it difficult to identify and measure the specific activity-regulating conformational changes in an enzymatic reaction. Recently, it has been demonstrated that spFRET is able to obtain a detailed characterization and analysis of the conformational change dynamics and energy landscape [12]. The limitation of the spFRET approach is that the enzymatic reaction turnovers can only be assured statistically but not assigned individually, since there is a statistical probability of nonproductive conformational motions in the reactive nuclear coordinates. Using a single-molecule fluorescence spectroscopic measurement, it is still extremely hard to probe simultaneously the single-molecule enzymatic reaction turnovers, the product generation or substrate consumption, and the specific conformational changes. There is typically a trade-off in probing one critical parameter. Nevertheless, enzymatic reactions typically involve an inhomogeneous environment and complex mechanism; it is often crucial to probe directly the specific conformational changes involved in an enzymatic reaction. For example, the hinge-bending motions associated with the T4 lysozyme hydrolysis enzymatic reaction are critical, but cannot be studied by either conventional ensemble-averaged measurements or alternative approaches, other than single-molecule intramolecular spFRET [12]. However, the approach described here has the advantage of characterizing the enzyme-activity-related conformational fluctuation without directly probing the product release. A combined approach to probe both parameters is still a great challenge for single-molecule enzymology due to the congestion of the fluorescence of the chromospheres. In recent years, using three FRET probe dyes to measure multiple dimensional conformational dynamics has made progresses [31–33], and a careful selection of the dye molecules may give a chance to circumvent the fluorescence spectral congestion problem.

Typically, single-molecule spectroscopy studies on enzymatic reactions and associated conformational change dynamics have time resolutions longer than sub-milliseconds. It is not only desirable, but also critical to probe the conformational change dynamics at an ultra-fast time-scale, nanoseconds or even picoseconds, because many important protein conformational motions are at nanosecond time-scale [34, 35]. Moreover, inhomogeneous protein conformational dynamics often show a power-law behavior extended in a wide time range from seconds to nanoseconds. Our group [22] and others [20, 21] have developed photon-stamping single-molecule fluorescence detection techniques that can measure single-molecule ultra-fast fluorescence anisotropy dynamics [21, 22]. The single-molecule nanosecond anisotropy is readily applicable to probing protein conformational dynamics by tethering bi-functional dyes [36, 37] or tetradentately-attached dyes [22, 38–40] to specific sites or domains of the proteins. For example, a mutant T4 lysozyme can be used with two cysteine residues at the same distance as the bi-functional dye can be labeled. A bi-functional fluorescent probe can be used to anchor it on the protein surface across two cysteine residues so that the dye molecule self-wobbling motions are fixed and only the protein matrix motion is probed. Therefore, the inter-domain and intra-domain conformational dynamics of a protein can be studied by probing the fluorescent dipole rotational motion without the complication of convoluted dye motions. Noticeably, the nanosecond single-molecule dynamics studied by single-molecule nanosecond anisotropy can be directly comparable and correlated with single-molecule dynamics from an MD simulation.

24.11 Concluding Remarks

In biological systems, enzymatic reactions typically involve multiple kinetic steps, complex molecular interactions, complex conformational changes, and confined local environments. These complexities, associated with intrinsic spatial and temporal inhomogeneities, often make a solution-phase ensemble-averaged measurement inadequate. In many systems, only the overall enzymatic reaction rates are measured without further characterization of kinetic mechanism, intermediate states, and step-specific reaction rates. Single-molecule spectroscopy, studying one molecule under a specific physiological condition at a time, is potentially a powerful and unique approach to characterize and analyze the complex enzymatic reaction dynamics and the correlated conformational change dynamics. An even more informative and powerful methodology is to combine single-molecule spectroscopy, computational MD, and theoretical modeling, an approach that provides molecular-level characterization of the enzymatic reaction dynamics, conformational dynamics, and energy landscape of specific conformational changes. It is the active complex formation processes $(E + S \rightarrow ES \rightarrow ES^*)$ that define the enzymatic reaction potential surfaces and contribute to the complexity and inhomogeneity of the enzymatic reactions. Understanding enzymatic reaction, conformational dynamics is intimately related to single-molecule studies of biomolecular interactions for the precise reason that the formation of an enzymatic reaction active complex involves biomolecular interactions. In recent years, mechanisms of protein conformation selection and induced conformational changes have been extensively explored [45–47], and it is anticipated that more single-molecule protein–protein interaction studies will also contribute to our fundamental understanding of enzymatic reaction dynamics and mechanisms. Applying the combined approaches, we have begun to obtain detailed mechanistic information about enzymatic conformational dynamics, including the intermediate enzyme–substrate complex structures and the associated energy landscape [41].

Acknowledgements

The author thanks Dehong Hu, Yu Chen, and Erich R. Vorpagel for their crucial contributions to the work discussed here; Brian Matthews and Walt Baas for providing us with T4 lysozyme proteins, the recipe for preparing the substrate, and helpful discussions; and Yuanmin Wang for computational simulation. We also acknowledge the support to our program from the Chemical Sciences Division of the Office of Basic Energy Sciences (BES) within the Office of Energy Research of the U.S. Department of Energy (DOE), The US Defense Advanced Research Projects Agency (DARPA), the Material Science Division of the US Army Research Office (ARO), National Science Foundation (NSF), National Institute of Environmental Health Sciences (NIEHS) of National Institute of Health (NIH), Pacific Northwest National Laboratory, and Bowling Green State University.

References

1. Q.F. Xue, E.S. Yeung, Nature 373, 681 (1995)
2. D.B. Craig, E.A. Arriaga, J.C.Y. Wong, H. Lu, N.J. Dovichi, J. Am. Chem. Soc. 118, 5245 (1996)
3. H.P. Lu, L.Y. Xun, X.S. Xie, Science 282, 1877 (1998); X.S. Xie, H.P. Lu, J. Biol. Chem. 274, 15967 (1999)
4. B.P. English, W. Min, A.M. van Oijen, K.T. Lee, G. Luo, H. Sun, B.J. Cherayil, S.C. Kou, X.S. Xie, Nat. Chem. Bio. 2, 87 (2006)
5. L. Edman, R. Rigler, Proc. Natl. Acad. Sci. USA 97, 8266 (2000); H. Lerch, R. Rigler, A. Mikhailov, Proc. Natl. Acad. Sci. USA 102, 10807 (2005)
6. R. Zwanzig, Accounts Chem. Res. 23, 148 (1990)
7. J. Wang, P. Wolynes, Phys. Rev. Lett. 74, 4317 (1995)
8. G.K. Schenter, H.P. Lu, X.S. Xie, J. Phys. Chem. A 103, 10477 (1999)
9. N. Agmon, J. Phys. Chem. B 104, 7830 (2000)
10. H.P. Lu, L.M. Iakoucheva, E.J. Ackerman, J. Am. Chem. Soc. 123, 9184 (2001)
11. A.M. van Oijen, P.C. Blainey, D.J. Crampton, C.C. Richardson, T. Ellenberger, X.S. Xie, Science 301, 1235 (2003)
12. Y. Chen, D. Hu, E.R. Vorpagel, H.P. Lu, J. Phys. Chem. B 107, 7947 (2003)
13. Part of the text appeared in a review article, Curr Pharm Biotech, 5, 261 (2004)
14. B.W. Matthews, Adv. Protein Chem. 46, 249 (1995)
15. X.J. Zhang, J.A. Wozniak, B.W. Matthews, J. Mol. Biol. 250, 527 (1995)
16. H.R. Faber, B.W. Matthews, Nature 348, 263 (1990)
17. R. Kuroki, L.H. Weaver, B.W. Matthews, Science 262, 2030 (1993)
18. G.E. Arnold, R.L. Ornstein, Biopolymers 41, 533 (1997)
19. S. Weiss, Science 283, 1676 (1999)
20. M. Bohmer, F. Pampaloni, M. Wahl, H. Rahn, R. Erdmann, J. Enderlein, Rev. Sci. Instrum. 72, 4145 (2001)
21. J.R. Fries, L. Brand, C. Eggeling, M. Kollner, C.A.M. Seidel, J. Phys. Chem. 102, 6601 (1998)
22. D. Hu, H.P. Lu, J. Phys. Chem. B 107, 618 (2003)
23. A. Tsugita, M. Inouye, E. Terzaghi, G. Streisinger, J. Biol. Chem. 243, 391 (1968)
24. I. Oppenheim, K.E. Shuler, G.H. Weiss, Stochastic Processes in Chemical Physics: The Master Equation (MIT, Cambridge, MA, 1977)
25. M. Vergeles, G. Szamel, J. Chem. Phys. 110, 6827 (1999)
26. J.E. Straub, M. Brokovec, B.J. Berne, J. Phys. Chem. 91, 4995 (1987)
27. D. Chandler, Introduction to Modern Statistical Mechanics (Oxford University Press, Oxford, 1987)
28. D. Hu, H.P. Lu, Biophys. J. 87, 656 (2004)
29. (a) A. Ishijima, H. Kojima, T. Funatsu, M. Tokunaga, H. Higuchi, H. Tanaka, T. Yanagida, Cell 92, 161 (1998); (b) H. Noji, R. Yasuda, M. Yoshida, K. Kinosita, Nature 386, 299 (1997)
30. T.J. Ha, A.Y. Ting, J. Liang, W.B. Caldwell, A.A. Deniz, D.S. Chemla, P.G. Schultz, S. Weiss, Proc. Natl. Acad. Sci. USA 96, 893 (1999)
31. M. Bates, B. Huang, G.T. Dempsey, X.W. Zhuang, Science 317, 1749 (2007)
32. S. Hohng, C. Joo, T. Ha, Biophys. J. 87, 1328 (2004)
33. N.K. Lee, et al., Biophys. J. 92, 303 (2007)
34. H. Frauenfelder, S.G. Sligar, P.G. Wolyne, Science 254, 1598 (1991)

35. C. Frieden, L.W. Nichol, Protein-Protein Interactions (Wiley, New York, 1981)
36. J.N. Forkey, M.E. Quinlan, Y.E. Goldman, Prog. Biophys. Mol. Biol. 74, 1 (2000)
37. E.J.G. Peterman, H. Sosa, L.S.B. Goldstein, W.E. Moerner, Biophys. J. 81, 2851 (2001)
38. R.Y. Tsien, A. Miyawaki, Science 280, 1954 (1998); R.Y. Tsien, Annu. Rev. Biochem. 67, 509 (1998)
39. R. Liu, D. Hu, X. Tan, H.P. Lu, J. Am. Chem. Soc. 128, 10034 (2006)
40. X. Tan, D. Hu, T.C. Squier, H.P. Lu, Appl. Phys. Lett. 85, 2420 (2004)
41. H.P. Lu, Acc. Chem. Res. 38, 557–565 (2005)
42. Y. Wang, H.P. Lu, Submitted
43. (a) I. Prigogine, The End of Certainty, Time, Chaos, and the New Laws of Nature (Fress Press, New York, 1997) (b) G. Nicolis, I. Prigogine, Exploring Complexity (W. H. Freeman, New York, 1989)
44. M.O. Vlad, J. Ross, Analysis of experimental observables and oscillations in single-molecule kinetics, The theory and evaluation of single-molecule signals, ed. by E. Barki, F. Brown, M. Orrit, H. Yang (World Scientific, New Jersey, 2008)
45. B. Ma, S. Kumar, C.J. Tsai, R. Nussinov, Protein Eng. Des. Sel. 12, 713 (1999); O.F. Lange, et al., Science 320, 1471 (2008)
46. D.E. Koshland, Proc. Natl. Acad. Sci. U.S.A. 44, 98 (1958)
47. R. Grunberg, J. Leckner, M. Nilges, Structure 12, 2125 (2004)

Watching Individual Enzymes at Work

Kerstin Blank, Susana Rocha, Gert De Cremer, Maarten B.J. Roeffaers,
Hiroshi Uji-i, and Johan Hofkens

Summary. Single-molecule fluorescence experiments are a powerful tool to analyze
reaction mechanisms of enzymes. Because of their unique potential to detect heterogeneities in space and time, they have provided unprecedented insights into the
nature and mechanisms of conformational changes related to the catalytic reaction.
The most important finding from experiments with single enzymes is the generally
observed phenomenon that the catalytic rate constants fluctuate over time (dynamic
disorder). These fluctuations originate from conformational changes occurring on
time scales, which are similar to or slower than that of the catalytic reaction. Here,
we summarize experiments with enzymes that show dynamic disorder and introduce
new experimental strategies showing how single-molecule fluorescence experiments
can be applied to address other open questions in medical and industrial enzymology, such as enzyme inactivation processes, reactant transfer in cascade reactions,
and the mechanisms of interfacial catalysis.

25.1 Introduction

Life is sustained by a complex network of chemical reactions. Enzymes, the
molecules that catalyze chemical reactions in biological systems, are one of
the most remarkable class of molecules generated by evolution. Their per-
formances, typically, far exceed those of man-made catalysts. Being able
to accelerate reactions by up to 10^{19}-fold relative to the uncatalyzed reaction [1], they allow reactions which would have half-lives of tens of millions
of years to occur in milliseconds. Furthermore, catalysis typically occurs at
ambient temperature and neutral pH, and is frequently exquisitely regio- and
stereo-selective.

Different theories have been proposed on how enzymes bind their specific
substrates and achieve these huge rate accelerations [1]. One key question that
remains to be answered is how dynamic effects contribute to the activity and
specificity of enzymes. The initially proposed "lock and key" mechanism was
supported by the large number of X-ray structures, which represent static
snapshots of one enzyme conformation. Along this line, the observation of

different conformations was explained by an "induced fit mechanism" which interprets conformational changes as the result of a specific binding event.

Although it has been known for a long time that proteins exist in an ensemble of slightly different and interconverting conformers which are defined by rugged energy landscapes [2], the "new view" of proteins was only recently introduced for the description of enzymatic reactions [3–5]: the concept of folding [6, 7] and binding funnels [7], which allows the existence of parallel reaction pathways [5, 8] is now being extended to reaction funnels that determine enzymatic reactions [5, 9]. Evidence for this "new view" of enzymatic reactions originates mainly from NMR experiments and molecular dynamics simulations, which suggest that reactions proceed via dynamic population shifts in the conformational ensemble [4].

Additional proof for this dynamic view was obtained from single-molecule studies. Experiments analyzing the activity of enzymes at the single-molecule level have shown that fluctuations between different conformations have a direct influence on the catalytic reaction [10–16]. Different conformations exhibit different rate constants for the catalytic process and the interconversion between these conformations results in time-dependent fluctuations of the measured rate constant. This effect, termed as dynamic disorder, has been shown for different enzymes and is considered to be a general property of enzymes.

In addition to the analysis of dynamic disorder, single-molecule approaches have the unique potential to reveal a number of processes that cannot be observed at the ensemble level. In the following, we wish to provide examples of how experiments with single enzymes can be extended beyond studies of dynamic disorder. We will show that single-molecule studies can identify rare events of enzyme inactivation, investigate cascade reactions, and determine the mechanism of interfacial catalysis at a phospholipid membrane.

25.2 Single Enzyme Experiments with Confocal Detection Schemes

Confocal detection schemes are ideally suited for the detection of individual fluorescent molecules with a high time resolution and have therefore been applied for a broad range of experiments designed to study the dynamics of individual biological systems. A confocal microscope, combined with a sensitive avalanche photodiode detector, allows for the observation of the catalytic activity of an individual surface-immobilized enzyme over long periods of time. For the experiments described in the following sections, the enzyme turns over a fluorogenic substrate into fluorescent product molecules. Each single turnover event results in a fluorescence burst. As the product diffuses away very quickly, it exits the confocal volume and cannot be detected anymore. Note that bleaching of the product has the same effect. The measured time intervals between two fluorescence bursts contain the relevant kinetic information.

25.2.1 Detection and Analysis of Dynamic Disorder

One example that has been analyzed with this approach is the enzyme lipase B from *Candida antarctica* (CalB) [13, 17]. Lipases and phospholipases are interfacial enzymes, which are found in most organisms in the microbial, plant, and animal kingdoms. They play a crucial role as catalysts in lipid metabolism and as mediators of cell signaling processes. As a consequence, their mode of action is the subject of extensive study. Moreover, a detailed understanding of the mechanism of CalB is also interesting because of its high stereoselectivity and its stability in organic solvents, making it an attractive biocatalyst for organic reactions.

Despite their different physical and biochemical properties, most lipases and phospholipases share a common structural element: an α-helical loop ("lid") that covers the active site. Since the opening of the lid exposes a large hydrophobic patch, the resulting open conformation is thermodynamically unfavorable in solution. In contrast, in the presence of a lipid interface the open conformation is stabilized by the interaction with lipids. Many lipases and phospholipases show higher activity on interfaces than with free lipids (interfacial activation). It has long been considered that interfacial activation and lid opening are correlated. However, a number of enzymes, such as CalB, possess a lid structure but do not show interfacial activation [18–20].

Since CalB does not show interfacial activation, its activity can be determined with soluble substrates. CalB has been used as a model enzyme for single-molecule experiments as it is active on fluorogenic substrates with short alkyl chains, such as $2', 7'$-bis-(2-carboxyethyl)-5-(and-6)-carboxyfluorescein, acetoxymethyl ester (BCECF-AM) (Fig. 25.1a). Experiments using this substrate were performed as follows. Fluorescently labeled molecules of CalB were adsorbed on a hydrophobically treated glass surface. After localizing one labeled enzyme in the focus of the laser, the fluorescence label was bleached and BCECF-AM was added into the reaction medium. Then time traces of fluorescence intensity were recorded at this position. These time traces show frequent spikes of high intensity, each corresponding to a single turnover event.

The time between two successive turnover events (waiting time) was determined and plotted into a histogram. According to Kramer's theory, the catalytic rate constant can be determined from an exponential fit to the histogram. However, the histogram of the waiting times showed a non-exponential decay; the distribution was stretched over several orders of magnitude. This observation clearly indicates that different processes, each described with an exponential distribution, contribute to the process and that the rate constant fluctuates over time. This is a clear manifestation of dynamic disorder. The subsequent analysis of successive waiting times showed a clustering of waiting times with a similar length, suggesting a correlation between

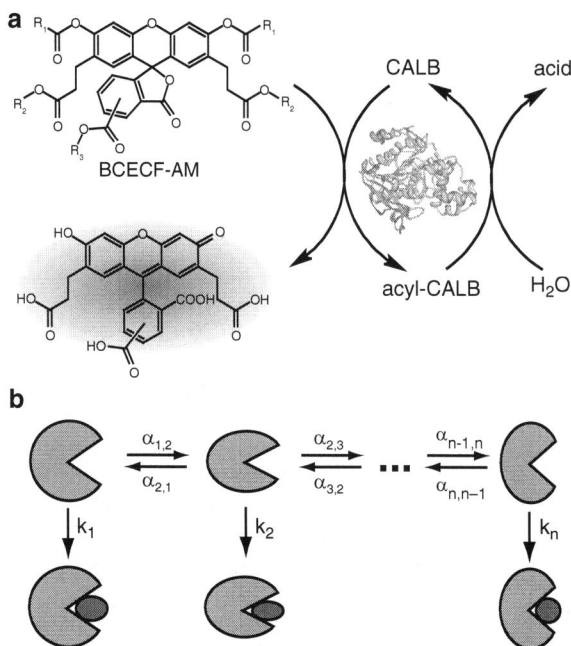

Fig. 25.1. Analysis of the catalytic activity of CalB at the single-molecule level. (**a**) Detection of single enzymatic turnover events of the enzyme CalB. The fluorogenic substrate BCECF-AM is hydrolyzed by CalB yielding the highly fluorescent dye BCECF. (**b**) Proposed reaction scheme explaining dynamic disorder. The enzyme interconverts between different conformations with the rate constants $\alpha_{a, b}$. Each conformation hydrolyzes the substrate with its own rate constant k_i. If conformational changes are slower than the catalytic reaction, a certain conformation performs several turnover cycles before it switches into another conformation. While subsequent turnovers in one conformation are correlated, the system loses its memory after a conformational change

waiting times. This observation was explained with the kinetic scheme shown in Fig. 25.1b, which assumes different activity states, which are related to different interconverting conformers.

Although static and dynamic disorder had been detected for other enzyme substrate systems before [10–12], one could argue that its observation in this case is an artifact of the way, how the experiments were performed. The nonspecific immobilization procedure might result in static disorder. Enzymes with different orientations on the surface might possess a different lid mobility and accessibility of the active site and, as a result, show different activities. And the use of the highly unnatural substrate might be a possible source of dynamic disorder. An alternative detection scheme for this class of enzymes, which solves these shortcomings, will be presented in Sect. 25.3.

25.2.2 Observation of Enzyme Inactivation

Experiments performed in a very similar way also showed dynamic disorder for the enzyme α-chymotrypsin [16]. α-Chymotrypsin is an endopeptidase acting on water-soluble polypeptides. The substrate (suc-AAPF)$_2$-rhodamine 110 was designed to interact optimally with the binding site of the enzyme. It consists of a rhodamine 110 core that is derivatized with a succinylated AlaAlaProPhe peptide sequence, known to bind very specifically at the enzyme's active site (Fig. 25.2a). To further avoid potential artifacts, the enzyme was immobilized by entrapment in an agarose matrix, which restricts enzyme diffusion while still allowing free rotation and conformational dynamics of the enzyme as well as the diffusion of the substrate.

The analysis of more than 100 individual enzymes under substrate saturation conditions yielded k_{cat} values between $1\,s^{-1}$ and $25\,s^{-1}$, again indicating static disorder. Furthermore, dynamic disorder was observed again: the histogram of the waiting times showed a non-exponential decay and a correlation between waiting times was observed for up to 15 turnovers.

For approximately 5% of the measured enzymes, a peculiar spontaneous inactivation pattern was observed (Fig. 25.2b and c). Rather than a one-step "all-or-nothing" inactivation or a continuous gradual decrease in activity, a transient phase was observed. During this phase, the enzyme switched between discrete active and inactive states. After each inactive period, only a fraction of the original activity was recovered until the enzyme was irreversibly deactivated.

Measurements of the temperature dependence of enzymatic activity suggested the presence of an intermediate step in enzyme inactivation. The proposed mechanism involves a reversible change to an inactive state, which precedes the final irreversible inactivation step [21–23]. This mechanism could now be refined based on experiments with single enzymes, which detected the intermediate steps directly. This more detailed information allowed the establishment of a tentative model for α-chymotrypsin inactivation (Fig. 25.2d).

In summary, the experiments with α-chymotrypsin have again proven the existence of dynamic disorder in the catalytic rate constant. Moreover, these experiments show that the analysis of single enzymes directly visualizes rare events, such as the described inactivation pathway. The approach can be extended to the analysis of induced denaturation processes, e.g. by adding denaturants or oxidizing agents. Besides being of basic interest, a more detailed understanding of enzyme inactivation and denaturation processes is of great importance for the optimization of biocatalytic processes whose performance is often limited by the inactivation of the biocatalyst.

25.2.3 Analysis of Cascade Reactions

Another important aspect in biocatalysis is the transfer of reactants in cascade reactions. In a cascade reaction, the product of the first reaction can be

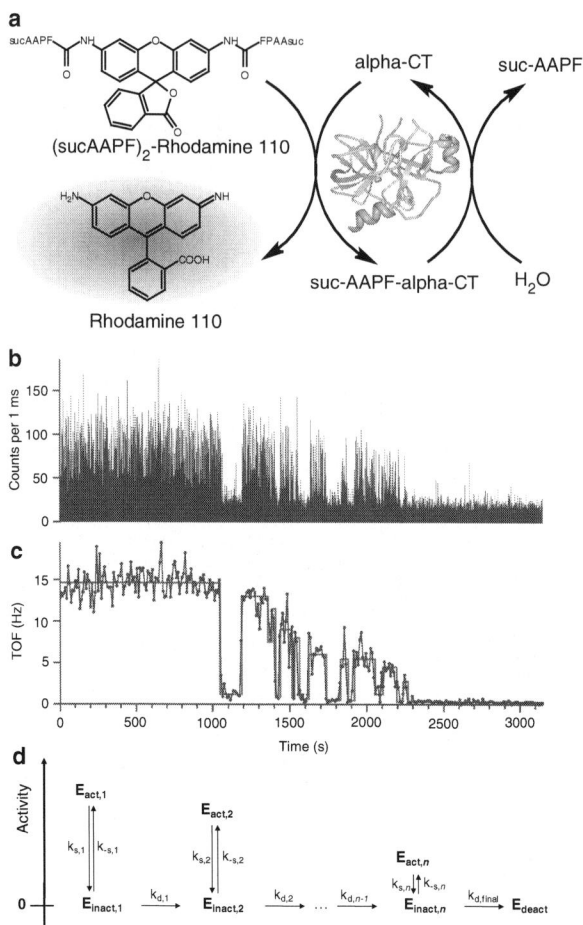

Fig. 25.2. Analysis of the catalytic activity and the inactivation of α-chymotrypsin at the single-molecule level. (**a**) Detection of single enzymatic turnover events of α-chymotrpysin. The fluorogenic substrate (suc-AAPF)$_2$-rhodamine 110 is hydrolyzed by α-chymotrypsin, yielding the highly fluorescent dye rhodamine 110. (**b**) Representative intensity time trace for an individual α-chymotrypsin molecule undergoing spontaneous inactivation under reaction conditions. (**c**) Inactivation trace for the intensity time transient in (**b**), obtained by counting the amount of turnover peaks in (**b**) in 10 s intervals. After approximately 1000 s, the enzyme deactivates through a transient phase with discrete active and inactive states. (**d**) Proposed model for the inactivation process. An initial active state is in equilibrium with an inactive state. This inactive state converts to another inactive state irreversibly whereby the corresponding active state has a lower activity than the previous one. All the transitions involved have energy barriers that can be overcome spontaneously at room temperature

released from the catalyst into the reaction medium where the second reaction takes place. Alternatively, the product of the first reaction might stay transiently bound to the first catalyst and the second reaction takes place on the surface of the first catalyst. With an ensemble of enzymes free in solution, it is difficult to ascertain the mechanism that determines a certain reaction, since no information about the localization of the individual reaction events can be obtained. In contrast, experiments with individual immobilized enzymes have the spatial precision to separate these events.

For our studies, we have chosen the haloperoxidase from *Curvularia verruculosa*. Haloperoxidases are typically extracted from marine microorganisms and algae, and are important industrial biocatalysts. In the presence of hydrogen peroxide and halides, they produce very reactive hypohalites which halogenate or oxidize organic compounds in a secondary reaction [24,25]. The mechanism of the secondary reaction is not fully understood. The presence of regio- and stereoselectively brominated compounds in haloperoxidase producing organisms suggests that the secondary reaction might be enzymatically catalyzed by the haloperoxidase itself [26, 27]. In this case, the hypohalite would remain bound to the enzyme while an organic compound binds to a secondary binding site on the enzyme allowing for the catalytic reaction to occur.

The use of aminophenyl fluorescein [28] as the organic compound allows the selective localization of the secondary oxidation reaction with confocal fluorescence microscopy (Fig. 25.3a) [29]. In a similar way as described for α-chymotrypsin, the enzyme was immobilized in an agarose matrix and fluorescence intensity time traces were recorded at a position where an enzyme was found. Time traces with exceptional signal-to-noise ratio were obtained (Fig. 25.3b) and a histogram of time-averaged single enzyme activities was constructed (Fig. 25.3c) which allowed the determination of the average activity of the analyzed enzymes.

By positioning the laser focus of the confocal fluorescence microscope at various distances with respect to the enzyme, a spatially resolved picture of the hypobromite reactivity in the surroundings of the enzyme was obtained (Fig. 25.3d). A high number of fluorescent product molecules of the secondary reaction was observed at distances larger than the spatial resolution of the measurement, leading to the conclusion that the major fraction of the hypobromite is released into the medium. However, the fact that the investigated enzyme provides a quasi-constant flux of highly reactive hypobromite species into the medium does not exclude that other haloperoxidases from other species might follow a different reaction mechanism. The ability of localizing individual reaction events shows the potential of confocal approaches for the study of cascade reactions. The transfer of reactants is not only of interest in biocatalytic reactions, but also in naturally existing cascade reactions where the transfer of intermediates might occur via transient-specific interactions between enzymes.

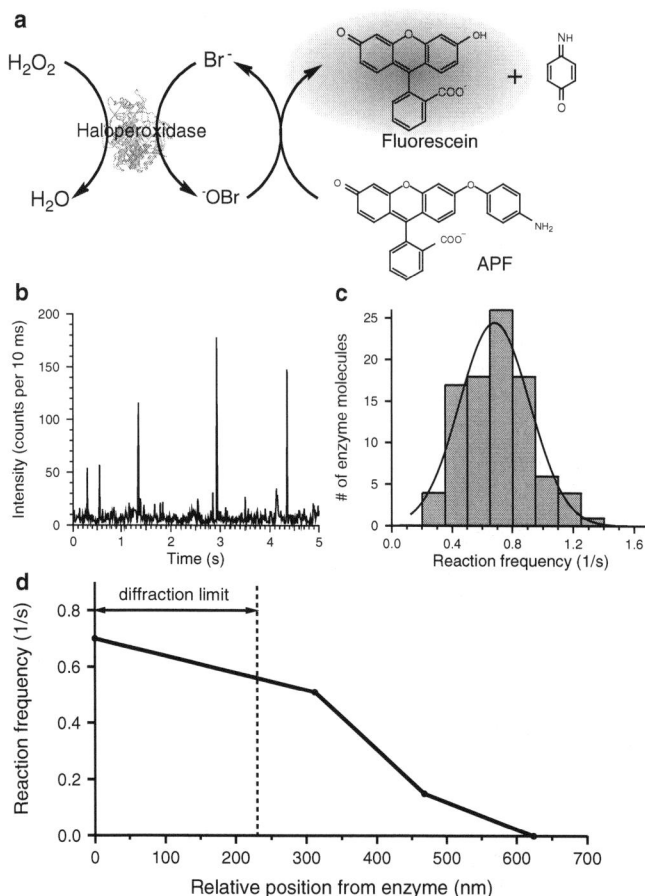

Fig. 25.3. Detection of a two-step cascade reaction by observing single hypobromite reaction events. (**a**) Reaction mechanism. The haloperoxidase enzyme produces hypobromite. These reactive oxygen species react with aminophenyl fluorescein (APF) in a secondary reaction to yield the highly fluorescent dye fluorescein. (**b**) Representative part of a fluorescence intensity time trace. Every intensity burst corresponds to one reaction event between hypobromite and APF. (**c**) Histogram of the time-averaged activities of approximately 100 individual enzymes, obtained by measuring the secondary reaction between hypobromite and APF. (**d**) Frequency of observing individual reaction events as a function of distance from the enzyme. At distances farther than the diffraction limit, a considerable number of reaction events occurs indicating that the secondary reaction takes place in solution following the release of most of the hypobromite from the enzyme

25.3 Single Enzyme Experiments with Wide Field Detection Schemes

Despite its broad applicability, the confocal approach is restricted to the observation of one enzyme immobilized at a specific position. In contrast, wide-field detection schemes allow for the observation of an area of up to $1\,mm^2$ and can therefore detect a number of enzymes in parallel. Moreover, the movement of individual molecules can be followed. This allows the study of processive enzymes moving on their natural substrates, as has been elegantly demonstrated for some DNA interacting enzymes [12, 30, 31].

Resulting from this unique potential, wide-field approaches are generally more appropriate for the study of enzymes that bind to high molecular weight substrates, such as DNA or carbohydrates. Furthermore, confocal approaches are less suitable for insoluble substrates, such as phospholipid bilayers. In this case the substrate can be considered as immobilized and the movement of the enzyme on the substrate needs to be detected. For these interfacial reactions, the catalytic step must be preceded by diffusion of the enzyme to an appropriate site on its target substrate.

25.3.1 Analysis of Interfacial Catalysis

In more detail, catalysis of an interfacial enzyme involves the following steps: (1) diffusion and adsorption of the enzyme to the (phospho)lipid surface, (2) opening of the lid/interfacial activation (3) penetration of the enzyme into the lipid phase, (4) lipid hydrolysis, and (5) either scrolling of the enzyme to the next substrate molecule or enzyme desorption [32]. This model implies that the preexisting structural ordering of the lipid and/or lipid–water interface influences the activity of lipolytic enzymes [20]. Indication for the validity of this assumption was obtained from atomic force microscopy (AFM) [33, 34] and fluorescence microscopy [35, 36] studies monitoring the desorption of the substrate layer upon hydrolysis.

Biological Model System

In contrast to the studies mentioned above, the new approach described here [37] monitors the behavior of the enzyme and/or the changes on the substrate layer during hydrolysis. The substrate used was a supported bilayer of the unsaturated phospholipid 1-palmitoyl-2-oleoyl-sn-glycero-3-phosphocholine (POPC). Supported bilayers with unsaturated fatty acid chains form more fluid bilayers and are often used for mimicking biological membranes [38].

Three different variants of *Thermomyces lanuginosus* lipase (TLL) were used as the model enzyme to study the hydrolysis of the phopsholipid bilayer. Being a lipase, TLL has low affinity for phospholipid bilayers [39, 40]. TLL is, therefore, of interest for the characterization of the diffusion and adsorption of an enzyme on phospholipid bilayers, without activation and penetration of the enzyme into the phospholipid layer, thereby mimicking step (1) of the described reaction mechanism. Furthermore, using mutagenesis TLL was converted into a phospholipase, which can perform the whole reaction cycle: After mutating both the lid and the C-terminal region, the new variant cleaves the ester bond of a phospholipid at the sn-1 position and is, therefore, designated phospholipase A1 (aPLA1). In addition, an inactive phospholipase (iPLA1) was generated based on aPLA1. This was achieved by replacing the active site serine with an alanine residue (S146A). Although this mutation causes some changes in the interaction between the substrate molecule and the enzyme's active site, it is often used to generate inactive mutants [39, 40]. The iPLA1 variant is thus able to mimic steps (1)–(3) of the reaction cycle. Therefore, the study of this set of closely related variants with different affinities and catalytic activities towards the substrate allows the detailed observation of the different steps of the reaction cycle.

Substrate Hydrolysis by the Active Phospholipase

In order to visualize the enzyme acting on the substrate, it was labeled with a water-soluble and highly photostable, fluorescent perylene diimide derivate (PDI) [41, 42]. When labeled aPLA1 ($\sim 10^{-7}$ M) was added to a non-labeled POPC multilayer (stacks of bilayers), enzymes could be visualized as bright spots and areas of high enzyme localization could be clearly seen (Fig. 25.4).

Despite the fact that the substrate layer itself is not directly visualized in this experiment, it is clear that the enzymes are localized primarily at the edge region of one bilayer. The position of the top bilayer of the POPC multilayer can easily be inferred as it is "outlined" by a high local concentration of enzymes, and it is this area that is reduced over time due to enzymatic activity. The experiments also show a non-uniform distribution of the labeled aPLA1 along the layer edge. Importantly, areas with a faster retraction of

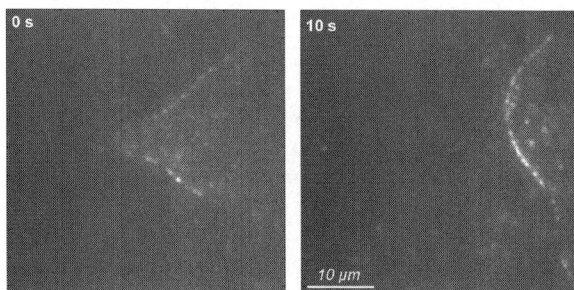

Fig. 25.4. Time-resolved fluorescence images of labeled aPLA1 hydrolyzing POPC multilayers. The preferential localization of the enzymes at the edge between two bilayers allows the discrimination between two consecutive bilayers. The speed of the retraction of the top layer is proportional to the amount of enzyme present

the top layer coincide with areas of higher local enzyme concentration. These results established for the first time the direct relationship between the local structural characteristics of the phospholipids and the activity of interfacial enzymes.

Unfortunately, the observation and the tracking of single aPLA1 molecules is only possible at lower enzyme concentrations ($\sim 10^{-9}$ M). However, when using this low concentration the position of the edge between two consecutive bilayers is no longer visible. This can be solved by fluorescently labeling both enzyme and bilayer, as will be described below.

Simultaneous Observation of Layer Desorption and Enzyme Mobility

While the enzymes were again labeled with PDI, the bilayer was doped with 3, 3′-dioctadecyloxacarbocyanine perchlorate (DiO). Two-color excitation was employed to excite both the PDI and the DiO label efficiently. By choosing an appropriate optical filter before the detection apparatus, it was possible to detect the majority of PDI emission and block most of the DiO emission (Fig. 25.5). Thus, emission from the enzymes and the substrate was discriminated based on their relative brightness, and single enzymes appear clearly as bright spots against the background of the low intensity bilayer (Fig. 25.6).

The observation of single enzymes hydrolyzing the POPC bilayer reveals that the diffusion of aPLA1 on the bilayer is a highly heterogeneous process (Figs. 25.6a and 25.7). Fast and slow-diffusing as well as immobilized enzyme molecules were observed, with the slower motion being observed mainly on the edge between two consecutive bilayers. The same experiment with the

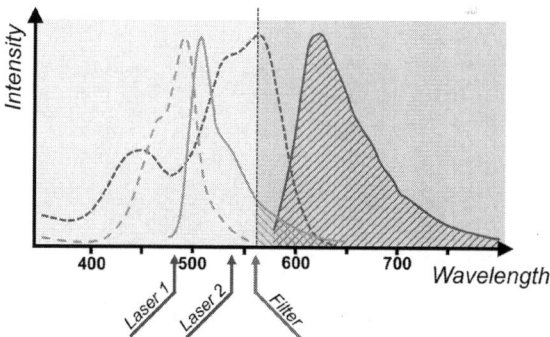

Fig. 25.5. Detection scheme for imaging labeled enzymes and multilayers simultaneously. Shown are the steady state absorption (*dashed line*) and emission (*solid line*) spectra for DiO (*light grey*) and PDI (*dark gray*). The excitation wavelengths used for each dye are indicated by arrows at the bottom. The best ratio of detected emission from the layers and from the enzyme molecules was accomplished by the use of an appropriate long pass cut-off filter (cut-off wavelength indicated)

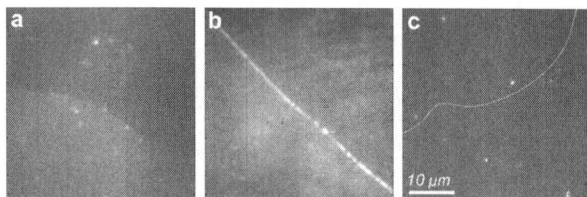

Fig. 25.6. Representative fluorescence images showing the localization of three different variants. (**a**) aPLA1, (**b**) iPLA1, and (**c**) TLL. The white line in (**c**) indicates the position of the layer edge

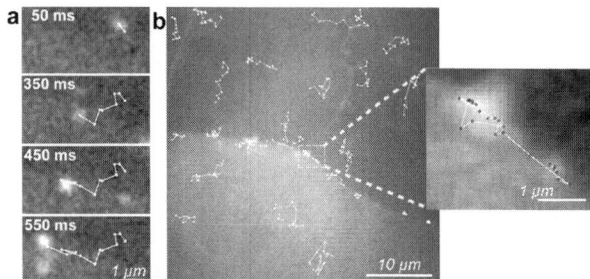

Fig. 25.7. Diffusion behavior of labeled aPLA1 molecules. (**a**) Snapshots of one enzyme diffusing on the layer edge and its corresponding trajectory. (**b**) Typical trajectories of individual aPLA1 molecules diffusing on the layer and on the edge (background image accumulated over 100 frames). The magnified trajectory shows hot spots where diffusion is slow

labeled inactive iPLA1 (Fig. 25.6b) showed a very different behavior of the enzyme. The image clearly shows the strong affinity of the inactive enzyme for the layer edge. When performing the experiment with TLL (Fig. 25.6c), the enzyme molecules again showed different behavior. Enzymes could only be visualized when using a 100 times higher enzyme concentration. Enzymes adsorbed poorly on the layer and diffused faster on the top of the layer, indicating only weak affinity for the phospholipid bilayer.

For a quantitative analysis of these dynamic heterogeneities in the diffusion behavior, individual enzymes were tracked (Fig. 25.7). The obtained trajectories were analyzed using a custom-written Matlab® routine. Instead of extracting an average diffusion coefficient for the individual trajectories, each individual step of the trajectory was determined and analyzed using cumulative distribution functions (CDFs) [43]. The CDFs describe the probability that an enzyme starting at an initial position is found within a circle of radius r at time τ. With the CDFs, the enzyme motions could be distinguished and quantified even in the presence of heterogeneous motion.

With separate CDFs for aPLA1 diffusing on top of the layer (996 trajectories) and on the layer edges (448 trajectories), the differences in behavior in each region could be quantified. Two distinct diffusion constants (D) were

Table 25.1. Diffusion constants detected for both active and inactive PLA1

Diffusion Constant	aPLA1		iPLA1	
$(10^{-8}\,\mathrm{cm}^2\,\mathrm{s}^{-1})$	Layer (%)	Edge (%)	Layer (%)	Edge (%)
5.1–5.2	58	–	62	–
1.7–2.3	–	41	38	–
0.7	42	–	–	–
0.07	–	52	–	–
~0	–	7	–	100

determined for enzymes moving on top of the layer ($D = 5.1 \times 10^{-8}\,\mathrm{cm}^2\,\mathrm{s}^{-1}$ and $7 \times 10^{-9}\,\mathrm{cm}^2\,\mathrm{s}^{-1}$) and for enzymes localized at the layer edge ($D = 1.7 \times 10^{-8}\,\mathrm{cm}^2\,\mathrm{s}^{-1}$ and $7 \times 10^{-10}\,\mathrm{cm}^2\,\mathrm{s}^{-1}$). For enzymes localized at the layer edge, 19% of these trajectories also included immobilized periods. The same analysis for iPLA1 (516 trajectories) yielded only one type of motion at the layer edge: immobilization. On top of the layer again two types of motions were detected with $D = 5.2 \times 10^{-8}\,\mathrm{cm}^2\,\mathrm{s}^{-1}$ and $2.3 \times 10^{-8}\,\mathrm{cm}^2\,\mathrm{s}^{-1}$. For TLL (531 trajectories), only one mode of motion was detected irrespective of the position on the substrate ($D = 3 \times 10^{-8}\,\mathrm{cm}^2\,\mathrm{s}^{-1}$). Table 25.1 summarizes the diffusion constants and the relative occurrence of each type of motion (individual steps) for aPLA1 and iPLA1.

A comparison of the different types of motion of the three different variants allows the correlation of the enzyme diffusion behavior with specific stages of the catalytic cycle. TLL, an enzyme which cannot interact strongly with phospholipid bilayers, was found to diffuse quickly on the POPC multilayers with no specific preference for the edge or the top of the layer. The motion detected is most likely associated with weak adsorption and desorption of the enzyme on the layer since the diffusion constant is 100 times slower than that expected for free diffusion in solution [42]. These motions correspond to parts A and eventually B of the catalytic cycle shown schematically in Fig. 25.8.

Similar fast diffusive motions ($D = 5.1$–$5.2 \times 10^{-8}\,\mathrm{cm}^2\,\mathrm{s}^{-1}$) were observed for both aPLA1 and iPLA1 diffusing on the POPC multilayers. Although the surfaces of TLL and PLA1 are different, resulting from the mutations, their diffusion constants cannot be compared directly and we attribute the fast motions of aPLA1 and iPLA1 on top of the layer to weak adsorption/desorption events.

The additional type of motion observed for both aPLA1 ($D = 0.7 \times 10^{-8}\,\mathrm{cm}^2\,\mathrm{s}^{-1}$) and iPLA1 ($D = 2.3 \times 10^{-8}\,\mathrm{cm}^2\,\mathrm{s}^{-1}$) on top of the layer most likely results from activated enzymes with an open lid. These enzymes are able to penetrate into the lipid phase (parts c and d of Fig. 25.8). The slight differences in the diffusion constants can be attributed to the structural differences originating from the S146A mutation, which results in a different lid mobility and different binding affinities [44].

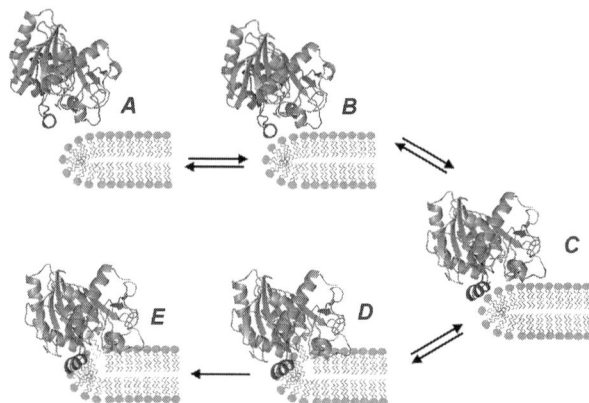

Fig. 25.8. Proposed catalytic cycle. While in solution, the enzyme remains in the closed form, with the lid covering the active site (**a**). Binding of the enzyme to the surface (**b**) promotes lid displacement and exposure of hydrophobic residues that interact with the phospholipid interface (**c**), thereby stabilizing the open form. Partitioned substrate accesses the active site (**d**), resulting in the formation of the enzyme-substrate and enzyme product complex (**e**). Hydrolysis is followed by product desorption and the enzyme diffuses along the substrate or into solution

Unlike TLL, both the active and inactive forms of PLA1 showed periods of immobilization at the layer edge with a much longer residence time for iPLA1. Enzymatic activity is thus not a prerequisite for strong enzyme intercalation at the layer edge, but is clearly a prerequisite for efficient desorption of the enzyme. The products of the hydrolysis reaction cause considerable reorganization and solubilization of the phospholipid bilayer, and either effect could trigger the desorption of enzyme.

In summary, this approach is much better suited for the analysis of lipases and phospholipases than the confocal approach described in 2.1 since all steps of the catalytic cycle can be observed. Interfacial enzymology is a growing field of research [32, 45] and the method described here can contribute to a more detailed understanding of catalysis at interfaces.

25.4 Summary

Despite enormous progress during the last 10 years, several important questions in enzymology are yet to be answered. The contribution of dynamic processes to the function of enzymes is still a matter of debate. In some cases, conformational changes contribute directly to the catalytic reaction [46] and in other cases they have shown to lead to dynamic disorder [10–17]. The connection between these seemingly opposing effects still needs to be established. Other open questions are related to inactivation processes, the transfer of reactants in cascade reactions, and the mechanisms of processive enzymatic reactions at interfaces, such as (phospho)lipid bilayers and high molecular weight substrates like carbohydrates.

Single-molecule experiments have unique properties and can contribute significantly to a number of different approaches. With the examples summarized in this chapter, we have shown that experiments at the single enzyme level can:

- Identify heterogeneities in time and space
- Follow the time series of events in enzyme-catalyzed reactions
- Identify rare events, such as inactivation processes
- Reveal parallel reaction pathways, and
- Localize reaction events in space

The examples summarized here are just a first demonstration of the potential of single-molecule experiments for the analysis of enzyme-catalyzed reactions. However, they provide a clear perspective that single-molecule experiments will continue to contribute to our detailed understanding of enzymes. A more detailed understanding is not only of fundamental scientific importance, but will also provide the basis for the design of better enzymes and enzyme inhibitors for a broad range of biomedical and industrial applications. Furthermore, the concepts outlined here are generic and can be applied to other systems, such as industrial catalysts [47].

Acknowledgments

The authors are grateful to the following organizations for individual fellowships: the Human Frontier Science Program HFSP (K.B), the Portuguese Foundation for Science and Technology FCT (S.R.), and the Fonds voor Wetenschappelijk Onderzoek FWO (G.D.C and M.B.J.R.). Furthermore, the authors acknowledge support from grants from FWO (G.0366.06 and G.0229.07) the KULeuven Research Fund (GOA 2006/2, Center of Excellence CECAT, CREA2007), the Federal Science Policy of Belgium (IAPVI/27), the European Union (NMP4-CT-2003–505211, Bioscope), and the Flemish government (Long term structural funding – Methusalem funding)

References

1. M. Garcia-Viloca, J. Gao, M. Karplus et al., Science **303**, 186–195 (2004)
2. H. Frauenfelder, S.G. Sligar, P.G. Wolynes, Science **254**, 1598–1603 (1991)
3. L.C. James, D.S. Tawfik, Trends Biochem. Sci. **28**, 361–368 (2003)
4. K. Henzler-Wildman, D. Kern, Nature **450**, 964–972 (2007)
5. L. Swint-Kruse, H.F. Fisher, Trends Biochem. Sci. **33**, 104–112 (2008)
6. K.A. Dill, H.S. Chan, Nat. Struct. Biol. **4**, 10–19 (1997)
7. S. Kumar, B. Ma, C.J. Tsai et al., Protein Sci. **9**, 10–19 (2000)
8. L.A. Wallace, C.R. Matthews, Biophys. Chem. **101–102**, 113–131 (2002)

9. W. Min, X.S. Xie, B. Bagchi, J. Phys. Chem. B. **112**, 454–466 (2008)
10. H.P. Lu, L. Xun, X.S. Xie, Science **282**, 1877–1882 (1998)
11. L. Edman, Z. Foldes-Papp, S. Wennmalm et al., Chem. Phys. **247**, 11–22 (1999)
12. A.M. van Oijen, P.C. Blainey, D.J. Crampton et al., Science **301**, 1235–1238 (2003)
13. K. Velonia, O. Flomenbom, D. Loos et al., Angew. Chem. Int. Ed. **44**, 560–564 (2005)
14. B.P. English, W. Min, A.M. van Oijen et al., Nat. Chem. Biol. **2**, 87–94 (2006).
15. N.S. Hatzakis, H. Engelkamp, K. Velonia et al, Chem. Commun. 2012–2014 (2006)
16. G. De Cremer, M.B.J. Roeffaers, M. Baruah et al., J. Am. Chem. Soc. **129**, 15458–15459 (2007)
17. O. Flomenbom, K. Velonia, D. Loos et al, Proc. Natl. Acad. Sci. U S A. **102**, 2368–2372 (2005)
18. J. Uppenberg, M.T. Hansen, S. Patkar et al., Structure **2**, 293–308 (1994)
19. M. Martinelle, M. Holmquist, K. Hult, Biochim. Biophys. Acta **1258**, 272–276 (1995)
20. R. Verger, Trends Biotechnol. **15**, 32–38 (1997)
21. R. Lumry, H. Eyring, J. Phys. Chem. **58**, 110–120 (1954)
22. C.L. Tsou, Biochim. Biophys. Acta **1253**, 151–162 (1995)
23. R.M. Daniel, M.J. Danson, R. Eisenthal, Trends Biochem. Sci. **26**, 223–225 (2001)
24. B. Sels, P. Levecque, R. Brosius et al., Adv. Synth. Catal. **347**, 93–104 (2005)
25. B.F. Sels, D.E. De Vos, P.A. Jacobs, Angew. Chem. Int. Ed. **44**, 310–313 (2005)
26. J.N. Carter-Franklin, A. Butler, J. Am. Chem. Soc. **126**, 15060–15066 (2004)
27. A. Yarnell, Chem. Eng. News **84**, 12–18 (2006)
28. K. Setsukinai, Y. Urano, K. Kakinuma et al., J. Biol. Chem. **278**, 3170–3175 (2003)
29. V.M. Martinez, G. De Cremer, M.B.J. Roeffaers et al., J. Am. Chem. Soc. **130**, 13192–13193 (2008)
30. J. Elf, G.W. Li, X.S. Xie, Science **316**, 1191–1194 (2007)
31. J.B. Lee, R.K. Hite, S.M. Hamdan et al., Nature **439**, 621–624 (2006)
32. A. Aloulou, J.A. Rodriguez, S. Fernandez et al., Biochim. Biophys. Acta **1761**, 995–1013 (2006)
33. M. Grandbois, H. Clausen-Schaumann, H. Gaub, Biophys. J. **74**, 2398–2404 (1998)
34. K. Balashev, J.N. DiNardo, T.H. Callisen et al., Biochim. Biophys. Acta **1768**, 90–99 (2007)
35. A.C. Simonsen, U.B. Jensen, P.L. Hansen, J. Colloid Interface Sci. **301**, 107–115 (2006)
36. A.C. Simonsen, Biophys. J. **94**, 3966–3975 (2008)
37. S. Rocha, J.A. Hutchinson, K. Peneva et al., Chem. Phys. Chem. **10**, 151–161 (2009)
38. O.G. Mouritsen, *Life – As a Matter of Fat.* (Springer, Berlin Heidelberg, 2004)
39. G.H. Peters, A. Svendsen, H. Langberg et al., Biochemistry **37**, 12375–12383 (1998)
40. Y. Cajal, A. Svendsen, V. Girona et al., Biochemistry **39**, 413–423 (2000)
41. F.C. De Schryver, T. Vosch, M. Cotlet et al., Acc. Chem. Res. **38**, 514–522 (2005)

42. K. Peneva, G. Mihov, F. Nolde et al., Angew. Chem. Int. Ed. **47**, 3372–3375 (2008)
43. G.J. Schütz, H. Schindler, T. Schmidt, Biophys. J. **73**, 1073–1080 (1997)
44. G.H. Peters, S. Toxvaerd, N.B. Larsen et al., Nat. Struct. Biol. **2**, 395–401 (1995)
45. F. Forneris, A. Mattevi, Science **321**, 213–216 (2008)
46. S. Hammes-Schiffer, S.J. Benkovic, Annu. Rev. Biochem. **75**, 519–541 (2006)
47. M.B.J. Roeffaers, G. De Cremer, H. Uji-i et al., Proc. Natl. Acad. Sci. U S A. **104**, 12603–12609 (2007)

The Influence of Symmetry on the Electronic Structure of the Photosynthetic Pigment-Protein Complexes from Purple Bacteria

Martin F. Richter, Jürgen Baier, Richard J. Cogdell, Silke Oellerich, and Jürgen Köhler

Summary. The primary reactions of purple bacterial photosynthesis take place in two pigment-protein complexes, the peripheral LH2 complex and the core RC-LH1 complex. In order to understand any type of excitation-energy transfer in the LH system detailed knowledge about the correlation between the geometrical structure and the nature of the electronically excited states is crucial. The interplay between the geometrical arrangement of the pigments and the transition probabilities of the various exciton states leads to key spectral features, such as narrow lines, that are clearly visible with single-molecule spectroscopy but are averaged out in conventional ensemble experiments. Combining low-temperature single-molecule spectroscopy with numerical simulations has allowed us to achieve a refined structural model for the bacteriochlorophyll a (BChl a) pigment arrangement in RC-LH1 core complexes of *Rps. palustris*. The experimental data are consistent with an equidistant arrangement of 15 BChl a dimers on an ellipse, where each dimer has been taken homologous to those from the B850 pigment pool of LH2 from *Rps. acidophila*.

26.1 Introduction

In photosynthesis, solar radiation is absorbed by a light-harvesting (LH) apparatus and the excitation energy is then transferred efficiently to a reaction center (RC), where it is used to create a charge-separated state that ultimately drives all the subsequent metabolic reactions. In our group, we study pigment-protein complexes from purple photosynthetic bacteria which have evolved an elegant modular LH system. These modules consist of pairs of hydrophobic, low molecular weight polypeptides, called α and β (usually 50–60 amino acids long) that noncovalently bind a small number of bacteriochlorophyll (BChl) and carotenoid (Car) molecules [1]. The modules then oligomerise to produce the native circular or elliptical complexes. Most purple bacteria have two main types of antenna complexes, called LH1 and LH2. The LH1 complexes surround the RCs to form the so-called core complex [1, 2]. The core complexes are surrounded by the LH2 complexes, which are also called the peripheral antenna complexes. Typically, light-energy is absorbed by LH2 and is then transferred via LH1 to the RC. There it is used to drive a series of electron transfer reactions that result in the reduction of ubiquinone (UQ). In the photosynthetic membrane, when UQ in the RC has been fully reduced to UQH_2, the quinol must leave the RC in order to transfer its reducing equivalents to the cytochrome b/c_1 complex, as part to a rather simple cyclic electron transport pathway. Unfortunately, it appears very difficult to obtain highly resolved structural information about the LH complexes. This reflects the fact that, though it is not too difficult to isolate the pigment-protein complexes from the membrane in high purity to allow for crystallization, obtaining highly resolving single crystals is still a major challenge. As a consequence of this, there is a reliance on atomic force- and electron-microscopy studies on 2d crystals where the resolution is in the range of 5–10 Å. This then still leaves many structural details obscured. Highly resolved X-ray structures have been obtained as yet only for the peripheral LH complexes for a few bacterial species [3–6].

The basic building block of LH2 is a protein heterodimer ($\alpha\beta$), which accommodates three BChl a pigments and one carotenoid molecule [3]. Depending on the bacterial species, the LH2 complex consists either of eight or nine copies of these heterodimers, which are arranged in a ring-like structure. The BChl a molecules are arranged in two pigment pools labeled B800 and B850, according to their room-temperature absorption maxima in the near infrared. The B800 assembly comprises nine well-separated BChl a molecules which have their bacteriochlorin plane aligned nearly perpendicularly to the symmetry axis whereas the B850 assembly comprises 18 BChl a molecules in close contact oriented with the plane of the molecules parallel to the symmetry axis, see Fig. 26.1.

In contrast to LH2, where highly resolved X-ray structures are available, the discussion in the literature about the structural properties of RC-LH1 complexes is much more controversial. A detailed discussion on this issue can be found in [1,7,8]. Initial structural models of the RC-LH1 complex pictured the RC completely surrounded by a closed LH1 ring. Similar to LH2, the basic structural unit of LH1 is an $\alpha\beta$-heterodimer, which binds two molecules of BChl a and one or two molecules of carotenoid. These dimers then oligomerize around the RC to form the closed ring. The pairs of BChl a molecules from each dimer interact together to form a strongly coupled ring of pigments giving rise to the strong absorption band in the 870–890 nm spectral region. The spectroscopic properties of this ring reflect the strong excitonic coupling among these BChl a molecules. Based on these models, an obvious question arises. How does UQH_2 escape from such a core complex? There are two possible solutions, either the LH1 ring is not complete, i.e., there is a gap, or the LH1 structure is inherently flexible enough to allow UQH_2 to diffuse through it [9].

a   b

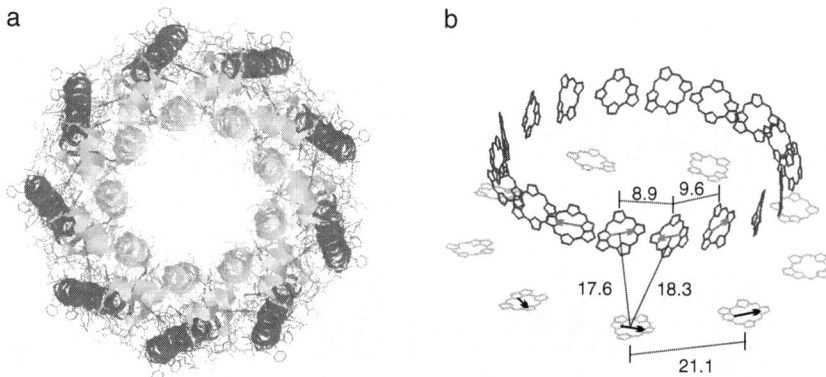

Fig. 26.1. Structure of the LH2 complex from *Rps. acidophila* as determined by X-ray diffraction. Part (**a**) shows the whole pigment-protein complex in a top view; part (**b**) displays the spatial arrangement of the BChl a in a tilted side-view. The B800 BChl a pigments are shown in light gray and the B850 BChl a molecules are shown in dark gray. The numbers indicate the centre to centre distances of the pigments in Å. The arrows indicate the direction of the Q_y transition moments. The phytol chains of the pigments are removed for clarity. Adapted from [3]

During the last decade, there have been reports that the RC-LH1 complexes are circular [10], square [11], "S" shaped [12–14], elliptical [15, 16], or even just arcs [17]. Indeed this field has become, and indeed still is, very confused. There is a further complication, especially in species such as *Rhodobacter (Rb.) sphaeroides*, which relates to a protein called PufX. When this protein is present in the RC-LH1 complex they are dimeric. Whereas in a PufX$^-$ phenotype, the RC-LH1 complex is monomeric [13]. PufX appears to be a member of the LH1 ring, replacing one of the $\alpha\beta$-dimers. This introduces a gap through which it has been proposed that the UQH$_2$ could pass. The current view is that there are at least two distinct classes of RC-LH1 complexes. One class is monomeric, i.e., consist of one RC surrounded by one LH1 complex. Examples of this class are the RC-LH1 complexes from *Rhodospirillum (Rsp.) rubrum* and *Rps. palustris* [18]. The second class is dimeric, i.e., consist of two RC-LH1 units. An example of this class is the RC-LH1 complex *Rb. sphaeroides* [13, 19].

Here, we focus on the RC-LH1 complex from *Rps. palustris* for which the first X-ray structure has been determined recently, see Fig. 26.2 [20]. Even though at this relatively low resolution the structure must be considered somewhat tentative, its main features seem quite clear. An elliptical LH1 complex surrounds the RC and adjacent to the RC-UQ-binding site, from where UQH$_2$ must leave, there is a gap in the LH1 ring. This gap is associated with a protein called W, which replaces an $\alpha\beta$-dimer and which is believed to be an orthologue of PufX. The presence or absence of such a gap and its functional significance has become rather controversial [21, 22].

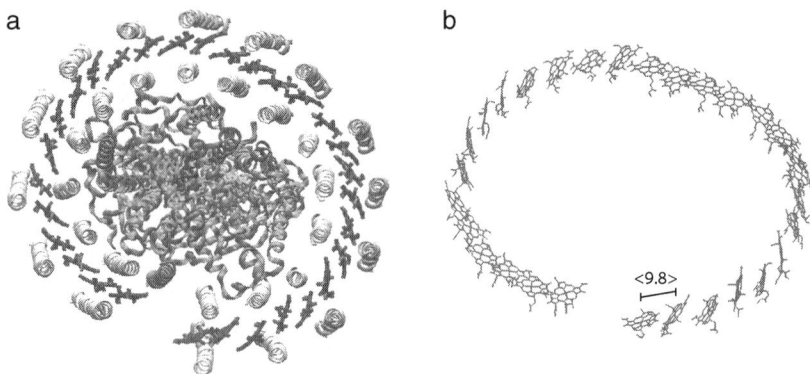

Fig. 26.2. Structure of the RC-LH1 complex from *Rps. palustris* as determined by X-ray diffraction. Part (**a**) shows the whole pigment-protein complex in a top view; part (**b**) displays the spatial arrangement of the BChl *a* pigments in LH1 of *Rps. palustris*, in a tilted side-view. The number indicates the average centre to centre distance of the pigments in Å. Adapted from [20]

26.2 Single-Molecule Spectroscopy on Light-Harvesting Complexes from Purple Bacteria

Important parameters that determine the character of the electronically excited states of the LH complexes are the transition energy of the BChl *a* molecules, E_0, the spread in transition energies, ΔE (diagonal disorder), and the intermolecular interaction strength V between neighboring BChl *a* molecules. V is mainly determined by the intermolecular distance and the relative orientation of the molecular transition-dipole moments. Variations in site energies, ΔE, can often be attributed to structural variations in the environment of the BChl *a* molecules, leading to changes in the electrostatic interaction with the surrounding protein. Generally, information about these parameters can be obtained by optical spectroscopy [23–26], however, the great difficulty encountered is the fact that the optical absorption lines are inhomogeneously broadened as a result of heterogeneity in the ensemble of absorbing pigments. To avoid these difficulties, single-molecule studies on pigment-protein complexes from purple bacteria were conducted, initially under ambient conditions. Fluctuations of the emission intensity as well as fluctuations of the polarization state of the emitted light were observed [27,28]. Although still in use [29–31], the information that can be extracted from these systems by single-molecule spectroscopy at room temperature is rather limited. First, photobleaching of the chromophores usually restricts the observation times to some 10 s. Since the main causes for photobleaching are photochemical reactions in the electronically excited state in the presence of oxygen, these processes play a negligible role at cryogenic temperatures simply due to the lack of (mobile) oxygen. Second, at room temperature the

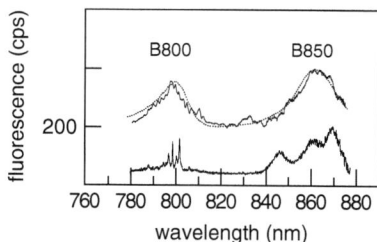

Fig. 26.3. Fluorescence-excitation spectra of LH2 complexes of *Rps. acidophila*. The top traces show the comparison between an ensemble spectrum (dotted line) and the sum of spectra recorded from nineteen individual complexes (full line). The lower trace displays the spectrum from a single LH2 complex. The spectra have been averaged over all polarizations of the incident radiation. All spectra were measured at 1.2 K at 20 W/cm^2 with LH2 dissolved in a PVA-buffer solution. Adapted from [39]

thermal broadening of the spectral features is so large that details in the optical spectra remain obscured.

Employing cryogenic temperatures allowed to retrieve valuable information about the character of the electronically excited states of the antenna complexes [10, 32–38]. As an example, Fig. 26.3 displays the fluorescence-excitation spectrum of an individual LH2 complex [39]. The upper trace shows, for comparison, the fluorescence-excitation spectrum taken from a bulk sample (dotted line) together with the spectrum that results from the summation of the spectra of 19 individual LH2 complexes (full line). The two spectra are in excellent agreement and both feature two broad structureless bands around 800 and 860 nm corresponding to the absorptions of the B800 and B850 pigments of the complex. By measuring the fluorescence-excitation spectra of the individual complexes, remarkable features become visible which are obscured in the ensemble average. The spectra around 800 nm show a distribution of narrow absorption bands, whereas in the B850 spectral region 2–3 broad bands are present. The striking differences between the B800 and B850 absorption bands of LH2 reflect that the ratio $V/\Delta E$ differs by about an order of magnitude in the two ring assemblies. In first approximation, the excitations of the B800 BChl a molecules can be described as being localized on an individual BChl a molecule whereas for the optical spectra of the B850 assembly excitonic interactions have to be considered [1, 2, 40–42].

Meanwhile, low-temperature single-molecule spectroscopy has become a versatile tool to study the properties of the electronically excited states of the LH complexes from purple bacteria. For example, the robustness of the LH process in purple bacteria has been demonstrated by the observation of the excitation-energy transfer within a single self-aggregated photosynthetic unit in a nonmembrane environment under cryogenic conditions [43]. In more recent studies, details about the electronic coupling between the BChl a chromophores in LH2 and the electron–phonon coupling between these chromophores and the protein backbone have been uncovered [44, 45]. Changing the point of view and considering the weakly coupled B800 BChl a molecules as local probes to monitor their local environment led to insights about the organization of the energy landscape within the binding pocket [46–50]. Reviews about the single-molecule work on bacterial LH complexes can be found in [1, 7, 51–53].

In Sect. 26.3, we address how the symmetry of the arrangement of the BChl a molecules affects the exciton states and their spectroscopic properties.

26.3 Excitons and Symmetry

The concept of Frenkel excitons (molecular excitons) [54] provides a good starting point for the description of the electronically excited states of the LH complexes. The respective model Hamiltonian reads in Heitler–London approximation

$$H = \sum_{n=1}^{N} (E_0 + \Delta E_n) |n\rangle \langle n| + \frac{1}{2} \sum_{n=1}^{N} \sum_{m \neq n} V_{nm} |n\rangle \langle m| \qquad (26.1)$$

where $|n\rangle$ and $|m\rangle$ correspond to excitations localized on molecules "n" and "m," respectively, $(E_0 + \Delta E_n)$ denotes the site energy of pigment "n" which is separated into an average, E_0, and a deviation from this average, ΔE_n (diagonal disorder), and V_{nm} denotes the interaction between molecules "n" and "m." The eigenstates of the Hamiltonian can be obtained by numerical diagonalization and correspond to the electronically excited states (exciton states $|k\rangle$) of the array of pigments. Once the $|k\rangle$-states are determined, the transition probabilities and polarization properties can be calculated for each exciton state.

In order to illustrate the influence of symmetry (or the lack of it) on the properties of the exciton states of the ring of BChl a molecules in RC-LH1, we present the calculated excited state manifold for three different geometries: (i) a circular symmetric assembly of 32 BChl a molecules, (ii) an elliptical assembly of 32 BChl a molecules, and (iii) an overall elliptical assembly of 30 BChl a molecules that feature a gap (as in the *Rps. palustris* structure, Fig. 26.2). For the circular symmetric arrangement, one obtains two nondegenerate and 15 pairwise degenerate exciton states, Fig. 26.4, top. The nondegenerate exciton states are labeled by the quantum numbers $k = 0$ and $k = 16$, and the degenerate exciton states are refered to as $k = \pm 1, \pm 2, \ldots, \pm 15$. Due to the high symmetry of the circular ring arrangement, the oscillator strength is concentrated in the lowest degenerate pair of the exciton states, i.e., $k = \pm 1$. The dominant effect of an elliptical distortion is that the pairwise degeneracies of the exciton states will be lifted and that oscillator strength from the $k = \pm 1$ states is redistributed to the $k = \pm 3$ states, leading to several spectral bands in the absorption spectrum

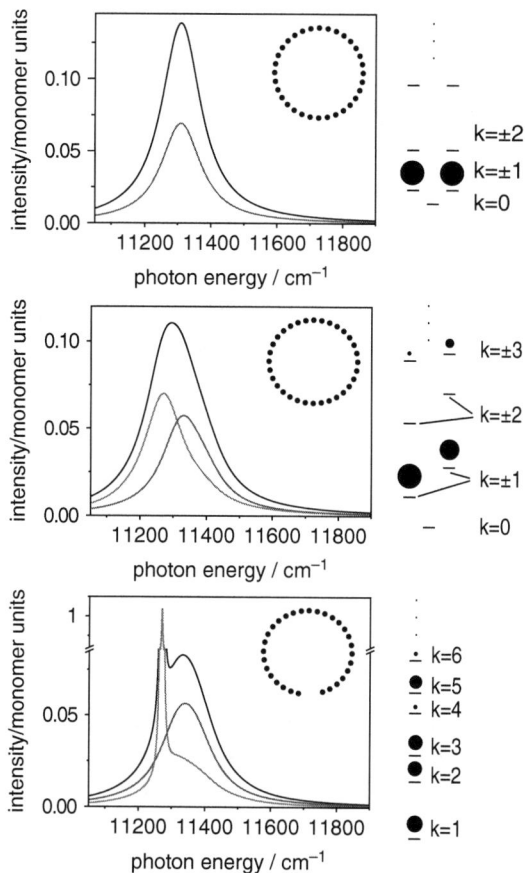

Fig. 26.4. Calculated spectra (*left*) and exciton manifold (*right*) for three qualitatively different BChl *a* arrangements, i.e., circular symmetric (*top*), elliptic (*centre*), and elliptic with a gap (*bottom*). The black circles indicate the oscillator strength of the respective exciton states. For each geometry two spectra for mutually orthogonal polarizations (*black, gray*) are shown. For the circular symmetric geometry the two spectra coincide. The upper black line corresponds to the sum of these spectra. Adapted from [55]

[42]. Since the relaxation of the higher exciton states occurs on an ultrafast timescale of about 100 fs [23, 26], the absorption spectrum for a closed structure, Fig. 26.4, top and centre, consists of either one or a few relatively broad spectral bands, respectively. For both cases, i.e., circular and elliptical arrangement, the transitions from the $k = \pm 1$ exciton states are polarized perpendicular with respect to each other. Moreover, the lowest exciton state is optically forbidden, because a C_2-type symmetry reduction alone, i.e., an ellipse, does not give rise to oscillator strength in the $k = 0$ state. This situation is reminescent to the electronically excited states of the B850 BChl *a*

molecules of LH2 [39,42]. Striking consequences for the absorption spectra are expected by introducing a symmetry-breaking gap in the BChl a assembly. The presence of such a gap in the LH1 ring is equivalent to the case of a linear excitonic system where due to the loss of symmetry the resulting exciton states are commonly referred to as $k = 1, 2, \ldots$. As it has been well described for linear excitonic systems, such as J-aggregates [56], now the lowest exciton state, i.e., $k = 1$ in this case, gains considerable oscillator strength. Since the lifetime of this state is longer than a few hundred picosecond, the associated optical transition will appear as a relatively narrow feature in the low-energy wing of the absorption spectrum, Fig. 26.4, bottom.

26.4 Results and Discussion

26.4.1 Fluorescence Excitation Spectra from Individual RC-LH1 Complexes

Since the details in the optical spectra are averaged out in a large ensemble of pigment-protein complexes, single-molecule spectroscopy has to be used to look for the spectroscopic signature of a gap in the electronic structure of LH1 BChl a molecules. In Fig. 26.5, we show a comparison of several fluorescence-excitation spectra from RC-LH1 core complexes from *Rps. palustris*. The top traces show an ensemble spectrum (black) and the sum spectrum (gray) from 41 individual complexes. The ensemble spectrum features a broad band at $11,322 \, \mathrm{cm}^{-1}$ with a width of $378 \, \mathrm{cm}^{-1}$ (FWHM) and two weak shoulders at $11,545 \, \mathrm{cm}^{-1}$ and $11,655 \, \mathrm{cm}^{-1}$, respectively. This spectrum is rather well reproduced by the spectrum that results from summing the 41 individual spectra, indicating that the selected individual complexes are a fair statistical representation of the ensemble. The lower traces display examples of fluorescence-excitation spectra recorded from individual RC-LH1 complexes. Common to these spectra are variations in the spectral positions and the widths of the observed bands. However, the most striking feature in the spectra from individual core complexes is indeed a narrow spectral line on the low-energy side. As pointed out earlier, the general features of these spectra can be understood on the basis of a simple exciton model for the lowest electronically excited states that takes a physical gap in the BChl a arrangement into account as has been observed in the X-ray structure.

26.4.2 The Narrow Spectral Feature

For the studied individual RC-LH1 complexes from *Rps. palustris*, the distribution of the spectral positions of the narrow feature is shown in Fig. 26.6a together with an ensemble absorption spectrum. The histogram is centered at $11,195 \, \mathrm{cm}^{-1}$ in the red wing of the ensemble spectrum and covers a spectral range of about $125 \, \mathrm{cm}^{-1}$, which illustrates the spectral heterogeneity

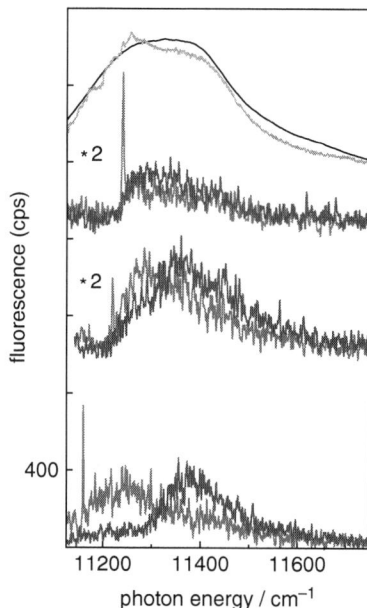

Fig. 26.5. Fluorescence-excitation spectra from RC-LH1 complexes from *Rps. palustris*. The top traces show the comparison between an ensemble spectrum (*black line*) and the sum of about 41 spectra recorded from individual complexes (*gray line*). The lower traces show spectra from single RC-LH1 complexes. For each individual complex two spectra, recorded with mutually orthogonal polarization of the excitation light, are displayed. The vertical scale is valid for the two lowest traces, the other spectra have been scaled by a factor of two and are offset for clarity. Adapted from [55]

(intercomplex disorder) of the ensemble of RC-LH1 complexes. These findings are consistent with hole-burning action spectra [57–60]. The spectral width of the narrow absorption showed a distribution as well, Fig. 26.6b, which ranges from 1 to $5\,\mathrm{cm}^{-1}$ where most of the entries cover the range between 1 and $2\,\mathrm{cm}^{-1}$. This width would correspond to a lifetime of the lowest exciton state of about 33 ps, which is significantly shorter than the actual lifetime of this state, indicating that these linewidths were determined by spectral diffusion rather than the lifetime limited value. This has been verified by using a single-mode laser of less than 1 MHz spectral bandwidth [61]. Figure 26.7a shows an expanded view of the narrow spectral feature for a sequence of spectra from an individual RC-LH1 complex from *Rps. palustris* in a two-dimensional representation where 23 individual scans are stacked on top of each other. The horizontal axis of the pattern in Fig. 26.7a corresponds to the relative photon energy, the vertical axis to the scan number, and the detected fluorescence intensity is coded by the gray scale. The sum spectrum of these scans is shown at the bottom of Fig. 26.7a and features a spectral line with

522   M.F. Richter et al.

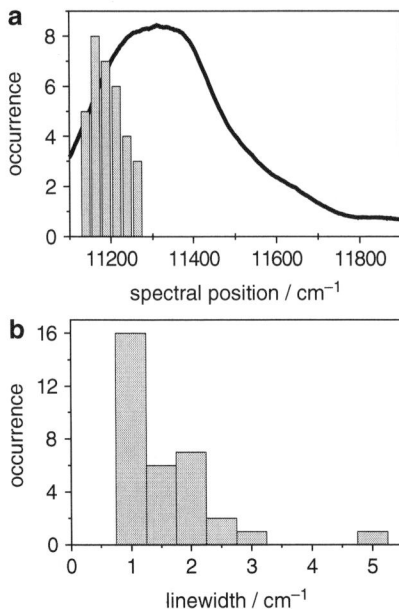

Fig. 26.6. (a) Distribution of the spectral positions of the narrow feature observed for individual RC-LH1 complexes from *Rps. palustris*. The bin width is $25\,\mathrm{cm}^{-1}$ and the bold line corresponds to the ensemble absorption spectrum, which has been overlaid for illustration. (b) Distribution of the spectral bandwidth of the narrow feature. The bin width is $0.5\,\mathrm{cm}^{-1}$. Adapted from [55]

a width of $1.8\,\mathrm{cm}^{-1}$ (FWHM). Figure 26.7b shows some individual scans of this feature together with a Lorentzian that has been fitted to the absorption profile. These spectra clearly demonstrate that both the spectral position and the width of the narrow feature vary with time.

26.4.3 Structural Models

In order to analyze the spectral bands from the individual core complexes in more detail, we recorded the fluorescence-excitation spectra as a function of the polarization of the incident radiation. The excitation spectra have been recorded in rapid succession and the polarization of the excitation light has been rotated by $6.4°$ between consecutive scans. An example of this protocol is shown in the top part of Fig. 26.8a in a two-dimensional representation where 312 individual scans are stacked on top of each other. The horizontal axis corresponds to photon energy, the vertical axis to the individual scans, or equivalently to the polarization of the excitation, and the detected fluorescence intensity is coded by the gray scale. The sum spectrum of these scans is presented at the bottom of Fig. 26.8a and shows two broad bands at $11{,}253$ and $11{,}398\,\mathrm{cm}^{-1}$ with a linewidth of 250 and $153\,\mathrm{cm}^{-1}$ (FWHM),

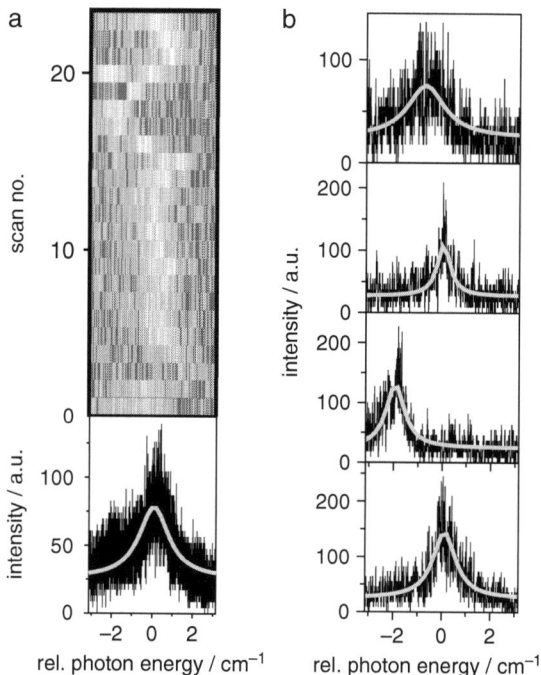

Fig. 26.7. Sequence of fluorescence-excitation spectra of the narrow spectral feature recorded with the single-mode laser. (**a**) Stack of 23 fluorescence-excitation spectra recorded at a scan speed of $0.2\,\mathrm{cm}^{-1}/\mathrm{s}$ ($5\,\mathrm{GHz/s}$) and an excitation intensity of $0.5\,\mathrm{W/cm}^2$. The fluorescence intensity is indicated by the gray scale. The averaged spectrum is shown in the lower panel and features a linewidth of $1.8\,\mathrm{cm}^{-1}$ (FWHM). (**b**) Individual fluorescence-excitation spectra together with Lorentzian fits (solid line). From top to bottom the linewidths (FWHM) are $1.8\,\mathrm{cm}^{-1}$, $0.7\,\mathrm{cm}^{-1}$, $0.9\,\mathrm{cm}^{-1}$ and $1.1\,\mathrm{cm}^{-1}$, respectively. Adapted from [61]

respectively. Again a narrow feature appears at the low-energy side, which is barely visible in the sum spectrum. In the two-dimensional representation of the data, however, the narrow feature is clearly observable as an intense stripe that undergoes spectral diffusion. The pattern clearly reveals the polarization dependence of the three absorptions. This becomes even more evident in Fig. 26.8b. The bottom part shows two individual scans that have been recorded with mutually orthogonal polarization, where the angle of polarization that yields the maximum intensity for the narrow spectral feature has been set arbitrarily to "horizontal" and provides the reference point. The top part of Fig. 26.8b shows the fluorescence intensity of the three bands as a function of the polarization of the excitation light (dots) and is consistent with a $\cos^2$ dependence (black line). As discussed earlier, the narrow spectral feature is assigned to the $k = 1$ exciton transition. Numerical simulations, that will be discussed in more detail later, show that the $k = 2$ exciton state

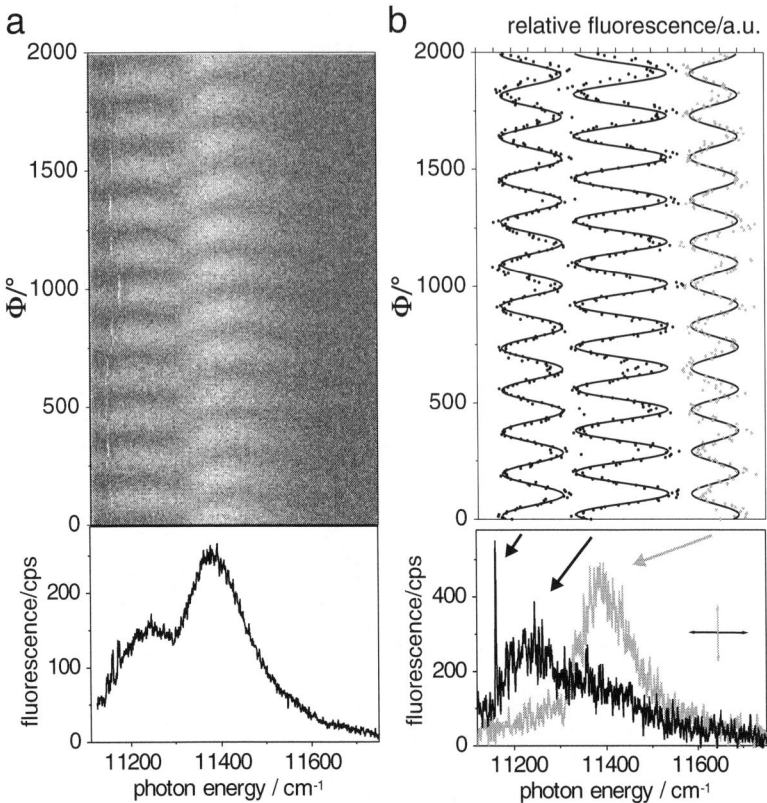

Fig. 26.8. Fluorescence-excitation spectrum from an individual RC-LH1 complex from *Rps. palustris* as a function of the polarization of the excitation light. (**a**) *Top*: Stack of 312 individual spectra recorded consecutively. Between two successive spectra the polarization of the incident radiation has been rotated by 6.4°. The horizontal axis corresponds to the photon energy, the vertical axis to the scan number or equivalently to the polarization angle and the intensity is given by the gray scale. The excitation intensity was $10\,\mathrm{W/cm^2}$. *Bottom*: Spectrum that corresponds to the average of the 312 consecutively recorded spectra. (**b**) *Top*: Fluorescence intensity of the three bands marked by the arrows in the lower part as a function of the polarization of the incident radiation (*dots*) together with $\cos^2$-type functions fitted to the data (*black*). *Bottom*: Two fluorescence-excitation spectra from the stack that correspond to mutually orthogonal polarization of the excitation light. The spectra where chosen such that the "horizontal" polarization yielded maximum intensity for the narrow feature at the low-energy side. Adapted from [62]

contributes preferentially to the adjacent broad absorption (at $11{,}253\,\mathrm{cm^{-1}}$ in Fig. 26.8b), whereas the other broad band (at $11{,}398\,\mathrm{cm^{-1}}$ in Fig. 26.8b) corresponds to a superposition of transitions from exciton states with quantum numbers $k \geqslant 3$. This is also reflected by the curvature of this band in the 2d representation in Fig. 26.8a, because the mutual orientations of the transition-dipole moments of the higher exciton states vary with respect to each other.

In the following, for a better comparison between the experimental data and our simulations, we focus on the data obtained for the $k = 1$, and $k = 2$ exciton transitions, since the individual contributions from the $k \geqslant 3$ exciton states to the second broad absorption band vary substantially as a function of the diagonal disorder. In the example shown in Fig. 26.8, we find for the spectral separation of the $k = 1$ and $k = 2$ exciton transitions $\Delta E = 90\,\mathrm{cm}^{-1}$ and for the respective angle between their transition-dipole moments $\Delta\alpha = 7°$. The histograms for ΔE and $\Delta\alpha$ obtained from 33 individual core complexes are displayed in the two lower rows of Fig. 26.9, together with the results from numerical simulations (black squares) that will be discussed later. The histogram for ΔE is centered at $50\,\mathrm{cm}^{-1}$ and has a width of about $60\,\mathrm{cm}^{-1}$. The distribution for $\Delta\alpha$ displays a maximum around $5°$, and only very few entries can be found for $\Delta\alpha > 20°$, indicating that the transition-dipole moments of the $k = 1$ and $k = 2$ exciton states are oriented nearly parallel with respect to each other. A correlation between the values for ΔE and $\Delta\alpha$ from an individual complex was not observed.

We compared these data with results from numerical simulations based on the Hamiltonian in (26.1) for various models of the pigment arrangement and thereby seek to test the crystallographic structure model. In order to keep things relatively simple, the average site energies were set to $E_0 = 11{,}570\,\mathrm{cm}^{-1}$ for all pigments and only diagonal disorder, taken randomly from a Gaussian distribution with a width of $100\,\mathrm{cm}^{-1}$ (FWHM), was considered. Further, we omitted off-diagonal disorder and restricted the interaction to nearest-neighbors only. The only variation that we allowed for the different models was the arrangement of the BChl a molecules, i.e., mutual distance and orientation, and the concomitant variation of their interaction strengths. As a prerequisite for this analysis, the tested structures had to be compatible with the X-ray structure, which we used as a starting point. As the simplest approach to evaluate the dependence of the interaction on the distance and the mutual orientation of the pigments, we used a dipole–dipole type approach. Accordingly, the proper eigenstates were calculated from diagonalizing the Hamiltonian (26.1), which provided the energy, the oscillator strength, and the orientation of the transition-dipole moment for each exciton state. Each model was run for 3,000 individual realizations of the diagonal disorder. More details about the simulation procedure can be found in [42, 62].

In model A, the mutual orientations and distances of the pigments correspond to the data provided by the protein database which reflects the X-ray structure (see top row Fig. 26.9, left). For models B and C, we distributed 15 BChl a dimers on an ellipse with an eccentricity of $\varepsilon = 0.5$, where each dimer was taken to be homologous to the B850 BChl a dimers in LH2 from *Rps. acidophila* [3]. For model B, the dimers were distributed around the ellipse such that their mutual angle was fixed to $\Phi = 22.5°$ (see top row

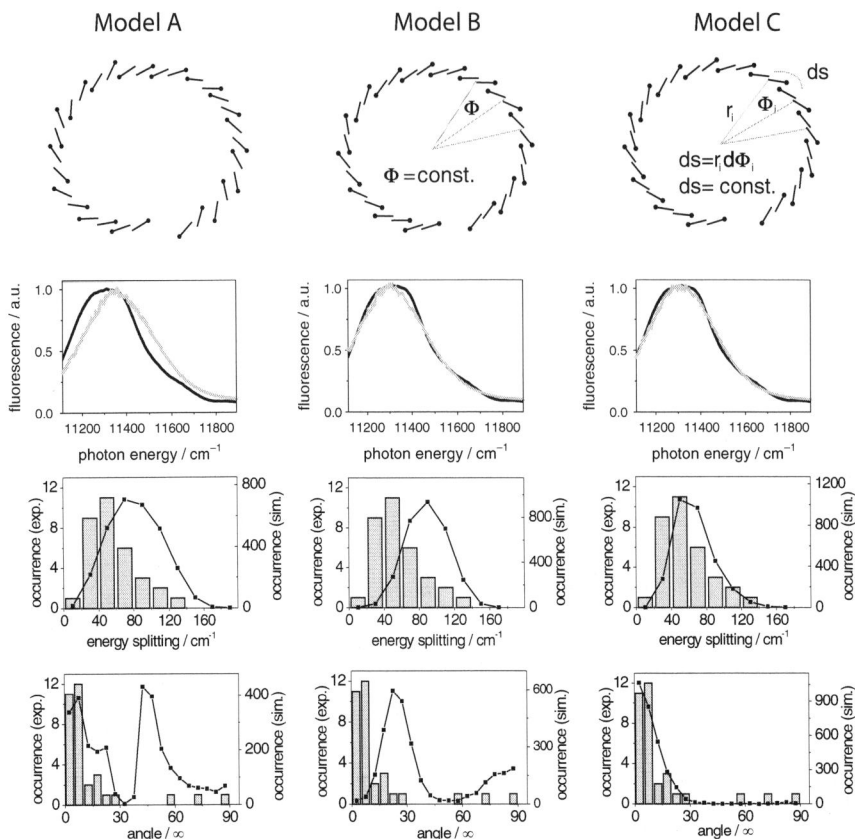

Fig. 26.9. Comparison of the energetic separation and the relative orientation of the transition-dipole moments of the $k = 1$ and $k = 2$ exciton states from individual RC-LH1 complexes with results from numerical simulations for three different arrangements of the BChl a molecules in the pigment-protein complex. The top row shows the model structures A–C that have been used for the numerical simulations. Details are given in the text. The second row compares the experimental ensemble absorption spectrum (*black line*) with the ensemble spectrum that results from numerical simulation (*gray line*) for the three model structures. The third row compares the experimentally obtained energetic separations between the $k = 1$ and $k = 2$ exciton states (*gray columns*) with numerical simulations (*black squares*) for the three model structures. The fourth row compares the experimentally obtained relative orientations of the transition-dipole of the $k = 1$ and $k = 2$ exciton states (*gray columns*) with numerical simulations (*black squares*) for the three model structures. Adapted from [62]

Fig. 26.9, centre, $\Phi = $ const.), whereas for model C the dimers were placed equidistantly around the ellipse, i.e., keeping the length of the arc $ds = r_i d\Phi_i$ between the dimers on the ellipse constant (see top row Fig. 26.9 right, $ds = $ const.). Although for both models the average interaction strengths between the BChl a molecules are about the same, the modulation of the interaction strengths between adjacent molecules differs significantly between models B and C around the ellipse. For model B, the inter-pigment distance is shortest along the long axis and longest along the short axis of the ellipse, and *vice versa* for model C. The black squares in Fig. 26.9 show, from left to right, the results of the simulations for the distributions of ΔE and $\Delta\alpha$ for models A–C together with the experimental data. The simulations based on the structure taken from the protein database (model A) are not able to reproduce the experimental histograms. The simulated distribution for ΔE is significantly broader than that observed experimentally. The simulation is even worse for $\Delta\alpha$ and yields a bimodal distribution with maxima around 8° and 43° which is clearly not compatible with the experimental data. For model B, the simulated distribution for ΔE shows a peak at $90 \, \text{cm}^{-1}$ (significantly shifted from that of the experimental data) with a width of $62 \, \text{cm}^{-1}$. Moreover, the distribution for $\Delta\alpha$ again shows a bimodal shape with maxima at 23° and 88°. Finally, with model C we calculate for ΔE a distribution that peaks at $50 \, \text{cm}^{-1}$ and features a width of $51 \, \text{cm}^{-1}$. The distribution predicted for $\Delta\alpha$ with model C shows a maximum at 0° and decreases rapidly toward larger values. Comparing the three models, it is obvious that neither model A nor model B describes the single-molecule data satisfactorily. Only in model C can we find a reasonable agreement between the simulations and the experimental data. The slight discrepancies still observed between simulation and experiment probably reflect the fact that we employed a rather simple exciton model, i.e., equal values for all site energies E_0, restricting the interaction to nearest neighbors, and taking only random diagonal disorder into account. Here, we have restricted the discussion of the model structures to three examples. However, we have tried many more models than those shown here. For structures reminiscent of models B and C, we also shifted the geometrical position of the gap around the ellipse. However, in all the models tested, only the one with the gap as presented in model C earlier (i.e., as shown in the X-ray structure, Fig. 26.2), allowed the simulations to reproduce the essential findings of the experimental data. For illustration, the second row of Fig. 26.9 compares the ensemble absorption spectrum (black line) with the resulting simulated ensemble spectrum (gray line) for the three model structures. Interestingly, the ensemble absorption spectrum is reproduced by all tested model structures. Obviously, agreement between a simulated and an experimental ensemble spectrum is a minimum prerequisite that has to be fulfilled by any proposed structural model, but this does not give much insight into the underlying structure of the pigment-protein complexes. In contrast, the use of the detailed single-molecule spectrocopic data provides a much more stringent test of these structural models.

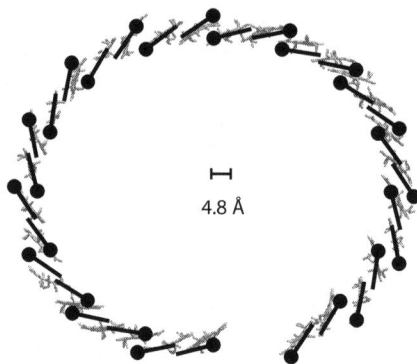

Fig. 26.10. Position and orientation of the BChl *a* molecules of the RC-LH1 complex from *Rps. palustris* as taken from the protein database (*gray*) and from model C (*black*). The scale bar indicates the resolution of the X-ray structure. The change in the positions of the BChl *a* molecules in the refined structure are well within the limits of accuracy of the X-ray model. Adapted from [62]

Finally, Fig. 26.10 displays the positions of the BChl *a* molecules as obtained from model A, the X-ray structure (gray), and according to model C (black), respectively. A comparison testifies that the difference between the two model structures corresponds only to very slight changes of the positions of the BChl *a* molecules, well within the limits of the accuracy of the X-ray structure. Hence, based on the X-ray structure as a starting point, the exploitation of low-temperature single-molecule spectroscopy allows us to propose a refined structural model for the RC-LH1 complex from *Rps. palustris*.

26.5 Conclusions

Employing low-temperature single-molecule spectroscopy allows to unmask subtle details in the fluorescence-excitation spectra from entire RC-LH1 complexes from *Rps. palustris*. Numerical simulations as a function of the geometrical arrangement of the BChl *a* molecules yield the best agreement with the experimental data for an equidistant arrangement of 15 BChl *a* dimers on an ellipse, where the BChl *a* molecules within a dimer are taken homologeous to those of the B850 BChl *a* molecules in LH2 from *Rps. acidophila*. Furthermore, successful modeling of the experimental data can only be achieved when the gap in the LH1 ellipse is placed in the same position as that shown in the X-ray structure. The single molecule data therefore allow us to propose a refined structural model of the RC-LH1 complex.

Acknowledgements

We thank June Southall for the preparation of the light-harvesting complexes. RJC thanks the BBSRC for funding.

References

1. R.J. Cogdell, A. Gall, J. Köhler, The architecture and function of purple bacteria: from single molecules to *in vivo* membranes. Quart. Rev. Biophys. 39, 227–324 (2006)
2. X. Hu, T. Ritz, A. Damjanovic, F. Autenrieth, K. Schulten, Photosynthetic apparatus of purple bacteria. Quart. Rev. Biophys. 35, 1–62 (2002)
3. G. McDermott, S.M. Prince, A.A. Freer, A.M. Hawthornthwaite-Lawless, M.Z. Papiz, R.J. Cogdell, N.W. Isaacs, Crystal structure of an integral membrane light-harvesting complex from photosynthetic bacteria. Nature 374, 517–521 (1995)
4. J. Koepke, X. Hu, C. Muenke, K. Schulten, H. Michel, The crystal structure of the light harvesting complex II (B800–B850) from *Rhodospirillum molischianum*. Structure 4, 581–597 (1996)
5. K. McLuskey, S.M. Prince, R.J. Cogdell, N.W. Isaacs, The crystallographic structure of the B800–820 Lh3 light-harvesting complex from the purple bacteria *Rhodopseudomonas acidophila* Strain 7050. Biochem. 40, 8783–8789 (2001)
6. M.Z. Papiz, S.M. Prince, T. Howard, R.J. Cogdell, N.W. Isaacs, The structure and thermal motion of the B800–850 LH2 complex from *Rps. acidophila* at 2.0 (A)over-circle resolution and 100 K: new structural features and functionally relevant motions. J. Mol. Biol. 326, 1523–1538 (2003)
7. T.J. Aartsma, J. Matysik, Advanced in Photosynthesis and Respiration. Vol. 26: Biophysical Techniques in Photosynthesis, Vol. II (Springer, Dordrecht, 2008)
8. B. Grimm, R.J. Porra, R. Wolfhart, H. Scheer, in Advances in Photosynthesis and Respiration Vol. 25: Chlorophylls and Bacteriochlorophylls: Biochemistry, Biophysics, Functions and Applications (Series ed. Govindjee) (Springer, Dordrecht, 2004)
9. A. Aird, J. Wrachtrup, K. Schulten, C. Tietz, Possible pathway for ubiquinone shuttling in *Rhodospirillum rubrum* revealed by molecular dynamics simulation. Biophys. J. 92, 23–33 (2007)
10. U. Gerken, D. Lupo, C. Tietz, J. Wrachtrup, R. Ghosh, Circular symmetry of the light-harvesting 1 complex from *Rhodospirillum rubrum* is not perturbed by interaction with the reaction center. Biochemistry 42, 10354–10360 (2003)
11. H. Stahlberg, J. Dubochet, H. Vogel, R. Gosh, Are the light harvesting I complexes from *Rhodospirillum rubrum* arranged around the reaction centre in a square geometry? J. Mol. Biol. 282, 819–831 (1998)
12. S. Scheuring, F. Francia, J. Busselez, B.A. Melandri, J.L. Rigaud, D. Levy, Structural role of Pufx in the dimerization of the photosynthetic core complex of *Rhodobacter sphaeroides*. J. Biol. Chem. 279, 3620–3626 (2004)
13. C.A. Siebert, P. Qian, D. Fotiadis, A. Engel, C.N. Hunter, P.A. Bullough, Molecular architecture of photosynthetic membranes in *Rhodobacter sphaeroides*: the Role of Pufx. EMBO J. 23, 690–700 (2004)

14. R.P. Goncalves, J. Busselez, D. Levy, J. Seguin, S. Scheuring, Membrane insertion of *Rhodopseudomonas acidophila* light harvesting complex 2 investigated by high resolution AFM. J. Struct. Biol. 149, 79–86 (2005)

15. S. Scheuring, J. Seguin, S. Marco, D. Levy, B. Robert, J.L. Rigaud, Nanodissection and high-resolution imaging of the Rhodopseudomonas viridis photosynthetic core complex in native membranes by AFM. Atomic force microscopy. Proc. Natl. Acad. Sci. 100, 1690–1693 (2003)

16. D. Fotiadis, P. Qian, A. Philippsen, P.A. Bullough, A. Engel, C.N. Hunter, Structural analysis of the reaction center light-harvesting complex I photosynthetic core complex of *Rhodospirillum rubrum* using atomic force microscopy. J. Biol. Chem. 279, 2063–2068 (2004)

17. S. Bahatyrova, R.N. Frese, C.A. Siebert, J.D. Olsen, K.O. van der Werf, R. van Grondelle, R.A. Niederman, P.A. Bullough, C. Otto, C.N. Hunter, The native architecture of a photosynthetic membrane. Nature 430, 1058–1062 (2004)

18. S. Karrasch, P.A. Bullough, R. Ghosh, The 8.5 Å projection map of the light harvesting complex I from *Rhodospirillum rubrum* reveals a ring composed of 16 subunits. EMBO J. 14, 631–638 (1995)

19. F. Francia, J. Wang, G. Venturoli, B.A. Melandri, W.P. Barz, D. Oesterhelt, The reaction center-LH1 antenna complex of *Rhodobacter sphaeroides* contains one pufx molecule which is involved in dimerization of this complex. Biochemistry 38, 6834–6845 (1999)

20. A.W. Roszak, T.D. Howard, J. Southall, A.T. Gardiner, C.J. Law, N.W. Isaacs, R. Cogdell, Crystal structure of the RC-LH1 core complex from *Rhodopseudomonas palustris*. Science 302, 1969–1971 (2003)

21. P. Qian, C. Neil Hunter, P.A. Bullough, The 8.5 A projection structure of the core RC-LH1-PufX dimer of *Rhodobacter sphaeroides*. J. Mol. Biol. 349, 948–960 (2005)

22. S. Scheuring, R.P. Goncalves, V. Prima, J.N. Sturgis, The photosynthetic apparatus of *Rhodopseudomonas palustris*: structures and organization. J. Mol. Biol. 358, 83–96 (2006)

23. T. Pullerits, V. Sundström, Photosynthetic light-harvesting pigment-protein complexes: toward understanding how and why. Acc. Chem. Res. 29, 381–389 (1996)

24. H.-M. Wu, M. Rätsep, I.-J. Lee, R.J. Cogdell, G.J. Small, Exciton level structure and energy disorder of the B850 ring of the LH2 antenna complex. J. Phys. Chem. B 101, 7654–7663 (1997a)

25. T.M.H. Creemers, C. de Caro, R.W. Visschers, R. van Grondelle, S. Völker, Spectral hole burning and fluorescence line narrowing in subunits of the light harvesting complex LH1 of purple bacteria. J. Phys. Chem. B 103, 9770–9776 (1999)

26. V. Sundström, T. Pullerits, R. van Grondelle, Photosynthetic light-harvesting: reconciling dynamics and structure of purple bacterial LH2 reveals function of photosynthetic unit. J. Phys. Chem. B 103, 2327–2346 (1999)

27. M.A. Bopp, Y. Jia, L. Li, R.J. Cogdell, R.M. Hochstrasser, Fluorescence and photobleaching dynamics of single light-harvesting complexes. Proc. Natl. Acad. Sci. 94, 10630–10635 (1997)

28. M.A. Bopp, A. Sytnik, T.D. Howard, R.J. Cogdell, R.M. Hochstrasser, The dynamics of structural deformations of immobilized single light-harvesting complexes. Proc. Natl. Acad. Sci. 96, 11271–11276 (1999)

29. D. Rutkauskas, R. Novoderezkhin, R.J. Cogdell, R. van Grondelle, Fluorescence spectral fluctuations of single LH2 complexes from *Rhodopseudomonas acidophila* strain 10050. Biochemistry 43, 4431–4438 (2004)

30. D. Rutkauskas, V. Novoderezhkin, R.J. Cogdell, R. van Grondelle, Fluorescence spectroscopy of conformational changes of single LH2 complexes. Biophys. J. 88, 422–435 (2005)

31. V.I. Novoderezhkin, D. Rutkauskas, R. Van Grondelle, Multistate conformational model of a single LH2 complex: quantitative picture of time-dependent spectral fluctuations. Chem. Phys. 341, 45–56 (2007)

32. C. Tietz, O. Cheklov, A. Draebenstedt, J. Schuster, J. Wrachtrup, Spectroscopy on single light-harvesting complexes at low temperature. J. Phys. Chem. B 103, 6328–6333 (1999)

33. A.M. van Oijen, M. Ketelaars, J. Köhler, T.J. Aartsma, J. Schmidt, Unraveling the electronic structure of individual photosynthetic pigment-protein complexes. Science 285, 400–402 (1999)

34. M. Ketelaars, C. Hofmann, J. Köhler, T.D. Howard, R.J. Cogdell, J. Schmidt, T.J. Aartsma, Spectroscopy on individual light-harvesting 1 complexes of *Rhodopseudomonas acidophila*. Biophys. J. 83, 1701–1715 (2002)

35. C. Hofmann, T.J. Aartsma, J. Köhler, Energetic disorder and the B850-exciton states of individual light-harvesting 2 complexes from *Rhodopseudomonas acidophila*. Chem. Phys. Lett. 395, 373–378 (2004a)

36. M. Ketelaars, J.M. Segura, S. Oellerich, W.P.F. de Ruijter, G. Magis, T.J. Aartsma, M. Matsushita, J. Schmidt, R.J. Cogdell, J. Köhler, Probing the electronic structure and conformational flexibility of individual light-harvesting 3 complexes by optical single-molecule spectroscopy. J. Phys. Chem. B 110, 18710–18717 (2006)

37. W.P.F. de Ruijter, J.M. Segura, R.J. Cogdell, A.T. Gardiner, S. Oellerich, T.J. Aartsma, Fluorescence-emission spectroscopy of individual LH2 and LH3 complexes: Ultrafast Dynamics of Molecules in the Condensed Phase: Photon Echoes and Coupled Excitations - A Tribute to Douwe A. Wiersma. Chem. Phys. 341, 320–325 (2007)

38. M.F. Richter, J. Baier, R.J. Cogdell, J. Köhler, S. Oellerich, Single-molecule spectroscopic characterization of light-harvesting 2 complexes reconstituted into model membranes. Biophys. J. 93, 183–191 (2007a)

39. M. Ketelaars, A.M. van Oijen, M. Matsushita, J. Köhler, J. Schmidt, T.J. Aartsma, Spectroscopy on the B850 band of individual light-harvesting 2 complexes of *Rhodopseudomonas acidophila*: I. Experiments and Monte-Carlo simulations. Biophys. J. 80, 1591–1603 (2001)

40. M.V. Mostovoy, J. Knoester, Statistics of optical spectra from single ring aggregates and its application to LH2. J. Phys. Chem. B 104, 12355–12364 (2000)

41. S. Jang, S.E. Dempster, R.J. Silbey, Characterization of the static disorder in the B850 band of LH2. J. Phys. Chem. B 105, 6655–6665 (2001)

42. M. Matsushita, M. Ketelaars, A.M. van Oijen, J. Köhler, T.J. Aartsma, J. Schmidt, Spectroscopy on the B850 band of individual light-harvesting 2 complexes of *Rhodopseudomonas acidophila*: II. Exciton states of an elliptically deformed ring aggregate. Biophys. J. 80, 1604–1614 (2001)

43. C. Hofmann, F. Francia, G. Venturoli, D. Oesterhelt, J. Köhler, Energy transfer in a single self-aggregated photosynthetic unit. FEBS Lett. 546, 345–348 (2003a)

44. C. Hofmann, M. Ketelaars, M. Matsushita, H. Michel, T.J. Aartsma, J. Köhler, Single-molecule study of the electronic couplings in a circular array of molecules:

Light-harvesting-2 complex from *Rhodospirillum molischianum*. Phys. Rev. Lett. 90, 013004 (2003b)

45. C. Hofmann, H. Michel, M. van Heel, J. Köhler, Multivariate analysis of single-molecule spectra: Surpassing spectral diffusion. Phys. Rev. Lett. 94, 195501 (2005)

46. C. Hofmann, T.J. Aartsma, H. Michel, J. Köhler, Direct observation of tiers in the energy landscape of a chromoprotein: A single-molecule study. Proc. Natl. Acad. Sci. 100, 15534–15538 (2003c)

47. C. Hofmann, T.J. Aartsma, H. Michel, J. Köhler, Spectral dynamics in the B800 band of LH2 from *Rhodospirillum molischianum*: A single-molecule study. New J. Phys. 6, 1–15 (2004b)

48. J. Baier, M.F. Richter, R.J. Cogdell, S. Oellerich,, J. Köhler, Do proteins at low temperature behave as glasses? A single-molecule study. J. Phys. Chem. B 111, 1135–1138 (2007)

49. J. Baier, M.F. Richter, R.J. Cogdell, S. Oellerich, J. Köhler, Determination of the spectral diffusion kernel of a protein by single molecule spectroscopy. Phys. Rev. Lett. 100, 8108–1–4 (2008)

50. H. Oikawa, S. Fujiyoshi, T. Dewa, M. Nango, M. Matsushita, How deep is the potential well confining a protein in a specific conformation? A single-molecule study on temperature dependence of conformational change between 5 and 18 K. J. Am. Chem. Soc. 130, 4580–4581 (2008)

51. T. Aartsma, J. Köhler, Optical spectroscopy of individual light-harvesting complexes, in Advanced in Photosynthesis and Respiration. Vol. 26: Biophysical Techniques in Photosynthesis, Vol. II, ed. by T.J. Aartsma, J. Matysik (Springer, Dordrecht, 2008), pp. 241–266

52. J. Köhler, T.J. Aartsma, Single molecule spectroscopy of pigment protein complexes from purple bacteria, in Advances in Photosynthesis and Respiration Vol. 25: Chlorophylls and Bacteriochlorophylls: Biochemistry, Biophysics, Functions and Applications (Series ed. Govindjee), ed. by B. Grimm, R.J. Porra, R. Wolfhart, H. Scheer (Springer, Dordrecht, 2004), pp. 309–321

53. Y. Berlin, A. Burin, J. Friedrich, J. Köhler, Low temperature spectroscopy of proteins. Part II: Experiments with single protein complexes. Phys. Life Rev. 4, 64–89 (2007)

54. A.S. Davidov, Theory of Molecular Excitons (Plenum, New York, 1971)

55. M.F. Richter, J. Baier, T. Prem, S. Oellerich, F. Francia, G. Venturoli, D. Oesterhelt, J. Southall, R.J. Cogdell, J. Köhler, Symmetry matters for the electronic structure of core complexes from *Rhodopseudomonas palustris* and *Rhodobacter sphaeroides* PufX. Proc. Natl. Acad. Sci. 104, 6661–6665 (2007b)

56. J. Knoester, V.M. Agranovich, Frenkel and charge-transfer excitons in organic solids, in Thin Films and Nanostructures Vol. 31, ed. by V.M. Agranovich, C.F. Bassani (Elsevier, San Diego, CA, 2003), pp. 1–96

57. H.-M. Wu, N.R.S. Reddy, G.J. Small, Direct observation and hole burning of the lowest exciton level (B870) of the LH2 antenna complex of *Rhodopseudomonas acidophila* (strain 10050). J. Phys. Chem. B 101, 651–656 (1997b)

58. H.-M. Wu, M. Rätsep, R. Jankowiak, R.J. Cogdell, G.J. Small, Hole burning and absorption studies of the LH1 antenna complex of purple bacteria: Effects of pressure and temperature. J. Phys. Chem. B 102, 4023–4034 (1998)

59. K. Timpmann, Z. Katiliene, N.W. Woodbury, A. Freiberg, Exciton self trapping in one-dimensional photosynthetic antennas. J. Phys. Chem. B 105, 12223–12225 (2001)

60. K. Timpmann, M. Rätsep, C.N. Hunter, A. Freiberg, Emitting excitonic polaron states in Core LH1 and peripheral LH2 bacterial light-harvesting complexes. J. Phys. Chem. B 108, 10581–10588 (2004)
61. M. Richter, J. Baier, J. Southall, R. Cogdell, S. Oellerich, J. Köhler, Spectral diffusion of the lowest exciton component in the core complex from *Rhodopseudomonas palustris* studied by single-molecule spectroscopy. Photosynth. Res. 95, 285–290 (2008)
62. M.F. Richter, J. Baier, J. Southall, R.J. Cogdell, S. Oellerich, J. Köhler, Refinement of the x-ray structure of the RC LH1 core complex from *Rhodopseudomonas palustris* by single-molecule spectroscopy. Proc. Natl. Acad. Sci. 104, 20280–20284 (2007c)

Part X

Fields and Outlook

Exploring Nanostructured Systems with Single-Molecule Probes: From Nanoporous Materials to Living Cells

Christoph Bräuchle

Summary. Molecular movement in confined spaces is of broad scientific and technological importance in areas ranging from molecular sieving and membrane separation to active transport along intracellular networks. Whereas measurements of ensemble diffusion provide information about the overall behavior of the guests in a nanoporous host, tracking of individual molecules provides insight into both the heterogeneity and the mechanistic details of the molecular diffusion, as well as into the structure of the host.

We first show how single dye molecules can be used as nanoscale probes to map the structure of nanoporous silica channel systems. These channel systems are prepared as thin films via cooperative self-assembly of surfactant molecules with polymerizable silicate species. In order to correlate the porous structure of the host with the diffusion dynamics of single molecules, we present a unique combination of transmission electron microscopic (TEM) mapping and optical single-molecule tracking experiments (SMT). With this approach, we can uncover how a single luminescent dye molecule travels through various defect structures in a thin film of nanoporous silica, how it varies its mobility in the channel structure, and how it bounces off a domain boundary having a different channel orientation. Additional polarization-dependent studies reveal simultaneous orientational and translational movements. In single-molecule measurements with very high positioning accuracy, we show how lateral motions between leaky channels allow a molecule to cross through defect structures into the neighboring channels and how the adsorption of single molecules at the walls of the nanoporous host can be observed as trapping events. Furthermore, a mechanism to switch on and off the diffusion of the guest molecules was discovered. These experiments reveal unprecedented details of the guest–host interactions and of the host's structure, its domains, defects, and the accessibility as well as the connectivity of the nanostructured channel systems. The knowledge of these details and the use of mesoporous nanoparticles with functionalized pore walls will finally lead to novel drug delivery systems. Another type of drug delivery systems is synthetic viruses. Live cell experiments have been performed and the targeting and uptake of DNA-polyplexes as synthetic viruses were investigated. Single-molecule techniques can be used to improve the efficiencies of such synthetic viruses in novel gene therapy applications.

27.1 Introduction

By viewing a movie of a single fluorescent dye molecule moving within the nanostructured channel system of a mesoporous silica host, we see incredible details of molecular motions, be it translation, rotation, trapping at specific sites, lateral motion between defect ("leaky") channels, or bouncing back from disordered regions, to mention only a few examples. In this way, single-molecule tracking experiments help us understand the dynamics and interactions of molecules in nano- or mesoporous silica structures [1–5]. This is of high importance for many applications of these attractive nanomaterials, which have been used as hosts for numerous molecular and cluster-based catalysts [6], for molecular sieving and chromatography [7], for the stabilization of conducting nanowires [8–10], as matrix for ultrasmall dye lasers [11], and for novel drug delivery systems [12, 13], to mention only some of them. In many of these cases, the transport and dynamics of guest molecules in the channels is of paramount importance for the successful functionality of these materials. They can be formed through cooperative self-assembly of surfactants and framework building blocks [12], with widely tuneable properties, like e.g. channel diameters (2–50 nm), topologies (hexagonal, cubic, or lamellar), and functionalized walls. In this chapter, we will first show how high resolution transmission electron microscopy maps can be overlaid with single-molecule trajectories, allowing a correlation between structural elements of the nanoporous host and the movement of the molecules in the nanometer-sized channels. This is followed by discussion on (1) the movement of oriented single molecules with switchable mobility in long unidimensional channels, (2) high localization accuracy experiments with single molecules down to the single channel limit, and (3) the development of functionalized mesoporous nanoparticles as novel drug delivery systems. The chapter concludes with a live cell imaging study of synthetic viruses as a further drug or gene delivery system.

27.2 Correlation of Structural and Dynamic Properties Using TEM and SMT

Mesoporous structures are commonly characterized with diffraction, electron microscopy methods [14], and gas sorption techniques. The ensemble diffusion behavior of small molecules has been examined with pulsed-field gradient NMR spectroscopy [15] and neutron scattering [16]. Here, we are interested in techniques which give a more direct access to the real structure of the mesoporous host and to the dynamics on a single-molecule basis, and thus reveal structural and dynamic features which are not obscured by ensemble or statistical averaging as in conventional techniques.

High-resolution transmission electron microscopy (TEM) offers a way to directly see the channel structure of a mesoporous host [17]. Fig. 27.1a shows

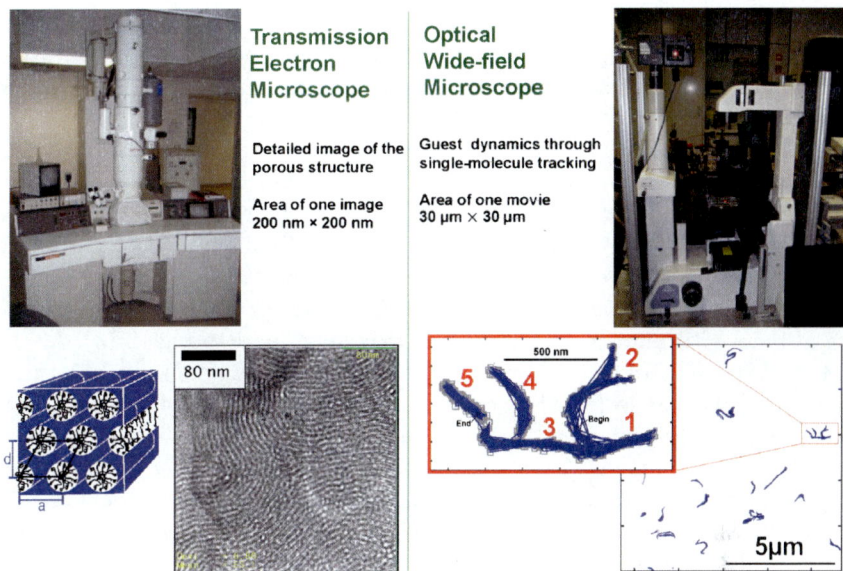

Fig. 27.1. Comparison between transmission electron microscopy (TEM) and single-molecule tracking (SMT) by optical wide-field microscopy. (**a**) High-resolution TEM gives the landscape of a channelar structure of a hexagonal mesoporous system and (**b**) SMT gives the trajectories of the movement of single molecules as guests in the nanoporous host

an example of a high-resolution TEM image of a M41S hexagonal phase prepared by spin-coating a mixture of a silica precursor (TEOS: tetraethyl-ortho-silicate), a template (Brij56: Polyoxyethylene10cetylether), and the probe dye molecule (a terrylendiimide derivative [18]: TDI) in an acidic water-ethanol solution resulting in a thin film (<100 nm) via evaporation-induced self-assembly. It is quiet clear that such an image gives us the landscape of the channels in which our dye molecule can move. Although the TEM image is an average over the thickness of the film (∼10–20 layers of channels), this is the most realistic picture we can get of the real structure of the host. The drawback, however, is that the size of such a high-resolution TEM image is limited to about 200 × 200 nm and the film has to be very thin.

Structural details like channel diameters on the scale of a few nanometers cannot be directly imaged with optical methods. However, single-molecule fluorescence microscopy can track individual dye molecules inside mesoporous silica thin films in an epifluorescence microscope and detect with a highly sensitive CCD camera in a wide-field imaging set-up [1,2]. As the films were much thinner than the focal depth of the microscope objective used, the images contain data from molecules at all heights inside and on the surface of the sample. Series of 1,000 images were acquired with a temporal resolution of 100 ms per frame. In each movie frame, single molecules show up as bright spots on a dark background. Their positions were obtained by fitting theoretical diffraction patterns to the spots, with a positioning accuracy of down to 15 nm. Single-molecule trajectories were then built up by tracking spots from frame to frame, and are shown in Fig. 27.1b.

The trajectories show that the molecules in the hexagonal phase travel in a highly structured manner over distances of several μm during the acquisition time of the movie (500 s). The inset shows an enlargement of one of these trajectories. This molecule travels first along the C-shaped structure on the right (1) and after 65 s enters the side-arm (2). Then, 100 s later, it passes into the linear structure at the bottom (3). After another 144 s, it enters region (4) and moves around there for 69 s before coming back to region (3). At the end it passes into region (5), where it moves back and forth for 109 s until the end of the movie (Movie, see supplementary material of [1]). We note that the molecule apparently probes the domain boundaries in this process, by repeatedly 'bouncing' back from dead ends of the channel regions. This is one of the many striking examples that show how a single molecule seems to explore the structure of the host. We argue that such non-random diffusion, which is repeatedly seen in the hexagonal phase, directly maps the alignment of the channels and the domain structure. Moreover, it seems that we "see" the structure of the host from the viewpoint of the molecule, i.e. we get information about the accessibility of the channels and the connectivity of the domains for the molecule in an unprecedented way, which is not possible with any other method. A proof, however, that the molecule really follows the channel system can only be given by a proper overlay of the structure of the channels as obtained by TEM with the trajectories of the single molecules as observed by SMT. Such a correlation can then clearly illuminate all the highly interesting aspects mentioned above, which can be summarized in one general question: How do structural elements correlate with, and influence the dynamics of the molecules in the nanoporous channels? Because the molecular movement in the pore system is the most important and defining characteristic of nanoporous materials and its various applications, it is of great interest to learn about this behavior as a function of the local structure.

For the overlay of the optical and electron microscopic images, we prepared a ca 50 nm thin spin-coated film directly onto the 30 nm thick Si_3N_4 membrane of the TEM sample holder [2]. Single polystyrene beads were incorporated into the film which can be imaged by TEM and optical microscopy. By first considering the trajectories with the optical wide-field microscope and then measuring the TEM images at the same sample, the correct correlation of both images could be achieved by overlapping the same pattern of the polystyrene beads. This tedious procedure is described in detail in [2] and gives an overlay accuracy between 4 and 30 nm, depending on the number of beads in the images. Fig. 27.2a shows the overlay of a single-molecule trajectory with a high resolution ($\times 40,000$ magnification) TEM map. The latter was obtained from many high-resolution TEM images (as shown in Fig. 27.1a), where adjacent square regions of 133 nm × 133 nm make up the whole map. In each square

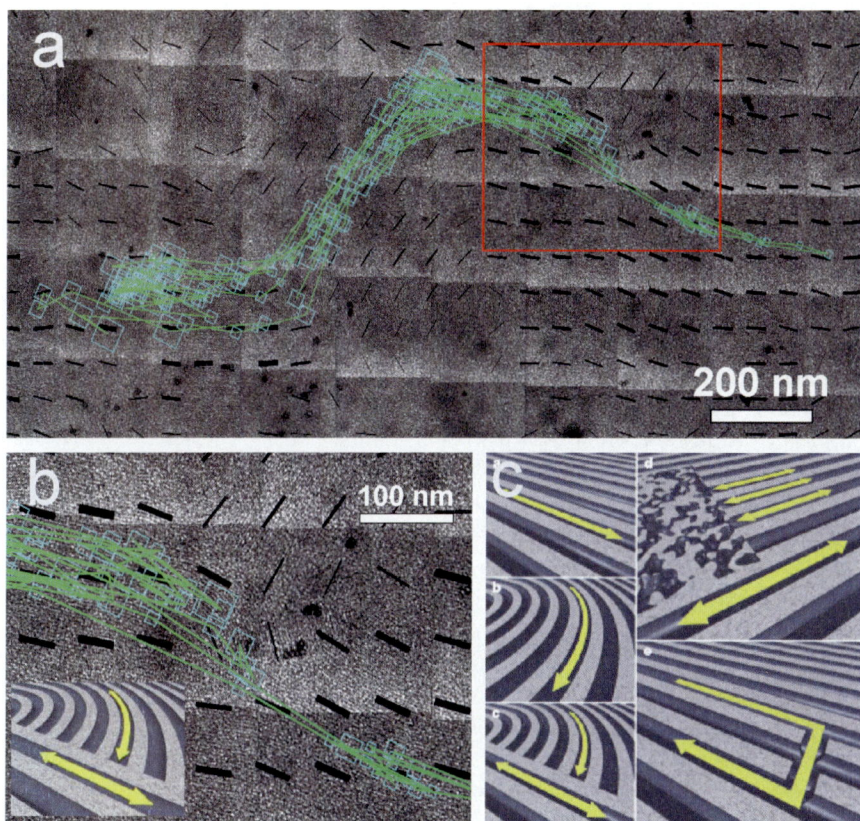

Fig. 27.2. Molecular trajectories and structural elements in a hexagonal meso-porous silica film. (**a**) Overlay of an S-shaped trajectory of a single molecule with a transmission electron microscopy map. The molecule exploring regions of parallel channels, strongly curved areas, and domain boundaries indicated by the FFT directors (black bars). (**b**) Enlarged area of (**a**) showing part of the trajectory of the single molecule which bounces back repeatedly from a domain boundary formed by channels having different orientations sketched in the inset. The light blue boxes in (**a**) and (**b**) depict the positioning accuracy. (**c**) Sketches of structural elements and molecular movements found in hexagonal mesoporous silica films

region, fast Fourier transformation (FFT) results in a FFT director which depicts the average orientation of the pores and the line thickness is a measure for the degree of structural order in this region. These directors are a good guide for the eye about the orientation of the channels and give an overview of the domain sizes. Figure 27.2a is an example of a molecule faithfully following the pores and mapping out specific elements of the host structure (movie, see supplementary material of [2]). The perfect overlay of the S-shaped trajectory on the pore system is shown well by the FFT directors. In Fig. 27.2b, a specific region of Fig. 27.2a is enlarged and shows the channel structure as well as the trajectories in more detail. In both figures, the light blue boxes of the trajectory indicate the position accuracy of the molecular positions determined. These are in the range of 15–30 nm, which means that the position of the molecules can therefore be assigned not to a single channel, but to an ensemble of about 3–6 parallel channels. Furthermore, one should keep in mind that we are sampling diffusion at discrete points in time and space. Thus, the connecting lines are just a method of visualizing the trajectories; they do not represent the molecules' exact path. The enlargement in Fig. 27.2b, however, clearly shows that the molecule in the upper part bounces back from the domain boundary with channels having different orientation as sketched in the inset of Fig. 27.2b. In the lower part of the trajectory, however, the molecule can find its way along the main domain which guides the pathway of the S-shaped trajectory. Many structural elements, as shown in Fig. 27.2c, are found and could be correlated with the dynamic behavior of single molecules.

In summary, the combination of the two techniques provides the first direct proof that the molecular diffusion pathway through the pore system correlates with the pore orientation of the hexagonal structure. In addition, the influence of specific structural features of the host on the diffusion behavior of the guest molecules can be clearly seen.

With this approach we can uncover, in an unprecedented detail, how a single fluorescent dye molecule travels through linear or strongly curved sections of the hexagonal channel system, how it changes speed, and how it bounces off a domain boundary with a different channel orientation. Furthermore, we can show how molecular travel is stopped at a less ordered region, or how lateral motions between 'leaky' channels allow a molecule to explore different parallel channels within an otherwise well-ordered periodic structure.

27.3 Oriented Single Molecules with Switchable Mobility in Long Unidimensional Nanochannels

Using CTAB (Cetyltrietylammoniumbromide) instead of Brij56 as surfactant in the template synthesis of the mesoporous M41S systems, we obtained hexagonally arranged channels where the diameter of the channel (2–3 nm) is smaller than the length of the TDI molecule (3 nm) used as fluorescent probe. Therefore, rotation of the TDI molecule should be impossible in well-ordered channels. Polarization-modulated confocal microscopy was performed to monitor simultaneously the diffusional and orientational behavior of the TDI molecules in such systems [4]. Figure 27.3a shows three fluorescence images taken at time 0, 2, and 4 min where single TDI dye molecules appear with a characteristic fluorescence-intensity profile (striped patterns) due to the polarization modulation during the scan.

From these patterns, we compute both the position of the molecule and the orientation of its transition dipole moment (shown as yellow bars). Following the molecule in the circle, we can obtain the trajectory of this single

a)

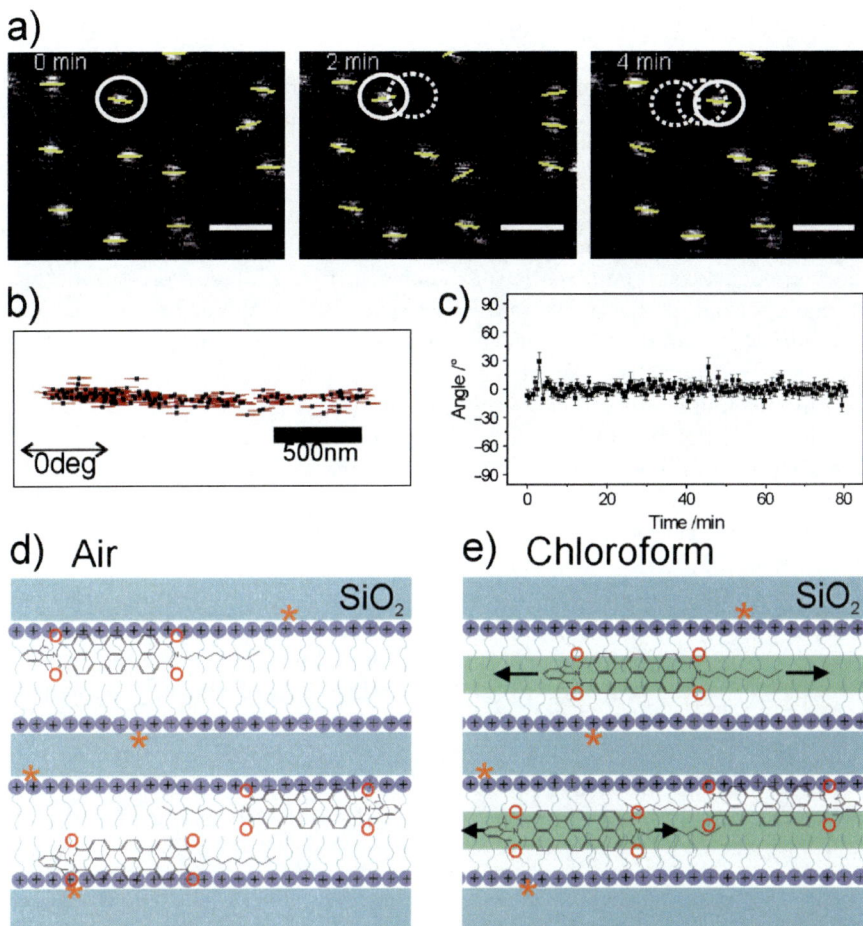

b)

c)

0deg   500nm

d) Air   e) Chloroform

SiO₂   SiO₂

Fig. 27.3. Parallel orientation and diffusion of single TDI molecules in a highly
ordered domain. (a) Sequence of fluorescence images showing linear diffusion of
single TDI molecules in a chloroform atmosphere extracted from a time series. Scale
bar: 2 μm. (b) Trajectory extracted from the molecule marked with the white circle
in (a). (c) Calculated angular time trajectory of the same molecule. (d) Sketch
of TDI molecules immobilized in the mesoporous film in air. The stars indicate
active silanol groups. (e) TDI molecules in the mesoporous film in the presence of
chloroform. The solvent provides a lubricant for the molecular movement

TDI molecule including position and orientation as shown in Fig. 27.3b. The
trajectory shows that the molecule is moving linearly back and forth over a
distance of about 2 μm, while it remains remarkably aligned with the direction
of the diffusion, which we assigned to the direction of the pores. Figure 27.3c
shows the alignment of the fluorophore within the pore which clearly indicates that free rotation is prevented by the well-ordered channel and the

geometrical constraints due to the size of the molecule and the diameter of the channel. Thus, the orientation of single TDI molecules and their trajectories map directly the direction of the channels. Furthermore, the rotational free translation of the single molecule indicates a structurally well-ordered area.

With this method at hand, we could improve the film preparation to produce highly ordered linear channels in domains up to $100\,\mu m$ in size. To our knowledge, such a high degree of order over long distances has not been reported before for mesoporous structures. It is, however, highly desirable for many applications. The observations shown in Fig. 27.3a–c were taken with the mesoporous film in a saturated chloroform atmosphere. In this case the molecules were mobile. Changing the atmosphere above the film with air (40% relative humidity), the molecular motion immediately stopped. This process is highly reversible and indicates that by changing the atmosphere around the porous film the diffusion of TDI guest molecules can be switched on and off reversibly in a very easy manner [4]. These observations lead us to a model for the immobility of the TDI molecules in the mesoporous host: The TDI molecule has a very high hydrophobic core and four oxygen atoms pointing to the side (Fig. 27.3d–e), whose lone-pair electrons can interact with the positively charged heads of the CTAB molecules. In addition, interactions are possible with active silanol groups or other defects in the channel walls. These interactions seem to be responsible for the immobilization of the molecule in air atmosphere (Fig. 27.3d). In contrast, when chloroform, a good solvent for TDI, is added to the system it is likely that the small solvent molecules form a lubricant-like phase inside the pores (Fig. 27.3f). As a result, the TDI molecules can be solvated and diffuse along the pores [4]. Hence, the solvent exchange allows an easy control of the diffusional behavior of the guest molecules.

27.4 High Localization Accuracy of Single Molecules Down to a Single Channel Limit

Our aim to achieve an extremely high positioning accuracy in the optical SMT experiments, which is better than the pore diameter of the mesoporous system, could be achieved in the CTAB templated M41S system [4]. Here, the TDI molecules move slower than in the Brij56 templated system and by increasing the laser power and improving other parameters of the experiment we could achieve a spatial resolution of 2–3 nm. This allows us to identify a moving molecule in a specific channel, and to observe jumps between neighboring channels. Figure 27.4a shows the trajectory of a TDI molecule (Movie, see supplementary material of [4]) which first moves in one channel (black trajectory) and then switches over into the neighboring channel (green trajectory). This can be clearly seen if we separate the movement along the pore and in a direction perpendicular to it in $x(t)$ and $y(t)$ graphs (Fig. 27.4c). By inspection of the $y(t)$ graph, a jump to a neighboring pore is observable after

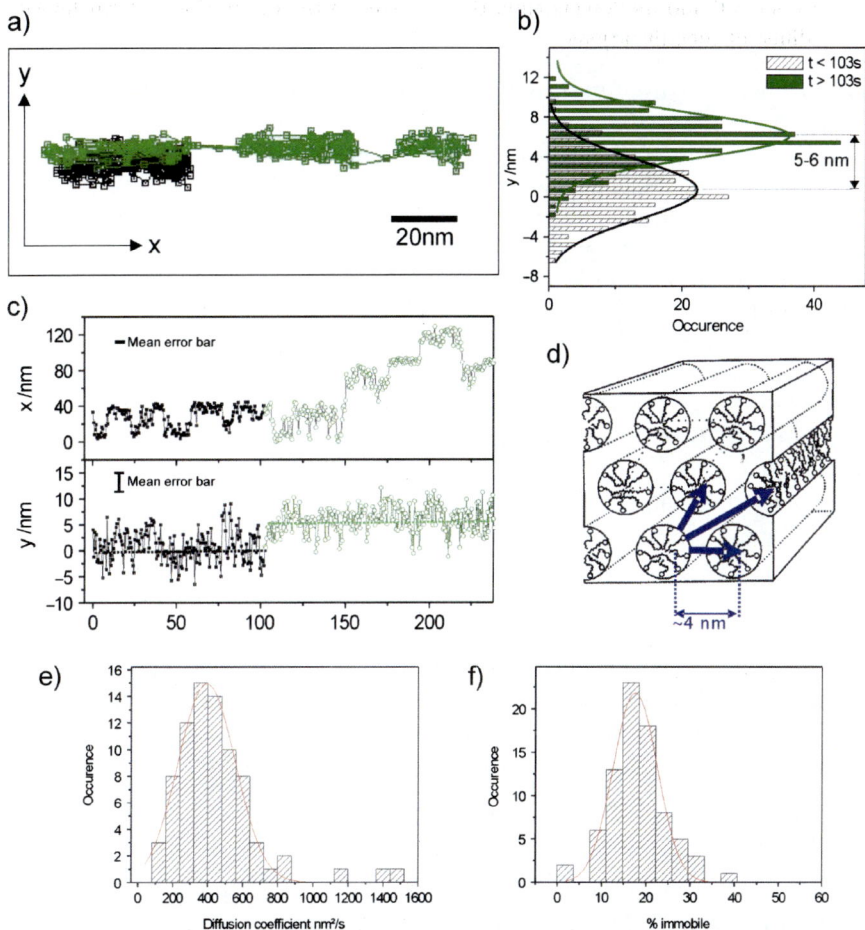

Fig. 27.4. Diffusion and switching in two distinct neighboring channels and trapping behavior. (**a**) Trajectory with high optical resolution of a single TDI molecule switching from one channel (*black*) to a neighboring channel (*green*). (**b**) Histogram with Gaussian distribution of the lateral (*y*) coordinate for the time intervals before (*black*) and after (*green*) the switching to the neighboring channel at time $t = 103$ s. (**c**) Projected $x(t)$ and $y(t)$ coordinates for the single TDI molecule in a) diffusing in two distinct neighboring pores. $y(t)$ clearly shows the switch into the neighboring channel at time $t = 103$ s. $x(t)$ shows repeated trapping of the single TDI molecule at the walls. (**d**) Hexagonal channel system of the CTAB templated mesoporous host with arrows indicating a switch to neighboring channels. (**e**) Histogram of the diffusion coefficients of 80 molecules in a CTAB templated mesoporous film. (**f**) Histogram of the percentage of adsorption time per trajectory of 80 molecules in the same system

103 s. Figure 27.4b displays the histogram of $y(t)$ before (green) and after (black) 103 s. The distributions are clearly distinct and can be fitted with two Gaussian curves with a maximum at 0.6 and 6.1 nm and with a half width of $\sigma = 2.9$ and $\sigma = 2.3$ nm, respectively. We attribute the observed dynamics to a TDI molecule which switches between two neighboring pores separated by 5–6 nm as indicated in Fig. 27.4d. In other cases, we observed molecules which explored even more distant pores by switching through defects from pore to pore. This seems to be an important process for a molecule to circumvent dead ends in one pore and to travel over larger distances within the mesoporous system. Figure 27.4a is also an example for this behavior because the molecule in the first channel is kept between two dead ends and can extend its pathway along the pores only by switching in the neighboring channel.

The $x(t)$ graph of this molecule as shown in Fig. 27.4b reveals another interesting property: the trapping of a molecule during its passage through the channel network. Whenever the molecule reaches one of the two dead ends of the first channel, it is trapped for some seconds. The same is true when it moves in the neighboring channel. A detailed analysis of the trajectories of 80 molecules reveals a Gaussian distribution of the diffusion coefficients (Fig. 27.4e) with a mean value of 390 nm^2 s^{-1} and a half width of $\sigma = 100$ nm^2 s^{-1}. In addition, the histogram of the percentage of the adsorption time per trajectory (Fig. 27.4f) is also Gaussian with a maximum at 18%. This means that a molecule spends on an average 18% of its walk immobilized at an adsorption site. Such kinetic data can give a very detailed picture of the dynamic behavior of molecules inside the channellar network of a mesoporous system.

27.5 Functionalized Mesoporous Silica Structures: Towards Novel Drug Delivery Systems

For many applications, the mesoporous materials are expected to show enhanced properties when their inner channel walls are functionalized with organic moieties to fine-tune host–guest interactions. This is particularly important for drug delivery systems which require the drug to be released at a slow rate in order to generate the desired depot-effect [19].

In our study, a cocondensation method [19] was used to enable the homogeneous incorporation of functional groups (Fig. 27.5a) like e.g. alkyl chains or cyano-propyl groups which differ strongly in polarity. This resulted in a strong influence on the diffusion coefficient of the probe molecule [20].

Furthermore, mesoporous nanoparticles with a diameter of about 100 nm can be produced as shown in the upper part of Fig. 27.5b. We intend to use them as the basis for a novel drug delivery system which is sketched in the lower part of Fig. 27.5b. This system has several functions for controlled release and targeting of specific cells. These functions are: (1) functionalized pore walls to create a depot effect and controlled release of the drug;

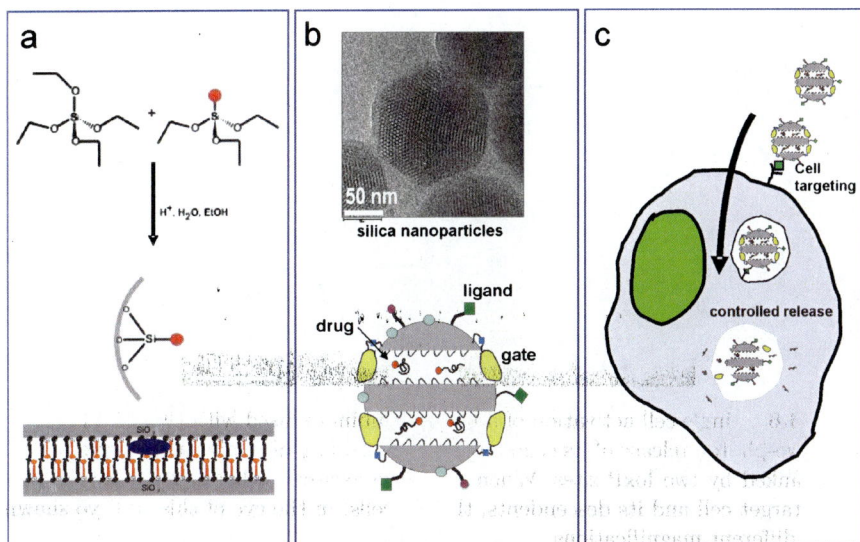

Fig. 27.5. Mesoporous silica nanoparticles as novel drug delivery systems. (a) Cocondensation method to form functionalized mesoporous silica structures in a surfactant template synthesis. (b) TEM image of mesoporous silica nanoparticles and sketch of a novel drug delivery particle which contains functionalized pores, closed by a gate, and is decorated with ligands for cell targeting. (c) Cell targeting by ligand-receptor interaction at the cell membrane, endosomal uptake and controlled release after pH change from early to late endosome

(2) a gate which closes the channels and can be opened by a pH drop from pH 7 to pH 5, taking place from an early to a late endosome within the cell; (3) ligands attached to the nanoparticle which serve for cell targeting using specific receptors at the cell surface. The nanoparticle attaches with its ligand to a specific receptor at the cell membrane. Then it is endocytosed and the early endosome transforms into a late endosome by changing its pH from 7 to 5. This will open the gates of the nanoparticle and a controlled release of the drug can take place (Fig. 27.5c). Such novel drug delivery systems will be especially useful for the delivery of toxic drugs like e.g. cytostatica in cancer chemotherapy, where the drug delivery system is targeted to the cancer cell and opens its toxic load not before it is inside the cancer cell. This novel drug delivery system using mesoporous nanoparticles with different functionalities is presently under development in our laboratory.

27.6 Synthetic Viruses as Gene Delivery Systems

Synthetic viruses are drug delivery systems in which the drug is a therapeutic gene [21–23]. They are used in gene therapy [24]. There are two different

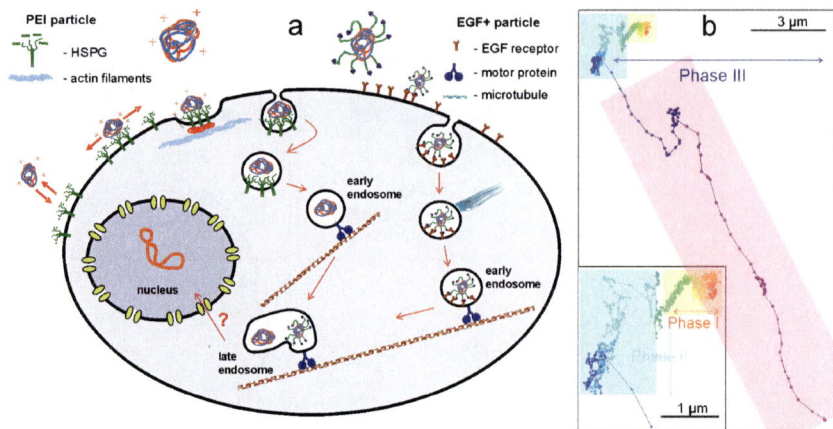

Fig. 27.6. Uptake and trafficking of synthetic viruses in a cell. (**a**) Uptake and trafficking processes of plain (*left*) and EGF + decorated (*right*) polyplexes in a cell shown in a schematic way. (**b**) Trajectory of a single EGF + polyplex showing three phases: phase I and phase III are directed movements whereas phase II is in most cases normal diffusion. Phase I and phase II are enlarged in the inset. The whole trajectory was recorded in 4:30 min at a frame rate of 300 ms

ways to introduce a gene into the cell nucleus: with the help of viruses [25] (viral vectors) or by means of synthetic viruses [22,23] (non-viral vectors). In general, the viral vectors induce a high expression of the target gene. Nevertheless, their disadvantage is the immunogenic response of the body upon administration. For this reason, a lot of research has been done on alternative vectors like synthetic viruses, which, however, are not as efficient as natural viruses. They consist of plasmid DNA condensed by e.g. a cationic polymer forming a so-called polyplex [22,23]. In order to induce gene expression, the DNA has to enter the cell nucleus. On its way to the cell nucleus there are several barriers to overcome like the cell membrane, the escape from the endosome, and the nuclear membrane. These processes are shown in Fig. 27.6a. First, the synthetic viruses have to enter the cell which can occur by different uptake processes like receptor-induced endocytosis and others [26,27]. Besides diffusion in the cell cytoplasm, transport processes with motor proteins are responsible to bring the polyplex closer to the nucleus [28,29]. In addition, the complex has to escape the endosome [30,31] and finally, the DNA has to be transported into the nucleus. Viruses have involved specialized mechanisms to overcome the barriers and to be transported very efficiently. Non-viral vectors or synthetic viruses, in contrast, have to be programmed chemically using e.g. a biomimetic approach. This can be based on the experience resulting from investigations of the infection pathway of single viruses in living cells [32,33]. In order to make synthetic viruses more efficient, the uptake and transport processes have to be characterized in detail in order to identify bottlenecks in the internalization pathway. This can be done with optical single particle tracking techniques, similar to the tracking of single molecules in nanoporous systems.

We used polyethylenimine (PEI) as polycation to bind nuclear acids and to form the synthetic viruses [28, 29]. In order to allow tissue selective delivery of these polyplexes, cell-specific ligands are attached to the polyplexes. To specifically target cancer cells, in which e.g. the epidermal growth factor receptor (EGFR) is overexpressed by a factor of more than 100, the corresponding EGF ligand is used. Moreover, the plasmid DNA condensed in the polyplex is labeled with a fluorescent dye. This allows us to observe individual synthetic viruses by means of highly sensitive fluorescence wide-field microscopy during different stages of their internalization process, starting with the initial binding to the cell membrane and followed by the internalization and intracellular transport.

Figure 27.6b shows the trajectory of an individual synthetic virus during such an internalization process [29] (Movie, see supplementary material of [29]). Three different phases can be identified: In phase I, binding to the plasma membrane is followed by a slow movement with drift, which can be deduced from the quadratic dependence of the mean square displacement $<r^2>$ as a function of time. Furthermore, a strong correlation between neighboring particles is seen and subsequent internalization is observed, and can be proven by quenching experiments. During this phase, the particles are subjected to actin-driven processes mediated by transmembrane proteins. Phase II is characterized by a sudden increase in particle velocity and random movement, often followed by confined movement.

This is indicated by a linear or saturating behavior of $<r^2>$ as a function of time. In this phase, the particles entrapped in intracellular vesicles diffuse in the cytosol, waiting for an encounter with a motor protein as a microtubule-dependent transporting system. Phase III started with a dramatic increase in particle velocity and directed movement over long distances (several micrometer) within the cell along microtubules. The directed movement is demonstrated by the quadratic dependence of $<r^2>$ as a function of time with transport velocities up to $4\,\mu m\ s^{-1}$.

As a result, the internalization of EGF + polyplexes in phase I was very rapid: 50% of the observed particles were internalized within 5 min. In sharp contrast, internalization of plain polyplexes which do not contain EGF was much slower and far less efficient. To our knowledge, this is the first study [31] describing the dynamics of targeted artificial viruses on a single-particle basis, in respect to cell binding and intracellular transport. From such results a very detailed knowledge can be obtained about the mechanistic processes of uptake and transport, which helps to improve the design and the functionality of synthetic viruses as a specific system for drug delivery.

Acknowledgments

I am very grateful to all coworkers and collaborators who are mentioned in our publications and who have shaped the understanding of molecular dynamics in nanoporous systems and drug delivery in living cells. The work was supported by the Excellence Clusters Nanosystems Initiative Munich (NIM) and Center for Integrated Protein Science Munich (CIPSM) as well as the collaborative research centers SFB 486 and SFB 749.

References

1. J. Kirstein, B. Platschek, C. Jung, R. Brown, T. Bein, C. Bräuchle, Nat. Mat. **6**, 303–310 (2007)
2. A. Zürner, J. Kirstein, M. Döblinger, C. Bräuchle*, T. Bein*, Nature **450**, 705–708 (2007)
3. C. Jung, C. Hellriegel, B. Platschek, D. Wöhrle, T. Bein, J. Michaelis, C. Bräuchle, J. Am. Chem. Soc. **129**, 5570–5579 (2007)
4. C. Jung, J. Kirstein, B. Platschek, T. Bein, M. Budde, I. Frank, K. Müllen, J. Michaelis, C. Bräuchle, J. Am. Chem. Soc. **130**, 1638–1648 (2008)
5. C. Jung, C. Hellriegel, J. Michaelis, C. Bräuchle, Adv. Mater. **19**, 956–960 (2007)
6. D.E. De Vos, M. Dams, B.F. Sels, P.A. Jacobs, Chem. Rev. **102**, 3615–3640 (2002)
7. V. Rebbin, R. Schmidt, M. Fröba, Angew. Chem. Int. Edn. Engl. **45**, 5210–5214 (2006)
8. D.J. Cott et al., J. Am. Chem. Soc. **128**, 3920–3921 (2006)
9. B. Ye, M.L. Trudeau, D.M. Antonelli, Adv. Mater. **13**, 561–565 (2001)
10. N. Petkov, N. Stock, T. Bein, J. Phys. Chem. B. **109**, 10737–10743 (2005)
11. I. Braun, G. Ihlein, F. Laeri, J.U. Nockel, G. Schulz-Ekloff, F. Schuth, U. Vietze, O. Weiss, D. Wohrle, Appl. Phys. B: Lasers Opt. **70**, 335–343 (2000)
12. C.J. Brinker, Y. Lu, A. Sellinger, H. Fan, Adv. Mater. **11**, 579–585 (1999)
13. I. Roy et al., Proc. Natl. Acad. Sci. U S A. **102**, 279–284 (2005)
14. O. Terasaki, T. Ohsuna, in *Handbook of Zeolite Science and Technology*, ed. by S.M. Auerbach, K.A. Carrado, P.K. Dutta 291–315 (Dekker, New York, 2003)
15. V. Kukla et al., Science **272**, 702–704 (1996)
16. N.E. Benes, H. Jobic, H. Verweij, Micropor. Mesopor. Mater. **43**, 147–152 (2001)
17. Y. Sakamoto et al., Nature **408**, 449–453 (2000)
18. C. Jung et al., J. Am. Chem. Soc. **128**, 5283–5291 (2006)
19. W.S. Han, Y. Kang, S.J. Lee, H. Lee, Y. Do, Y.A. Lee, J.H. Jung, J. Phys. Chem. **109**, 20661–20664 (2005)
20. T. Lebold, L.A. Mühlstein, J. Blechinger, M. Riederer, H. Amenitsch, R. Köhn, K. Peneva, K. Müllen, J. Michaelis, C. Bräuchle, T. Bein, Chem. Eur. J., **15**, 1661–1672 (2009, submitted 2008)
21. S.M. Sullivan, in *Pharmaceutical Gene Delivery Systems*, ed. by A. Rolland (Dekker, New York, 2003), pp. 1–16
22. T.G. Park, J.H. Jeong, S.W. Kim, Adv Drug Deliv Rev. **58**, 467–486 (2006)
23. D. Schaffert, E. Wagner, Gene Ther. 1–8 (2008)
24. T.R. Flotte, J. Cell Physiol. **213**, 301–305 (2007)

25. R. Waehler, S.J. Russell, D.T. Curiel, Nat. Rev. Genet.. **8**, 573–587 (2007)
26. I. Kopatz, J.S. Remy, J.P. Behr, J. Gene Med. **6**, 769–776 (2004)
27. J. Rejman, A. Bragonzi, M. Conese, Mol. Ther. **12**, 468–474 (2005)
28. R. Bausinger, K. von Gersdorff, K. Braeckmans, M. Ogris, E. Wagner, C. Brauchle et al., Angew. Chem. Int. Ed. Engl. **45**, 1568–1572 (2006)
29. K. de Bruin, N. Ruthardt, K. von Gersdorff, R. Bausinger, E. Wagner, M. Ogris, C. Bräuchle, Mol. Ther. **15**, 1297–1305 (2007)
30. Y.W. Cho, J.D. Kim, K. Park, J. Pharm. Pharmacol. **55**(6), 721–734 (2003)
31. K. de Bruin, C. Fella, M. Ogris, E. Wagner, N. Ruthardt, C. Bräuchle, J. Control. Release, **130**, 175–182 (2008)
32. G. Seisenberger, M.U. Ried, T. Endreß, H. Büning, M. Hallek, C. Bräuchle, Science **294**, 1929–1932 (2001)
33. M. Lakadamyali, M.J. Rust, H.P. Babcock, X. Zhuang, Proc. Natl. Acad. Sci. USA. **100**, 9280–9285 (2003)

Gene Regulation: Single-Molecule Chemical Physics in a Natural Context

Peter G. Wolynes

Summary. Single-molecule studies provide us with greater insights into the mechanistic details of chemical processes, but in natural biological systems, the deterministic formalism of ensemble kinetics has been considered to provide a sufficient phenomenology. The most striking exception to this view of the basis of systems biology is gene regulation. Most cells possess only one or two copies of any given gene. The proteins regulating these genes are also present in small numbers. Temporal averaging over DNA-protein binding events would return the problems of gene regulation to one of macroscopic kinetics, but this temporal averaging is not always adequate. The stability of genetic switches depends on the dynamics of individual gene binding events. A non-adiabatic formalism is required. In some models of gene switches, even more dramatically stochastic attractors apparently exist that have no deterministic counterparts. These attractors arise directly from the single-molecule nature of the gene and are analogous to extinction events in population biology.

Advances in technology have converted single-molecule experiments from being primarily thrillingly explicit demonstrations of the atomic hypothesis to powerful methodologies for exploring molecular systems. For the simpler chemical systems, such experiments generally recapitulate what can be learned from bulk ensemble measurements. On the contrary, when there is statistical complexity arising from a rugged energy landscape [1] or where there is mechanistic complexity in a multistep process, the path to understanding is made considerably easier by being able to avoid averaging and directly see within the ensemble, and to monitor fleeting intermediates that do not accumulate to a significant degree macroscopically. Single-molecule probes have thus given direct evidence for the cooperatively rearranging regions in supercooled liquids as they approach the glass transition [2,3] and evidence for the intrinsic complexity of the so-called two-level systems found in low temperature glasses [4]. Single-molecule kinetics has revealed that the conformational substates on the folded protein energy landscape can be the source of enzymic kinetic intermittency [5] at physiological temperatures [6,7]. Many aspects of protein folding concerning the complexity and malleability of the partially denatured state are coming into focus using single-molecule fluorescence measurements of distance distributions [8]. Very complex sequential mechanisms of biomolecular processing are also coming into sight [9].

Single-molecule physical chemistry, clearly, is here to stay! The focus of this chapter moves away from these analytical uses of single-molecule physics in the laboratory. Instead of exploiting single-molecule dynamics as an exploratory tool, I wish to highlight an exciting part of biology where the properties of a single molecule have huge macrolevel consequences and where the inevitable stochasticity of molecular dynamics cannot be neglected, if the natural system is to be understood. This area is gene regulation, a central part of systems biology [10].

While some of what follows may appear to be obvious, the viewpoint that single-molecule statistical physics enters gene regulation, actually is a bit controversial – most systems biologists today talk as if macroscopic chemical kinetics will eventually prove to be sufficient to explain cellular biology. These systems biologists argue that we only are lacking knowledge of the relevant macroscopic kinetic parameters but that standard physical chemistry will eventually be enough to describe the responses of a cell. Instead I think the single-molecule nature of the gene already forces (or at least strongly encourages!) a fresh approach: The statistics of individual reaction events really do matter. The reaction events in any biochemical system are themselves complex, as has already been established. They are not simply Poisson processes, in general, although when finally analyzed into their individual steps of conformational change and intracellular transport, such simplicity must eventually emerge.

The first point is that no matter what the complexity of its regulatory machinery, the gene itself is a single molecule. In fact, the earliest experiments leading to today's molecular biology were studies of the statistics of the macroscopic results of mutations [11]: development is a wonderful amplifier! These statistics established that the gene was a single molecule long before its structural characterization via biophysical measurements. Many of Delbrück's mathematical results that he developed in that study are still used in single-molecule data analysis today. Mutations are, however, dramatic molecular modifications. The more refined question facing us today is "do the regulatory processes in ordinary cells have significant statistical variations due to the molecular nature of the gene and is this variation relevant to the proper functioning of the cell?" It seems increasingly likely that this question must be answered in the affirmative. In addition, facing the single-molecule nature of the gene straight on provides an especially convenient language for discussing gene regulation: cell biologists have long spoken of a gene being either "off" or "on." This linguistic convenience carries over to the mathematics: In what follows, I shall argue that mathematical techniques used to describe the dynamics of molecules in fluctuating environments can be borrowed from solid state and chemical physics, and that these ideas are especially useful.

28.1 A Single-Molecule Gene Switch as a Many Body Problem

From the point of view of regulation, a gene is a binding site for proteins. These proteins are called transcription factors. Since a single site is either occupied or not, the state of a gene has essentially two values: "on" or "off". A gene is thus like a quantum spin (or perhaps better said, a semi-classical spin) which is either up or down [12]. When the gene is on, the DNA begins to produce proteins through a series of steps starting with the serial synthesis of mRNA followed by transcription of the mRNA into protein. These synthesis events build up a "proteomic" atmosphere of transcription factor molecules that can act on the original gene itself or other genes. The number of proteins in the atmosphere takes on discrete integer values. These discrete levels of protein expression resemble the discrete levels of excitation of a quantum mechanical harmonic oscillator. These two points of verbal analogy can be made mathematically precise. Sasai and Wolynes showed how the Master equation describing the joint probability of a gene being turned on or off, along with protein number can in fact be written like the Schrödinger equation for a spin interacting with an oscillator – the so-called "spin-boson problem" [13].

In spin-boson problems, one either has a strongly fluctuating situation in which the two state entities rapidly go back and forth which can be described by a single average activity or the system exhibits a pair of very long-lived steady states, in which the environment stabilizes the two state objects in one or the other of its options. These options for two state tunneling systems are molecular configurations. For gene switches, the states correspond with two different levels of gene expression. On sufficiently long time scales, stochastic fluctuations always eventually take the system from one state to the other. For a molecular spin boson system, this change is a chemical reaction. For gene switches, this event is called a spontaneous epigenetic change.

Figure 28.1 shows an "effective potential" that describes the probability of finding a certain number of protein molecules in the cell as a function of the gene's occupancy state. Alongside the figure is a picture of the crossing potential surfaces for the nonbiological spin-boson problem. In both cases, we have a situation familiar to chemists of two-state kinetics. The spin-boson problem in chemistry is especially useful for the theory of electron transfer reactions where it forms the basis of the Marcus theory [14]. This picturesque analogy also motivates mathematical treatments using variational methods like those used for many-body problems like the polaron in solid state physics and Hartree methods in quantum chemistry.

A phase diagram for a simple gene switch emerges as a function of the molecular parameters characterizing the gene using such a variational approach. This is shown in Fig. 28.2. The two key quantities are the number of protein molecules needed to flip the switch (X_{ad}) and the ratio of the unbinding rate to the protein degradation rate, ω. The latter can be called an adiabaticity parameter. When it is large, the binding/unbinding process has

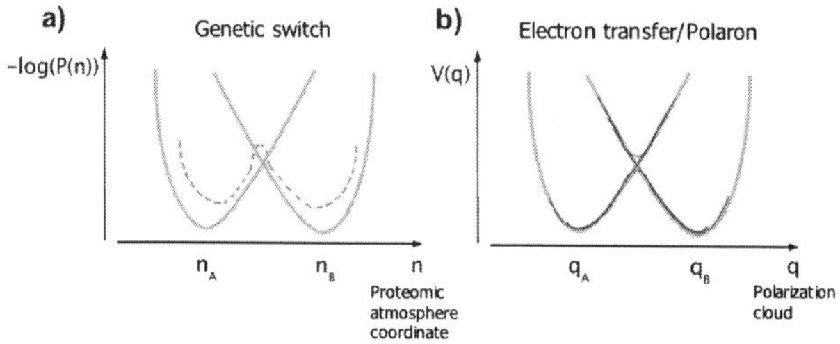

Fig. 28.1. In a genetic switch, the total state of the system depends on two variables: whether the DNA site is occupied or not and the number of copies of the transcription factor. In the left panel, the logarithm of the steady state probability for the occupancy state and protein number is plotted. This probability acts like an effective potential. In the right panel, the effective potential for a charge transfer or two site polaron is plotted as a function of the environment polarization for the two electronic states. The governing time-dependent equations for the two problems share many similarities

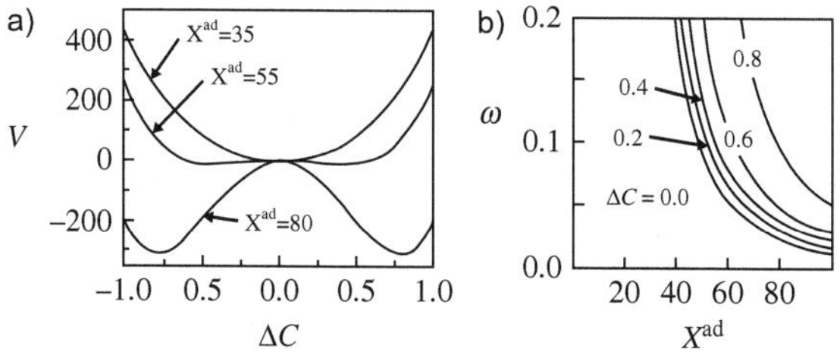

Fig. 28.2. The left panel shows the effective potential for a simple gene switch analogous to that shown in Fig. 28.1. ΔC is a variable that scales with the protein number. X^{ad} gives the value of the number of transcription factor molecules needed (at equilibrium) to give an equal likelihood of the site being occupied or not. The right panel shows the demarcation lines for two-state switch behavior (for large ω and x^{ad} versus one-state averaged behavior for small ω or x^{ad})

time to come to equilibrium. Switches are more stable at large ω and large X_{ad}. It seems likely that achieving the simultaneous limits can be costly to the cell. Thus, many switches may be left by evolution to function near the border of switch stability.

Most gene switches have complex molecularity [15], leading to the necessity of the approximate solution of the Master equation. One particularly simple switch has a Master equation that can be solved without making significant approximations [16]. This exactly solvable switch involves a gene regulated by a protein which binds as a monomer. In Fig. 28.3, I show the probability distribution for protein number obtained from the exact solution. When the binding rate is high, a strongly adiabatic situation, we see a single-peaked probability distribution. This is what would be predicted by macroscopic kinetics with mild noise.

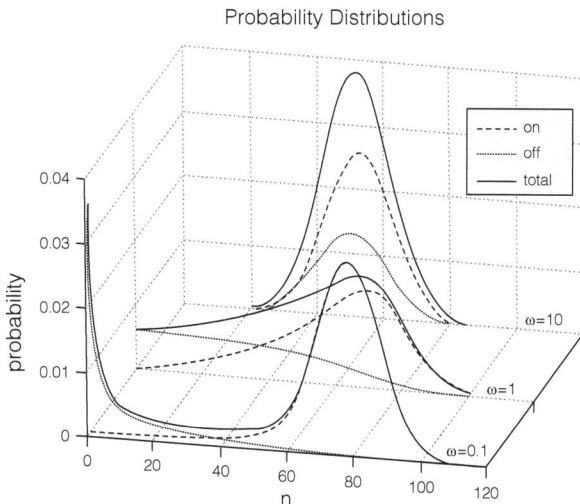

Fig. 28.3. The steady state probability distribution for a switch modulated by a monomeric transcription factor. The two peak behavior at low ω (a highly "nonadiabatic" system) comes from the extinction of the transcription factor population

The single-molecule nature of the gene, however, allows a second peak to form at low ω when the system is non-adiabatic. This extra peak exhibited by the exact solution essentially corresponds to the extinction of the repressor protein population. This second peak is not predicted by macroscopic modeling accompanied by Gaussian noise – it is a clear single-molecule effect. (Once the last whale is gone, they all are!) Sunny Xie has observed a highly analogous distribution recently in a simple gene circuit [17].

The single-molecule nature of the gene allows the gene in steady state to behave much differently from what the macroscopic laws postulate. Even the law of mass action is modified as seen in Fig. 28.4. Again, most models of gene regulation have assumed macroscopic equilibrium behavior and would obey the law of mass action but this simple picture breaks down due to single-molecule effects.

The single-molecule nature of the gene clearly modifies the landscape of possible attractors – that is, the set of switching states of a gene network. Single-molecule event also can greatly modify the stability of individual network attractors. It is again possible to look at this in a way consonant with the treatment of reaction events in chemical physics. We follow the thinking of transition state theory. Once the steady state probability distributions for a switch are known, escape from an attractor can be understood as occurring by moving the system past a stochastic separatrix or transition state.

The escape can happen either via one or a few binding/unbinding events or via the more continuous fluctuating growth and death of the populations of molecules in atmosphere. In Fig. 28.5, I show a variety of scenarios by which a

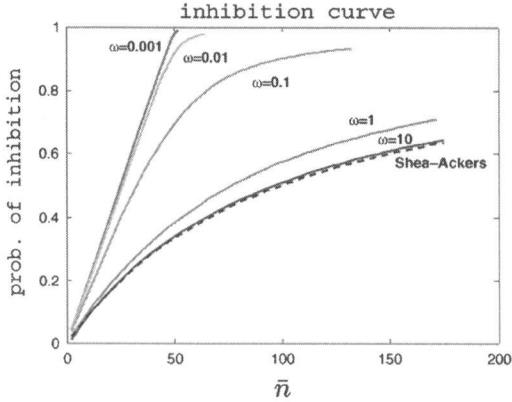

Fig. 28.4. The apparent "renormalized" law of mass action for a gene switch is shown in the outer panel. The probability of the gene being off versus average population of transcription factor is shown. The "Shea-Ackers" curve is the conventional macroscopic law of mass action. Single-molecule physical chemistry reigns!

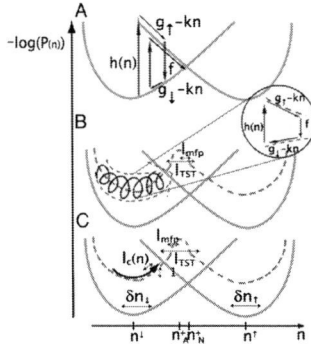

Fig. 28.5. The effective potential surfaces for a simple gene switch are shown – h gives the binding rate, f the unbinding rate, and $g_\uparrow$ and $g_\downarrow$ are the synthesis rates when the gene is on or off, respectively. The different diagrams correspond with different sequences of binding/synthesis/unbinding events. The upper plot shows the typical trajectory at high non-adiabaticity. The lowest plot shows the adiabatic case. A churning process gives an enhanced rate of protein number fluctuations in the intermediate "weakly non-adiabatic" case B

simple gene switch can change its state. Simple formulae have been found for the escape rates in the various regimes by Walczak, Onuchic, and Wolynes [18]. In Fig. 28.6, I show the typical escape rates from either attractor as a function of the ratio of the rate of single molecule binding to the growth/death rate parameters. It is interesting that there is a "stochastic resonance" when these two input time scales are close to each other.

In other words, switches are most agile in the weakly non-adiabatic regime, corresponding to the case B, binding/growth/unbinding scenario. It is

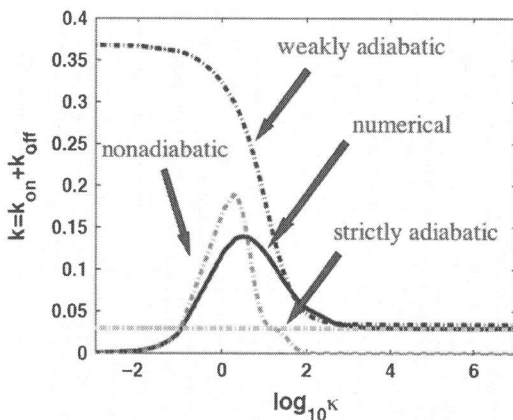

Fig. 28.6. Comparison of the rate of spontaneous epigenetic switching for different values of the non-adiabaticity K. The exact numerical results are shown as a solid curve. Various approximations reflecting the mechanisms described in Fig. 28.5 are also plotted. A stochastic resonance occurs near the weakly non-adiabatic regime

interesting that protein-DNA binding is indeed a very rapid process – one may argue unnaturally so. Eigen [19] and Berg and von Hippel [20] have drawn attention to this fact very early and it is now clear that a variety of factors contribute to this rapidity – one-dimensional search via sliding, transfer between DNA chains, and flycasting dynamics of repressors assisted by electrostatics [21]. An interesting question must, therefore, be – is this speed an adaptation to achieve the stochastic resonance? As the experimental study of gene regulation at the single-molecule level blossoms, I am sure we will soon have an answer to this question.

Acknowledgments

This work was supported by the NSF grant to the Center for Theoretical Biological Physics. The wonderful collaborations with Masaki Sasai, Aleksandra Walczak, José Onuchic, Daniel Schultz, and J.E. Hornos are much appreciated.

References

1. H. Frauenfelder, S. Sligar, P.G. Wolynes, Science **254**, 1598–1603 (1991)
2. E.V. Russel, N.E. Israeloff, Nature **408**, 695–698 (2000)
3. R. Zondervan et al., Proc. Nat. Acad. Sci. USA. **104**, 12628–12633 (2007)
4. A.M. Boiron, P. Tamarat, B. Loomis, R. Brown, M. Orrit, Chem. Phys. **247**, 119–132 (1999)
5. J. Wang, P.G. Wolynes, Phys. Rev. Lett. **74**, 4317–4320 (1995)
6. H.P. Lu, L.Y. Xun, X.S. Xie, Science **282**, 1877–1882 (1998)

7. L. Edman, R. Rigler, Proc. Natl. Acad. Sci. USA. **97**, 8266–8271 (2000)
8. B. Schuler, W. Eaton, Curr. Opin. Struct. Biol. **18**, 16–26 (2008)
9. W.J. Greenleaf, M.T. Woodside, S.M. Block, Annu. Rev. Biophys. Biomolec. Struct. **36**, 171–190 (2007)
10. F. Jacob, Mol. Biol. **3**, 318 (1961)
11. M.L. Delbrück, J. Chem. Phys. **8**, 120 (1940)
12. M. Sasai, P.G. Wolynes, Proc. Natl. Acad. Sci. USA. **100**, 2374–2379 (2003)
13. D. Chandler, P.G. Wolynes, J. Chem. Phys. **74**, 4078–4095 (1981)
14. R.A. Marcus, Angew. Chem. **32**, 1111–1121 (1993)
15. N.E. Buchler, U. Gerland, T. Hwa, Proc. Natl. Acad. Sci. USA. **100**, 5136–5141 (2003)
16. J.E.M. Hornos, D. Schultz, G.C.P. Innocentini, A.M. Walczak, J. Wang, J.N. Onuchic, P.G. Wolynes, Phys. Rev. E. **72**, 051907/1–5 (2005)
17. P.J. Choi, L. Cai, K. Frieda, S. Xie, Science **232**, 442–446 (2008)
18. A.M. Walczak, J.N. Onuchic, P.G. Wolynes, Proc. Natl. Acad. Sci. USA. **102**, 18926–18931 (2005)
19. P.H. Richter, M. Eigen, Biophys. Chem. **2**, 255–263 (1974)
20. O.G. Berg, R.G. Winter, P.H. von Hippel, Biochemistry. **20**, 6929–6948 (1981)
21. Y. Levy, J.N. Onuchic P.G. Wolynes, J. Am. Chem. Soc. **3**, 738–739 (2007)

Index